I0036299

Textile Engineering

Textile Engineering

Statistical Techniques, Design of Experiments and Stochastic Modeling

Anindya Ghosh, Bapi Saha,
and Prithwiraj Mal

CRC Press
Taylor & Francis Group
Boca Raton London New York

CRC Press is an imprint of the
Taylor & Francis Group, an **informa** business

MATLAB® is a trademark of The MathWorks, Inc. and is used with permission. The MathWorks does not warrant the accuracy of the text or exercises in this book. This book's use or discussion of MATLAB® software or related products does not constitute endorsement or sponsorship by The MathWorks of a particular pedagogical approach or particular use of the MATLAB® software.

First edition published 2022
by CRC Press
6000 Broken Sound Parkway NW, Suite 300, Boca Raton, FL 33487-2742

and by CRC Press
4 Park Square, Milton Park, Abingdon, Oxon, OX14 4RN

© 2022 Taylor & Francis Group, LLC

CRC Press is an imprint of Taylor & Francis Group, LLC

Reasonable efforts have been made to publish reliable data and information, but the author and publisher cannot assume responsibility for the validity of all materials or the consequences of their use. The authors and publishers have attempted to trace the copyright holders of all material reproduced in this publication and apologize to copyright holders if permission to publish in this form has not been obtained. If any copyright material has not been acknowledged please write and let us know so we may rectify in any future reprint.

Except as permitted under U.S. Copyright Law, no part of this book may be reprinted, reproduced, transmitted, or utilized in any form by any electronic, mechanical, or other means, now known or hereafter invented, including photocopying, microfilming, and recording, or in any information storage or retrieval system, without written permission from the publishers.

For permission to photocopy or use material electronically from this work, access www.copyright.com or contact the Copyright Clearance Center, Inc. (CCC), 222 Rosewood Drive, Danvers, MA 01923, 978-750-8400. For works that are not available on CCC please contact mpkbookspermissions@tandf.co.uk

Trademark notice: Product or corporate names may be trademarks or registered trademarks and are used only for identification and explanation without intent to infringe.

Library of Congress Cataloging-in-Publication Data

Names: Ghosh, Anindya, author. | Saha, Bapi, author. | Mal, Prithwiraj, author.
Title: Textile engineering : statistical techniques, design of experiments and stochastic modeling / Anindya Ghosh, Bapi Saha, and Prithwiraj Mal.
Description: First edition. | Boca Raton, FL : CRC Press, 2022. | Includes bibliographical references and index. | Summary: "Focusing on the importance of application of statistical techniques, design of experiments and stochastic modeling in textile engineering, this book endeavors to cover all these areas in one book. Practical examples and end-of-chapter exercises selected from textile industry are included. In order to solve various numerical problems relevant to statistics, design of experiment and stochastic modeling, computer codes in MATLAB® are provided at the end of each chapter for better comprehension. Overall theme of the book is analysis and interpretation of the textile data for improving the quality of textile process and product using various statistical techniques"-- Provided by publisher.
Identifiers: LCCN 2021035520 (print) | LCCN 2021035521 (ebook) | ISBN 9780367532741 (hbk) | ISBN 9780367532765 (pbk) | ISBN 9781003081234 (ebk)
Subjects: LCSH: Textile fabrics.
Classification: LCC TS1765 .G485 2022 (print) | LCC TS1765 (ebook) | DDC 677--dc23
LC record available at https://lccn.loc.gov/2021035520
LC ebook record available at https://lccn.loc.gov/2021035521

ISBN: 978-0-367-53274-1 (hbk)
ISBN: 978-0-367-53276-5 (pbk)
ISBN: 978-1-003-08123-4 (ebk)

DOI: 10.1201/9781003081234

Typeset in Times
by KnowledgeWorks Global Ltd.

Contents

Foreword

I am happy to write the foreword of the book entitled *Textile Engineering: Statistical Techniques, Design of Experiments and Stochastic Modeling* authored by Dr. Anindya Ghosh, Dr. Bapi Saha and Dr. Prithwiraj Mal. Textile materials have variability galore in their dimension and properties which often creates quality problems in textile products. The globalization and free-trade agreements have resulted in stringent quality standards for all products and textiles are no exception. Besides, the wafer-thin profit margin of commodity textiles further adds to the complexity. Therefore, use of statistics is of paramount importance for the quality and process control in textile industry.

Statistics is an integral part of textile engineering programs. However, there is scarcity of good textbooks in this area with theoretical underpinning and customised problems. This books presents a perfect amalgamation of theory and practice of textile statistics. The fundamental aspects of probability and various probability distributions have laid the foundation of the subject. The sampling distribution, test of significance, analysis of variance, correlation and regression have been covered with a good number of numerical examples. Topics like design of experiments and statistical quality control will immensely benefit research students and practicing engineers. The key strength of the book is that all the mathematical expressions have been derived from the basics. This will trigger the young minds to understand the fundamentals before applying them in practice. The book also contains a vast collection of solved problems and practice exercises. Overall, it is a perfect textbook for the undergraduate and postgraduate students of textile engineering programs. The authors have shown remarkable acumen and patience to construct each of the chapters. I am sure that this book is going to serve the engineering community in general and the textile community in particular in a big way.

Abhijit Majumdar
Institute Chair Professor
Department of Textile and Fibre Engineering
Indian Institute of Technology
Delhi, India

Preface

Among the plethora of books making rounds in the world of statistics are well-known titles that stand out, continuing to hold their sway with enduring appeal and utility. It justifiably then raises the question, will the publication of *Textile Engineering: Statistical Techniques, Design of Experiment and Stochastic Modeling* be just another addition to the common run of the ordinary or carry with it some novelty that justifies the labour the authors have poured into it?

Traditionally textile industries have wielded statistical techniques as an indispensable tool playing the role of quality watchdog, guiding and deciding upon process parameters that are to be fine-tuned so that the best possible quality products can be offered to the consumers.

It is the felt realization of the authors that a great deal of whys in science and technology at the fundamental level have remained either unanswered or ignored as too trivial to be considered. This is the inescapable truth even authors of great books must confront. To address this fundamental grey area, authors have chosen to start from the very beginning – the first principle. All the statistical theories and relevant mathematical rigors have boiled down to threadbare lucidities through step-by-step derivations.

Authors have gleaned a large body of real-world problems mostly drawn from the diverse areas of textiles and catered them with solutions for the readers. Each problem is aptly illustrated in an approach rarely found in traditional books. Again, equally large number of problems are given as exercises to stimulate the readers' comprehension about the subject they learn as theory.

MATLAB® coding for solving a wide range of statistical problems has been provided, which will surely bring out students' practical difficulties during problem solving. These MATLAB® coding will definitely enhance the scope of their application manifold in real world problems.

Probability is the cardinal subject in statistics. It is the centre point around which a great deal of statistical theories gravitate. Therefore, a large portion of the book is devoted to this subject. Authors lay great emphasis upon probability distributions for discrete and continuous variables as they are the bedrock over which the edifice of statistics stands. Statistical inference through estimation and hypothesis testing takes on instrumental role in making meaningful predictions about population on the basis of sample information. Many statistical investigations seek interrelationship among the variables, which correlation and regression theories elicit. Of great practical significance is statistical quality control, without which, product quality may go wrong. Authors therefore have focused their wits upon these areas in a reader friendly way.

In fulfilling their desire to write a broad-spectrum text, authors at length reached out to niche scholars and researchers with a clear and methodical approach to the essentials of experimental design and stochastic modelling. It is expected that young researchers will be well-guided through the little explored areas of stochastic modelling.

Authors, also, ambitiously stepped out their way to make this book useful for learners and practitioners in other science and technological domains.

The author trio will regard their effort amply rewarded once this book secures an prominent place upon the study table of students, scholars and professionals alike. In spite of all the care, if any error creeps into this book, authors are ready to amend and are open to every constructive criticism for further improvement.

Anindya Ghosh
Bapi Saha
Prithwiraj Mal

MATLAB® is a registered trademark of The MathWorks, Inc. For product information, please contact:

The MathWorks, Inc.
3 Apple Hill Drive
Natick, MA, 01760-2098 USA
Tel: 508-647-7000
Fax: 508-647-7001
E-mail: info@mathworks.com
Web: www.mathworks.com

Authors

Anindya Ghosh, PhD, is an Associate Professor at the Government College of Engineering and Textile Technology, Berhampore, West Bengal, India. He earned a BTech in textile technology in 1997 at the College of Textile Technology, Berhampore (Calcutta University), India. He then worked in a textile spinning industry for one year as a shift engineer. He earned an MTech and PhD in textile engineering at the Indian Institute of Technology, Delhi, India, in 2000 and 2004, respectively. He received the Career Award for Young Teacher–2009 from the All India Council for Technical Education (AICTE). He has more than 17 years of teaching experience. His research involves yarn manufacturing, yarn and fabric structures, modelling and simulation, statistical quality control, optimization and decision-making techniques. He has published more than 80 papers in various referred journals. He coauthored *Advanced Optimization and Decision-Making Techniques in Textile Manufacturing* (CRC Press).

Bapi Saha, PhD, is an Assistant Professor of mathematics at the Government College of Engineering and Textile Technology, Berhampore, West Bengal, India. He earned a BSc in mathematics and an MSc in applied mathematics in 2004 and 2006, respectively, at Jadavpur University, Kolkata, India. He received the Junior Research Fellowship from the Council of Scientific and Industrial Research, MHRD, Government of India, and worked as a Junior Research Fellow at the Indian Statistical Institute. He earned a PhD in applied mathematics at the University of Calcutta, India, in 2016. He has more than ten years of teaching experience. His research interests include dynamical system, mathematical biology, biostatistics, and stochastic modelling. Dr. Saha has published ten research articles, including a single-author article, in peer-reviewed international journals. He has also published articles in the proceedings of national and international conferences.

Prithwiraj Mal, PhD, is an Associate Professor in the Department of Textile Design at the National Institute of Fashion Technology (NIFT), Hyderabad, India. He earned a BTech in textile technology in 2000 at the College of Textile Technology, Berhampore (Calcutta University) and an MTech in textile engineering in 2004 at the Indian Institute of Technology Delhi. He has almost 19 years of cumulative experience in both industry and academics. He earned a PhD at Jadavpur University, Kolkata, in 2017.

Dr. Mal joined NIFT in 2008. His research work involves comfort, optimization, decision-making techniques, and product development. He has published more than 20 papers in referred journals, chapters in books, and presented or published papers at national and international conferences. He coauthored *Advanced Optimization and Decision-Making Techniques in Textile Manufacturing* (CRC Press).

1 Introduction

1.1 INTRODUCTION

Statistics is dealing with the collection of data, their subsequent description, summarization and analysis, which often leads to the drawing of conclusions. In order to draw a conclusion from the data, we must consider the possibility of chance or probability. For example, suppose that a finishing treatment was applied to a fabric to reduce its bending rigidity. The average reduction in bending rigidity was found to be lower for the fabric samples receiving the finishing treatment than that of untreated fabric samples. Can we conclude that this result is due to the finishing treatment? Or is it possible that the finishing treatment is really ineffective and that the reduction in bending rigidity was just a chance occurrence?

In statistics, we are interested in finding information about a total collection of elements, which refers to population. Nevertheless, the population is often too huge to inspect each of its members. For example, suppose that a garment manufacturer marks the size of men's trousers according to the waist size. For the purpose of designing the trousers, the manufacturer needs to know the average waist size of the men in the population to whom the trousers will be sold. In this example, in order to find out the accurate average, the waist size of every man in the population would have to be measured. However, there may be millions of men in the population and hence, it is quite impracticable to measure the waist size of every man in the population. In such cases, we try to learn about the population by examining a subgroup of its elements. This subgroup of a population is called a sample.

Definition 1.1: The total collection of all the elements is called a population. A subgroup of the population is called a sample.

The sample must be representative as well as informative about the population. For example, suppose that we are interested about the length distribution of a given variety of cotton fibre and we examine the lengths of the 300 fibres from a combed sliver. If the average length of these 300 fibres is 23.4 mm, are we justified in concluding that this is approximately the average length of the entire population? Probably not, for we could certainly argue that the sample chosen in this case is not representative of the total population because usually combed sliver contains a smaller number of short fibres. The representation does not mean that the length distribution of fibre in the sample is exactly that of the total population, but the sample should be chosen in such a way that all parts of the population had an equal chance to be included in the sample. In general, a given sample cannot be considered to be representative of a population unless that sample has been chosen in a random manner. This is because any specific non-random selection of a sample often results in biasing toward some data values as opposed to others.

DOI: 10.1201/9781003081234-1

Definition 1.2: A sample of k members of a population is said to be a random sample, if the members are chosen in such a way that all possible choices of the k members are equally likely.

It is unrealistic and impossible to examine every member of the population. Once a random sample is chosen, we can use statistical inference to say something meaningful about the entire population by studying the elements of the sample.

It is inevitable that the natural and man-made products vary one from another. For example, a cotton plant produces fibres with varying lengths. In a mass-producing manufacturing process, it is impossible to produce all the articles which are absolutely identical. Had there been no variation in a population, then it would have been sufficient to examine only one member to know everything about the population. But if there is variation in a population, the examination of only a few members contained in a random sample may provide incomplete and uncertain information about the population. The random or chance sources of variation give rise to this uncertainty. For example, the growth habit of the cotton plant, genetics and environmental conditions during cotton fibre development are the probable sources of random variation of cotton fibre length. One of the important tasks of statistics is to measure the degree of this uncertainty. Although, the values of a random variable may vary arbitrarily, there is generally an underlying pattern in the variation. The presence of such pattern in the data enables us to draw some meaningful conclusions about the population from the results of the sample. The recognition and description of patterns in the data are of fundamental importance in statistics.

Statistics is dealing with the quantities that vary. Such quantities are known as variables. The variables that are affected by random sources of variation are called random variables. Random variables are of two kinds, viz., discrete and continuous. A random variable is called discrete random variable if the set of its possible values is either finite or countable infinite. The number of warp breaks in a loom shed per shift, the number of defective ring frame bobbins produced in a day and the number of defects per 100 square meters of a fabric are some examples of discrete random variables. A random variable is called a continuous random variable if the set of its possible values is not finite and uncountable. A continuous random variable takes on value in an interval. The fineness of cotton fibre, tenacity of a yarn and bending rigidity of a fabric are some examples of continuous random variables.

1.2   ORGANIZATION OF THE BOOK

This book is divided into 12 chapters which discuss various statistical methods for the analysis of the data pertaining to the domain of textile engineering. A brief view of these chapters is given below.

This chapter is an elementary overview of statistics. A brief idea of the population, sample, random sample, random variation, uncertainty and variables is given in this chapter.

Frequency distribution and histograms are among the few techniques which are largely used in summarizing the data. In addition, a suitable representation of the data which is simple in nature is helpful in many situations. Chapter 2 discusses the representation and summarization of the data. It deals with the frequency distribution, relative frequency, histogram, probability density curve, mean, median, mode, range, mean deviation and variance.

Chapter 3 deals with the concept of probability. Firstly, some basic terminologies, viz., random experiment, sample space and events have been introduced. It is followed by the various definitions of probability, set theoretic approach to probability and conditional probability. A brief idea of random variable, probability mass function, probability density function and probability distribution function has been furnished in this chapter. It then explains the expectation, variance, moment, moment generating function, characteristic function, multivariate distribution and transformation of random variables.

Chapter 4 covers discrete probability distribution. Some important discrete probability distributions, viz., Bernoulli distribution, binomial distribution, Poisson distribution and hypergeometric distribution have been explained in this chapter with numerical examples and MATLAB® coding.

Chapter 5 contains the discussion on continuous probability distribution. Some important univariate continuous probability distributions, viz., uniform distribution, exponential distribution, Gaussian or normal distribution and lognormal distribution have been explained in this chapter with numerical examples and MATLAB® coding. Normal approximation to the binomial distribution as well as Poisson distribution has also been explained in this chapter. In addition, this chapter provides a brief view of bivariate normal distribution.

Statistical inference is an important decision-making tool to say something meaningful about population on the basis of sample information. The statistical inference is divided into two parts, viz. estimation and hypothesis testing. Chapter 6 is dealing with the sampling distribution and estimation. It begins with the explanation of distribution of sample mean, central limit theorem, chi-square distribution, Student's t-distribution and F-distribution. It then discusses on the point estimation and interval estimation. The section of point estimation is comprised of the discussion of unbiased estimator, consistency, minimum variance unbiased estimator, sufficiency and maximum likelihood estimator. The interval estimation of mean, difference between two means, proportion, difference between two proportions, variance and the ratio of two variances have been explained in the section of interval estimation with numerical examples and MATLAB® coding.

Chapter 7 treats the statistical test of significance. At the outset, the concept of null hypothesis, alternative hypothesis, type-I and type-II errors have been explained. Then various statistical tests concerning mean and difference between two means with small and large sample sizes, proportions, variance and difference between two variances, difference between expected and observed frequencies have been described with numerical examples and MATLAB® coding.

Analysis of variance (ANOVA) is a common procedure for comparing multiple population means across different groups while the number of groups is more than two. In Chapter 8, one-way ANOVA, two-way ANOVA with and without replication have been explained with numerical examples and MATLAB® coding. This chapter also discusses the multiple comparisons of treatment means.

A key objective in many statistical investigations is to establish the inherent relationships among the variables. This is dealt with the regression and correlation analysis.

In Chapter 9, firstly, simple linear regression, coefficient of determination, correlation coefficient and rank correlation coefficient have been explained. It then explains quadratic regression and multiple linear regression. In addition, a matrix approach of regression has been discussed with reference to simple linear regression, quadratic regression, multiple linear regression and multiple quadratic regression. Finally, the test of significance of regression coefficients has been discussed. A substantial number of numerical examples and MATLAB® coding have been furnished in this chapter.

It is of utmost importance to conduct experiments in a scientific and strategic way to understand, detect and identify the changes of responses due to change in certain input factors in complex manufacturing process. Hence, a statistical design of experiment (DOE) is required to layout a detailed experimental plan to understand the key factors that affect the responses and optimizing them as required. Chapter 10 discusses the need of designing such experiments to investigate and find the significant and insignificant factors affecting the responses. Initially, the basic principle of the design of experimentations is discussed. The importance of randomization, replication, different types of blocking and their effects are explained along with examples. The need of factorial design, fractional factorial design and response surface designs are explicated in details with examples. However, none of these designs of experiments consider the noise factor, which might have a significant role affecting the response. The Taguchi design of experiment considers noise factors and helps to overcome such limitations, which is explained towards the end of this chapter along with examples. The advantages and limitations of each design are also thoroughly discussed. Lastly, numerical examples along with MATLAB® coding are furnished in this chapter.

Chapter 11 includes sampling inspection and various means to check whether a lot is acceptable or rejected. Acceptance sampling, which is a statistical technique to deal with inspection of raw materials or products and to conclude whether to accept or reject in context to consumer's and producer's risk along with examples is initially discussed. Acceptance sampling for both the attributes and variance, average outgoing quality, average total inspection, assurance about a minimum/maximum/mean value are explained with suitable examples. Various control charts, which are a statistical technique of dealing with sampling inspection of materials at the conversion stage, are also described with appropriate examples. The need and concepts of control charts are explained initially. The centre line, warning limits and action limits and their interpretation with regard to control charts for mean, range, fraction defectives and number of defects are discussed with suitable examples of each. Finally, numerical examples with MATLAB® coding are given in this chapter.

In many situations, when a system is time dependent, a sequence of random variables which are functions of time is termed as a stochastic process. Chapter 12 deals with the stochastic modelling and its application in textile manufacturing. It begins with a brief discussion on the Markov chain followed by an application of it on the mechanism of a carding machine. The process of formation of the stochastic differential equation for a random process is discussed subsequently with its application in resolving the fibre breakage problem in the yarn manufacturing process. Finally, a numerical example and corresponding MATLAB® coding is given in this chapter.

2 Representation and Summarization of Data

2.1 INTRODUCTION

The science of statistics deals with the numerical data. Statistical data obtained from experiments and surveys are often so large that their significance is not readily comprehended. In such situations, it becomes necessary to summarize the data into a more suitable form. Grouping of data and representation of such groupings in graphical form can often be the most effective way to visualize the overall pattern in the data. It is also important to characterize the data by a few numerical measures which describe the nature of the data in a general way. The two most common measures are central tendency and dispersion.

Quite often there is a general tendency for the values of a data set to cluster around a central value. In such a situation, it is reasonable to use the central value to represent the whole data set. Such a representation is called the measure of central tendency or average. The most popular among central tendencies are mean, median and mode.

The measure of central tendency does not give proper idea about the overall nature of the variation of a set of data. In some cases, the values of a data set are closely bunched about the average, while in others they may be scattered widely about the average. Therefore, it is necessary to measure the dispersion which states how scattered the values of a data set are about their average. Three types of dispersions are in general used, viz., range, mean deviation and standard deviation.

2.2 FREQUENCY DISTRIBUTION AND HISTOGRAM

If a variable property is measured on a fairly large sample of individuals drawn from a population, it is usually expected that the results provide a meaningful perception about that property in the population. Consider the data in Table 2.1 which gives the strengths in cN/tex of 200 tests taken from a random sample of 200 cops of 20 tex cotton yarn, one test being done from each cop. The significance of this large mass of data is quite difficult to appreciate. Thus, it becomes necessary to summarize the data to a more comprehensive form. This is accomplished by frequency distribution.

A frequency distribution is a table that divides a data set into a suitable number of groups or classes, displaying also the frequency or number of the occurrences to each class. From the data in Table 2.1, we observe that the smallest is 17.5, the largest is 22.8, and the range is 5.3. As the data has been recorded to one decimal place, we might choose 12 classes having the limits 17.1–17.5, 17.6–18, …, 22.6–23, or we might choose 6 classes as 17–17.9, 18–18.9, …, 22–22.9, such that the classes do not

DOI: 10.1201/9781003081234-2

TABLE 2.1

Data of Yarn Strength in cN/tex of 200 Tests

19.8	19.9	19.7	18.7	20.6	20.4	20.5	21.2	20.0	18.7
19.6	18.4	21.1	19.9	18.3	20.2	20.4	21.1	20.3	19.9
21.0	20.8	19.4	20.4	20.3	19.3	20.6	19.7	19.6	21.2
19.7	19.9	20.2	21.0	19.1	19.6	20.8	20.3	18.1	20.6
20.3	18.9	20.6	20.1	18.4	19.4	21.1	20.2	21.0	19.9
20.6	18.8	19.0	20.6	19.8	20.2	19.0	21.0	19.3	19.6
19.1	19.0	18.5	21.8	20.6	19.3	18.7	21.3	20.2	18.9
19.6	19.2	20.8	19.9	21.0	20.3	21.3	20.8	21.0	20.9
21.9	19.9	20.4	19.4	21.2	19.1	21.0	20.0	20.6	20.9
20.1	19.9	19.5	18.6	19.6	20.6	19.2	19.1	19.2	19.8
20.9	19.4	21.1	20.9	19.0	19.8	20.0	18.7	19.7	20.4
21.3	19.9	20.3	19.5	22.1	17.9	21.9	21.1	19.8	20.3
20.6	19.3	19.6	21.4	19.4	20.2	18.8	22.0	21.0	20.9
20.1	20.2	19.6	20.1	19.3	21.0	20.3	20.3	21.3	21.6
21.3	21.6	19.5	19.9	20.5	21.3	18.6	19.9	18.9	18.2
18.7	20.9	18.9	19.4	18.9	20.2	20.2	20.9	19.0	19.7
21.2	19.6	22.8	19.5	19.3	21.5	20.3	18.3	20.1	19.9
17.9	20.1	17.5	18.7	18.5	20.7	19.8	19.7	20.0	18.7
18.7	19.8	20.3	20.6	19.8	22.1	20.9	18.0	20.2	19.7
19.6	19.6	20.0	19.6	18.9	20.3	19.9	20.4	18.9	20.8

overlap and they are all of the same size. Table 2.2 is formed with 6 classes. Note that had the original data been given to two decimal places, we would have used class limits 17–17.99, 18–18.99, ..., 22–22.99.

In this example the yarn strength is a continuous variable, but if we use classes such as 17–18, 18–19, ..., 22–23, an ambiguous situation may arise. For example, 18.0 could go into the first class or into the second, 19.0 could go into the second or into the third, and so on and so forth. This difficulty can be avoided if we can let the classes as 16.95–17.95, 17.95–18.95, ..., 21.95–22.95. These new limits are called

TABLE 2.2

Frequency Distribution for the Data of Table 2.1

Class Limits	Class Mid-point	Tally Marks	Frequency
17–17.9	17.45	///	3
18–18.9	18.45	THL THL THL THL THL ///	28
19–19.9	19.45	THL THL THL THL THL THL THL THL THL THL THL THL THL ///	68
20–20.9	20.45	THL THL THL THL THL THL THL THL THL THL THL THL THL /	66
21–21.9	21.45	THL THL THL THL THL THL /	31
22–22.9	22.45	////	4

class boundaries. Further, the data is recorded to only one decimal, for example, 17.0 represents any value between 16.95 and 17.05; similarly, 17.9 represents any value between 17.85 and 17.95. Thus, class 17–17.9 really stands for class boundaries 16.95–17.95. In practice, we use class boundaries rather than the original class limits, while drawing up the frequency distribution of a continuous variable. However, for the sake of simplicity we have used class limits as shown in Table 2.2.

Next, we take each value of the data one by one, and for each value we place a tally mark against the appropriate class. In order to simplify the method of counting, the tally marks are arranged in blocks of five by drawing the fifth tally mark across the preceding four. Following this procedure, we now tally 200 observations, and by counting the number of tally marks in each class we obtain the frequency distribution of the variable, which is shown in Table 2.2. The frequency distribution reveals that there is more concentrated data in classes 19–19.9 and 20–20.9, being positioned in the centre of the distribution, whereas the data tend to sparsely distribute in the classes at extreme ends of the distribution. The mid-points of the classes are also given in Table 2.2 which are calculated as the average of the upper-class and lower-class limits. For example, mid-point of the last class is $\frac{(22+22.9)}{2} = 22.45$. Often a class is represented by its mid-point.

A diagrammatic representation of the frequency distribution is called histogram. In a histogram, the horizontal axis is divided into segments corresponding to widths of the classes, and above each segment a rectangle is erected whose area represents the corresponding class frequency. When all classes are of equal width, as in this example, the height of the rectangle is proportional to the class frequency. Figure 2.1 depicts the histogram corresponding to the frequency distribution of yarn strength in Table 2.2, where the classes are represented by their mid-points.

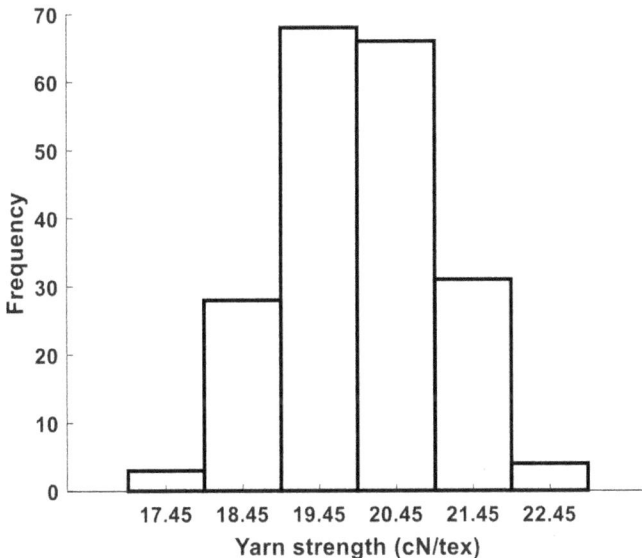

FIGURE 2.1   Histogram.

TABLE 2.3

Relative Frequency Distribution for the Data of Table 2.1

Class Mid-point	Relative Frequency	Cumulative Relative Frequency
17.45	0.015	0.015
18.45	0.140	0.155
19.45	0.340	0.495
20.45	0.330	0.825
21.45	0.155	0.980
22.45	0.020	1

2.2.1 RELATIVE FREQUENCY AND DENSITY HISTOGRAM

In order to find the relative frequency, we divide the actual frequency of a class by the total frequency. The sum of the frequencies in Table 2.2 is equal to 200. Thus, by dividing the frequency of each class by 200, we obtain the relative frequencies as shown in Table 2.3. The sum of the relative frequencies is 1. The last column of Table 2.3 shows the cumulative relative frequency. Now, if we construct a histogram for which the height of each rectangle is equal to the relative frequency divided by width, the resulting histogram is called density histogram. The ratios of relative frequencies to the corresponding widths of class intervals are called relative frequency densities. Figure 2.2 depicts the density histogram of the yarn strength. For a density histogram, the area of rectangle represents the relative frequency for the corresponding class, and the total area of all the rectangles is equal to 1.

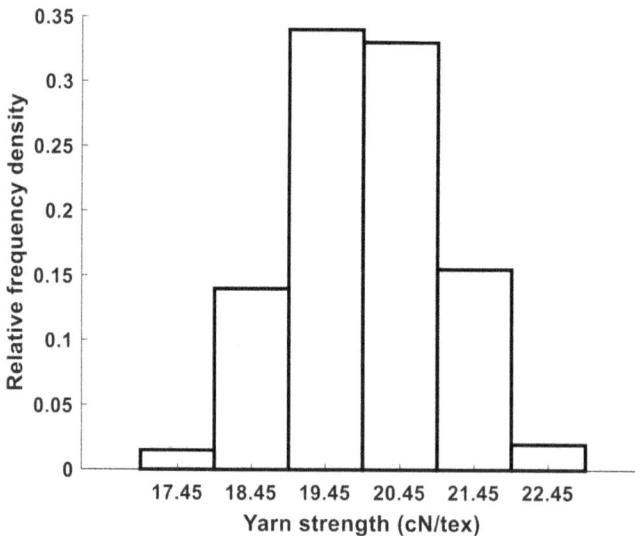

FIGURE 2.2 Density histogram.

2.3 PROBABILITY DENSITY CURVES

For a continuous variable, it is instructive to imagine the form of the density histogram if the number of observations tends to infinity and the measurement accuracy is considerably increased. In practice, we generally deal with the finite number of observations, and the frequency distribution is formed with a small number of classes. As in the previous example, the data is recorded only to the first decimal place, and the frequency distribution of yarn strength has been made with 6 classes from 200 observations. Nevertheless, theoretically, a continuous variable like yarn strength can be measured to any degree of accuracy. Imagine that the number of observations and the measuring accuracy are gradually increased and simultaneously the width of each class is gradually decreased so that the number of classes goes on increasing. In such a case, rectangles of the density histogram gradually become narrower and smoother, as depicted in Figure 2.3(a) to (c), even though the total area of each density histogram remains the same, viz., unity. If we further imagine that the number of observations tends to infinity and the measuring accuracy also increases, the outline of the density histogram will become indiscernible from the smooth curve, as shown in Figure 2.3(d). This smooth curve is called probability density curve. When the number of observations tends to infinity, the sample tends to become the population. Thus, a probability density curve describes the population from which the sample represented by the histogram is drawn.

It has been discussed in the previous section that the areas of the rectangles generating a density histogram are the relative frequencies. Let X be the continuous variable denoting yarn strength which takes the value x. The shaded area in Figure 2.3(a) indicates the relative frequency or proportion of values in the sample lying between x_1 and x_2.

When the sample size tends to become the population, the shaded area under the probability density curve in Figure 2.3(d) indicates the proportion of values in population lying between x_1 and x_2. In other words, the shaded area under the probability density curve in Figure 2.3(d) gives the probability that X is lying between x_1 and x_2, which is denoted as $P(x_1 \leq X \leq x_2)$. If the height of the probability density curve is denoted by $y = f(x)$, as shown in Figure 2.4, then

$$P(x_1 \leq X \leq x_2) = \int_{x_1}^{x_2} f(x)\,dx$$

If the maximum and minimum values of x are x_{min} and x_{max}, respectively, then

$$P(x_{min} \leq X \leq x_{max}) = \int_{x_{min}}^{x_{max}} f(x)\,dx = 1$$

Theoretically, x_{min} and x_{max} can be $-\infty$ and ∞, respectively. The probability is of utmost importance in statistics, and a thorough discussion on this topic is made in Chapter 3.

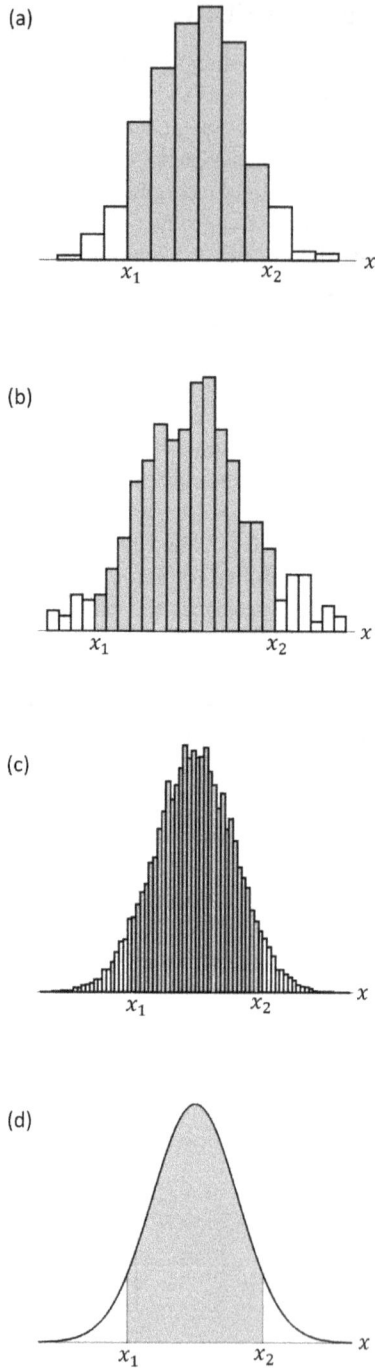

FIGURE 2.3 Development of the probability density curve.

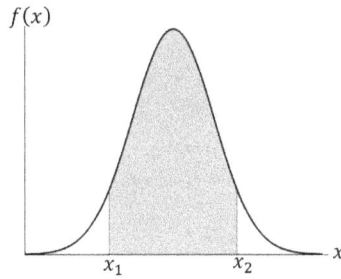

FIGURE 2.4 Probability density curve.

2.4 MEASURES OF CENTRAL TENDENCY

In many cases, we need to draw some inferences from a given data set. In such situations, a single representation of the given data set is useful. For example, if we are assigned a task to assess the quality of students in a school, we may consider the marks obtained by the students. If the number of students is large in the school, then the data set of marks obtained will also be large. In this situation, a single representation of the whole data is often proved to be useful. For this example, one intuitive approach is to take the average of those marks and, depending on the value of this average, we could infer about the student quality of the school. Apart from average or mean, several other representations of a given data are used, depending on the distribution of the data. This single representation of a given data is termed as central tendency. Mean, median and mode are among the frequently used central tendencies that are discussed as follows.

2.4.1 MEAN

Mean is one of the most frequently used and popular measures of the central tendency. This has gained popularity because it is simple in nature. Mean is found to be an appropriate measure when the data is symmetric. The general definition of mean is given as follows:.

Suppose $x_1, x_2, \ldots, x_n$ are the observed values of a random variable. The arithmetic mean of these observed values are given by

$$\overline{x} = \frac{\sum_{i=1}^{n} x_i}{n}$$

For a frequency data, the mean is given by

$$\overline{x} = \frac{\sum_{i=1}^{n} f_i x_i}{\sum_{i=1}^{n} f_i}$$

where x_i's are the observed values of a random variable with corresponding frequencies f_i's.

For a grouped frequency data, the mean is obtained by replacing x_i by the mid-point of the ith class interval. It is illustrated in the following example.

Example 2.1:

The ranges of the wages of 100 workers assigned to different jobs are given in the following table:

Range of Wages (in Rs.)	Frequency of Workers
100–150	5
150–200	12
200–250	25
250–300	30
300–350	18
350–400	10

Find the mean wage of the workers.

SOLUTION

For every class interval, the mid-point of the upper-class boundary and the lower-class boundary is considered. The frequency of each of the class mid-point is the same as the corresponding class and is depicted in the following table:

Range of Wages (in Rs.)	Mid-point of Class	Frequency of Workers
100–150	125	5
150–200	175	12
200–250	225	25
250–300	275	30
300–350	325	18
350–400	375	10

Therefore, the mean is given by

$$\bar{x} = \frac{125 \times 5 + 175 \times 12 + 225 \times 25 + 275 \times 30 + 325 \times 18 + 375 \times 10}{100} = 262$$

2.4.2 MEDIAN

The mean is obviously a good choice as a measure of central tendency when the data is symmetric. However, there exists some cases where mean is not appropriate and can be misleading. This situation may arise when the data set is not symmetric, i.e. skewed. In such cases, the median or mode are the alternative choices for the central tendency. In particular, if the data contains outliers, median is preferred to mean because the value of mean in this case is greatly influenced by the outliers.

The median of a data is a value for which the number of data points having value less than it and the number of data points having values greater than it are same.

For a given data set, say x_1, x_2, x_3,...,x_n, we arrange the values of the data either in increasing order or decreasing order. Depending on the nature of the sample size n, the median will be determined. For example, if n is odd, say $n = 2m+1$ for some $m \in N$, where N is the set of natural numbers, the $(m+1)$th observation is the median of the distribution; whereas, if n is even, say $n = 2k$ for some $k \in N$, then the average of the kth and $(k+1)$th observed value can be treated as the median. As an example, for the odd numbered data set $\{0, 9, 3, 1, 6, 11, 5\}$, we arrange the observed values in increasing order as 0, 1, 3, 5, 6, 9, 11. Here, $m = 3$ and therefore $3+1 = 4$th observation which is 5, is the median. Now, in the case of an even numbered data set $\{0, 9, 3, 1, 6, 11, 5, 14\}$, the arrangement in the increasing order yields 0, 1, 3, 5, 6, 9, 11, 14. Here, $k = 4$ and hence the median is the average of 4th and 5th observations, which is $(5+6)/2 = 5.5$.

For a grouped frequency data, the median M is given by

$$M = L + \frac{\dfrac{n}{2} - F}{f} \times c$$

where L is the lower boundary point of the median class, n is the total frequency, F is the cumulative frequency of the class preceding the median class, f is the frequency of the median class, and c is the class length of the median class. This is illustrated in the following example.

Example 2.2:

The marks obtained (out of 50) by 100 students in a class in the subject statistics are depicted in the following table. Find the median marks.

Marks Obtained	Frequency
0–10	5
10–20	10
20–30	22
30–40	33
40–50	30

SOLUTION

The grouped frequency data with cumulative frequency is given in the following table.

Marks Obtained	Frequency	Cumulative Frequency
0–10	5	5
10–20	10	15
20–30	22	37
30–40 ←	33	70
40–50	30	100

Here, the total frequency n is 100 and the frequency $\frac{n}{2} = 50$ corresponds to class 30–40, which is highlighted by left arrow. This is because the cumulative frequency of the preceding class is 37, which is less than 50, indicating that class 20–30 cannot contain the median. In this example, $L = 30$, $n = 100$, $F = 37$, $f = 33$ and $c = 10$. Hence, the median is calculated as follows:

$$M = 30 + \frac{\dfrac{100}{2} - 37}{33} \times 10$$

$$= 33.94$$

In the above example, the upper boundary of any class is the same as the lower boundary of the next class. In many situations, these two boundaries are not same and continuity correction is needed. This is illustrated in the following example:

Example 2.3:

The marks obtained (out of 50) by 100 students in a class in the subject statistics are depicted in the following table. Find the median marks.

Marks Obtained	Frequency
0–9	5
10–19	10
20–29	22
30–39	33
40–50	30

SOLUTION

We need to modify the data by performing the continuity correction of the marks obtained. In order to do that, we add and subtract $\frac{1}{2} = 0.5$ to the upper boundary point and from the lower boundary point, respectively. Thus, we have the following table.

Marks Obtained	Marks after Continuity Correction	Frequency	Cumulative Frequency
0–9	(−0.5)–9.5	5	5
10–19	9.5–19.5	10	15
20–29	19.5–29.5	22	37
30–39	29.5–39.5←	33	70
40–50	39.5–50.5	30	100

For this grouped frequency data, we can take 0 also as the lower-class boundary of the first class instead of –0.5. This is because the number of students who obtain marks less than 0 is 0 and thus do not contribute to the frequency of the corresponding class. Similar argument is applicable to justify that 50.0 can also be taken as the upper-class boundary of the last class instead of 50.5. In this example, the total frequency n is 100 and the frequency $\frac{n}{2} = 50$ corresponds to the class 29.5–39.5 which is highlighted by left arrow. Therefore, in this case $L = 29.5$, $n = 100$, $F = 37$, $f = 33$ and $c = 10$. Hence, the median is calculated as

$$M = 29.5 + \frac{\dfrac{100}{2} - 37}{33} \times 10$$

$$= 33.44$$

2.4.3 MODE

Apart from the mean and median, mode is also one of the frequently used central tendencies. The mode of a data is the value which occurs most frequently.

For example, let us consider the data {110, 125, 120, 110, 50, 120, 80, 110, 90}. In this data set, the value 110 occurs thrice which is highest among the frequencies with which the remaining data values appear and so 110 is the mode for this data set. A special case may arise when two or more data points occur with the same highest frequency. In such cases, all the data values having the same highest frequency are modes and, in this case, the data is said to be multi-modal. For example, in the previous data set, had there been one more data value 120, i.e. if the data set was {110, 125, 120, 110, 50, 120, 80, 110, 90, 120}, then 110 and 120 would have occurred with the same highest frequency 3. So, both 110 and 120 would be modes and hence, the given data set would be bimodal.

In the case of grouped frequency data, we cannot identify readily the value which occurs most frequently. In this case, the following formula is used to find the mode of the given data set.

If M_O is the mode of a given data set, then it can be expressed as

$$M_O = L + \frac{f_1 - f_0}{2f_1 - f_0 - f_2} \times c$$

where:

$L =$ lower boundary point of the modal class
$f_1 =$ frequency of the modal class
$f_0 =$ frequency of the class preceding the modal class
$f_2 =$ frequency of the class succeeding the modal class
$c =$ class length of the modal class

Example 2.4:

The following data shows the frequency distribution of cotton fibre lengths.

Length Class (mm)	Frequency
0–4	32
4–8	796
8–12	1056
12–16	3980
16–20	5774
20–24 ←	6800
24–28	4282
28–32	2015
32–36	846
36–40	228
40–44	72

Calculate the mode using the above data.

SOLUTION

From the given table, we observe that the highest frequency is 6800 and it corresponds to class 20–24, as indicated by left arrow. In this example, $L = 20$, $f_1 = 6800$, $f_0 = 5774$, $f_2 = 4282$ and $c = 4$. Hence, the mode is calculated as

$$M_O = L + \frac{f_1 - f_0}{2f_1 - f_0 - f_2} \times c$$

$$= 20 + \frac{6800 - 5774}{2 \times 6800 - 5774 - 4282} \times 4$$

$$= 21.16$$

2.5   MEASURES OF DISPERSIONS

The central tendency alone does not give proper representation of the data when the data values are largely dispersed from each other. For example, consider the monthly income of 10 Indian families in rupees are 8000, 7000, 90,000, 110,000, 9800, 9000, 120,000, 15,000, 150,000 and 81,000. The average of these incomes is found to be Rs. 59,980. This shows that it will not be reasonable to represent the monthly incomes of the families by the mean or average as the monthly incomes of 5 out of 10 families are less than or equal to Rs. 15,000, which is much less than the average income. Nevertheless, 4 out of 10 families have income equal to Rs. 90,000 or above, which is far above the average income. Thus, in order to judge the reliability of the central tendency of a given data, it is necessary to know how the data points are scattered around it. This can be accomplished by the measure of dispersion. The commonly used measures of dispersions are range, variance, standard deviation etc.

2.5.1 RANGE

The range of a data is defined as the difference between its maximum and minimum values. Thus, for a given data x_1, x_2, ..., x_n, the range is given by

$$R = \max_i x_i - \min_i x_i, \text{ where } i = 1, 2, ..., n.$$

If we consider the above example of the monthly income of ten families, the range is $150,000 - 7000 = 143,000$.

As a measure of dispersion, the range may be found to be inappropriate in many cases due to several flaws. The presence of very high values or very low values (which are known as outliers) greatly influence range. It may happen that the range estimate is found to be very high although most of the data are closely located to the central tendency. For example, for the data set $\{25, 0, 25, 25, 25, 100, 25\}$, the range is $100 - 0 = 100$ even if all the data points except 2nd and 6th, are 25. So, it is misleading to assign such a high value to the dispersion. Besides this, range does not involve all the data points. Mean deviation about mean or median, variance or standard deviation are some suitable choices to avoid these discrepancies.

2.5.2 MEAN DEVIATION ABOUT MEAN

Mean deviation about mean is the average absolute distance of the data points from the mean of the data. If x_1, x_2, ..., x_n are the observations of a data and $\bar{x}$ is the arithmetic mean of the data, the mean deviation about mean of the data is given by

$$MD = \frac{\sum_{i=1}^{n} |x_i - \bar{x}|}{n}$$

For a frequency data, the mean deviation about mean is given by

$$MD = \frac{\sum_{i=1}^{n} f_i |x_i - \bar{x}|}{\sum_{i=1}^{n} f_i}$$

where x_i's are the observed values of a random variable with corresponding frequencies f_i's. For a grouped frequency data, the mean deviation about mean is obtained by replacing x_i by the mid-point of the ith class interval.

Example 2.5:

In the following table, the daily wages of 100 workers are given.

Daily Wages (in $)	Number of Workers
20	10
30	15
45	25
50	20
60	15
70	10
75	5

Find the mean deviation about mean of the wages.

SOLUTION

The mean of the wages is

$$\bar{x} = \frac{20 \times 10 + 30 \times 15 + 45 \times 25 + 50 \times 20 + 60 \times 15 + 70 \times 10 + 75 \times 5}{100} = 47.50$$

Thus, we have the following table.

Daily Wages (in $)	Number of Workers (f_i)	$d_i = \|x_i - \bar{x}\|$	$d_i \times f_i$
20	10	27.5	275
30	15	17.5	262.5
45	25	2.5	62.5
50	20	2.5	50
60	15	12.5	187.5
70	10	22.5	225
75	5	27.5	137.5

From the above table, we have $\sum d_i f_i = 1200$ and $\sum f_i = 100$.

Therefore, the mean deviation about mean is $\dfrac{\sum d_i f_i}{\sum f_i} = 12$.

2.5.3   MEAN DEVIATION ABOUT MEDIAN

Following the same argument as in the case of mean deviation about mean, the mean deviation about median is defined as the average absolute distance of the median from the data points. In order to estimate the mean deviation about median, we replace $\bar{x}$ by the median in the expression for the mean deviation about mean. It is illustrated in the following example.

Example 2.6:

Find the mean deviation about median of the following data.

9, 5, 11, 3, 7, 6, 13, 15, 0, 1, 20

SOLUTION

Verify that the median (M) of the given data is 7. The mean deviation about median is calculated using the following table.

x_i	$d_i = \lvert x_i - M \rvert$
0	7
1	6
3	4
5	2
6	1
7	0
9	2
11	4
13	6
15	8
20	13

From the above table, we have, $\Sigma d_i = 53$, $n = 11$.

Hence, the mean deviation about median is $\dfrac{\Sigma d_i}{11} = 4.818$.

2.5.4 VARIANCE AND STANDARD DEVIATION

The variance and standard deviation are widely used measures of dispersion. In the following, we illustrate how the population variance and sample variance can be obtained.

2.5.4.1 Variance and Standard Deviation of a Population

Suppose, $x_1, x_2, \ldots, x_n$ are the observed values of a random variable. The population variance of these observed values is given by

$$\sigma^2 = \frac{1}{n} \sum_{i=1}^{n} (x_i - \bar{x})^2$$

For a frequency data, the variance is given by

$$\sigma^2 = \frac{1}{n} \sum_{i=1}^{n} f_i (x_i - \bar{x})^2$$

where:

x_i = observed values of a random variable
f_i = corresponding frequencies

$$n = \sum_{i=1}^{n} f_i$$

For a grouped frequency data, the x_i is replaced by the mid-points of the upper-class boundary and the lower-class boundary.

The square root of the population variance is called population standard deviation (σ).

2.5.4.2 Variance and Standard Deviation of a Sample

The process of finding the sample variance differs from that for finding the population variance. The process of finding the sample variance will be evident in the following definition.

Suppose x_1, x_2, ...,x_n are the observed values of a random sample of size n from a population. Then the sample variance of this sample is given by

$$s^2 = \frac{1}{n-1}\sum_{i=1}^{n}(x_i - \bar{x})^2$$

One should note that we have divided the sum of squared difference by $n-1$ instead of n to make the sample variance unbiased (For detail see Section 6.6.1).

The square root of the sample variance is called the sample standard deviation.

Example 2.7:

Find the sample variance of the sample 5, 9, 11, 7, 2, 3, 12.

SOLUTION

The mean of the given sample is

$$\bar{x} = \frac{5+9+11+7+2+3+12}{7} = \frac{49}{7} = 7$$

Therefore, the sample variance is given by,

$$s^2 = \frac{1}{n-1}\sum_{i=1}^{n}(x_i - \bar{x})^2$$

$$= \frac{(5-7)^2 + (9-7)^2 + (11-7)^2 + (7-7)^2 + (2-7)^2 + (3-7)^2 + (12-7)^2}{7-1} = \frac{90}{6} = 15$$

Example 2.8:

Find the variance of the marks obtained by the students in statistics in Example 2.2.

SOLUTION

The mean of the marks is

$$\bar{x} = \frac{5\times5 + 15\times10 + 25\times22 + 35\times33 + 45\times30}{100} = 32.30$$

From the given information of the mark's distribution, we construct the following table.

Class Interval of the Marks (x)	Mid-point of the Class Interval (m_i)	Frequency (f_i)	$m_i - \bar{x}$	$(m_i - \bar{x})^2$	$f_i(m_i - \bar{x})^2$
0–10	5	5	−27.3	745.29	3726.45
10–20	15	10	−17.3	299.29	2992.90
20–30	25	22	−7.3	53.29	1172.38
30–40	35	33	2.7	7.29	240.57
40–50	45	30	12.7	161.29	4838.70

From the above table, we have $\displaystyle\sum_{i=1}^{5} f_i(m_i - \bar{x})^2 = 12971$.

Thus, the variance $= \dfrac{1}{100}\left(\displaystyle\sum_{i=1}^{5} f_i(m_i - \bar{x})^2\right) = 129.71$.

To reduce the complexity of the calculation we can take suitable linear transformation of the given data. Suppose Z is the transformed random variable obtained as $Z = \frac{X-A}{d}$, where A and d are chosen suitably. Using Proposition 3.3 (Chapter 3), it can be shown that

$$var(X) = d^2 var(Z)$$

where:
$var(X)$ = variance of X
$var(Z)$ = variance of Z

In this example, $A = 25$ which lies in the middle of column 2 of the above table and $d = 10$. Thus, we get the following modified table.

Class Interval of the Marks (x)	Mid-value of the Class Interval (m_i)	Frequency (f_i)	$z_i = \dfrac{m_i - A}{d}$	$z_i - \bar{z}$	$(z_i - \bar{z})^2$	$f_i(z_i - \bar{z})^2$
0–10	5	5	−2	−2.73	7.4529	37.2645
10–20	15	10	−1	−1.73	2.9929	29.9290
20–30	25	22	0	−0.73	0.5329	11.7238
30–40	35	33	1	0.27	0.0729	2.4057
40–50	45	30	2	1.27	1.6129	48.3870

The mean of z_i is

$$\bar{z} = \frac{(-2)\times 5 + (-1)\times 10 + 0\times 22 + 1\times 33 + 2\times 30}{100} = \frac{73}{100} = 0.73.$$

From the above table, we have

$$\sum_{i=1}^{5} f_i(z_i - \bar{z})^2 = 129.71.$$

Thus, the variance of Z is $\frac{129.71}{100} = 1.2971$. Therefore, the variance of the marks (X) is $d^2 var(Z) = 100 \times 1.2971 = 129.71$. The corresponding standard deviation of the marks is $\sqrt{129.71} = 11.39$.

2.5.5   Coefficient of Variation

The coefficient of variation ($CV\%$), also known as relative standard deviation, is defined by expressing the standard deviation (s) as a percentage of the mean ($\bar{x}$). Thus, the $CV\%$ is given by

$$CV\% = \frac{s}{\bar{x}} \times 100$$

$CV\%$ is often used to compare the variation in several sets of data. In Example 2.7, $CV\%$ can be calculated as follows:

$$CV\% = \frac{\sqrt{15}}{7} \times 100 = 55.33\%.$$

Exercises

2.1 The following values are the fabric areal densities $\left(\text{g/m}^2\right)$ of a sample of ten fabrics chosen at random from a large batch of similar fabrics:

110, 113, 112, 108, 115, 110, 111, 109, 110, 114

Calculate the mean, median and mode.

2.2 The following table gives the frequency distribution of the lengths of the 200 fibres. Find the mean, median and mode from this grouped frequency data of fibre length distribution.

Length Class (in mm)	Frequency
15–20	12
20–25	15
25–30	30
30–35	28
35–40	45
40–45	35
45–50	25
50–55	10

2.3 The marks of 100 students in mathematics out of 100 is given in the following table. Find the mean and standard deviation of the marks obtained by
the students.

Marks	Frequency
0–10	4
10–20	5
20–30	7
30–40	14
40–50	20
50–60	14
60–70	21
70–80	7
80–90	5
90–100	3

2.4 Find the mean deviation about the mean of the following distribution.

x	10	15	25	30	40	55	60
Frequency (f)	5	7	20	25	20	10	13

3 Probability

3.1 INTRODUCTION

We start this chapter with examples which will give the readers a feel for the concept of probability. Suppose Ram and Shyam are two friends and they are gambling. They will toss a coin and if head occurs Ram will get 10 USD form Shyam and if tail occurs, Shyam will get 10 USD from Ram. Now if it is asked: What is the chance that Ram will get 10 USD from Shyam? We consider another example. Suppose a die is thrown once. Again, the question is asked, what is the probability of getting a 4 or 6? In the first case, the chance of Ram winning is the same as the chance of getting head. There is one head out of two possible cases viz., 'head' or 'tail'. So the chance of getting a head is 1/2 and the probability that Ram will get 10 USD from Shyam is also 1/2. In the next case, the number of possible outcomes of a dice is 6. Hence the chance of getting a 4 or 6 is $2/6 = 1/3$.

In the above two examples it is noted that in both the cases the probabilities are the ratios of the number of cases we are interested in and the total number of possible cases. Before giving the formal definition of probability we must mention some useful terminologies.

3.2 BASIC TERMINOLOGIES

3.2.1 RANDOM EXPERIMENT

Definition 3.1: An experiment or observation that can be repeated a large number of times under almost the same conditions, and the possible outcomes are known but unpredictable is known as a random experiment.

As an example, the tossing of a coin is a random experiment. The possible outcomes are 'head' or 'tail' which is known prior to performing the toss, but we cannot say definitely which one will occur prior to the tossing.

3.2.2 SAMPLE SPACE

Definition 3.2: The set of all possible outcomes of an experiment is called sample space.

Sometimes, the elements of sample space are called sample points. For example, let us consider the simplest experiment of coin tossing. The possible outcomes are 'head' (say H) or 'tail' (say T). So the set $S = \{H, T\}$ is the sample space. Similarly, if we throw a die the possible outcomes are $1, 2, 3, \ldots, 6$. In this case the sample space is given by $S = \{1, 2, 3, \ldots, 6\}$.

DOI: 10.1201/9781003081234-3

Now we need the idea of event. An event is an outcome of an experiment. The sample points are necessarily events but an event may not be a sample point. In order to explain this, let us suppose that we throw a die. The sample space here is $S = \{1, 2, 3, ..., 6\}$. The occurrence of 1, 2, 3, etc. are sample points which belong to S. But the event of the occurrence of say, '2 or 6' is not in the set S. Hence, we need to construct a bigger set. This set is called the event space which is defined as follows:

Definition 3.3: Suppose E is an experiment and S is the corresponding sample space. A set ω is called event space if

    i. the sample space $S \in \omega$
    ii. an event $A \in \omega$ implies $A^c \in \omega$
    iii. $A_1, A_2, A_3, ... \in \omega$ implies $A_1 \cup A_2 \cup A_3 \cup ... \in \omega$

These three conditions imply that the set ω contains the null event and it is closed under a countable intersection.

3.2.3 MUTUALLY EXCLUSIVE EVENTS

Definition 3.4: Two or more events are called mutually exclusive if any one of the events occur then the possibility of occurrence of any other event from the remaining set of events is ruled out.

If the events are considered as sets then two events, say A and B, are mutually exclusive if $A \cap B = \phi$. Suppose a coin is tossed. Then the outcomes H and T are mutually exclusive events because if H appears then T cannot appear simultaneously and vice versa. Analogously, if a die is thrown the outcomes 1, 2, 3,....,6 are mutually exclusive since any two numbers from 1 to 6 cannot occur simultaneously in a single throw.

3.2.4 EXHAUSTIVE EVENTS

Definition 3.5: A collection of events is said to be exhaustive if they constitute the whole sample space.

If we consider the experiment of throwing of a die then the events of appearing 1, 2, ..., 6 are exhaustive events. Note that they are also mutually exclusive events.

3.2.5 INDEPENDENT EVENTS

Definition 3.6: Two or more events are said to be independent if they are not dependent.

For example, suppose two friends A and B are tossing coins. In this case the event that 'a head will appear in the toss of friend A' is independent of the event that 'a head will appear in the toss of friend B'.

3.3   CONCEPT OF PROBABILITY

3.3.1   Classical Definition of Probability

Definition 3.7: The classical definition of probability is given by

$$\text{Probability of an event} = \frac{\text{Number of cases favourable to the event in the experiment}}{\text{Total number of possible cases in the experiment}}$$

Though the classical definition finds its application in many areas, it is not free from defects. To explain, suppose we are performing the toss of a coin. According to the classical definition of probability given above, the probability of getting a head or tail is 1/2. The answer is true as long as the coin is unbiased, that is when both the sides of the coin are equally likely to occur. But if the coin is not unbiased (we can add some extra weight on one of the sides to make it a biased coin), the chance of getting a head or tail will no longer be the same. In this case, the classical definition will not work to have the exact probability of obtaining a head or tail. If we wish to get the exact probability of getting say, head, we need to perform the experiment repeatedly. Each time we record the relative frequency of head, that is, the number of heads appearing divided by the number of repetitions of the experiment. Initially, the value of the relative frequency fluctuates and the fluctuation decreases as we go on increasing the number of experiments. Theoretically, the value of the relative frequency will converge to a fixed value if the repetition is performed infinitely many times. This is known as statistical regularity. Under this circumstance the axiomatic definition of probability is more appropriate which is given next.

3.3.2   Axiomatic Definition of Probability

Definition 3.8: Let $S(\neq \varnothing)$ be a sample space corresponding to an experiment and ω be the set of all events. A function $P: \omega \rightarrow R$ is called probability of an event, A (say) if it satisfies the following conditions:

 i. $P(A) \geq 0 \,\forall\, A \in \omega$
 ii. $P(S) = 1$
 iii. If $A_1, A_2, ..., A_n \in \omega$ is a collection of pairwise disjoint events then

$$P\left(\bigcup_{i=1}^{n} A_i\right) = \sum_{i=1}^{n} P(A_i)$$

From the definition given above we have the following properties:

 i. $P(\varnothing) = 0$
 ii. $P(A^C) = 1 - P(A)$

3.3.3   Frequency Definition of Probability

In many cases, it may not be possible to find the exact probability of an event by using the classical definition or the axiomatic definition. For an example, if someone is assigned a job of finding the probability of occurrence of heads with a biased coin, the classical definition cannot be applied. What one can do in this case is to toss the biased coin a large number, say N times, and observe the number of heads occurred, say n. Then, the estimate of the probability of getting head is

$$P(H) = \frac{n}{N}$$

If the number of tosses increases, the estimated probability will come closer to the actual probability. Theoretically, the estimated probability will converge to the actual probability as $N \to \infty$ if the tosses are carried out under identical conditions. Thus, the frequency definition of the probability of an event reads as follows

$$P = \frac{\text{Frequency of the occurence of the event}}{\text{Total frequency}}$$

For another example, suppose a yarn cross-section consists of 200 fibres among which 75 fibres have lengths greater than 30 mm and less than 40 mm. Then an estimate of the probability that a chosen fibre will have a length greater than 30 mm and less than 40 mm is $\frac{75}{200} = \frac{3}{8}$.

3.4 SET THEORETIC APPROACH TO PROBABILITY

The outcomes of an experiment can be treated as sets. The usual combinations of events such as simultaneous occurrence of events, the occurrence of at least one event etc., can be managed by using the usual set operations such as union, intersection etc. In particular, the occurrence of either an event A or an event B is denoted by $A \cup B$ and we use $A \cap B$ to denote that the events A and B occurs simultaneously. For better understanding, suppose two friends among which one is tossing a coin and the other is throwing a die simultaneously. Let A denotes the event of the occurrence of a 'head' in the first friend's toss and B denotes the event of occurring 6 in the second friend's throwing of die. Then $A \cup B$ represents the event that either a head will occur in the toss or number 6 will appear in the throwing of the die. The Venn diagram can help us represent the events such as $A \cup B$ and $A \cap B$. The shaded regions of the Venn diagrams in Figures 3.1 and 3.2 represent $A \cup B$ and $A \cap B$, respectively.

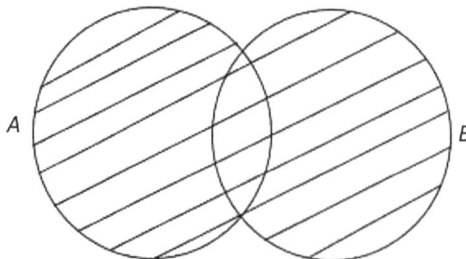

FIGURE 3.1   Venn diagram showing $A \cup B$.

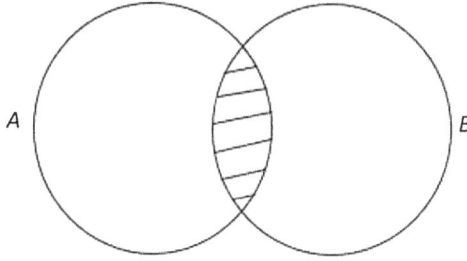

FIGURE 3.2 Venn diagram showing $A \cap B$.

The following theorems are useful.

Theorem 3.1: If A and B are two events, then $P(A \cup B) = P(A) + P(B) - P(A \cap B)$

Proof: Suppose that A and B are two events as shown in Figure 3.3. From Figure 3.3, we can write that

$$A \cup B = \left(A \cap B^c\right) \cup \left(B \cap A^c\right) \cup \left(A \cap B\right)$$

Thus, the set $A \cup B$ is written as the union of three disjoint sets viz. $A \cap B^c$, $B \cap A^c$ and $A \cap B$. Now, from the axiom (iii) of Section 3.3.2 we have

$$P(A \cup B) = P\left(A \cap B^c\right) + P\left(B \cap A^c\right) + P(A \cap B) \tag{3.1}$$

It can be shown that

$$P\left(A \cap B^c\right) = P(A) - P(A \cap B) \tag{3.2}$$

$$P\left(B \cap A^c\right) = P(B) - P(A \cap B) \tag{3.3}$$

So from Equation (3.1), using Equations (3.2) and (3.3), we get

$$P(A \cup B) = P(A) + P(B) - P(A \cap B)$$

This completes the proof.

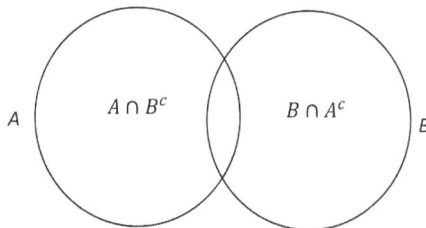

FIGURE 3.3 Venn diagram showing $A \cup B = \left(A \cap B^c\right) \cup \left(B \cap A^c\right) \cup \left(A \cap B\right)$.

For three events A, B and C, the above method can be applied to show that

$$P(A \cup B \cup C) = P(A) + P(B) + P(C) - P(A \cap B) - P(B \cap C) - P(A \cap C)$$
$$+ P(A \cap B \cap C)$$

We can generalize it further by the repeated application of the above theorem and hence obtain the following theorem.

Theorem 3.2: If A_i $(i = 1, 2, \ldots, n)$ are n events, then

$$P\left(\bigcup_{i=1}^{n} A_i\right) = \sum_{i=1}^{n} P(A_i) - \sum_{\substack{i_1, i_2 \\ i_1 < i_2}}^{n} P(A_{i_1} \cap A_{i_2}) + \sum_{\substack{i_1, i_2, i_3 \\ i_1 < i_2 < i_3}}^{n} P(A_{i_1} \cap A_{i_2} \cap A_{i_3})$$

$$- \cdots + (-1)^{r+1} \sum_{1 \le i_1 < i_2 < i_3 < \cdots < i_r \le n} P(A_{i_1} \cap A_{i_2} \cap A_{i_3} \cap \ldots \cap A_{i_r})$$

$$+ \cdots + (-1)^{n+1} P\left(\bigcap_{i=1}^{n} A_i\right)$$

Proposition 3.1: For any event A, $P(A^c) = 1 - P(A)$

Proof: If A be an event, then $A \cup A^c = S$ where S is the sample space. A and A^c are disjoint events. Hence $P(A \cup A^c) = P(S) = 1$ or $P(A) + P(A^c) = 1$ which implies $P(A^c) = 1 - P(A)$.

Proposition 3.2: If $E \subseteq F$, then $P(E) \le P(F)$

Proof: If $E \subseteq F$, $F = E \cup (F \cap E^c)$.

Therefore, $P(F) = P(E) + P(F \cap E^c)$ (Since, E and $F \cap E^c$ are disjoint)
Hence, $P(F) \ge P(E)$ since $P(F \cap E^c) \ge 0$

3.5 CONDITIONAL PROBABILITY

From the classical definition of probability, we have for any event A, the probability of A is given by

$$P(A) = \frac{\text{Number of cases favouring the event } A}{\text{Total number of cases}}$$

Now let us consider two events, say A and B. We define conditional probability of A, given B as

$$P(A \mid B) = \frac{n_{A \cap B}}{n_B} \tag{3.4}$$

where $n_{A \cap B}$ is the number of cases favouring the event $A \cap B$ that is the number of cases in which the events A and B occurs simultaneously. Similarly, n_B represents the number of cases in which the event B occurs. If we divide the numerator and denominator of the right-hand side of Equation (3.4) by n, the total number of possible outcomes of the corresponding random experiment, we have

$$P(A \mid B) = \frac{\dfrac{n_{A \cap B}}{n}}{\dfrac{n_B}{n}}$$

$$= \frac{P(A \cap B)}{P(B)}$$

This gives

$$P(A \cap B) = P(B)P(A \mid B) \tag{3.5}$$

It should be noted that if A and B are independent of each other, then $P(A \mid B) = P(A)$. In this case

$$P(A \cap B) = P(A)P(B) \tag{3.6}$$

So if two events are independent then the joint probability of the events is the product of the probabilities of the two events. This can be generalized for any finite number of events, i.e. if A_1, A_2, $\cdots$, A_n are n mutually independent events, then

$$P(A_1 \cap A_2 \cap \cdots \cap A_n) = \prod_{i=1}^{n} P(A_i) \tag{3.7}$$

Example 3.1:

Suppose there are 3 identical bags containing red and white balls. Bag 1 contains 3 white, 4 red balls. Bag 2 contains 2 white, 5 red balls and bag 3 contains 3 white and 5 red balls. A die is thrown to choose the bag. If 1 or 2 appears bag 1 is selected. If 3, 4 or 5 appears bag 2 is selected, else bag 3 is selected. A bag is selected by throwing the die and a pair of balls are selected randomly from the chosen bag. What is the probability that one ball is red and one ball is white among the drawn balls?

SOLUTION

Suppose E_i (i = 1, 2, 3) denotes the event that the ith bag is chosen. Now,

$$P(E_1) = P(\text{getting 1 or 2 when the die is thrown}) = 1/3$$

Similarly,

$$P(E_2) = \frac{1}{2} \text{ and } P(E_3) = \frac{1}{6}$$

Let A be the event that one ball is red and one ball is white among the drawn balls. It is evident that if no bag is chosen, the event A cannot occur. So, we can write

$$A \subseteq E_1 \cup E_2 \cup E_3$$

From the knowledge of set theory, we can write

$$A = (A \cap E_1) \cup (A \cap E_2) \cup (A \cap E_3) \tag{3.8}$$

Since E_i's are mutually exclusive events, $A \cap E_i$'s are also mutually exclusive. From Equation (3.8), we have

$$P(A) = P(A \cap E_1) + P(A \cap E_2) + P(A \cap E_3)$$
$$= P(E_1)P(A|E_1) + P(E_2)P(A|E_2) + P(E_3)P(A|E_3) \tag{3.9}$$

Now, $P(A|E_i)$ indicates the probability of getting one red and one white ball given that the ith bag is selected and two balls are selected randomly from it. We can see that if bag 1 is selected, then the probability of getting one red and one white ball is

$$P(A|E_1) = \frac{\binom{3}{1} \times \binom{4}{1}}{\binom{7}{2}} = \frac{4}{7}$$

Similarly, we can show that

$$P(A|E_2) = \frac{\binom{2}{1} \times \binom{5}{1}}{\binom{7}{2}} = \frac{10}{21} \text{ and } P(A|E_3) = \frac{\binom{5}{1} \times \binom{3}{1}}{\binom{8}{2}} = \frac{15}{28}$$

Putting these values in Equation (3.9) we get

$$P(A) = \frac{29}{56}$$

3.5.1 Baye's Rule

To have a better idea about the Baye's rule we consider Example 3.1. In this example we have evaluated the probability of getting one red and one white ball among the two drawn balls which is dependent on the event of selection of bags. In this case we are to evaluate the probability of an event based on the information about a past event. But if we are interested to find the probability that the ith $(i=1,2,3)$ bag is chosen given that among the drawn balls one is red and one is white, we actually need to find the probability of a past event based on the information given for the future event. This kind of problems can be solved using Baye's theorem which is discussed in the following.

Theorem 3.3 (Baye's theorem): If E_i $(i=1,2,3,\ldots,n)$ are n mutually exclusive events with $P(E_i)\neq 0$ for $i=1,2,3,\ldots,n$ and $A\subseteq\bigcup_{i=1}^{n}E_i$ with $P(A)\neq 0$, then

$$P(E_i\mid A)=\frac{P(E_i)P(A\mid E_i)}{\sum_{i=1}^{n}P(E_i)P(A\mid E_i)}$$

Proof: We have

$$P(E_i\mid A)=\frac{P(A\cap E_i)}{P(A)}=\frac{P(E_i)P(A\mid E_i)}{P(A)} \qquad (3.10)$$

Since $A\subseteq\bigcup_{i=1}^{n}E_i$, we have, $A=\bigcup_{i=1}^{n}(E_i\cap A)$. It is given that all E_i's are mutually exclusive and hence all $E_i\cap A$'s are also so. Therefore,

$$P(A)=\sum_{i=1}^{n}P(E_i\cap A)=\sum_{i=1}^{n}P(E_i)P(A\mid E_i).$$

Therefore, from Equation (3.10) we have

$$P(E_i\mid A)=\frac{P(E_i)P(A\mid E_i)}{\sum_{i=1}^{n}P(E_i)P(A\mid E_i)}$$

To demonstrate the use of Bayes' theorem we consider the last example with a minor change.

Example 3.2:

Suppose there are 3 identical bags containing red and white balls. Bag 1 contains 3 white, 4 red balls. Bag 2 contains 2 white, 5 red balls and bag 3 contains 3 white and 5 red balls. A die is thrown to choose the bag. If 1 or 2 appears bag 1 is selected. If 3, 4 or 5 appears bag 2 is selected, else bag 3 is selected. A bag is selected by throwing the die and a pair of balls are selected randomly from the chosen bag. What is the probability that bag 2 was chosen given that one ball is red and one ball is white among the drawn balls?

SOLUTION

One should note that in this example the outcome of a future event (in this case drawing of the 2 balls) is known and based on this information we need to find the probability of a past event that is selection of bag 2.

Let $E_i (i = 1, 2, 3)$ be the event that the ith bag is chosen. Therefore, we have

$$P(E_1) = P\left(\text{getting 1 or 2 when the die is thrown}\right) = \frac{1}{3}$$

Similarly,

$$P(E_2) = \frac{1}{2} \text{ and } P(E_3) = \frac{1}{6}.$$

Let A be the event that one ball is red and one ball is white among the drawn balls. Here, we need to find $P(E_2 \mid A)$ which can be obtained by using Bayes' theorem as follows.

$$P(E_2 \mid A) = \frac{P(E_2)P(A \mid E_2)}{P(E_1)P(A \mid E_1) + P(E_2)P(A \mid E_2) + P(E_3)P(A \mid E_3)}$$

$$= \frac{\dfrac{1}{2} \times \dfrac{10}{21}}{\dfrac{1}{3} \times \dfrac{4}{7} + \dfrac{1}{2} \times \dfrac{10}{21} + \dfrac{1}{6} \times \dfrac{15}{28}}$$

$$= \frac{40}{87}.$$

3.6 RANDOM VARIABLE

Definition 3.9: A random variable X is a function from the sample space S to the set of real numbers.

For an example, let us consider that a coin is tossed 2 times and X denotes the number of heads. The sample space consists of the events HH, HT, TH and TT. We call these events as ω_1, ω_2, ω_3 and ω_4, respectively. So, the sample space is $S = \{\omega_1, \omega_2, \omega_3, \omega_4\}$. Therefore, we have $X(\omega_1) = 2$. So we can write,

$$X(\omega) = 2 \text{ if } \omega = \omega_1$$
$$= 1 \text{ if } \omega \in \{\omega_2, \omega_3\}$$
$$= 0 \text{ if } \omega = \omega_4$$

Note that the random variable X acts on the events of the sample space giving some real numbers. This establishes the fact that X is a function on S. Furthermore, we can write

$$P(X = 2) = P(\omega_1) = \frac{1}{4}$$

$$P(X = 1) = P(\omega_2 \cup \omega_3) = P(\omega_2) + P(\omega_3) = \frac{1}{4} + \frac{1}{4} = \frac{1}{2} \text{ and so on.}$$

Random variables are of two kinds, viz., discrete random variable and continuous random variable. A random variable is called discrete random variable if the set of its possible values is either finite or countably infinite. On the contrary X is a continuous random variable if the set of its possible values is not finite and uncountable, in particular, the continuous random variable assumes its values in an interval. In the above example the random variable X is a discrete random variable. The body weight of the students of a college may be considered to be a continuous random variable.

3.6.1 PROBABILITY MASS FUNCTION

Definition 3.10: For a discrete random variable X, if a function $f(.)$ exists such that $P(X = x) = f(x)$, then $f(x)$ is called probability mass function (p. m. f.) of X.

Thus, $f(x)$ represents the probability that the random variable X assumes value x. The probability mass function has the following properties:

i. $f(x) \geq 0 \ \forall x$

ii. $\displaystyle\sum_x f(x) = 1$

In the aforementioned example of the toss of two coins, the probability mass function can be written as,

$$f(x) = \frac{1}{4} \text{ if } x \in \{0, 2\}$$

$$= \frac{1}{2} \text{ if } x = 1$$

Example 3.3:

Following table shows the probability mass function of a discrete random variable X.

$X = x$	-2	-1	0	1	2
$f(x) = P(X = x)$	0.2	0.1	k	$2k$	0.4

a. Find the value of k.
b. Find $P(X < 0)$ and $P(-2 < X \leq 1)$.

SOLUTION

a. Since $f(x)$ is a probability mass function, so we must have

$$0.2 + 0.1 + k + 2k + 0.4 = 1.$$

Thus,

$$k = 0.1$$

b. $P(X < 0) = P(X = -2) + P(X = -1) = 0.2 + 0.1 = 0.3$
   Again,

$$P(-2 < X \leq 1) = P(X = -1) + P(X = 0) + P(X = 1) = 0.1 + 0.1 + 0.2 = 0.4$$

3.6.2   PROBABILITY DENSITY FUNCTION

In the previous section we have observed that in case of discrete random variable we can define probability mass function. So it is natural to expect a similar kind of function for a continuous random variable. A continuous random variable takes values in an interval. For an example, we choose a point randomly from the closed interval $[0, 1]$. Suppose X is the random variable which represents the number chosen randomly from the said interval. In this case, the probability that X will assume a value in the interval $[0, 1]$ is actually zero since all the points in the interval are equally likely to be selected. But the probability that X will take value in a subinterval contained in the given interval $[0, 1]$ will be nonzero. In fact, it can be shown that for a continuous random variable X, $P(X = x) = 0$ for any x, showing that the probability mass function for a continuous random variable is identically zero. To resolve this issue probability density function is used in place of probability mass function.

Definition 3.11: For a continuous random variable X, if there exists a function $f(.)$ such that $P(x \leq X \leq x + dx) = f(x)dx$, then $f(.)$ is called probability density function (p. d. f.) of X.

The probability density function has the following useful properties:

   i. $f(x) \geq 0 \ \forall x$

   ii. $\int_{-\infty}^{\infty} f(x)dx = 1$

   iii. $P(a \leq X \leq b) = \int_{a}^{b} f(x)dx$

Using (iii), we observe that $P(a \leq X \leq a) = P(X = a) = \int_{a}^{a} f(x)dx = 0$ for any a contained in the range of X.

3.6.3   PROBABILITY DISTRIBUTION FUNCTION

Definition 3.12: For a given random variable X(discrete or continuous), a function $F(.)$ is called cumulative distribution function (c. d. f.) or probability distribution function if $F(x) = P(X \leq x)$ where $-\infty < x < \infty$.

So $F(x)$ denotes the probability that the value of the random variable X is less or equal to x. The probability distribution function has the following properties:

   i. $F(\infty) = 1$ and $F(-\infty) = 0$
   ii. $F(x)$ is a nondecreasing function; that is, if $a < b$, $F(a) \leq F(b)$
   iii. F is right continuous. That is for any point c $\lim_{x \to c+} F(x) = F(c)$.

For a discrete random variable X, the distribution function $F(x)$ has a jump discontinuity at every point at which the probability is non-zero. Suppose for a discrete

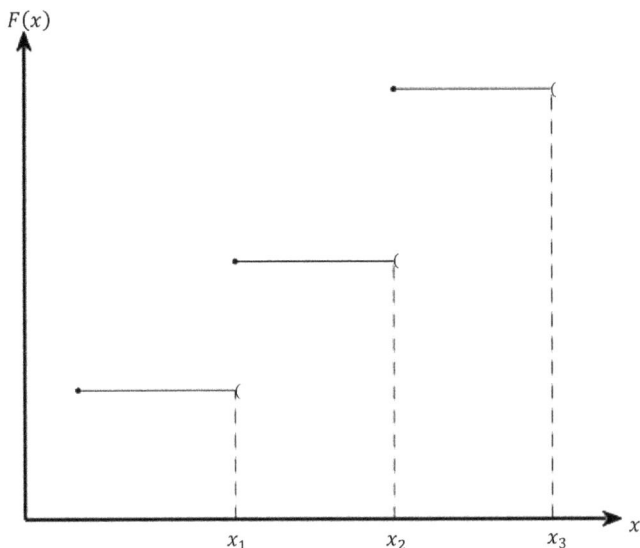

$F(x)$

x_1 x_2 x_3 x

FIGURE 3.4   Probability distribution function of a discrete random variable X.

random variable X the probability mass function and the probability distribution function are $f(x)$ and $F(x)$, respectively. Then

$$f(x) = F(x) - \lim_{t \to x-} F(t)$$

On the contrary, the probability distribution function of a continuous random variable is continuous. For better understanding, the probable graph of a distribution function of a discrete random variable is shown in Figure 3.4.

Example 3.4:

Show that

$$P(a < X \leq b) = F(b) - F(a) \text{ for } a < b.$$

SOLUTION

The event $\{X \leq b\} = \{X \leq a\} \cup \{a < X \leq b\}$
  Since the events $\{X \leq a\}$ and $\{a < X \leq b\}$ are disjoint, we can write

$$P(\{X \leq b\}) = P(\{X \leq a\}) + P(\{a < X \leq b\})$$

or,
$$F(b) = F(a) + P(\{a < x \leq b\})$$
  Thus,

$$P(\{a < x \leq b\}) = F(b) - F(a)$$

Example 3.5:

Find the probability distribution function of the random variable X whose probability density function is given by

$$f(x) = x, \text{ for } 0 < x < 1$$

$$= 2 - x, \text{ for } 1 \le x < 2$$

$$= 0, \text{ elsewhere.}$$

SOLUTION

From the definition of probability distribution function we have

$$F(x) = P(X \le x) = \int_{-\infty}^{x} f(t)\, dt$$

We consider the following cases.

Case 1: $-\infty < x < 0$

In this case, $F(x) = P(X \le x) = \int_{-\infty}^{x} 0\, dt = 0$

Case 2: $0 \le x < 1$

$$F(x) = P(X \le x) = \int_{-\infty}^{0} f(t)\, dt + \int_{0}^{x} f(t)\, dt = \int_{0}^{x} t\, dt = \frac{x^2}{2}$$

Here we consider the interval $0 \le x < 1$ instead of $0 < x < 1$. This is admissible since the given distribution is a continuous distribution and hence addition or deletion of a single point does not make difference in the cumulative distribution.

Case 3: $1 \le x < 2$

$$F(x) = P(X \le x) = \int_{-\infty}^{0} f(t)\, dt + \int_{0}^{1} f(t)\, dt + \int_{1}^{x} f(t)\, dt$$

$$= 0 + \int_{0}^{1} t\, dt + \int_{1}^{x} (2 - t)\, dt$$

$$= 1 - \frac{(2 - x)^2}{2}$$

Case 4: $2 \le x < \infty$

For this case $F(x) = 1$ since for $x \ge 2$

$$F(x) = P(X \le x) = \int_{-\infty}^{0} f(t)\, dt + \int_{0}^{1} f(t)\, dt + \int_{1}^{2} f(t)\, dt + \int_{2}^{x} f(t)\, dt = 1.$$

It should be noted that the first and the last integrals in the above expression are zero.

Thus, the distribution function can be written as

$$F(x) = 0, \text{ if } -\infty < x < 0$$

$$= \frac{x^2}{2}, \text{ if } 0 \le x < 1$$

$$= 1 - \frac{(2-x)^2}{2}, \text{ if } 1 \le x < 2$$

$$= 1, \text{ if } 2 \le x$$

Example 3.6:

If the probability distribution function of a random variable is given by

$$F(x) = 0 \text{ if } x \le 1$$

$$= x - 1 \text{ if } 1 < x < 2$$

$$= 1 \text{ if } x \ge 2,$$

then find the probability density function of the random variable.

SOLUTION

The given distribution function is differentiable in the open intervals $(-\infty, 1)$, $(1, 2)$ and $(2, \infty)$. Taking differentiation in the respective intervals, we get

$$\frac{dF}{dx} = 0 \text{ if } x \in (-\infty, 1)$$

$$= 1 \text{ if } x \in (1, 2)$$

$$= 0 \text{ if } x \in (2, \infty)$$

We need to check the differentiability at the point 1. Now, we have

$$\lim_{h \to 0+} \frac{F(1+h) - F(1)}{h} = \lim_{h \to 0+} \frac{1+h-1-0}{h} = 1$$

This is the right-hand limit at the point 1.

Again,

$$\lim_{h \to 0+} \frac{F(1-h) - F(1)}{h} = \lim_{h \to 0+} \frac{0-0}{h} = 0$$

So the left-hand limit is 0 at the point 1 which shows that the right-hand limit and the left-hand limit are not same and hence the given distribution function is not differentiable at $x = 1$. Using the same argument, it is easy to verify that the given distribution function is not differentiable at $x = 2$. Hence the required probability density function can be written as

$$f(x) = 0 \text{ for } x \in (-\infty, 1)$$

$$= 1 \text{ for } x \in (1, 2)$$

$$= 0 \text{ for } x \in (2, \infty)$$

We can assume the values of $f(1)$ and $f(2)$ to be 0 since the given distribution is a continuous one. Thus, we can write,

$$f(x) = 1 \text{ for } x \in (1, 2)$$

$$= 0 \text{ elsewhere}$$

3.7 EXPECTATION OF A RANDOM VARIABLE

The expectation of a random variable X is nothing but the mean or average value of the random variable and it is an important concept in probability theory. The expected value of a random variable X is the weighted average of the possible values that X can take on. For a discrete random variable X, the expectation is defined as

$$E(X) = \sum_{x \in S} x f(x)$$

where $f(x)$ is the probability mass function of the discrete random variable X and S is the set of all possible values which X can take on. Whereas, for a continuous random variable X, the expectation is defined as

$$E(X) = \int_{-\infty}^{\infty} x f(x) dx$$

where $f(x)$ is the probability density function of the continuous random variable X.

Example 3.7:

A discrete random variable X can take values {0, 1, 2} with $P(X = 0) = \frac{2}{3}$, $P(X = 1) = \frac{1}{6}$ and $P(X = 2) = \frac{1}{6}$. Find the expectation of X.

SOLUTION

Here, the random variable X is a discrete random variable and hence the expectation of X is given by

$$E(X) = 0 \times \frac{2}{3} + 1 \times \frac{1}{6} + 2 \times \frac{1}{6} = \frac{1}{2}.$$

Example 3.8:

If the probability density function of a continuous random variable X is given by

$$f(x) = 3x^2, \text{ if } 0 \le x \le 1$$

$$= 0, \text{ elsewhere.}$$

Determine the expectation of X.

SOLUTION

$$\text{Here, } E(X) = \int_{-\infty}^{\infty} x\, f(x)\, dx = \int_{-\infty}^{0} x\, f(x)\, dx + \int_{0}^{1} x\, f(x)\, dx + \int_{1}^{\infty} x\, f(x)\, dx$$

$$= 0 + \int_{0}^{1} x.3x^2 dx + 0 = 3.\left[\frac{x^4}{4}\right]_0^1 = \frac{3}{4}.$$

3.7.1 EXPECTATION OF THE FUNCTION OF A RANDOM VARIABLE

In many applications we are not only interested in the expectation of a random variable X, but also in the expectation of $g(X)$ which is the function of the random variable X. If X is a discrete random variable and $f(x)$ is the value of its probability mass function at x, the expected value of $g(X)$ is given by

$$E(g(X)) = \sum_{x \in S} g(x) f(x)$$

where S is the set of all possible values which the random variable X can take on. Similarly, if X is a continuous random variable and $f(x)$ is the value of its probability density function at x, the expected value of $g(X)$ is given by

$$E(g(X)) = \int_{-\infty}^{\infty} g(x) f(x)\, dx$$

The expectation of the function of a random variable has the following properties:

1. $E(c) = c$ where c is any real constant.
2. $E(a + bX) = a + bE(X)$ where a and b are constants.
3. $E(XY) = E(X)E(Y)$ if X and Y are independent random variables.

Example 3.9:

For a continuous random variable X having the same probability density function as given in the Example 3.8, if $g(x) = x^2 + 1$, determine the expected value of $g(X)$.

SOLUTION

We have

$$E\big(g(X)\big) = \int_{-\infty}^{\infty} g(x)f(x)\,dx$$

$$= \int_{-\infty}^{0} g(x)f(x)\,dx + \int_{0}^{1} g(x)f(x)\,dx + \int_{1}^{\infty} g(x)f(x)\,dx$$

$$= 0 + \int_{0}^{1}\big(1+x^2\big)\cdot 3x^2\,dx + 0 = 3\cdot\left[\frac{x^5}{5}\right]_0^1 + 3\cdot\left[\frac{x^3}{3}\right]_0^1 = \frac{8}{5}$$

3.8   COVARIANCE AND VARIANCE

3.8.1   COVARIANCE

Definition 3.13: The covariance of two random variables X and Y is defined as

$$cov(X,Y) = E\big[(X - E(X))(Y - E(Y))\big]$$

We can show that

$$cov(X,Y) = E(XY) - E(X)E(Y)$$

The covariance of two random variables expresses the relationship between the random variables. For example, if two random variables are such that the increase in one of the random variables results in decrease in the other, then covariance between the two random variables will be negative. Whereas, if the increase in one random variable also results in an increase in the other, the covariance between them will be positive. Consider the volume and pressure of a gas. We know that the volume and pressure of a gas are inversely proportional and hence the covariance between them is expected to be negative. From this analogy, we must say that if two random variables are independent, the covariance between them will be zero. In fact, if X and Y are two independent random variables then,

$$cov(X,Y) = E(XY) - E(X)E(Y) = E(X)E(Y) - E(X)E(Y) = 0$$

3.8.2   VARIANCE

The variance of a random variable is the average of the squared distances of all possible values of the random variable from its mean or expectation. The variance gives the idea of how the data points are scattered in a given sample of data. Intuitively, we can say that if the data points are scattered far apart from each other or in other words variance in the data is high, it is not justified to represent the data by a single point like mean. The estimation of the variance gives us a clue as to how much we can rely on the mean or average of a random variable. The mathematical form of variance of a random variable X is given by

$$var(X) = E\left[\left(X - E(X)\right)^2\right]$$

$$= E\left[X^2 - 2XE(X) + E^2(X)\right]$$

$$= E(X^2) - 2E[XE(X)] + E[E^2(X)]$$

$$= E(X^2) - 2E^2(X) + E^2(X)$$

$$= E(X^2) - E^2(X)$$

Here we use $E[E(X)] = E(X)$ and $E[E^2(X)] = E^2(X)$ as $E(X)$ and $E^2(X)$ are merely constants. The square root of $var(X)$ is called the standard deviation of X.

Proposition 3.3: $var(aX + b) = a^2 \ var(X)$

Proof: We have

$$var(aX + b) = E\left[aX + b - a\mu - b\right]^2$$

where μ is the mean of X.
   Therefore,

$$var(aX + b) = E\left[aX - a\mu\right]^2$$

$$= E\left[a^2(X - \mu)^2\right]$$

$$= a^2 E\left[(X - \mu)^2\right]$$

$$= a^2 var(X)$$

Proposition 3.4:

If $X_1, X_2, \cdots, X_n$ are n random variables then,

$$var\left(\sum_{i=1}^{n} X_i\right) = \sum_{i=1}^{n} var(X_i) + 2\sum_{i<j} cov(X_i, X_j)$$

In particular if $X_1, X_2, \cdots, X_n$ are n mutually independent random variables,

$$var\left(\sum_{i=1}^{n} X_i\right) = \sum_{i=1}^{n} var(X_i)$$

since $cov(X_i, X_j) = 0$ for all $i \neq j$ (see Section 3.8.1)

As we mentioned earlier, mean is one of the most widely used central tendencies, so naturally we expect that the average distance of the all-possible values of the random variable from its mean cannot be infinitely large. This assurance is provided in the Chebyshev's theorem. Moreover, this theorem reveals the fact that the variance of a random variable is an indicator of the spread of the random variable. In the following we discuss the theorem.

Theorem 3.4 (Chebyshev's theorem): If μ and σ are the mean and standard deviation of a random variable X, then for any positive constant k

$$P\left(|X - \mu| < k\sigma\right) \geq 1 - \frac{1}{k^2}, \ \sigma \neq 0$$

Proof: According to the definition of variance, we have

$$\sigma^2 = E\left[(X - \mu)^2\right]$$

$$= \int_{-\infty}^{\infty} (x - \mu)^2 f(x)dx$$

$$= \int_{-\infty}^{\mu - k\sigma} (x - \mu)^2 f(x)dx + \int_{\mu - k\sigma}^{\mu + k\sigma} (x - \mu)^2 f(x)dx + \int_{\mu + k\sigma}^{\infty} (x - \mu)^2 f(x)dx$$

As the integrand $(x - \mu)^2 f(x)dx$ is positive, we can write the following inequality

$$\sigma^2 \geq \int_{-\infty}^{\mu - k\sigma} (x - \mu)^2 f(x)dx + \int_{\mu + k\sigma}^{\infty} (x - \mu)^2 f(x)dx$$

For $x \geq \mu + k\sigma$ and $x \leq \mu - k\sigma$, we can write

$$(x - \mu)^2 \geq k^2 \sigma^2$$

Hence,

$$\sigma^2 \geq \int_{-\infty}^{\mu - k\sigma} k^2 \sigma^2 f(x)dx + \int_{\mu + k\sigma}^{\infty} k^2 \sigma^2 f(x)dx$$

Or,

$$\sigma^2 \geq k^2 \sigma^2 \left[\int_{-\infty}^{\mu - k\sigma} f(x)dx + \int_{\mu + k\sigma}^{\infty} f(x)dx\right]$$

Therefore,

$$\frac{1}{k^2} \geq \int_{-\infty}^{\mu - k\sigma} f(x)dx + \int_{\mu + k\sigma}^{\infty} f(x)dx$$

provided $\sigma^2 \neq 0$. Since the sum of the two integrals on the right-hand side of the above inequality is the probability that X will take on a value less than or equal to $\mu - k\sigma$ or greater than or equal to $\mu + k\sigma$, we can write that

$$\frac{1}{k^2} \geq P(|X - \mu| \geq k\sigma)$$

or,

$$\frac{1}{k^2} \geq 1 - P(|X - \mu| < k\sigma).$$

Thus,

$$P\big[|X - \mu| < k\sigma\big] \geq 1 - \frac{1}{k^2}$$

For example, when $k = 2$, $1 - \frac{1}{k^2} = 1 - \frac{1}{2^2} = 0.75$, hence the probability that a random variable X will take on a value within $\mu \pm 2\sigma$ is at least 0.75. Similarly, when $k = 3$, $1 - \frac{1}{k^2} = 1 - \frac{1}{3^2} = 0.89$, hence the probability that a random variable X will take on a value within $\mu \pm 3\sigma$ is at least 0.89. Therefore, the Chebyshev's theorem confirms that the probability that a random variable X will take on a value within $\mu \pm k\sigma$ is at least $1 - \frac{1}{k^2}$. Furthermore, from Chebyshev's theorem we have $\frac{1}{k^2} \geq P(|X - \mu| \geq k\sigma)$ for a given positive constant k. Now, $\lim_{k \to \infty} P(|X - \mu| \geq k\sigma) \leq \lim_{k \to \infty} \frac{1}{k^2} = 0$. This establishes our earlier claim that the possibility that the absolute difference between any value of the random variable and its mean will be infinitely large is negligible.

3.9   MOMENT OF A RANDOM VARIABLE

In statistics the moments of a random variable are important since it gives significant information about the nature of the distribution like symmetry, peaked-ness etc. of the concerned random variable. The moment about origin and the moment about mean are two important factors in this direction. A detail description of the moment about origin and the moment about mean of a random variable is furnished below.

Definition 3.14: The rth order moment about the origin of a random variable is the expected value of X^r and is denoted by μ_r'.

Thus, $\mu_r' = E(X^r)$ for $r = 0, 1, 2 \cdots$. If $r = 0$, $\mu_0' = E(X^0) = E(1) = 1$ and if $r = 1$, $\mu_1' = E(X^1) = E(X) = \mu$, which is the mean of the given random variable X.

We now define another important moment of a random variable which is useful to describe the shape of the distribution of the random variable about its mean.

Definition 3.15: The rth order moment about the mean of a random variable X, denoted by μ_r, is the expected value of $(X - \mu)^r$.

Therefore,

$\mu_r = E\big((X - \mu)^r\big)$ for $r = 0, 1, 2, \cdots$. If $r = 0$, $\mu_0 = 1$ and if $r = 1$, $\mu_1 = E(X - \mu) = E(X) - \mu = \mu - \mu = 0$.

If $r = 2$, $\mu_2 = E\big((X - \mu)^2\big) = var(X)$. Thus, the 2nd order moment about the mean gives the variance.

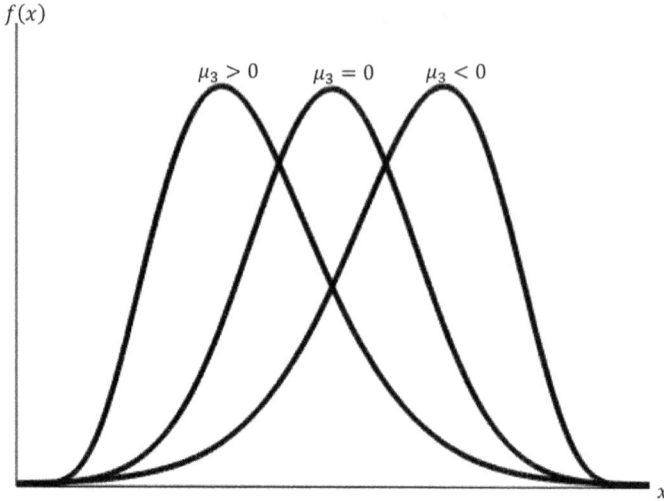

FIGURE 3.5   Positively skewed ($\mu_3 > 0$), symmetric ($\mu_3 = 0$) and negatively skewed ($\mu_3 < 0$) distribution.

If $r = 3$, $\mu_3 = E\left((X - \mu)^3\right)$. The 3rd order moment about the mean is important to measure the asymmetry about the mean of a distribution. If $\mu_3 = 0$, it means that the distribution is symmetric about the mean or, in other words, the graph of the probability density function on the right of the mean is the mirror image of that on the left of the mean. The violation of this implies that the distribution is not symmetric about the mean. For $\mu_3 > 0$, the distribution is termed as positively skewed and for $\mu_3 < 0$, the distribution is termed as negatively skewed (see Figure 3.5).

3.10   MOMENT GENERATING FUNCTION AND CHARACTERISTIC FUNCTION

3.10.1   Moment Generating Function

Definition 3.16: The moment generating function (MGF) of a random variable X is given by

$$M_X(t) = E\left(e^{tX}\right) = \sum_x e^{tx} f(x) \text{ when } X \text{ is discrete}$$

and

$$M_X(t) = \int_{-\infty}^{\infty} e^{tx} f(x) dx \text{ if } X \text{ is continuous}$$

where t is a real parameter.

This is called the moment generating function because the moments of all order of the random variable X can be obtained by differentiating the moment generating function with respect to t at $t = 0$. The details are discussed in the following.

We observe that for a discrete random variable X,

$$M_X(t) = E\left(e^{tX}\right) = \sum_x e^{tx} f(x)$$

By differentiating both sides with respect to t, we have

$$\frac{dM_X(t)}{dt} = \sum_x e^{tx} x f(x) \tag{3.11}$$

Therefore,

$$\left(\frac{dM_X(t)}{dt}\right)_{t=0} = \sum_x x f(x) = \mu_1'$$

Differentiating both sides of Equation (3.11) with respect to t again, we get

$$\frac{d^2 M_X(t)}{dt^2} = \sum_x e^{tx} x^2 f(x)$$

Thus,

$$\left(\frac{d^2 M_X(t)}{dt^2}\right)_{t=0} = \sum_x x^2 f(x) = \mu_2'$$

We can proceed further to show that $\left[\frac{d^r M_X(t)}{dt^r}\right]_{t=0} = \mu_r'$

For a continuous random variable, the same can be shown in a similar way.

Example 3.10:

Find the moment generating function of the random variable X whose probability density function is

$$f(x) = \frac{1}{2} e^{-|x|}, \ -\infty < x < \infty$$

SOLUTION

According to the definition of the moment generating function, we have

$$M_X(t) = \int_{-\infty}^{\infty} e^{tx} f(x) \, dx$$

$$= \int_{-\infty}^{0} \frac{1}{2} e^{tx} e^{-|x|} \, dx + \int_{0}^{\infty} \frac{1}{2} e^{tx} e^{-|x|} \, dx$$

$$= \int_{-\infty}^{0} \frac{1}{2} e^{tx} e^{x} \, dx + \int_{0}^{\infty} \frac{1}{2} e^{tx} e^{-x} \, dx$$

$$= \int_{-\infty}^{0} \frac{1}{2} e^{(t+1)x} \, dx + \int_{0}^{\infty} \frac{1}{2} e^{(t-1)x} \, dx$$

The 2nd integral of the above equation exists if $t < 1$. Under this assumption, we can write

$$M_X(t) = \frac{1}{2}\left(\frac{1}{1+t}\right) - \frac{1}{2}\left(\frac{1}{t-1}\right) = \frac{1}{1-t^2} ; \, t < 1$$

Theorem 3.5: If $M_X(t)$ is the moment generating function of the random variable X, then the moment generating function of the random variable $\frac{X+a}{b}$ is given by

$$M_{\frac{X+a}{b}}(t) = e^{\frac{a}{b}t} M_X\left(\frac{t}{b}\right)$$

where a and $b(\neq 0)$ are constants.

Proof: We have

$$M_X(t) = E\left(e^{tX}\right)$$

It therefore shows that

$$M_{\frac{X+a}{b}}(t) = E\left(e^{t\left(\frac{X+a}{b}\right)}\right) = E\left(e^{\frac{t}{b}X + t\frac{a}{b}}\right) = E\left(e^{\frac{t}{b}X} e^{t\frac{a}{b}}\right)$$

Note that $e^{t\frac{a}{b}}$ does not involve any randomness and so it assures that

$E\left(e^{\frac{t}{b}X} e^{t\frac{a}{b}}\right) = e^{t\frac{a}{b}} E\left(e^{\frac{t}{b}X}\right) = e^{t\frac{a}{b}} M_X\left(\frac{t}{b}\right)$. This completes the proof.

This theorem leads to the following corollaries depending on the different choices of the constants a and b.

Corollary 3.1: If $M_X(t)$ is the moment generating function of a random variable X, then $M_{X+a}(t) = e^{at} M_X(t)$

Proof: The proof is immediate from Theorem 3.5 if we put $b = 1$.

Corollary 3.2: If $M_X(t)$ is the moment generating function of a random variable X, then $M_{bX}(t) = M_X(bt)$.

Proof: In order to prove this corollary, we put $a = 0$, and replace b by $\frac{1}{b}$ in Theorem 3.5.

Theorem 3.6: If X_1, X_2, $\cdots, X_n$ are n independent random variables and a_1, $a_2 \cdots, a_n$ are n constants, then $M_{a_1 X_1 + a_2 X_2 + \cdots + a_n X_n}(t) = \prod_{i=1}^{n} M_{X_i}(a_i t)$.

Proof: From the definition of moment generating function (MGF), we have

$$M_{a_1 X_1 + a_2 X_2 + \cdots + a_n X_n}(t) = E\left[e^{t(a_1 X_1 + a_2 X_2 + \cdots + a_n X_n)}\right]$$

$$= E\left(e^{ta_1 X_1} e^{ta_2 X_2} \cdots e^{ta_n X_n}\right)$$

$$= E\left(e^{ta_1 X_1}\right) E\left(e^{ta_2 X_2}\right) \cdots E\left(e^{ta_n X_n}\right)$$

$$\left[\because X_1, X_2, ..., X_n \text{ are independent random variables}\right]$$

$$= M_{a_1 X_1}(t) M_{a_2 X_2}(t) \cdots M_{a_n X_n}(t)$$

$$= M_{X_1}(a_1 t) M_{X_2}(a_2 t) \cdots M_{X_n}(a_n t) \left[\text{Using Corollary 3.2}\right]$$

This completes the proof.

3.10.2 CHARACTERISTIC FUNCTION

Moment generating function is useful and finds its applications in many areas. Despite its usefulness, the main limitation of moment generating function is that it may not exist for a given random variable. This limitation is avoided in characteristic function which is defined as below.

Definition 3.17: The characteristic function of a random variable X is defined as

$$Q_X(t) = E\left(e^{itX}\right) \text{ where } i = \sqrt{-1} \text{ and } t \in \mathbb{R}.$$

The characteristic function of a random variable always exists finitely. We will show the existence of characteristic function for a discrete random variable. One can verify it for a continuous random variable as well.

For a discrete random variable X having probability mass function $f(x)$, the characteristic function can be written as

$$Q_X(t) = E\left(e^{itX}\right) = \sum_x e^{itx} f(x)$$

Therefore,

$$\left|Q_X(t)\right| = \left|\sum_x e^{itx} f(x)\right| \le \sum_x \left|e^{itx} f(x)\right| = \sum_x \left|e^{itx}\right| \left|f(x)\right| = \sum_x f(x) = 1$$

This shows that $\left|Q_X(t)\right| \le 1$ and hence it exists finitely. In Example 3.10 we have shown that the moment generating function for the given random variable does not exist for $t \ge 1$. We will show that the characteristic function of the said random variable exists as illustrated in the following.

The probability density function of the random variable is $f(x) = \frac{1}{2}e^{-|x|}$, $-\infty < x < \infty$. Therefore, the characteristic function is given by

$$Q_X(t) = \int_{-\infty}^{\infty} e^{itx} f(x) dx = \frac{1}{2} \int_{-\infty}^{0} e^{itx} e^x dx + \frac{1}{2} \int_{0}^{\infty} e^{itx} e^{-x} dx = \frac{1}{1+t^2}$$

which exists for all real t.

3.11 MULTIVARIATE DISTRIBUTION AND TRANSFORMATION OF RANDOM VARIABLES

So far, we have considered the issues related to one dimensional random variables. In many cases like resolving some industrial problems, multiple random variables are associated. For example, in manufacturing yarn, the simultaneous distribution of the fineness of fibres and the length of fibres may give useful information regarding the quality of the yarn. In such situations we need some knowledge about the distribution of multivariate random variables.

3.11.1 JOINT DISTRIBUTION AND TRANSFORMATION OF DISCRETE RANDOM VARIABLES

Definition 3.18: If X_1, X_2, $\cdots$, X_n are n discrete random variables, the function given by $f(x_1, x_2, \cdots, x_n) = P(X_1 = x_1, X_2 = x_2, \cdots, X_n = x_n)$, for each n tuple $(x_1, x_2, \cdots, x_n)$ is called the joint probability mass function of $(X_1, X_2, \cdots, X_n)$.

The joint probability distribution of two or more random variables are primarily related to two probability distributions functions viz., marginal probability distribution function and conditional probability distribution function.

3.11.1.1 Marginal Probability Distribution

Definition 3.19: Suppose X_1, X_2, $\cdots$, X_n are n discrete random variables having joint probability mass function $f(x_1, x_2, \cdots, x_n)$, then the marginal probability mass function of X_i is given by

$$g(x_i) = \sum_{\substack{x_j, j=1,2,\cdots,n \\ j \ne i}} f(x_1, x_2, \cdots, x_n)$$

3.11.1.2   Conditional Probability Distribution

Definition 3.20: Suppose $X_1, X_2, \cdots, X_n$ are n discrete random variables having probability mass function $f(x_1, x_2, \cdots, x_n)$, then the conditional probability mass function of the random variables X_j, $j = 1, 2, \ldots, i-1, i+1, \ldots, n$ given X_i is

$$P(x_1, x_2, \ldots, x_{i-1}, x_{i+1}, \ldots, x_n \mid X_i = x_i) = \frac{f(x_1, x_2, \cdots, x_n)}{g(x_i)}$$

where $g(x_i)$ is the marginal probability mass function of the random variable X_i.

In particular for two random variables X and Y, we can write

$$P(X = x, Y = y) = P(X = x) P(Y = y | X = x) = f_X(x) f_{Y|X}(y)$$

In the above expression $f_X(x)$ is the marginal probability mass function of X and $f_{Y|X}(y)$ is the conditional probability mass function of Y given X. Alternatively,

$$P(X = x, Y = y) = P(Y = y) P(X = x | Y = y) = f_Y(y) f_{X|Y}(x)$$

Similar to the previous case, $f_Y(y)$ is the marginal probability mass function of Y and $f_{X|Y}(x)$ is the conditional probability mass function of X given Y. Furthermore, we can also define the joint probability distribution function of X and Y as follows:

$$F(x, y) = P(X \leq x, Y \leq y) = \sum_{t \leq y} \sum_{s \leq x} f(s, t)$$

where $f(s, t)$ is the value of the joint probability mass function of X and Y at (s, t).

We extend our study by defining the conditional probability distribution function. The conditional probability distribution function of X given Y is given by,

$$F_{X|Y}(x) = P(X \leq x | Y = y)$$

Similarly, the conditional probability distribution function of Y given X can also be defined.

If X and Y are independent random variables, then

$$f(x, y) = P(X = x) P(Y = y) = f_X(x) f_Y(y)$$

and

$$F(x, y) = F_X(x) F_Y(y)$$

where $F_X(x)$ and $F_Y(y)$ are the marginal distribution function of X and Y, respectively.

Example 3.11:

The joint distribution of the random variables X and Y is given in the following table.

x \ y	0	1	2	3
0	0.1	0.05	0.025	0.025
1	0.015	0.045	0.1	0.05
2	0.19	0.15	0.1	0.15

Find

a. $P(X = 1, \ Y = 2)$
b. $P(X < Y)$
c. $P(X = Y)$
d. Find the marginal probability mass function of X and Y.
e. Find conditional distribution function of X given $Y = 1$.

SOLUTION

a. From the given table we have $P(X = 1, \ Y = 2) = 0.15$
b. $P(X < Y) = P(X = 0, Y = 1) + P(X = 0, Y = 2) + P(X = 1, Y = 2)$
   $= 0.015 + 0.19 + 0.15 = 0.355$
c. $P(X = Y) = P(X = 0, Y = 0) + P(X = 1, Y = 1) + P(X = 2, Y = 2)$
   $= 0.1 + 0.045 + 0.1 = 0.245$
d. From the given table, we have

x \ y	0	1	2	3	$P(Y = y)$ $= \sum_x P(X = x, Y = y)$
0	0.1	0.05	0.025	0.025	0.2
1	0.015	0.045	0.1	0.05	0.21
2	0.19	0.15	0.1	0.15	0.59
$P(X = x)$	0.305	0.245	0.225	0.225	
$= \sum_y P(X = x, Y = y)$					

In the above table the last row and the last column gives the marginal probability mass function of the random variable X and Y, respectively.
e. We write the conditional probability mass function of X given $Y = 1$ as

$$f(x \mid y = 1) = \frac{f(x,1)}{P(Y = 1)} = \frac{0.015}{0.21} = 0.0714 \text{ if } x = 0$$

Therefore, for $x = 0$, $f(x \mid y = 1) = \frac{f(x,1)}{P(Y = 1)} = \frac{0.015}{0.21} = 0.0714$

For $x = 1$, $\frac{f(x,1)}{P(Y=1)} = \frac{0.045}{0.21} = 0.2142$

For $x = 2$, $\frac{f(x,1)}{P(Y=1)} = \frac{0.1}{0.21} = 0.4762$

Finally, for $x = 3$, $\frac{f(x,1)}{P(Y=1)} = \frac{0.05}{0.21} = 0.2381$

Thus, the conditional probability mass function of X given $Y = 1$ is

x	0	1	2	3
$f(x \mid y = 1)$	0.0714	0.2142	0.4762	0.2381

3.11.1.3 Transformation of Discrete Random Variables

In case of discrete random variable, there is no definite rule of transformation. The only thing which can be helpful is that the transformation is one to one. Suppose X is a discrete random variable having the mass function $P(X = x_i) = f(x_i)$. Now if the transformation is given by $Y = h(X)$, then the probability mass function of Y is given by

$$P(Y = y_i) = P(h(X) = y_i) = P(X = h^{-1}(y_i))$$

The transformation process of discrete random variable is illustrated in the following examples.

Example 3.12:

Let X be a discrete random variable having probability mass function as follows.

$X_i = x_i$	0	1	2	3	4	5
$P(X = x_i)$	0.1	0.2	0.15	0.25	0.12	0.18

Find the distribution of the transformed random variable $Y = 1 + 2X$.

SOLUTION

The values of Y given the values of $X \in \{0, 1, 2, 3, 4, 5\}$ are obtained following the rule $y_i = h(x_i)$ and the probability $P(Y = y_i)$ will be same as that of $X = x_i$. Thus, the probability mass function of Y will be as follows.

$Y = y_i$	1	3	5	7	9	11
$P(Y = y_i)$	0.1	0.2	0.15	0.25	0.12	0.18

Example 3.13:

Suppose X is discrete random variable having the following probability mass function

$$P(X = x) = \binom{5}{x}\left(\frac{1}{2}\right)^5, \quad x = 0, 1, 2, 3, 4, 5$$

$$= 0, \text{ elsewhere.}$$

If Y is a random variable as given in Example 3.12, find the probability mass function of Y.

SOLUTION

Given that

$$P(X = x) = \binom{5}{x}\left(\frac{1}{2}\right)^5, x = 0, 1, 2, 3, 4, 5$$

The transformation $y = 2x + 1$ is a one-to-one transformation and the inverse transformation is given by $x = \frac{y-1}{2}$. Therefore,

$$P(Y = y) = P(1 + 2X = y) = P\left(X = \frac{y-1}{2}\right) = \binom{5}{\frac{y-1}{2}}\left(\frac{1}{2}\right)^5, y = 1, 3, 5, 7, 9, 11$$

Hence, the probability mass function of Y is

$$P(Y = y) = \binom{5}{\frac{y-1}{2}}\left(\frac{1}{2}\right)^5, y = 1, 3, 5, 7, 9, 11$$

The probability mass function of a discrete random variable can be obtained even if the transformation is not a one-to-one transformation as illustrated in the following example.

Example 3.14:

Suppose X is a discrete random variable, the probability mass function of which is given in Example 3.12. If we take the transformed random variable $Y = (X - 3)^2$, find the probability mass function of Y.

SOLUTION

In this case, the given transformation is not a one to one. Now, $P(Y = y) = P(X = 3 + \sqrt{y}$ or $X = 3 - \sqrt{y}) = P(X = 3 + \sqrt{y}) + P(X = 3 - \sqrt{y})$. This shows that the admissible values of y are 0, 1, 4 and 9. The corresponding probabilities are as follows.

$$P(Y = 0) = P(X = 3) = 0.25$$

$$P(Y = 1) = P(X = 4) + P(X = 2) = 0.27$$

$$P(Y = 4) = P(X = 1) + P(X = 5) = 0.38$$

$$P(Y = 9) = P(X = 0) = 0.1$$

Therefore, the probability mass function of Y is given by

$Y = y_i$	0	1	4	9
$P(Y = y_i)$	0.25	0.27	0.38	0.1

3.11.2 JOINT DISTRIBUTION AND TRANSFORMATION OF CONTINUOUS RANDOM VARIABLES

3.11.2.1 Joint Probability Density Function

Definition 3.21: Suppose X_1, X_2, ..., X_n are n continuous random variables. A function $f:\mathbb{R}^n \rightarrow \mathbb{R}$ where $\mathbb{R}$ is the set of real numbers, is called the joint probability density function of the random variables X_1, X_2, ..., X_n if

$$P(x_1 < X_1 < x_1 + dx_1, \ x_2 < X_2 < x_2 + dx_2, ..., x_n < X_n < x_n + dx_n)$$

$$= f(x_1, \ x_2, \ ..., x_n) dx_1 dx_2 ... dx_n.$$

As in the discrete case, here also we define the marginal probability density function and conditional probability density function.

3.11.2.2 Marginal Probability Density Function

Definition 3.22: Suppose X_1, X_2, ..., X_n are n continuous random variables having the value of the joint probability density $f(x_1, \ x_2, \ ..., x_n)$ at $(x_1, \ x_2, \ ..., x_n)$, then the marginal probability density of the random variable x_i is given by,

$$g(x_i) = \int\limits_{-\infty}^{\infty} \int\limits_{-\infty}^{\infty} \cdots \int\limits_{-\infty}^{\infty} f(x_1, \ x_2, \ \cdots, x_n) dx_1 dx_2 \cdots dx_{i-1} dx_{i+1} \cdots dx_n.$$

3.11.2.3 Conditional Probability Density Function

Suppose X_1, X_2, $\cdots$, X_n are n continuous random variables having the value of the joint probability density $f(x_1, \ x_2, \ \cdots, x_n)$ at $(x_1, \ x_2, \ \cdots, x_n)$ and A_i be the event that $x_{i1} \leq X_i \leq x_{i2}$ for $i = 1, 2, \cdots, n$. Then, we have

$$P(A_1 \cap A_2 \cap \cdots \cap A_{i-1} \cap A_{i+1} \cap \cdots \cap A_n \mid A_i) = \frac{P(A_1 \cap A_2 \cap \cdots \cap A_i \cap \cdots \cap A_n)}{P(A_i)}$$

The joint conditional probability of $A'_j s \, (j = 1, 2, \ ..., i-1, i+1, ..., n)$ given A_i is

$$P(A_1 \cap A_2 \cap \cdots \cap A_{i-1} \cap A_{i+1} \cap \cdots \cap A_n \mid A_i)$$

$$= \frac{\displaystyle\int_{x_{11}}^{x_{12}} \int_{x_{21}}^{x_{22}} \cdots \int_{x_{i1}}^{x_{i2}} \cdots \int_{x_{n1}}^{x_{n2}} f(x_1, \ x_2, \ \cdots, x_n) dx_1 dx_2 \cdots dx_i \cdots dx_n}{\displaystyle\int_{x_{i1}}^{x_{i2}} g(x_i) dx_i}$$

where $g(x_i)$ is the value of the marginal probability density of X_i at x_i. It can be generalized by conditioning on some arbitrary number (say, $1 \leq r < n$) of events among the events A_i, $i = 1, 2, \cdots, n$. But here we consider r to be 1 to avoid unwanted mathematical complexity.

In particular, for two continuous random variables X and Y having joint probability density function $f(x, y)$, the conditional probability density of X given $Y = y$ is

$$h(x|y) = \frac{f(x,y)}{g(y)}$$

where $g(y)$ is the marginal probability density of Y. Analogously, the conditional probability density of Y given $X = x$ is

$$w(y|x) = \frac{f(x,y)}{\varphi(x)}$$

where $\varphi(x)$ being the marginal probability density of X.

Example 3.15:

Suppose $f(x,y) = 2(x + 2y)$, $x > 0; y > 0; x + y < 1$
$$= 0, \text{ elsewhere}$$
a. Find the marginal distribution of X
b. Find $P\left(0 < X < \frac{1}{3} \middle| \frac{1}{2} < Y < 1\right)$
c. Find $P\left(0 < X < \frac{1}{4} \middle| Y = \frac{2}{3}\right)$

SOLUTION

a. Suppose $g(x)$ represents the marginal probability density function of the random variable X. Then from Section 3.11.2.2, we have

$$g(x) = \int_{-\infty}^{\infty} f(x,y)dy = \int_{0}^{1-x} f(x,y)dy = 2\int_{0}^{1-x} (x+2y)dy$$

$$= 2\left[xy + y^2\right]_0^{1-x} = 2(1-x), \ 0 < x < 1$$

Therefore,

$$g(x) = 2(1-x), \ 0 < x < 1$$

$$= 0 \text{ elsewhere}$$

b. According to Section 3.11.2.3, we have

$$P\left(0 < X < \frac{1}{3}\bigg|\frac{1}{2} < Y < 1\right) = \frac{P\left(0 < X < \frac{1}{3}, \frac{1}{2} < Y < 1\right)}{P\left(\frac{1}{2} < Y < 1\right)}$$

$$P\left(0 < X < \frac{1}{3}\bigg|\frac{1}{2} < Y < 1\right) = \frac{\iint f(x,y)\,dx\,dy}{\int_{\frac{1}{2}}^{1} h(y)\,dy} \qquad (3.12)$$

Here the double integral in the numerator is taken over the region S as shown in Figure 3.6, where S consists of the regions S_1 and S_2. It can be shown that the marginal probability density function of y, say $h(y)$, is given by

$$h(y) = 1 + 2y - 3y^2, \text{ if } 0 < y < 1$$

$$= 0 \text{ elsewhere}$$

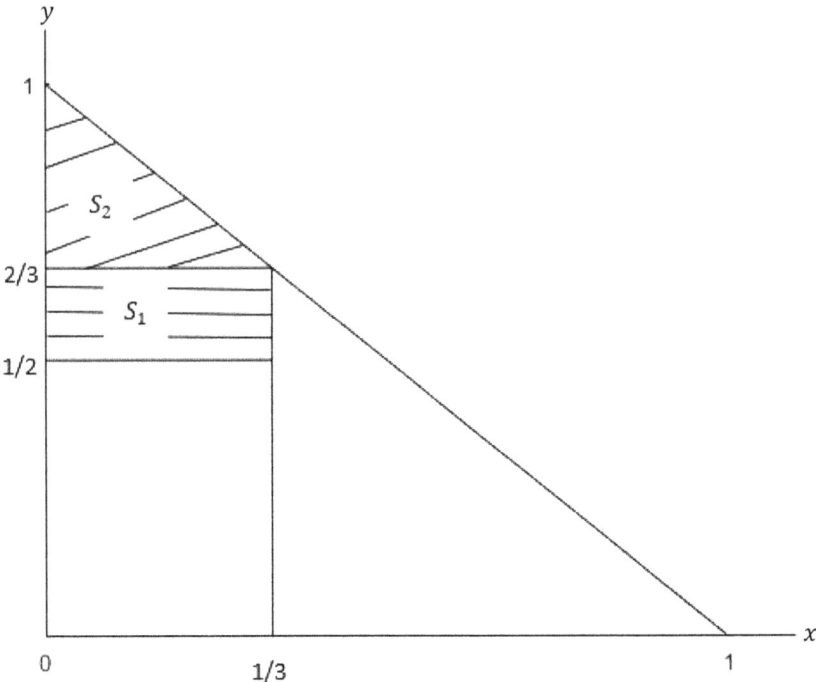

FIGURE 3.6 The region of integration for the multiple integral in Equation (3.12).

Therefore, Equation (3.12) gives

$$P\left(0 < X < \frac{1}{3}\bigg|\frac{1}{2} < Y < 1\right) = \frac{\displaystyle\int_0^{\frac{1}{3}}\int_{\frac{1}{2}}^{\frac{2}{3}} f(x,y)\,dy\,dx + \int_0^{\frac{1}{3}}\int_{\frac{2}{3}}^{1-x} f(x,y)\,dy\,dx}{\displaystyle\int_{\frac{1}{2}}^1 (1+2y-3y^2)\,dy}$$

$$= \frac{\displaystyle 2\int_0^{\frac{1}{3}}\int_{\frac{1}{2}}^{\frac{2}{3}} (x+2y)\,dy\,dx + 2\int_0^{\frac{1}{3}}\int_{\frac{2}{3}}^{1-x} (x+2y)\,dy\,dx}{\displaystyle\int_{\frac{1}{2}}^1 (1+2y-3y^2)\,dy} = \frac{\frac{1}{3}}{\frac{3}{8}} = \frac{8}{9}$$

c. It is observed that the marginal probability density function of Y is given by

$$h(y) = 1 + 2y - 3y^2 \text{ if } 0 < y < 1$$

$$= 0 \text{ elsewhere}$$

So, the conditional probability density of X given $Y = y$ is

$$\varphi(x|y) = \frac{f(x,y)}{h(y)}$$

Or, $\varphi(x|y) = \dfrac{2(x+2y)}{1+2y-3y^2}$, $0 < x < 1-y$

$$= 0 \text{ elsewhere}$$

Therefore,

$$\varphi\left(x|y=\frac{2}{3}\right) = \frac{2\left(x+2.\frac{2}{3}\right)}{1+2.\frac{2}{3}-3\left(\frac{2}{3}\right)^2} = 2x + \frac{8}{3} \text{ for } 0 < x < \frac{1}{3}$$

This shows that

$$P\left(0 < X < \frac{1}{4}\bigg|Y=\frac{2}{3}\right) = \int_0^{\frac{1}{4}} \varphi\left(x|y=\frac{2}{3}\right)dx$$

$$= \int_0^{\frac{1}{4}} \left(2x+\frac{8}{3}\right)dx = \frac{35}{48}$$

In continuation of the marginal and conditional probability density of two random variables X and Y, we define the joint distribution function of two random variables X and Y in the following.

Definition 3.23: If X and Y are two continuous random variables, the function given by $F(x, y) = P(X \le x, Y \le y) = \int_{-\infty}^{y} \int_{-\infty}^{x} f(s, t) ds\, dt, -\infty < x < \infty, -\infty < y < \infty$ is called the joint distribution function of X and Y, where $f(s, t)$ is the value of the joint probability density of X and Y at (s, t).

The repeated partial differentiation of $F(x, y)$ with respect to x and y leads to the joint probability density function of X and Y. Therefore, we have

$$\frac{\partial^2 F(x, y)}{\partial x\, \partial y} = f(x, y)$$

if this partial derivative exists.

So far, we have discussed the different relevant aspects of joint distribution of two or more random variables. We now extend our study with the joint distribution of transformed random variables. In order to avoid mathematical complexity, which does not suit the scope of this book, we confine our discussion within the random variables of dimensions two and three only.

Theorem 3.7: Let X_1 and X_2 be two continuous random variables with joint probability density function $f(x_1, x_2)$. Suppose $y_1 = u_1(x_1, x_2)$ and $y_2 = u_2(x_1, x_2)$ are the transformations of x_1, x_2 such that u_1 and u_2 are partially differentiable with respect to x_1, x_2 and the mapping u_1 and u_2 are one to one so that for a given value of y_1 and y_2; x_1, x_2 can be solved uniquely as $x_1 = w_1(y_1, y_2)$ and $x_2 = w_2(y_1, y_2)$. Then the joint probability density function of $Y_1 = u_1(X_1, X_2), Y_2 = u_2(X_1, X_2)$ is given by

$$g(y_1, y_2) = f\big(w_1(y_1, y_2), w_2(y_1, y_2)\big)|J| \text{ if } f\big(w_1(y_1, y_2), w_2(y_1, y_2)\big) \ne 0$$

$$= 0 \text{ elsewhere}$$

Here, J is the Jacobian of the transformation and it is given by

$$J = \begin{vmatrix} \dfrac{\partial x_1}{\partial y_1} & \dfrac{\partial x_1}{\partial y_2} \\[2mm] \dfrac{\partial x_2}{\partial y_1} & \dfrac{\partial x_2}{\partial y_2} \end{vmatrix}$$

In Theorem 3.7 we have considered the transformation of two random variables. This theorem can be generalised to the distribution of three or more number of random variables. For example, if the joint probability density function of three random variables X_1, X_2 and X_3 is given as $f(x_1, x_2, x_3)$, then the joint probability density function of $Y_1 = u_1(x_1, x_2, x_3)$, $Y_2 = u_2(x_1, x_2, x_3)$ and $Y_3 = u_3(x_1, x_2, x_3)$ is given by

$$g(y_1, y_2, y_3) = f(w_1, w_2, w_3) . |J| \text{ if } f(w_1, w_2, w_3) \neq 0$$

$$= 0 \text{ elsewhere}$$

and x_1, x_2 and x_3, for the given values of y_1, y_2, y_3, are uniquely given by $x_1 = w_1(y_1, y_2, y_3)$, $x_2 = w_2(y_1, y_2, y_3)$ and $x_3 = w_3(y_1, y_2, y_3)$. Here, the Jacobian of the transformation is given by

$$J = \begin{vmatrix} \dfrac{\partial x_1}{\partial y_1} & \dfrac{\partial x_1}{\partial y_2} & \dfrac{\partial x_1}{\partial y_3} \\[2mm] \dfrac{\partial x_2}{\partial y_1} & \dfrac{\partial x_2}{\partial y_2} & \dfrac{\partial x_2}{\partial y_3} \\[2mm] \dfrac{\partial x_3}{\partial y_1} & \dfrac{\partial x_3}{\partial y_2} & \dfrac{\partial x_3}{\partial y_3} \end{vmatrix}$$

Example 3.16:

Let X and Y be two continuous random variables having the joint probability density function

$$f(x, y) = 24xy; \text{ if } 0 < x < 1, 0 < y < 1, x + y < 1$$

$$= 0 \text{ elsewhere}$$

Find the joint probability density function of $Z = X + Y$ and $W = X$.

SOLUTION

Here, $z = x + y$ and $w = x$ which implies that $x = w$ and $y = z - w$. Therefore,

$$J = \begin{vmatrix} \dfrac{\partial x}{\partial z} & \dfrac{\partial x}{\partial w} \\[2mm] \dfrac{\partial y}{\partial z} & \dfrac{\partial y}{\partial w} \end{vmatrix} = \begin{vmatrix} 0 & 1 \\ 1 & -1 \end{vmatrix} = -1 \text{ and } |J| = 1.$$

Now, the important part is to identify the range of z and w. The range of w is same as that of x as $w = x$. Now the infimum of z is zero and the supremum of it is 1 since it is given that $0 < x + y < 1$ and $z = x + y$. Note that $y = z - w > 0$. This shows that $z > w$. Hence, applying Theorem 3.7 we get the joint probability density function of Z and W as

$$g(z, w) = 24w(z - w) \text{ if } w < z < 1, 0 < w < 1$$

$$= 0 \text{ elsewhere}$$

The region $R : w < z < 1, 0 < w < 1$ is the shaded region in Figure 3.7.

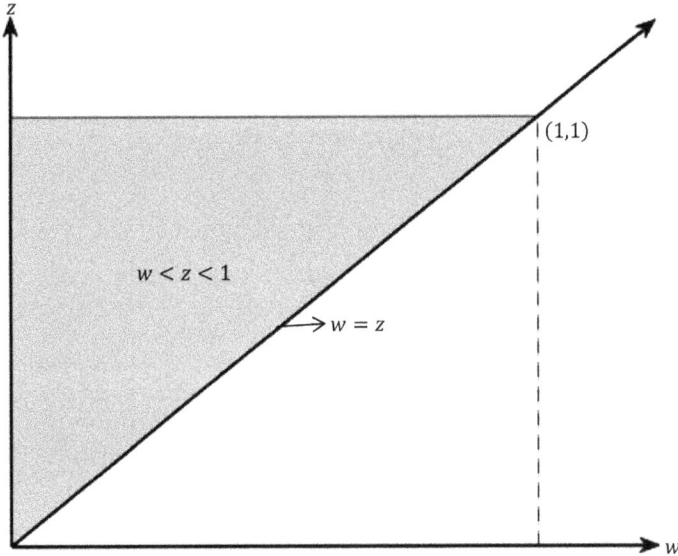

FIGURE 3.7   Diagram showing region R.

In the case of one dimensional random variables the formation of probability density function of a transformed random variable is a special case of the general method described in the previous section. The following theorem is used to serve the purpose.

Theorem 3.8: Let $f(x)$ be the probability density function of the continuous random variable X at x. We consider a function $y = u(x)$ is differentiable and a one-to-one function within the domain of X for which $f(x) \neq 0$. Then for those values of x, the probability density function of Y at the point $y = u(x)$ is given by

$$g(y) = f[w(y)].|w'(y)| \text{ where } u'(x) \neq 0$$

$$= 0 \text{ elsewhere}$$

where the function $w(y)$ uniquely determines x for a given value of y.
   The use of this theorem is illustrated in the following example.

Example 3.17:

The probability density of a random variable X is given by

$$f(x) = 2xe^{-x^2} \text{ for } x > 0$$

$$= 0 \text{ elsewhere}$$

and $Y = X^2$. Find the probability density function and the distribution function of Y.

SOLUTION

Here, $y = x^2$
   Thus,

$$x = \sqrt{y}, \text{ since } x > 0.$$

Therefore,

$$\frac{dx}{dy} = \frac{1}{2\sqrt{y}}$$

Hence,

$$\left|\frac{dx}{dy}\right| = \frac{1}{2\sqrt{y}}$$

Using Theorem 3.8, the probability density function of Y is given by

$$f_Y(y) = 2\sqrt{y}e^{-y}\left|\frac{dx}{dy}\right| = e^{-y} \text{ if } y > 0$$

$$= 0 \text{ elsewhere}$$

The distribution function of Y is given by

$$F(y) = P(Y \le y) = P(X^2 \le y) = P\left(-\sqrt{y} \le X \le \sqrt{y}\right) = P\left(0 < X \le \sqrt{y}\right), \text{ since } x > 0.$$

Therefore,

$$F(y) = \int_0^{\sqrt{y}} 2xe^{-x^2}dx = \left(1 - e^{-y}\right), y > 0$$

We can also obtain the probability density function by differentiating $F(y)$. Thus, in this case the probability density function of Y is given by $\frac{dF(y)}{dy} = \frac{d\left(1-e^{-y}\right)}{dy} = e^{-y}$, $y > 0$ and for $y \le 0$, $f_Y(y) = \frac{d(0)}{dy} = 0$.

Exercises

3.1 For any two events A and B, show that
    i. $P(A \cup B) \ge P(A) + P(B) - 1$
    ii. $P(A \cap B) \le P(A)$
3.2 Show that if A and B are two independent events then A^c and B^c are also independent.
3.3 An urn contains 3 white and 4 black balls. A ball is drawn and the colour of it is noticed and put back in the urn. A second ball is drawn from the urn. What is the probability that the second drawn ball is black?
3.4 In Exercise 3.3, if the first drawn ball is kept aside after noticing its colour, what is the probability that the second drawn ball is black?

3.5 Two cards are drawn from a pack of 52 cards. What is the probability that
    i. one card is the queen of spades and the other is a heart?
    ii. at least one of the cards is a queen?

3.6 Due to an outbreak of a new virus in a city, 30% of the total population of the city are infected. The probability that a certain kit correctly diagnoses the infection for a patient when the patient is indeed infected by the new virus is 0.80. Whereas, the kit gives a positive result for a given patient when the patient is actually not infected by the virus with probability 0.15. Suppose that for a particular patient the test result comes out to be positive. What is the probability that the patient is not infected by the new virus?

3.7 For what value of k, the following function

$$f(x) = \frac{k}{2^x} ; x = 0, 1, 2, \ldots$$

will serve as a probability mass function?

3.8 Find the value of k for which the following distribution of the random variable X represents a probability distribution.

x	-2	-1	0	1	2
$F(x)$	0.1	0.25	$2k$	0.15	0.30

Also determine
    i. $P(X < 0)$
    ii. $P(-1 < X \le 2)$
    iii. the distribution function of X.

3.9 Suppose a random variable X has mean 3 and variance 3/2. Show the probability that the value of X will be greater than 0 and less than 6 is at least 5/6.

3.10 Find the moment generating function of the following uniform distribution.

$$f(x) = \frac{1}{3} \text{ if } 0 < x < 3$$

$$= 0 \text{ elsewhere}$$

Also determine the mean and variance of the random variable.

3.11 The joint probability mass function of two discrete random variables is given in the following table.

x / y	0	1	2
0	0.05	0.15	0.075
1	0.045	0.05	0.025
2	0.1	0.04	0.065
3	0.2	0.05	0.15

Find

   i. $P(X=0, Y=2)$

  ii. Marginal probability mass function of X and Y

 iii. $P(X>Y)$

 iv. Conditional probability mass function of Y given $X=1$.

3.12 i. Show that

$$f(x, y) = k \text{ if } 0 < x < 1; 0 < y < 1 \text{ and } x+y < 1$$

$$= 0 \text{ elsewhere}$$

is a joint probability density function of two random variables X and Y for some suitable value of k.

 ii. Find the joint probability density function of the transformed random variables $U = X+Y$ and $V = Y$.

iii. Find the marginal probability density function of the random variable $X+Y$.

4 Discrete Probability Distribution

4.1 INTRODUCTION

A random variable is called discrete random variable if the set of its possible values is either finite or countable infinite. The number of imperfections per Km length of yarn, the number of end breakages per 100 spindles per shift of a ring frame, the number of defective garments produced in a day are few examples of discrete random variables. Let X be a discrete random variable. The probability that the random variable X assume value x is expressed as $P(X = x) = f(x)$, where $f(x)$ is called the probability mass function. For example, $f(3)$ is the probability that the random variable X assumes the numerical value of 3. A discrete random variable must satisfy the conditions that $f(x) \geq 0$ and $\sum f(x) = 1$. Well-known discrete probability distributions include Bernoulli distribution, binomial distribution, Poisson distribution and hypergeometric distribution.

4.2 BERNOULLI DISTRIBUTION

A trial which has only two outcomes, one is called success and the other is called failure, is called a Bernoulli trial. For example, suppose we toss a coin. It has two outcomes, head or tail. Hence it is a Bernoulli trial. Suppose for a particular Bernoulli trial, the probability of success is p and the probability of failure is $1 - p$. Then the probability mass function for success is given by

$$f(x) = p^x (1-p)^{1-x}; \text{ for } x = 0, 1.$$

The distribution of a discrete random variable X with probability mass function given above is called Bernoulli distribution.

4.3 BINOMIAL DISTRIBUTION

Let there be n number of independent Bernoulli trials and we are interested to know the probability that there will be exactly x number of occurrences of a particular event such that when this event occurs we call it as 'success' and if this event does not occur it will be called as 'failure'. In order to find out exactly x number of successes in n independent trials, let p be the probability of success or the probability that a particular event will occur in any single trial and q be the probability of failure or the probability that a particular event will fail to occur in any single trial. As the total probability of both success and failure in a single trial is one,

DOI: 10.1201/9781003081234-4

obviously, $q = 1 - p$. Let us begin with one arrangement of the occurrences of x number of consecutive successes and then n-x number of consecutive failures in the following way:

$$\overbrace{p.p.p. \ldots .p}^{x} \overbrace{q.q. \ldots q}^{n-x} = p^x q^{n-x}$$

This is the expression of the probability of exactly x number of successes in n independent trials for the above particular arrangement. However, there may be several other arrangements of p and q which will be attained by interchanging their respective positions. The total number of ways in which we can arrange x successes and $n - x$ failures out of n trials is $\frac{n!}{x!(n-x)!}$ or $\binom{n}{x}$. Each of these arrangements will have the same probability of occurrence, i.e. $p^x q^{n-x}$. It thus follows that the desired probability of getting exactly x number of successes in n independent trials is $\binom{n}{x} p^x q^{n-x}$. If X is a discrete random variable that equals to the number of successes in n trials, the probability mass function of binomial distribution is expressed by

$$P(X = x) = f(x) = \binom{n}{x} p^x (1 - p)^{n-x}; \text{ for } x = 0, 1, 2, \ldots, n$$
(4.1)
$$= 0, \text{ otherwise.}$$

Thus, for binomial distribution, the probability mass function $f(x)$ designates the probability of finding exactly x number of successes in n independent trials and the corresponding discrete random variable X which represents the number of success in n trials is called a binomial random variable with parameters (n, p). Note that, a Bernoulli random variable is also a binomial random variable with parameters $(1, p)$.

The term binomial distribution originates from the fact that the values of $f(x)$ for $x = 0, 1, 2, 3, \ldots, n$ are the successive terms of the binomial expansion $(q + p)^n$ as follows:

$$(q + p)^n = \binom{n}{0} q^n + \binom{n}{1} q^{n-1} p + \binom{n}{2} q^{n-2} p^2 + \binom{n}{3} q^{n-3} p^3 + \cdots + \binom{n}{n} p^n$$

$$= \sum_{x=0}^{n} \binom{n}{x} p^x q^{n-x}$$
(4.2)

$$= \sum_{x=0}^{n} f(x)$$

$$= 1 \text{ (because, } q + p = 1)$$

This also shows that the sum of the probabilities equals 1.

It should be noted that a random variable X which follows binomial distribution with parameters (n, p) can be thought of as a sum of n Bernoulli random variables. Suppose n independent Bernoulli trials are performed and X_i denotes the number of success in each trial. Then each X_i follows Bernoulli distribution. If $X := \sum_{i=1}^{i=n} X_i$, then X denotes the total number of success in n trials and X follows binomial distribution with parameters (n, p).

Theorem 4.1: The mean (μ) and the variance (σ^2) of the binomial distribution are $\mu = np$ and $\sigma^2 = np(1-p)$.

Proof: Mean (μ) is the first moment about the origin, hence it becomes

$$\mu = \mu_1'$$

$$= \sum_{x=0}^{n} x f(x)$$

$$= \sum_{x=0}^{n} x \frac{n!}{x!(n-x)!} p^x q^{n-x}$$

$$= \sum_{x=1}^{n} x \frac{n!}{x!(n-x)!} p^x q^{n-x} \quad (\text{Because the term corresponding to } x = 0 \text{ is } 0)$$

Now, $n! = n(n-1)!$ and $\frac{x}{x!} = \frac{1}{(x-1)!}$

Therefore, $\mu = \sum_{x=1}^{n} \frac{n(n-1)!}{(x-1)!(n-x)!} p^x q^{n-x}$

$$= np \sum_{x=1}^{n} \frac{(n-1)!}{(x-1)!(n-x)!} p^{x-1} q^{n-x}$$

$$= np \sum_{x-1=0}^{n-1} \frac{(n-1)!}{(x-1)!(n-x)!} p^{x-1} q^{n-x}$$

By putting $y = x - 1$, and $m = n - 1$, it becomes

$$\mu = np \sum_{y=0}^{m} \frac{m!}{y!(m-y)!} p^y q^{m-y}$$

As the term $\sum_{y=0}^{m} \frac{m!}{y!(m-y)!} p^y q^{m-y} = (p+q)^m = 1$

$$\therefore \mu = np \qquad (4.3)$$

The variance $\left(\sigma^2\right)$ is the second moment about mean, thus it becomes

$$\sigma^2 = \mu_2$$

$$= \sum_{x=0}^{n}(x-\mu)^2 f(x)$$

$$= \sum_{x=0}^{n}x^2 f(x) - 2\mu \sum_{x=0}^{n} xf(x) + \mu^2 \sum_{x=0}^{n} f(x)$$

$$= \sum_{x=0}^{n}x^2 f(x) - 2\mu^2 + \mu^2$$

$$= \mu_2' - \mu^2 \tag{4.4}$$

Now, $x^2 = x(x-1) + x$

$$\therefore \mu_2' = \sum_{x=0}^{n}\left\{x(x-1) + x\right\} f(x)$$

$$= \sum_{x=0}^{n}\left\{x(x-1) + x\right\}\frac{n!}{x!(n-x)!}p^x q^{n-x}$$

$$= \sum_{x=0}^{n}x(x-1)\frac{n!}{x!(n-x)!}p^x q^{n-x} + \sum_{x=0}^{n}x\frac{n!}{x!(n-x)!}p^x q^{n-x}$$

Now, $\frac{x(x-1)}{x!} = \frac{1}{(x-2)!}$, $n! = n(n-1)(n-2)!$, and $\sum_{x=0}^{n} x\frac{n!}{x!(n-x)!}p^x q^{n-x} = \mu$, hence

$$\mu_2' = \sum_{x=2}^{n}\frac{n(n-1)(n-2)!}{(x-2)!(n-x)!}p^x q^{n-x} + \mu$$

$$= n(n-1)p^2 \sum_{x=2}^{n}\frac{(n-2)!}{(x-2)!(n-x)!}p^{x-2} q^{n-x} + \mu$$

$$= n(n-1)p^2 \sum_{x-2=0}^{n-2}\frac{(n-2)!}{(x-2)!(n-x)!}p^{x-2} q^{n-x} + \mu$$

By putting $z = x - 2$, and $t = n - 2$ it becomes

$$\mu_2' = n(n-1)p^2 \sum_{z=0}^{t}\frac{t!}{z!(t-z)!}p^z q^{t-z} + \mu$$

As the term $\sum_{z=0}^{t} \frac{t!}{z!(t-y)!} p^z q^{t-z} = 1$, thus

$$\mu_2' = n(n-1)p^2 + \mu$$

Now, from Equation (4.4), we have

$$\sigma^2 = \mu_2' - \mu^2$$
$$= n(n-1)p^2 + \mu - \mu^2$$
$$= n(n-1)p^2 + np - n^2 p^2$$
$$= np - np^2$$
$$\therefore \sigma^2 = np(1-p) \tag{4.5}$$

Some generalization can be made about binomial distribution as follows:

1. When $p = 0.5$, the binomial distribution is symmetric. This is depicted in Figure 4.1.
2. When p is smaller than 0.5, the binomial distribution is skewed to the right, that is it is positively skewed as shown in Figure 4.2.
3. When p is larger than 0.5, the binomial distribution is skewed to the left, that is it is negatively skewed as shown in Figure 4.3.
4. As n increases, the binomial distribution becomes symmetric. Figure 4.4 depicts that the shape of the binomial distribution is skewed at $n = 10$ and 20, however it approaches to a symmetric shape at $n = 150$.

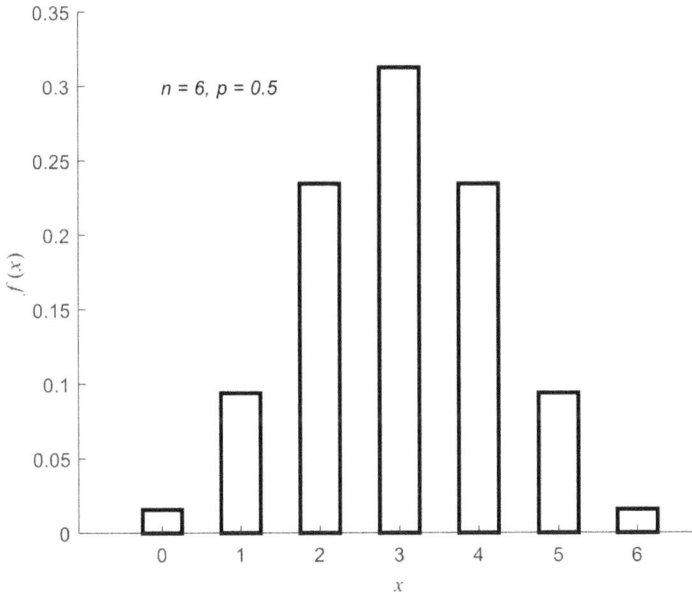

FIGURE 4.1 Symmetric binomial distribution with $n = 6$ and $p = 0.5$.

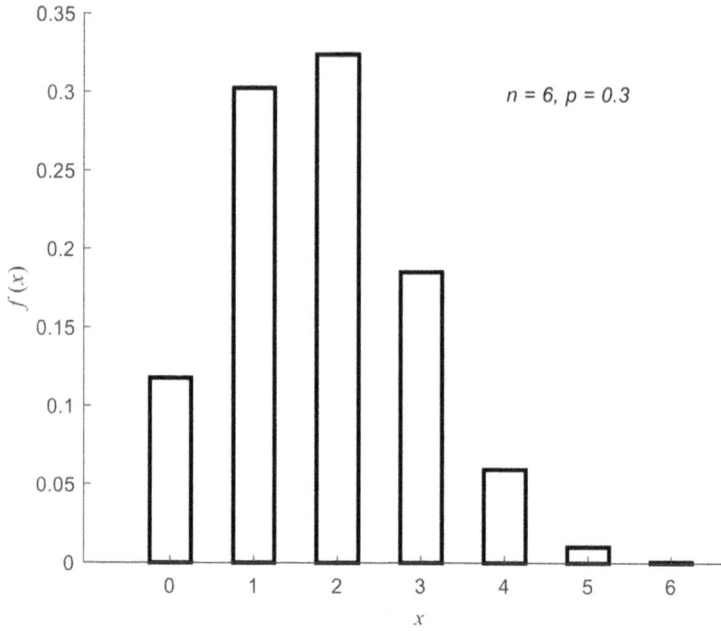

FIGURE 4.2 Positively skewed binomial distribution with $n = 6$ and $p = 0.3$.

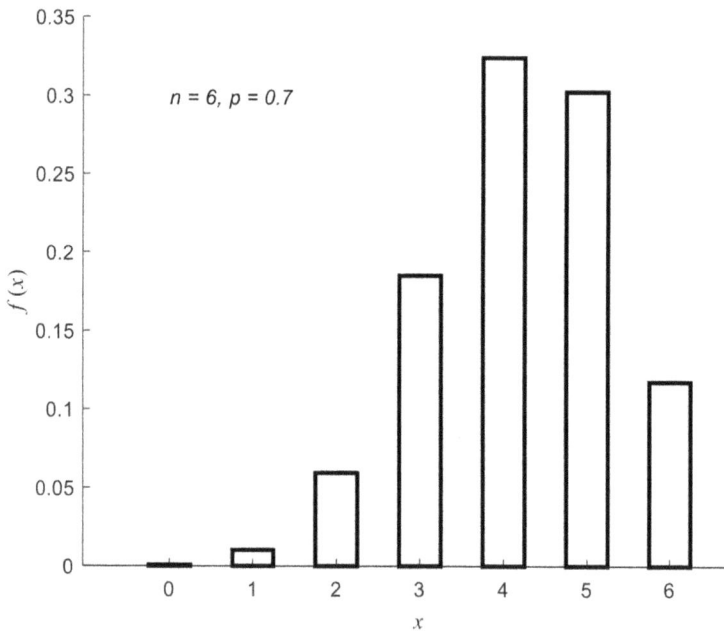

FIGURE 4.3 Negatively skewed binomial distribution with $n = 6$ and $p = 0.7$.

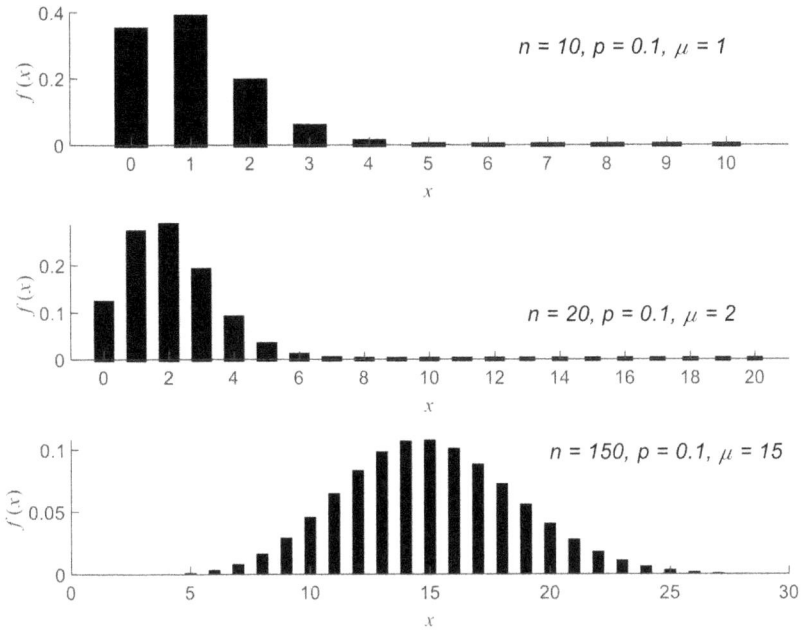

FIGURE 4.4 Change of the shape of binomial distribution as *n* increases.

Example 4.1:

If a coin is tossed for five times, find the probability of getting exactly 3 heads.

SOLUTION

Probability of getting a head in a single toss is 0.5. Substituting $x = 3$, $n = 5$, and $p = 0.5$ into the Equation (4.1), we get

$$P(X = 3) = f(3) = \binom{5}{3} 0.5^3 (1-0.5)^{5-3} = 0.3125$$

Example 4.2:

If 1.2% of the cones produced by an autoconer winding machine are defective, determine the probability that out of 10 cones chosen at random (a) 0, (b) 1, (c) less than 2, cones will be defectives.

SOLUTION

a. Substituting $x = 0$, $n = 10$, and $p = 0.012$ into the Equation (4.1), we get

$$P(X = 0) = f(0) = \binom{10}{0} 0.012^0 (1-0.012)^{10-0} = 0.8862$$

b. Putting $x = 1$,

$$P(X = 1) = f(1) = \binom{10}{1} 0.012^1 (1 - 0.012)^{10-1} = 0.1076$$

c.

$$\begin{aligned}
P(X < 2) &= P(X = 0 \text{ or } X = 1) \\
&= P(X = 0) + P(X = 1) \\
&= f(0) + f(1) \\
&= 0.8862 + 0.1076 \\
&= 0.9938
\end{aligned}$$

Example 4.3:

If the probability of producing a defective garment is 0.01, find the probability that there will be more than 2 defectives in a total of 20 garments.

SOLUTION

We have

$$\begin{aligned}
P(X > 2) &= 1 - P(X \leq 2) \\
&= 1 - P(X = 0 \text{ or } X = 1 \text{ or } X = 2) \\
&= 1 - \{P(X = 0) + P(X = 1) + P(X = 2)\} \\
&\quad 1 - f(0) - f(1) - f(2)
\end{aligned}$$

Substituting $n = 20$, $p = 0.01$ for $x = 0$, 1, and 2 into the Equation (4.1), we get

$$P(X > 2) = 1 - \binom{20}{0} 0.01^0 (1 - 0.01)^{20-0} - \binom{20}{1} 0.01^1 (1 - 0.01)^{20-1}$$

$$- \binom{20}{2} 0.01^2 (1 - 0.01)^{20-2}$$

$$= 1 - 0.8179 - 0.1652 - 0.0159$$

$$= 0.001$$

4.4   POISSON DISTRIBUTION

For large value of n, the estimation of binomial probabilities using Equation (4.1) will involve a cumbersome task. For example, to find out the probability that 25 of 4000 shirt buttons are defectives, we need to determine $\binom{4000}{25}$, and if the probability is 0.001 that a single button is defective, we also have to esti-mate the value of $(0.001)^{25}(1-0.001)^{4000-25}$. Therefore, in such cases, a suitable approximation of the binomial distribution would be more convenient. If n is very large, but p is very small, while np remains constant, then for such kind of rare event the Poisson distribution gives a better approximation to the bino-mial distribution. The binomial distribution has the following probability mass function

$$P(X=x)=f(x)=\frac{n!}{x!(n-x)!}p^x(1-p)^{n-x} \text{ for } x=0, 1, 2,\ldots,n.$$

When $n\to\infty$ and $p\to 0$, and letting the constant value of np be λ, which gives $p=\frac{\lambda}{n}$, we can write

$$f(x)=\frac{n!}{x!(n-x)!}\left(\frac{\lambda}{n}\right)^x\left(1-\frac{\lambda}{n}\right)^{n-x}$$

$$=\frac{n(n-1)(n-2)\cdots\{n-(x-1)\}}{x!n^x}\lambda^x\left(1-\frac{\lambda}{n}\right)^{n-x}$$

$$=\frac{\frac{n}{n}\left(\frac{n-1}{n}\right)\left(\frac{n-2}{n}\right)\cdots\left\{\frac{n-(x-1)}{n}\right\}}{x!}\lambda^x\left(1-\frac{\lambda}{n}\right)^n\left(1-\frac{\lambda}{n}\right)^{-x}$$

$$=\frac{1\left(1-\frac{1}{n}\right)\left(1-\frac{2}{n}\right)\cdots\left(1-\frac{x-1}{n}\right)}{x!}\lambda^x\left\{\left(1-\frac{\lambda}{n}\right)^{-\frac{n}{\lambda}}\right\}^{-\lambda}\left(1-\frac{\lambda}{n}\right)^{-x} \quad (4.6)$$

As $n\to\infty$, $\frac{1}{n}\to 0$, while x and λ remain fixed, we obtain that

$$1\left(1-\frac{1}{n}\right)\left(1-\frac{2}{n}\right)\cdots\left(1-\frac{x-1}{n}\right)\to 1$$

and

$$\left(1-\frac{\lambda}{n}\right)^{-x}\to 1$$

By putting, $-\frac{n}{\lambda} = a$, we have

$$\left(1-\frac{\lambda}{n}\right)^{-\frac{n}{\lambda}} = \left(1+\frac{1}{a}\right)^{a}$$

$$= 1 + a\left(\frac{1}{a}\right) + \frac{a(a-1)}{2!}\left(\frac{1}{a}\right)^2 + \frac{a(a-1)(a-2)}{3!}\left(\frac{1}{a}\right)^3 + \cdots \text{up to } \infty$$

$$= 1 + \left(\frac{a}{a}\right) + \frac{1}{2!}\left(\frac{a}{a}\right)\left(\frac{a-1}{a}\right) + \frac{1}{3!}\left(\frac{a}{a}\right)\left(\frac{a-1}{a}\right)\left(\frac{a-2}{a}\right) + \cdots \text{up to } \infty$$

$$= 1 + 1 + \frac{1}{2!}\left(1-\frac{1}{a}\right) + \frac{1}{3!}\left(1-\frac{1}{a}\right)\left(1-\frac{2}{a}\right) + \cdots \text{up to } \infty$$

As $n \to \infty$, $-\frac{1}{a} \to 0$, therefore it becomes

$$\left(1-\frac{\lambda}{n}\right)^{-\frac{n}{\lambda}} = 1 + 1 + \frac{1}{2!} + \frac{1}{3!} + \cdots \text{up to } \infty$$

$$= e$$

Thus Equation (4.6) becomes

$$f(x) = \frac{\lambda^x e^{-\lambda}}{x!}$$

Therefore, the probability mass function of Poisson distribution is expressed as

$$f(x) = \frac{\lambda^x e^{-\lambda}}{x!} \text{ for } x = 0, 1, 2, \ldots, \infty$$

$$= 0, \text{ otherwise}$$

(4.7)

Equation (4.7) indicates the probability that there will be exactly x number of successes with the given value of λ. Thus, the Poisson distribution is a limiting form of binomial distribution of a discrete random variable X when $n \to \infty$ and $p \to 0$, while np remains constant. Poisson distribution is named after the French mathematician Simeon Poisson (1781–1840).

Theorem 4.2: The mean (μ) and the variance (σ^2) of the Poisson distribution are both equal to λ.

Proof: Mean (μ) is expressed as the first moment about the origin, therefore we can write

$$\mu = \mu_1'$$

$$= \sum_{x=0}^{\infty} x f(x)$$

$$= \sum_{x=0}^{\infty} x \frac{\lambda^x e^{-\lambda}}{x!}$$

$$= \sum_{x-1=0}^{\infty} \lambda \frac{\lambda^{x-1} e^{-\lambda}}{(x-1)!}$$

If we now let $y = x - 1$, we obtain

$$\mu = \lambda \sum_{y=0}^{\infty} \frac{\lambda^y e^{-\lambda}}{y!}$$

As the term $\sum_{y=0}^{\infty} \frac{\lambda^y e^{-\lambda}}{y!} = 1$

$$\therefore \mu = \lambda \qquad\qquad (4.8)$$

The variance $\left(\sigma^2\right)$ is the second moment about mean, therefore we can write

$$\sigma^2 = \mu_2$$

$$= \mu_2' - \mu^2$$

$$= \sum_{x=0}^{\infty} x^2 \frac{\lambda^x e^{-\lambda}}{x!} - \lambda^2$$

$$= \sum_{x=0}^{\infty} \{x(x-1) + x\} \frac{\lambda^x e^{-\lambda}}{x!} - \lambda^2$$

$$= \sum_{x=0}^{\infty} x(x-1) \frac{\lambda^x e^{-\lambda}}{x!} + \sum_{x=0}^{\infty} x \frac{\lambda^x e^{-\lambda}}{x!} - \lambda^2$$

$$= \sum_{x=2}^{\infty} \lambda^2 \frac{\lambda^{x-2} e^{-\lambda}}{(x-2)!} + \lambda - \lambda^2$$

$$= \lambda^2 \sum_{x-2=0}^{\infty} \frac{\lambda^{x-2} e^{-\lambda}}{(x-2)!} + \lambda - \lambda^2$$

If we now let $z = x - 2$, we obtain

$$\sigma^2 = \lambda^2 \sum_{z=0}^{\infty} \frac{\lambda^z e^{-\lambda}}{z!} + \lambda - \lambda^2$$

$$= \lambda^2 + \lambda - \lambda^2$$

$$\therefore \sigma^2 = \lambda \qquad (4.9)$$

Some generalization can be made about Poisson distribution as follows:

In general, Poisson distribution is a good approximation to the binomial distribution when $n \geq 50$ and $np \leq 5$, however for smaller value of n, this approximation may not be valid. This is illustrated in Figures 4.5 and 4.6 where the empty and filled bins represent the binomial and Poisson probabilities, respectively. Figure 4.5 shows that for $n = 100$ and $p = 0.02$, i.e., $np = 2$, Poisson distribution provides an excellent approximation to the binomial distribution. Figure 4.6 depicts that for $n = 9$ and $p = 1/3$, i.e., $np = 3$, Poisson distribution does not approach the binomial distribution.

As mean (μ) increases, the Poisson distribution becomes symmetric. Figure 4.7 illustrates that the shape of the Poisson distribution gradually becomes more and more symmetric as the value of μ increases from 1 to 15.

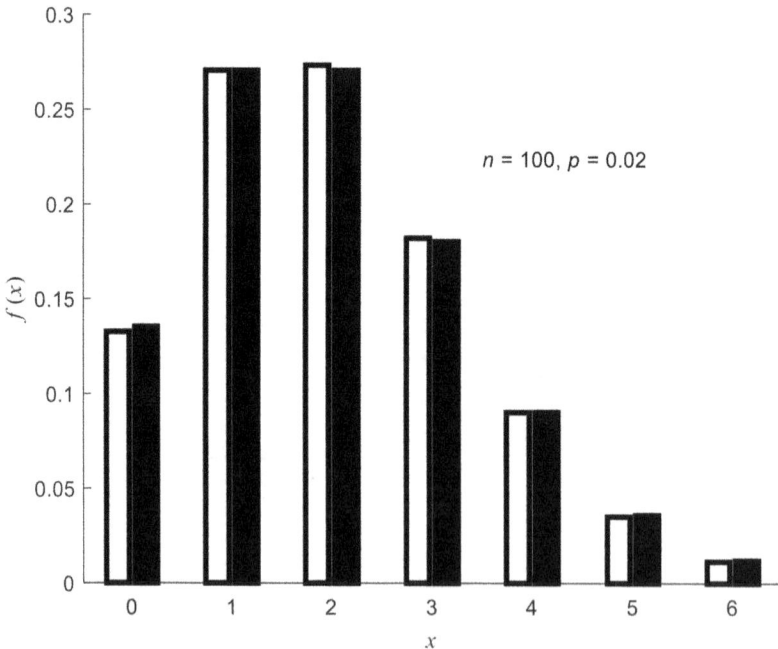

FIGURE 4.5 Binomial and Poisson distributions for $n = 100$ and $p = 0.02$.

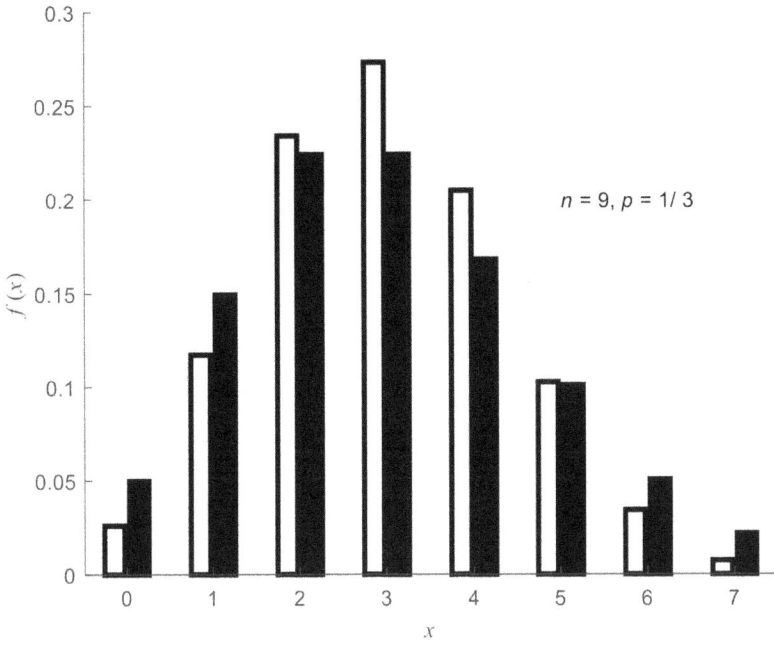

FIGURE 4.6   Binomial and Poisson distributions for $n = 9$ and $p = 1/3$.

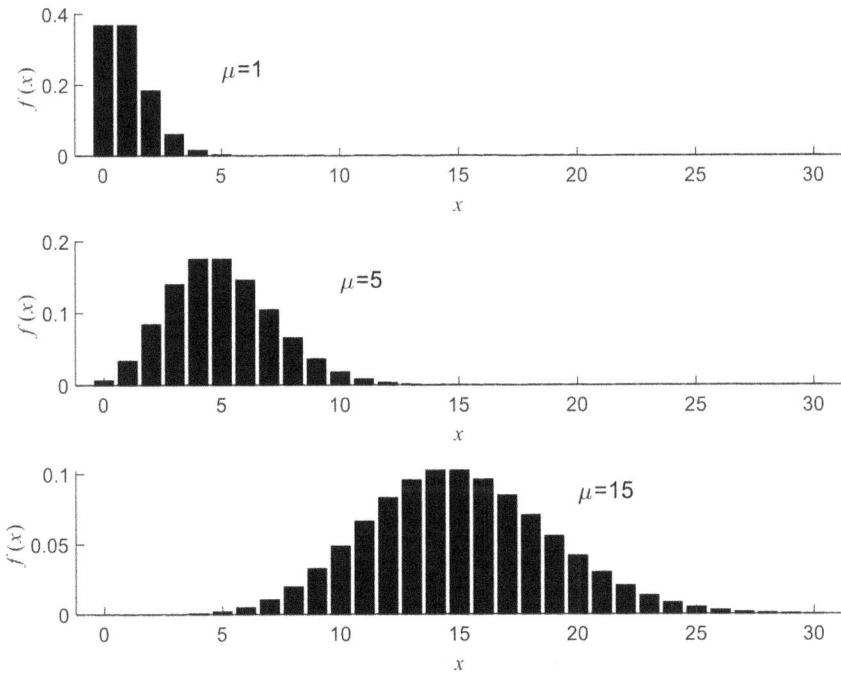

FIGURE 4.7   Change of the shape of Poisson distribution as mean increases.

Example 4.4:

While inspecting a fabric in a continuous roll, the mean number of defects per meter is 0.25. Determine the probability that 10 m length of fabric will contain (a) 0, (b) 1, (c) 2, (d) less than or equal to 2 defects.

SOLUTION

Let X be the defects per 10 m length of fabric. The mean number of defects present in 10 m length of fabric = $0.25 \times 10 = 2.5$. Therefore, the discrete random variable X has a Poisson distribution with $\lambda = 2.5$.

a. Substituting $x = 0$ and $\lambda = 2.5$ into the Equation (4.7), we get

$$P(X = 0) = f(0) = \frac{2.5^0 e^{-2.5}}{0!} = 0.0821$$

b. Putting $x = 1$,

$$P(X = 1) = f(1) = \frac{2.5^1 e^{-2.5}}{1!} = 0.2052$$

c. Putting $x = 2$,

$$P(X = 2) = f(2) = \frac{2.5^2 e^{-2.5}}{2!} = 0.2565$$

d. $P(X \le 2) = P(X = 0 \text{ or } X = 1 \text{ or } X = 2)$

$$= P(X = 0) + P(X = 1) + P(X = 2)$$
$$= f(0) + f(1) + f(2)$$
$$= 0.0821 + 0.2052 + 0.2565$$
$$= 0.5438$$

Example 4.5:

The average number of monthly breakdowns of an auto leveller in a draw frame machine is 1.4. Determine the probability that in a given month the auto leveller will function

a. without a breakdown
b. with only one breakdown

SOLUTION

Let X be the number of monthly breakdowns of an auto leveller in a draw frame machine. The mean number of monthly breakdowns of the auto leveller = 1.4.

Therefore, X is a discrete random variable having Poisson distribution with $\lambda = 1.4$.

a. Substituting $x = 0$ and $\lambda = 1.4$ into the Equation (4.7), we get

$$P(X = 0) = f(0) = \frac{1.4^0 e^{-1.4}}{0!} = 0.2466$$

b. Putting $x = 1$,

$$P(X = 1) = f(1) = \frac{1.4^1 e^{-1.4}}{1!} = 0.3452$$

Example 4.6:

Neps in a card web are estimated by placing over the web template with 34 round holes, each one having 6.45 cm² in area, and counting the proportion of holes that are free from neps. In an experiment, if 17 holes are free from neps, determine the mean number of neps per m² of card web.

SOLUTION

Let X be the number of neps present in a hole and λ be the mean number of neps present in a hole. Therefore, X is a discrete random variable having Poisson distribution with mean = λ. Substituting $x = 0$ into the Equation (4.7), we get

The probability that a hole is free from neps = $P(X = 0) = f(0) = \frac{\lambda^0 e^{-\lambda}}{0!} = e^{-\lambda}$

From the experiment, the probability that a hole is free from neps = $\frac{17}{34} = 0.5$.

Hence, $e^{-\lambda} = 0.5$

or, $\lambda = -\ln 0.5 = 0.6931$

Thus, the average number of neps per 6.45 cm² = 0.6931

Therefore, the average number of neps per 10,000 cm² = $\frac{0.6931 \times 10000}{6.45} = 1075$ (rounded to nearest integer).

4.5 HYPERGEOMETRIC DISTRIBUTION

We observed that, the number of successes in a sample of n independent trials follows binomial distribution as long as the sampling is done with replacement. If such a sampling is done without replacement, the trials are no longer independent. In this case the number of success does not follow the binomial distribution. To understand it better let us consider the following example.

Example 4.7:

A bag contains 4 red and 5 black balls. 3 balls are drawn without replacement. Then what will be the probability that out of the three drawn balls 2 balls are black?

SOLUTION

The required probability can not be obtained by binomial distribution since the balls are drawn without replacement. The actual probability will be obtained as follows: Out of the 5 black balls, 2 black balls can be drawn in $\binom{5}{2}$ ways and the remaining 1 red ball can be drawn in $\binom{4}{1}$ ways. So, the total number of ways of getting 2 black balls out of the drawn 3 balls is $\binom{5}{2} \times \binom{4}{1}$. The total number of ways 3 balls can be drawn (without replacement) is $\binom{9}{3}$. Hence the required probability is $\dfrac{\binom{5}{2} \times \binom{4}{1}}{\binom{9}{3}} = 0.476$.

This can be generalised to formulate another type of distribution called hypergeometric distribution.

Suppose we are sampling m objects from a source of N objects out of which M objects belong to one category, say category 1, and remaining $N - M$ belong to the category 2. Furthermore, assume that the sample is taken without replacement and hence the probability of success or failure does not remain same for each trial of the sampling process. Let X be the discrete random variable representing the number of objects belonging to the category 1 in the drawn sample. Then,

$$P(X = x) = \frac{\binom{M}{x}\binom{N-M}{m-x}}{\binom{N}{m}}; \ x = 0, 1, \ldots, \min(m, M) \tag{4.10}$$

$$= 0, \text{ otherwise.}$$

This distribution is called hypergeometric distribution whose probability mass function is given by (4.10).

4.6 MATLAB® CODING

4.6.1 MATLAB® CODING OF EXAMPLE 4.2

```
clc
clear
close all
n=10;
p=0.012;
pr_1=binopdf(0,n,p)
pr_2=binopdf(1,n,p)
pr_3=binocdf(1,n,p)
```

4.6.2   MATLAB® CODING OF EXAMPLE 4.3

```
clc
clear
close all
n=20;
p=0.01;
pr=1-binocdf(2,n,p)
```

4.6.3   MATLAB® CODING OF EXAMPLE 4.4

```
clc
clear
close all
lambda=2.5;
pr_1=poisspdf(0,lambda)
pr_2=poisspdf(1,lambda)
pr_3=poisspdf(2,lambda)
pr_4=poisscdf(2,lambda)
```

Exercises

4.1 An unbiased coin is tossed multiple times. What is the probability that the first head will occur at the 5th trial?

4.2 The market research section of a large textile retail shop has noticed that the probability that a customer who is just browsing the shop's mobile apps will buy something is 0.4. If 30 customers browse the apps in one hour, determine the probability that at least 10 browsing customers will buy something during that specified hour.

4.3 In the ring frame section of a spinning mill, it was observed that 1 in 10 end breakages occurs due to the problem of the traveller flying off. What is the probability that 2 out of 5 end breakages are due to the cause of the traveller flying off?

4.4 A company manufactures soft body armours and 90% of the armours satisfy the bullet proof test. If 8 armours are tested at random, find the probability that all armours will satisfy the test.

4.5 The average number of complaints that a dry-cleaning establishment receives per day is 2.7. Find the probability that it will receive only 3 complaints on any given day.

4.6 A roll of fabric contains on an average 5 defects scattered at random in 100 m². Pieces of fabric, 5 m by 2 m, are cut from the roll. Determine the probability that

   i. a piece contains no defects

   ii. a piece will contain more than one defects

   iii. a piece will contain fewer than two defects

   iv. five pieces selected at random will all be free from defects

4.7 Neps in a card web are estimated by placing over the web template with 34 round holes, each one having 6.45 cm² in area. In an experiment, if the mean number of neps per m² of card web is 1800, determine the probability that a hole is free from neps.

4.8 In a lot of 50 cloths 10 are defectives. Twelve cloths are chosen at random. What is the probability that 3 cloths will be defectives out of these 12 cloths?

4.9 A group of applicants consisting of 5 females and 10 males are called for an interview for a job and 5 are selected for the job. What is the probability that out of 5 selected candidates 2 will be female and 3 will be male?

5 Continuous Probability Distributions

5.1 INTRODUCTION

In Chapter 4, we introduced the discrete random variables and considered a few useful discrete distributions. In many cases, the total probability mass may be distributed over an interval instead of being distributed among a finite number of points or countably infinite number of points. The random variables which take on values in an interval are called continuous random variables. The length of cotton fibres, fineness of a yarn and strength of a fabric are few examples of continuous random variables. Sometimes it is more convenient to use the approximate continuous distribution of a discrete distribution. In this chapter we will consider some important continuous distributions which is useful in many statistical problems.

5.2 UNIFORM DISTRIBUTION

We start with uniform distribution which is one of the simple continuous distributions.

Definition 5.1: A continuous random variable X is said to follow uniform distribution in the interval $[a, b]$ if its probability density function is given by

$$f(x) = \frac{1}{b-a} \text{ if } a \leq x \leq b$$

$$= 0 \text{ elsewhere}$$

(5.1)

Figure 5.1 depicts the graph of the uniform probability density function given by Equation (5.1). The mean of the uniformly distributed random variable is given by

$$E(X) = \int_a^b x f(x) dx = \int_a^b \frac{x}{b-a} dx = \frac{a+b}{2}$$

(5.2)

The variance of the uniform random variable is obtained as

$$E(X^2) - E^2(X) = \int_a^b x^2 f(x) dx - \left(\frac{a+b}{2}\right)^2 = \frac{(a-b)^2}{12}$$

(5.3)

DOI: 10.1201/9781003081234-5

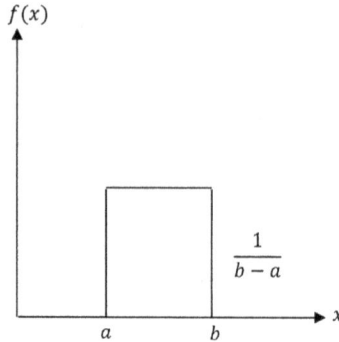

FIGURE 5.1   Uniform distribution.

5.3 EXPONENTIAL DISTRIBUTION

Definition 5.2: A continuous random variable X is said to follow exponential distribution if its probability density function is

$$f(x) = \alpha e^{-\alpha x} \text{ for } 0 < x < \infty$$
$$= 0 \text{ elsewhere}$$

(5.4)

with $\alpha > 0$.

The mean and variance of a random variable following exponential distribution can be obtained using the following theorem.

Theorem 5.1: The mean and variance of an exponential distribution given in Equation (5.4) are $\frac{1}{\alpha}$ and $\frac{1}{\alpha^2}$ respectively.

Proof: Let X be the random variable following exponential distribution given in Equation (5.4). The mean of X is given by

$$E(X) = \int_{-\infty}^{\infty} x\alpha e^{-\alpha x} dx = \int_{-\infty}^{0} x\alpha e^{-\alpha x} dx + \int_{0}^{\infty} x\alpha e^{-\alpha x} dx = 0 + \int_{0}^{\infty} x\alpha e^{-\alpha x} dx.$$

Now,

$$\int_{0}^{\infty} x\alpha e^{-\alpha x} dx = \lim_{a \to \infty} \int_{0}^{a} x\alpha e^{-\alpha x} dx$$

$$= \alpha \left[\lim_{a \to \infty} \left(\frac{xe^{-\alpha x}}{-\alpha} \right)_{0}^{a} + \frac{1}{\alpha} \lim_{a \to \infty} \int_{0}^{a} e^{-\alpha x} dx \right]$$

(5.5)

$$= \frac{1}{\alpha}$$

Therefore, we have, $E(X) = \frac{1}{\alpha}$.

The variance of X is given by

$$var(X) = E(X^2) - (E(X))^2$$

$$= \int_{-\infty}^{\infty} x^2 f(x) dx - \frac{1}{\alpha^2}$$

$$= \int_{-\infty}^{0} x^2 f(x) dx + \int_{0}^{\infty} x^2 f(x) dx - \frac{1}{\alpha^2}$$

$$= 0 + \int_{0}^{\infty} x^2 \alpha e^{-\alpha x} dx - \frac{1}{\alpha^2} \qquad (5.6)$$

Now,

$$\int_{0}^{\infty} x^2 \alpha e^{-\alpha x} dx = \lim_{a \to \infty} \int_{0}^{a} x^2 \alpha e^{-\alpha x} dx$$

$$= \alpha \left[\lim_{a \to \infty} \left(\frac{x^2 e^{-\alpha x}}{-\alpha} \right)_{0}^{a} + \frac{2}{\alpha} \lim_{a \to \infty} \int_{0}^{a} x e^{-\alpha x} dx \right]$$

$$= 2 \lim_{a \to \infty} \int_{0}^{a} x e^{-\alpha x} dx$$

$$= \frac{2}{\alpha^2} \qquad (5.7)$$

Therefore, from Equation (5.6) we have, $var(X) = \frac{2}{\alpha^2} - \frac{1}{\alpha^2} = \frac{1}{\alpha^2}$.

Note: In Equations (5.5) and (5.7) we have used the fact that $xe^{-\alpha x}$ and $x^2 e^{-\alpha x}$ both tends to 0 as $x \to \infty$. This is because $x^n e^{-\alpha x} \to 0$ as $x \to \infty$ for any finite value of n with $\alpha > 0$.

Example 5.1:

The hair length of rotor spun yarn has an exponential distribution with a mean of 2.5 mm. Find the probability that the hair length is

a. less than 3 mm
b. between 4 to 6 mm
c. greater than 8 mm

SOLUTION

Let the random variable X be the hair length following exponential distribution with the parameter $\alpha = \frac{1}{2.5}$.

a. The probability that the hair length is less than 3 mm is given by

$$P(X < 3) = \frac{1}{2.5} \int_0^3 e^{-\frac{x}{2.5}} dx$$

$$= -e^{-\frac{x}{2.5}} \Big|_0^3$$

$$= -e^{-\frac{3}{2.5}} + 1$$

$$= -0.301 + 1 = 0.699$$

This is indicated in Figure 5.2(a).

b. The probability that the hair length is between 4 to 6 mm is given by

$$P(4 < X < 6) = \frac{1}{2.5} \int_4^6 e^{-\frac{x}{2.5}} dx$$

$$= -e^{-\frac{x}{2.5}} \Big|_4^6$$

$$= -e^{-\frac{6}{2.5}} + -e^{-\frac{4}{2.5}}$$

$$= -0.091 + 0.202 = 0.111$$

This is indicated in Figure 5.2(b).

c. The probability that the hair length is greater than 8 mm is given by

$$P(X > 8) = \frac{1}{2.5} \int_8^\infty e^{-\frac{x}{2.5}} dx$$

$$= -e^{-\frac{x}{2.5}} \Big|_8^\infty$$

$$= 0 + e^{-\frac{8}{2.5}} = 0.041$$

This is indicated in Figure 5.2(c).

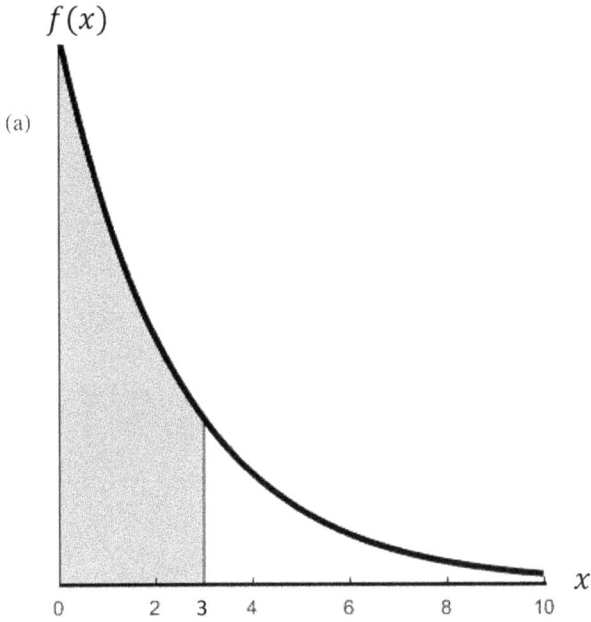

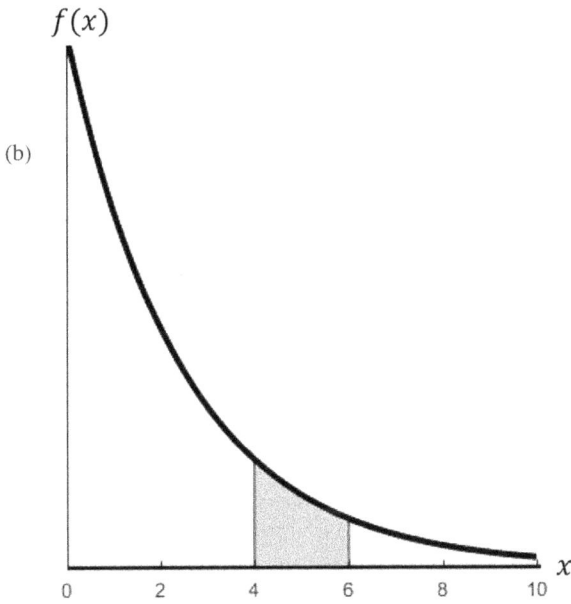

FIGURE 5.2 (a) Exponential distribution showing $P(X < 3)$ (b) Exponential distribution showing $P(4 < X < 6)$ (c) Exponential distribution showing $P(X > 8)$. *(Continued)*

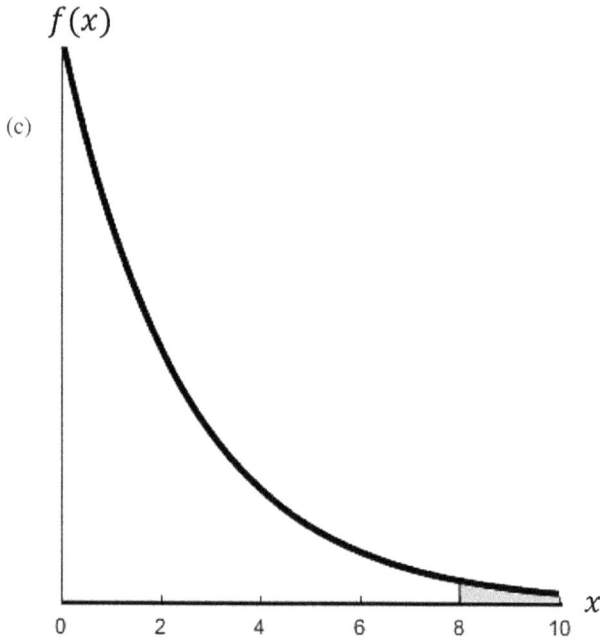

FIGURE 5.2   *(Continued)*

5.4   GAUSSIAN/NORMAL DISTRIBUTION

One of the most popular and useful distributions in statistics is the Normal distribution. It finds its application in estimation related issues (see Chapter 6) apart from many other areas of statistics. Normal distribution is a symmetric distribution about the mean, where the left side of the distribution mirrors the right side. In this section we will discuss some important properties of this distribution.

Definition 5.3: A continuous random variable X is said to follow Gaussian or normal distribution if the probability density function of X is given by

$$f(x) = \frac{1}{\sigma\sqrt{2\pi}} e^{\left[-\frac{1}{2}\left(\frac{x-\mu}{\sigma}\right)^2\right]}; \ -\infty < x < \infty \tag{5.8}$$

where the parameters μ and σ are the mean and standard deviation of the random variable X. The notation $X \sim N(\mu,\sigma^2)$ is often used to denote that the random variable X follows a normal distribution with mean μ and variance σ^2.

In the following theorem it is established that the mean and variance of a normal random variate having probability density function given by Equation (5.8) is indeed μ and σ^2 respectively.

Theorem 5.2: The mean and variance of a normal distribution given in Equation (5.8) are μ and σ^2, respectively.

Proof: The mean of a random variable $X \sim N\left(\mu, \sigma^2\right)$ is given by

$$E(X) = \int\limits_{-\infty}^{\infty} x f(x)\,dx = \frac{1}{\sigma\sqrt{2\pi}} \int\limits_{-\infty}^{\infty} x e^{-\frac{1}{2}\left(\frac{x-\mu}{\sigma}\right)^2}\,dx$$

Let us substitute $\frac{x-\mu}{\sigma} = z$ which leads to $dx = \sigma\,dz$
 Therefore,

$$\frac{1}{\sigma\sqrt{2\pi}} \int\limits_{-\infty}^{\infty} x e^{-\frac{1}{2}\left(\frac{x-\mu}{\sigma}\right)^2}\,dx = \frac{1}{\sigma\sqrt{2\pi}} \int\limits_{-\infty}^{\infty} (\sigma z + \mu) e^{-\frac{1}{2}z^2}\sigma\,dz$$

$$= \frac{1}{\sqrt{2\pi}} \int\limits_{-\infty}^{\infty} \sigma z e^{-\frac{1}{2}z^2}\,dz + \frac{\mu}{\sqrt{2\pi}} \int\limits_{-\infty}^{\infty} e^{-\frac{1}{2}z^2}\,dz \qquad (5.9)$$

The first integrand in Equation (5.9) is an odd function and hence

$$\frac{1}{\sqrt{2\pi}} \int\limits_{-\infty}^{\infty} z e^{-\frac{1}{2}z^2}\,dz = 0.$$

Thus,

$$\frac{1}{\sigma\sqrt{2\pi}} \int\limits_{-\infty}^{\infty} x e^{-\frac{1}{2}\left(\frac{x-\mu}{\sigma}\right)^2}\,dx = \frac{\mu}{\sqrt{2\pi}} \int\limits_{-\infty}^{\infty} e^{-\frac{1}{2}z^2}\,dz$$

Since $e^{-\frac{1}{2}z^2}$ is an even function, we have

$$\frac{\mu}{\sqrt{2\pi}} \int\limits_{-\infty}^{\infty} e^{-\frac{1}{2}z^2}\,dz = 2\frac{\mu}{\sqrt{2\pi}} \int\limits_{0}^{\infty} e^{-\frac{1}{2}z^2}\,dz$$

$$= \frac{\mu}{\sqrt{\pi}} \int\limits_{0}^{\infty} e^{-u} u^{\frac{1}{2}-1}\,du \qquad \left[\text{on substituting } \frac{1}{2}z^2 = u\right]$$

$$= \frac{\mu}{\sqrt{\pi}} \Gamma\left(\frac{1}{2}\right)$$

$$= \frac{\mu}{\sqrt{\pi}} \sqrt{\pi} = \mu \qquad \left[\begin{array}{l}\text{Because } \Gamma\left(\frac{1}{2}\right) = \sqrt{\pi} \text{ where } \Gamma(\cdot) \text{ is the gamma}\\ \text{function}\end{array}\right]$$

Now, $var(X) = E\left(X^2\right) - \left(E(X)\right)^2 = E\left(X^2\right) - \mu^2$ \qquad (5.10)

and

$$E\left(X^2\right)=\frac{1}{\sigma\sqrt{2\pi}}\int_{-\infty}^{\infty}x^2 e^{-\frac{1}{2}\left(\frac{x-\mu}{\sigma}\right)^2}dx$$

On substitution of $z=\frac{x-\mu}{\sigma}$ it becomes

$$\frac{1}{\sigma\sqrt{2\pi}}\int_{-\infty}^{\infty}x^2 e^{-\frac{1}{2}\left(\frac{x-\mu}{\sigma}\right)^2}dx$$

$$=\frac{1}{\sqrt{2\pi}}\int_{-\infty}^{\infty}\left(\mu+\sigma z\right)^2 e^{-\frac{1}{2}z^2}dz$$

$$=\mu^2\frac{1}{\sqrt{2\pi}}\int_{-\infty}^{\infty}e^{-\frac{1}{2}z^2}dz+\frac{2\mu\sigma}{\sqrt{2\pi}}\int_{-\infty}^{\infty}ze^{-\frac{1}{2}z^2}dz+\frac{\sigma^2}{\sqrt{2\pi}}\int_{-\infty}^{\infty}z^2 e^{-\frac{1}{2}z^2}dz \qquad (5.11)$$

It is already shown that $\frac{1}{\sqrt{2\pi}}\int_{-\infty}^{\infty}e^{-\frac{1}{2}z^2}dz=1$ and $\frac{1}{\sqrt{2\pi}}\int_{-\infty}^{\infty}ze^{-\frac{1}{2}z^2}dz=0$. Therefore, from Equation (5.11) we have,

$$E\left(X^2\right)=\frac{1}{\sigma\sqrt{2\pi}}\int_{-\infty}^{\infty}x^2 e^{-\frac{1}{2}\left(\frac{x-\mu}{\sigma}\right)^2}dx=\mu^2+\frac{\sigma^2}{\sqrt{2\pi}}\int_{-\infty}^{\infty}z^2 e^{-\frac{1}{2}z^2}dz$$

On substituting $\frac{1}{2}z^2=u$, we have,

$$E\left(X^2\right)=\mu^2+\frac{2\sigma^2}{\sqrt{\pi}}\int_{0}^{\infty}e^{-u}u^{\frac{1}{2}}du$$

$$=\mu^2+\frac{2\sigma^2}{\sqrt{\pi}}\Gamma\left(\frac{3}{2}\right)$$

$$=\mu^2+\frac{2\sigma^2}{\sqrt{\pi}}\frac{1}{2}\Gamma\left(\frac{1}{2}\right)$$

$$=\mu^2+\sigma^2$$

Therefore, from Equation (5.10) we have,

$$var\left(X\right)=\mu^2+\sigma^2-\mu^2=\sigma^2$$

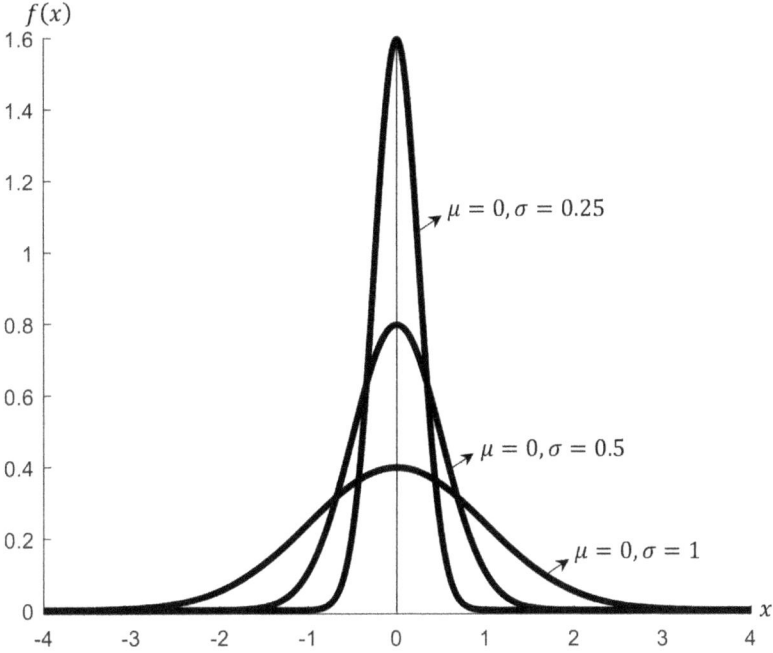

FIGURE 5.3   Normal distribution for different values of σ.

Figure 5.3 depicts the graph of normal probability density. The curvature of the graph of normal distribution remains unaltered if the variance remains same. However, the graph gradually flattens if the variance is increased (Figure 5.3). The normal distribution with $\mu = 0$ and $\sigma = 1$ is called the standard normal distribution.

Theorem 5.3: If the random variable X follows a normal distribution with mean μ and variance σ^2, then $Z = \frac{X-\mu}{\sigma}$ follows the standard normal distribution.

Proof: Since the relation between X and Z is linear, whenever X assumes a value x_1, the corresponding value of Z is $z_1 = \frac{x_1-\mu}{\sigma}$. Hence, we may write

$$P(X > x_1) = \int_{x_1}^{\infty} f(x)\,dx = \frac{1}{\sigma\sqrt{2\pi}} \int_{x_1}^{\infty} e^{-\frac{1}{2}\left(\frac{x-\mu}{\sigma}\right)^2}\,dx$$

$$= \frac{1}{\sqrt{2\pi}} \int_{z_1}^{\infty} e^{-\frac{1}{2}z^2}\,dz = P(Z > z_1)$$

where Z is seen to be a standard normal variate with $\mu = 0$ and $\sigma = 1$.

The distributions of X and Z are illustrated in Figure 5.4, from which it is evident that the shaded area under the X-curve between the ordinates $x = x_1$ and $x = \infty$ is equal to the shaded area under the Z-curve between the ordinates $z = z_1$ and $z = \infty$.

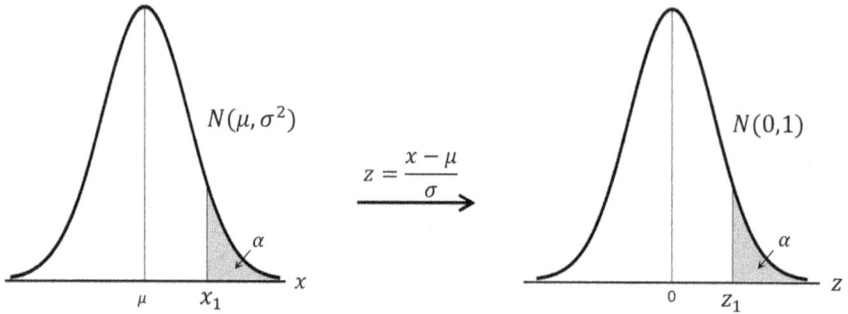

FIGURE 5.4 $P(X > x_1)$ and corresponding $P(Z > z_1)$.

The shaded area is denoted by α in Figure 5.4. Table A1 in Appendix A shows the values of areas under the standard normal curve corresponding to $P(Z > z_\alpha) = \alpha$ for the given values of z_α ranging from 0 to 3.09, while Table A2 in Appendix A shows the values of z_α for the given values of $P(Z > z_\alpha) = \alpha$.

Theorem 5.4: If the random variable X has a normal distribution with mean μ and standard deviation σ, then

$$P(\mu - \sigma < X < \mu + \sigma) = 0.6826$$

$$P(\mu - 2\sigma < X < \mu + 2\sigma) = 0.9544$$

$$P(\mu - 3\sigma < X < \mu + 3\sigma) = 0.9974$$

Proof: Using Theorem 5.3, we have

$$P(X > \mu + \sigma) = P\left(Z > \frac{\mu + \sigma - \mu}{\sigma}\right)$$

$$= P(Z > 1)$$

$$= 0.1587 \text{ (From Table A1)}$$

Again,

$$P(X > \mu - \sigma) = P\left(Z > \frac{\mu - \sigma - \mu}{\sigma}\right)$$

$$= P(Z > -1)$$

$$= 1 - P(Z < -1)$$

$$= 1 - P(Z > 1) \left(\text{Because the distribution is symmetric}\right)$$

$$= 1 - 0.1587 \text{ (From Table A1)}$$

$$= 0.8413$$

Therefore,

$$P(\mu-\sigma < X < \mu+\sigma) = P(X > \mu-\sigma) - P(X > \mu+\sigma) = 0.8413 - 0.1587 = 0.6826.$$

The other results given in the theorem can be proved similarly.

Hence, it can be inferred that for a normally distributed random variable X, 68.26% of the observed values should lie within one standard deviation of its mean, 95.44% of the observed values should lie within two standard deviations of its mean, and 99.74% of the observed values should lie within three standard deviations of its mean. Figure 5.5 depicts the results of the Theorem 5.4, as it applies to the standard normal distribution.

Some important properties of the normal distribution are listed below.

- The probability density function of the normal distribution is symmetric about mean μ.
- The mean, median and mode of a normal distribution are same.
- If $X \sim N(\mu, \sigma^2)$, then $Z = \frac{X-\mu}{\sigma} \sim N(0, 1)$.

Example 5.2:

The lengths of 1000 cotton fibres are found to be normally distributed with a mean of 28.5 mm and standard deviation of 8 mm. How many of these fibres are expected to have lengths

    a. greater than 40 mm?
    b. less than 12.5 mm?
    c. between 20 to 35 mm?

Assume that the fibre length can be measured to any degree of accuracy.

SOLUTION

Let the random variable X be the fibre length following normal distribution with mean $\mu = 28.5$ mm and standard deviation $\sigma = 8$ mm.

    a. The proportion of fibres having lengths greater than 40 mm is shown by the shaded area in Figure 5.6(a). This can be calculated as

$$P(X > 40) = P\left(Z > \frac{40 - 28.5}{8}\right) = P(Z > 1.44) = 0.075$$

$$\text{(From Table A1).}$$

Thus, around $1000 \times 0.075 = 75$ fibres have lengths greater than 40 mm.

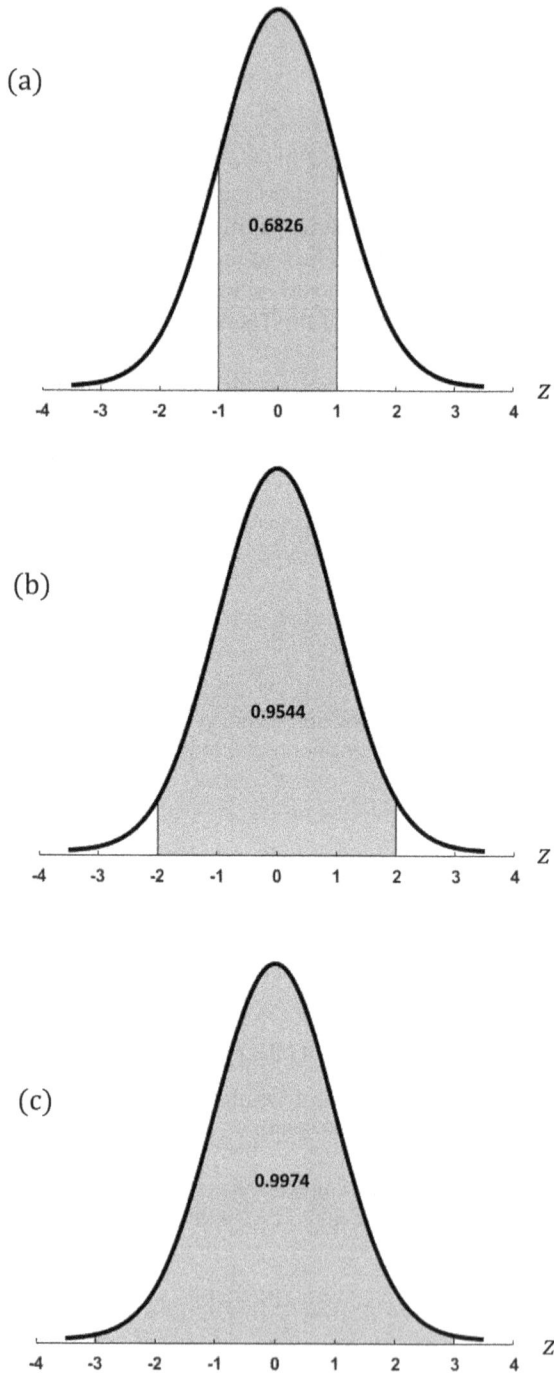

FIGURE 5.5   (a) $P(-1 < Z < 1)$, (b) $P(-2 < Z < 2)$, (c) $P(-3 < Z < 3)$.

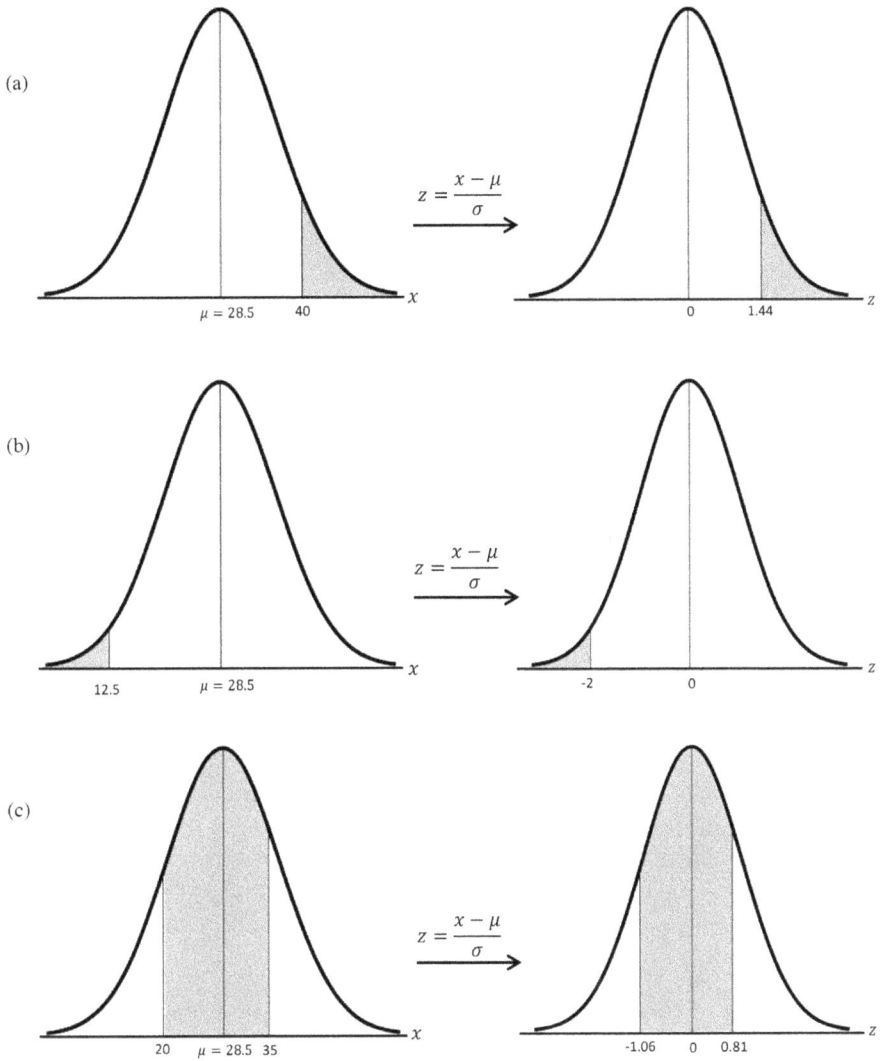

FIGURE 5.6 (a) $P(X > 40)$ and corresponding $P(Z > 1.44)$; (b) $P(X < 12.5)$ and corresponding $P(Z < -2)$; (c) $P(20 < X < 35)$ and corresponding $P(-1.06 < Z < 0.81)$.

b. The proportion of fibres having lengths less than 12.5 mm is shown by the shaded area in Figure 5.6(b). Hence,

$$P(X < 12.5) = P\left(Z < \frac{12.5 - 28.5}{8}\right) = P(Z < -2) = P(Z > 2) = 0.023$$

(From Table A1).

Thus, around $1000 \times 0.023 = 23$ fibres have lengths less than 12.5 mm.

c. The proportion of fibres having lengths between 20 to 35 mm is shown by the shaded area in Figure 5.6(c). Hence,

$$P(20 < X < 35) = P(X > 20) - P(X > 35)$$

$$= P\left(Z > \frac{20 - 28.5}{8}\right) - P\left(Z > \frac{35 - 28.5}{8}\right)$$

$$= P(Z > -1.06) - P(Z > 0.81)$$

$$= 1 - P(Z < -1.06) - P(Z > 0.81)$$

$$= 1 - P(Z > 1.06) - P(Z > 0.81)$$

$$= 1 - 0.145 - 0.209 \text{ (From Table A1)}$$

$$= 0.646$$

Thus, around $1000 \times 0.646 = 646$ fibres have lengths between 20 to 35 mm.

Example 5.3:

Find the number of fibres exceeding 40 mm length in Example 5.2 if the length is measured to the nearest 0.1 mm.

SOLUTION

As the length is measured to the nearest 0.1 mm, the fibres having lengths greater than 40 mm would have at least 40.1 mm length. In this case, we approximate a discrete distribution by means of a continuous normal distribution. Hence,

$$P(X > 40.1) = P\left(Z > \frac{40.1 - 28.5}{8}\right) = P(Z > 1.45) = 0.074 \text{ (From Table A1)}.$$

Thus around $1000 \times 0.074 = 74$ fibres have lengths greater than 40 mm.

Example 5.4:

The mean and standard deviation of yarn count are 20.1 tex and 0.9 tex, respectively. Assuming the distribution of yarn count to be normal, how many leas out of 300 would be expected to have counts

a. 21.5 tex
b. less than 19.6 tex
c. between 19.65 to 20.55 tex

SOLUTION

Let the random variable X be the yarn count following normal distribution with mean $\mu = 20.1$ tex and standard deviation $\sigma = 0.9$ tex.

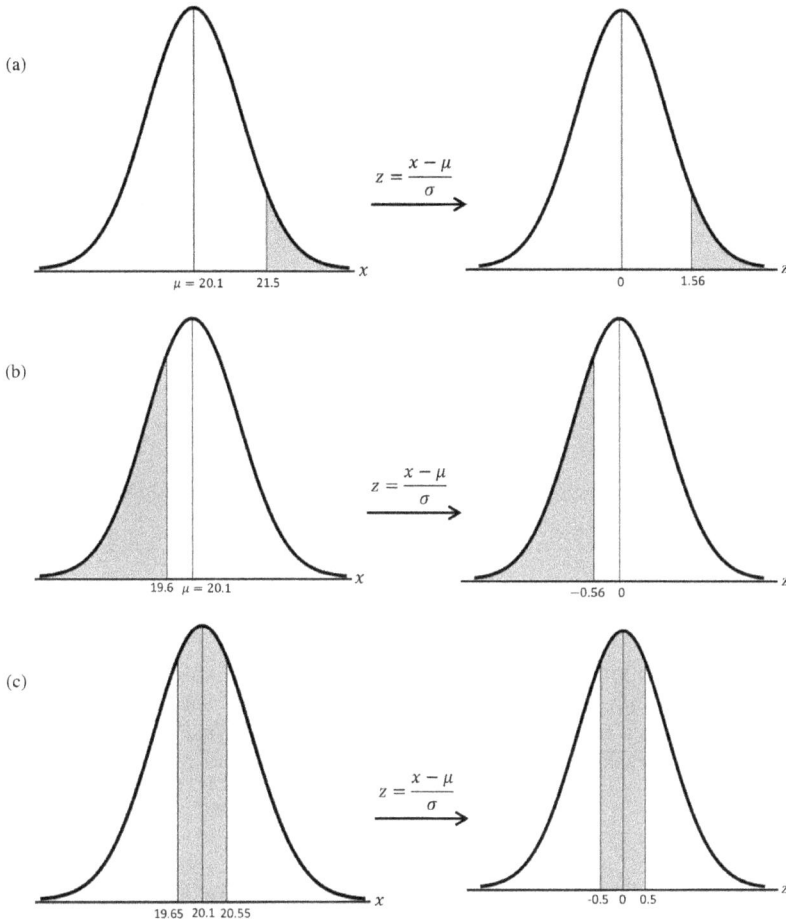

FIGURE 5.7 (a) $P(X > 21.5)$ and corresponding $P(Z > 1.56)$; (b) $P(X < 19.6)$ and corresponding $P(Z < -0.56)$; (c) $P(19.65 < X < 20.55)$ and corresponding $P(-0.5 < Z < 0.5)$.

a. The proportion of leas having counts greater than 21.5 tex is depicted by the shaded area in Figure 5.7(a). This can be calculated as

$$P(X > 21.5) = P\left(Z > \frac{21.5 - 20.1}{0.9}\right) = P(Z > 1.56) = 0.0594 \text{ (From Table A1).}$$

Thus, approximately $300 \times 0.0594 \cong 18$ leas have counts greater than 21.5 tex.

b. The proportion of leas having counts less than 19.6 tex shown by the shaded area in Figure 5.7(b). Hence,

$$P(X < 19.6) = P\left(Z < \frac{19.6 - 20.1}{0.9}\right) = P(Z < -0.56) = P(Z > 0.56) = 0.2877$$

$$\text{(From Table A1).}$$

Thus, approximately $300 \times 0.2877 \cong 86$ leas have counts less than 19.6 tex.

c. The proportion of leas having counts between 19.65 to 20.55 tex is shown by the shaded area in Figure 5.7(c). Hence,

$$P(19.65 < X < 20.55) = P(X > 19.65) - P(X > 20.55)$$

$$= P\left(Z > \frac{19.65 - 20.1}{0.9}\right) - P\left(Z > \frac{20.55 - 20.1}{0.9}\right)$$

$$= P(Z > -0.5) - P(Z > 0.5)$$

$$= 1 - P(Z > 0.5) - P(Z > 0.5)$$

$$= 1 - 2 \times P(Z > 0.5)$$

$$= 1 - 2 \times 0.3085 \text{ (From Table A1)}$$

$$= 0.383$$

Thus, approximately $300 \times 0.383 \cong 115$ leas have counts between 19.65 to 20.55 tex.

Example 5.5:

When a large number of single threads of cotton sized yarn are stretched by 2% of their length, 1.3% of them break, and when they are stretched by 5% of their length, a further 80.3% of them break. Assuming the distribution of single thread breaking elongation to be normal, what are the mean and standard deviation?

SOLUTION

Let the random variable X be the yarn breaking elongation following normal distribution with mean μ and standard deviation σ. Figures 5.8(a) and (b) depict two areas under the normal curve by shaded regions. From the data, the probability of breaking elongation less than 2% is 0.013. Since it is less than 0.5, it corresponds to a negative value of variable Z. Thus, we can write

$$P(X < 2) = P\left(Z < \frac{2 - \mu}{\sigma}\right)$$

$$= 0.013$$

$$= P(Z < -2.226) \text{ (From Table A2)}$$

Therefore,

$$\frac{2 - \mu}{\sigma} = -2.226$$

or, $\mu - 2.226\sigma = 2$

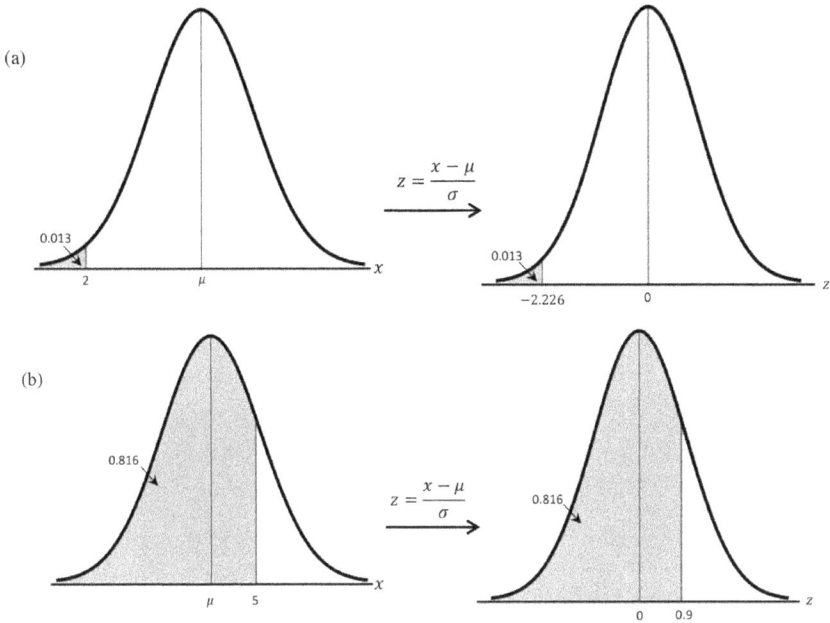

FIGURE 5.8 (a) $P(X < 2)$ and corresponding $P(Z < -2.226)$ (b) $P(X < 5)$ and corresponding $P(Z < 0.9)$.

The probability of breaking elongation less than 5% is $0.803 + 0.013 = 0.816$. Therefore, the probability of breaking elongation greater than 5% is $1 - 0.816 = 0.184$. Hence, we can write

$$P(X > 5) = P\left(Z > \frac{5 - \mu}{\sigma} \right)$$

$$= 0.184$$

$$= P(Z > 0.9) \text{ (From Table A2)}$$

Therefore,

$$\frac{5 - \mu}{\sigma} = 0.9$$

or, $\mu + 0.9\sigma = 5$

By solving the above two equations, the mean and standard deviation are found to be $\mu = 4.136$ and $\sigma = 0.96$.

5.4.1 NORMAL APPROXIMATION TO BINOMIAL DISTRIBUTION

In many situations it becomes difficult or cumbersome to find the probability of an event when the underlying distribution is binomial. To illustrate we consider the following example.

Example 5.6:

Suppose 10% of the items produced by a machine are defective. 100 items are selected at random. What is the probability that at least 70 items are non-defective?

SOLUTION

The number of non-defective items (say X) follows binomial distribution with probability 0.90.

Therefore,

$$P(X \geq 70) = \sum_{n=70}^{100} \binom{100}{n} (0.90)^n (0.10)^{100-n}$$

One can note that the evaluation of the above expression is very difficult. A suitable way to overcome this difficulty is to approximate the distribution of X to normal distribution.

If a random variable X follows binomial distribution with parameters n and p, the distribution of X can be approximated by normal distribution if the parameter n, the total number of trials is sufficiently large. The approximation is more accurate if the values of the random variable under consideration lie near the mean that is the standard deviation $\sqrt{npq}$ is not very large. The details of normal approximation to binomial distribution, when the number of trials n is large, is given below.

For binomial distribution we have

$$P(X = x) = f(x) = \frac{n!}{x!(n-x)!} p^x q^{n-x} \qquad (5.12)$$

We now use the Stirling approximation of $n!$ which is given by

$$n! \sim n^n e^{-n} \sqrt{2\pi n} \left(1 + O\left(\frac{1}{n}\right)\right)$$

where $O(\cdot)$ is the big-O notation that describes the limiting behaviour of a function when n tends towards a particular value or infinity.

Therefore, from Equation (5.12) we have,

$$f(x) = \frac{n^n e^{-n} \sqrt{2\pi n} \left(1 + O\left(\frac{1}{n}\right)\right)}{x^x e^{-x} \sqrt{2\pi x} (n-x)^{(n-x)} e^{-(n-x)} \sqrt{2\pi (n-x)} \left(1 + O\left(\frac{1}{n}\right)\right)} p^x q^{n-x}$$

$$= \left(\frac{np}{x}\right)^x \left(\frac{nq}{n-x}\right)^{n-x} \sqrt{\frac{n}{2\pi x(n-x)}} \left(1 + O\left(\frac{1}{n}\right)\right) \qquad (5.13)$$

We set, $x - np = \varepsilon$ so that $x = np + \varepsilon$ and $n - x = n - np - \varepsilon = nq - \varepsilon$

Now,

$$\log\left(\frac{np}{x}\right) = \log\left(\frac{np}{np+\varepsilon}\right) = -\log\left(1+\frac{\varepsilon}{np}\right) \text{ and } \log\left(\frac{nq}{n-x}\right) = \log\left(\frac{nq}{nq-\varepsilon}\right)$$

$$= -\log\left(1-\frac{\varepsilon}{nq}\right)$$

Thus,

$$\log\left(\frac{np}{x}\right)^{x}\left(\frac{nq}{n-x}\right)^{n-x}$$

$$= x\log\left(\frac{np}{x}\right) + (n-x)\log\frac{nq}{(n-x)}$$

$$= -(\varepsilon+np)\left[\frac{\varepsilon}{np} - \frac{1}{2}\frac{\varepsilon^2}{n^2p^2} + O\left(\frac{\varepsilon^3}{n^3}\right)\right] - (nq-\varepsilon)\left[-\frac{\varepsilon}{nq} - \frac{1}{2}\frac{\varepsilon^2}{n^2q^2} + O\left(\frac{\varepsilon^3}{n^3}\right)\right]$$

$$= -\frac{\varepsilon^2}{2npq} + O\left(\frac{\varepsilon^3}{n^2}\right)$$

Hence,

$$\left(\frac{np}{x}\right)^{x}\left(\frac{nq}{n-x}\right)^{n-x} = e^{-\frac{\varepsilon^2}{2npq}}\left[1+O\left(\frac{\varepsilon^3}{n^2}\right)\right] \tag{5.14}$$

Again,

$$\sqrt{\frac{n}{2\pi x(n-x)}} = \sqrt{\frac{n}{2\pi(np+\varepsilon)(nq-\varepsilon)}}$$

$$= \sqrt{\frac{1}{2\pi npq}}\left(1+O\left(\frac{\varepsilon}{n}\right)\right) \tag{5.15}$$

Thus, from Equations (5.13), (5.14) and (5.15) we have

$$f(x) = \sqrt{\frac{1}{2\pi npq}}e^{-\frac{\varepsilon^2}{2npq}}\left[1+O\left(\frac{\varepsilon^3}{n^2}\right)\right]\left(1+O\left(\frac{\varepsilon}{n}\right)\right)$$

$$= \frac{1}{\sqrt{2\pi npq}}e^{-\frac{\varepsilon^2}{2npq}}\left(1+O\left(\frac{\varepsilon}{n}\right)\right) \tag{5.16}$$

Note that the average deviation of the probable values of the random variable is actually the standard deviation. Therefore, we may argue that $\varepsilon = x - np = O\left(\sqrt{n}\right)$

at most and thus Equation (5.16) gives $f(x)=\frac{1}{\sqrt{2\pi npq}}e^{-\frac{\varepsilon^2}{2npq}}\left(1+O\left(\frac{1}{\sqrt{n}}\right)\right)$. So,

$f(x)=\frac{1}{\sqrt{2\pi npq}}e^{-\frac{(x-np)^2}{2npq}}$ as $n\to\infty$. This shows that X follows normal distribution with mean np and variance npq.

For normal approximation to binomial distribution, we should use the continuity correction according to which each non-negative integer x is represented by the interval from $x-0.5$ to $x+0.5$. The continuity correction is needed due to the fact that a discrete distribution is being approximated by a continuous distribution.

Example 5.7:

If 10% of the laps produced by a blow room line are defectives, what is the probability that in a lot of 250 randomly chosen laps for inspection

a. at most 20 will be defective?
b. exactly 20 will be defective?
c. 20 or more will be defective?

SOLUTION

Let the random variable X be the number of defective laps in a lot. The variable X has binomial distribution with mean $\mu=np=250\times0.1=25$ and standard deviation $\sigma=\sqrt{np(1-p)}=\sqrt{250\times0.1\times(1-0.1)}=4.74$. We obtain the following normal approximation to the binomial distribution for finding the desired probabilities with the continuity correction.

a. $P(X\le20)=P(X<20.5)$

$=P\left(Z<\frac{20.5-25}{4.74}\right)$

$=P(Z<-0.95)$

$=P(Z>0.95)$

$=0.171$ (From Table A1).

This is indicated in Figure 5.9(a).

b. $P(X=20)=P(19.5<X<20.5)$

$=P\left(Z<\frac{20.5-25}{4.74}\right)-P\left(Z<\frac{19.5-25}{4.74}\right)$

$=P(Z<-0.95)-P(Z<-1.16)$

$=P(Z>0.95)-P(Z>1.16)$

$=0.171-0.123$ (From Table A1)

$=0.048$

This is indicated in Figure 5.9(b).

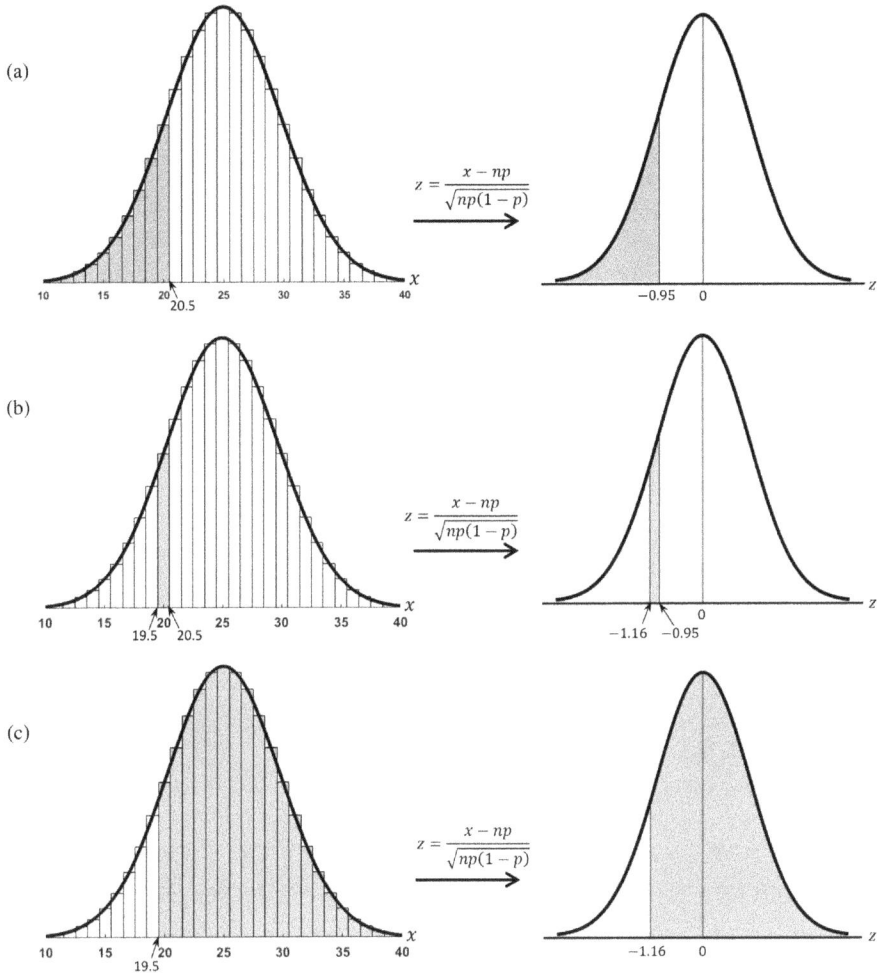

FIGURE 5.9 (a) $P(X < 20.5)$ and corresponding $P(Z < -0.95)$; (b) $P(19.5 < X < 20.5)$ and corresponding $P(-1.16 < Z < -0.95)$; (c) $P(X > 19.5)$ and corresponding $P(Z > -1.16)$.

c. $P(X \geq 20) = P(X > 19.5)$

$$= P\left(Z > \frac{19.5 - 25}{4.74}\right)$$

$$= P(Z > -1.16)$$

$$= 1 - P(Z < -1.16)$$

$$= 1 - P(Z > 1.16)$$

$$= 1 - 0.123 \text{ (From Table A1)}$$

$$= 0.877$$

This is indicated in Figure 5.9(c).

Example 5.8:

A garment manufacturer usually produces 2% defectives. What is the probability that in a lot of 1000 randomly chosen garments

a. less than 15 will be defective?
b. greater than 30 will be defective?
c. between 15 to 30 inclusive will be defective?

SOLUTION

Let the random variable X be the number of defective garments in a lot. The variable X has binomial distribution with mean $\mu = np = 1000 \times 0.02 = 20$ and standard deviation $\sigma = \sqrt{np(1-p)} = \sqrt{1000 \times 0.02 \times (1-0.02)} = 4.43$. We obtain the following normal approximation to the binomial distribution for finding the desired probabilities with the continuity correction.

a. $P(X < 15) = P(X < 14.5)$

$$= P\left(Z < \frac{14.5 - 20}{4.43}\right)$$

$$= P(Z < -1.24)$$

$$= P(Z > 1.24)$$

$$= 0.1075 \text{ (From Table A1)}$$

This is indicated in Figure 5.10(a).

b. $P(X > 30) = P(X > 30.5)$

$$= P\left(Z > \frac{30.5 - 20}{4.43}\right)$$

$$= P(Z > 2.37)$$

$$= 0.0089 \text{ (From Table A1)}$$

This is indicated in Figure 5.10(b).

c. $P(15 \le X \le 30) = P(14.5 < X < 30.5)$

$$= P(X > 14.5) - P(X > 30.5)$$

$$= P\left(Z > \frac{14.5 - 20}{4.43}\right) - P\left(Z > \frac{30.5 - 20}{4.43}\right)$$

$$= P(Z > -1.24) - P(Z > 2.37)$$

$$= 1 - P(Z < -1.24) - 0.0089 \text{ (From Table A1)}$$

$$= 1 - P(Z > 1.24) - 0.0089$$

$$= 1 - 0.1075 - 0.0089 \text{ (From Table A1)}$$

$$= 0.8836$$

This is indicated in Figure 5.10(c).

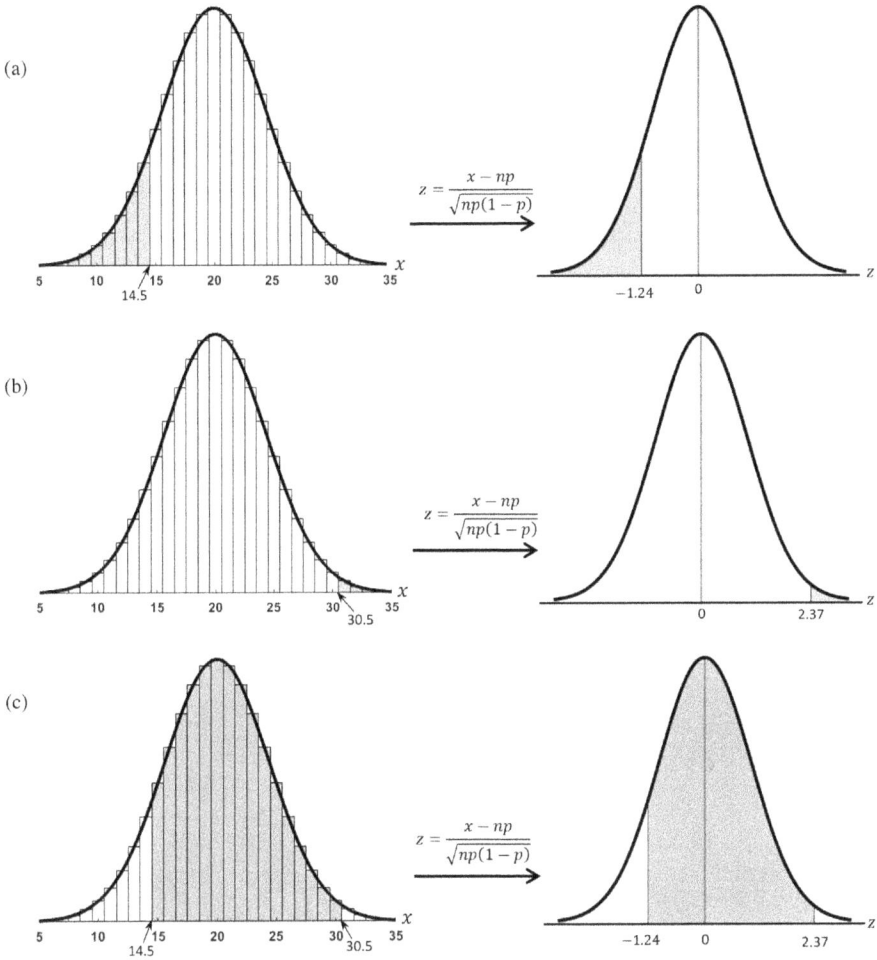

FIGURE 5.10   (a) $P(X < 14.5)$ and corresponding $P(Z < -1.24)$; (b) $P(X > 30.5)$ and corresponding $P(Z > 2.37)$; (c) $P(14.5 < X < 30.5)$ and corresponding $P(-1.24 < Z < 2.37)$.

5.4.2   NORMAL APPROXIMATION TO POISSON DISTRIBUTION

We have already shown that binomial distribution can be approximated by normal distribution under the restriction that the number of trials n is large. We also know that Poisson distribution is obtained from binomial distribution imposing the condition that the number of trials is large and the probability of success is very small. So it is legitimate to expect that the Poisson distribution can be converted to normal distribution under some suitable conditions. We furnish the details in the following:

Suppose X follows Poisson distribution with parameter λ. The moment generating function (MGF) of X is given by

$$M_X(t) = e^{\lambda(e^t - 1)}$$

We know that for Poisson distribution both the mean and variance of X is λ. Let $Z = \frac{X-\lambda}{\sqrt{\lambda}}$ and therefore,

$$M_Z(t) = e^{-\frac{\lambda t}{\sqrt{\lambda}}} M_X\left(\frac{t}{\sqrt{\lambda}}\right) = e^{-\sqrt{\lambda}t} e^{\lambda\left(e^{\frac{t}{\sqrt{\lambda}}}-1\right)}$$

Taking the logarithm of $M_Z(t)$ we have,

$$\ln M_Z(t) = -\sqrt{\lambda}t + \lambda\left(e^{\frac{t}{\sqrt{\lambda}}}-1\right)$$

$$= -\sqrt{\lambda}t + \lambda\left(1 + \frac{t}{\sqrt{\lambda}} + \frac{t^2}{2!\lambda} + \frac{t^3}{3!\lambda^{\frac{3}{2}}}\cdots -1\right)$$

$$= \frac{t^2}{2} + \frac{t^3}{\sqrt{\lambda}3!} + \cdots$$

Therefore,

$$\lim_{\lambda\to\infty} \ln M_Z(t) = \frac{t^2}{2}$$

This shows that the moment generating function of Z converges to that of a standard normal variate. So, from the uniqueness of moment generating function we can conclude that the random variable Z converges to standard normal variate as $\lambda \to \infty$. Therefore, we can infer that $X = \sqrt{\lambda}Z + \lambda$ will follow normal distribution with mean λ and variance λ, when λ is large.

Example 5.9:

Average number of complaints that a dry-cleaning establishment receives per month is 25. What is the probability that in a given month there will be

a. less than 20 complaints?
b. 30 or more complaints?

SOLUTION

Let the random variable X be the number of complaints received per month. The variable X has Poisson distribution with $\mu = 25$ and $\sigma = \sqrt{25} = 5$. We obtain the following normal approximation to the Poisson distribution for finding the desired probabilities with the continuity correction.

a. $P(X < 20) = P(X < 19.5)$

$$= P\left(Z < \frac{19.5 - 25}{5}\right)$$

$$= P(Z < -1.1)$$

$$= P(Z > 1.1)$$

$$= 0.1357 \text{ (From Table A1)}$$

This is indicated in Figure 5.11(a).

b. $P(X \geq 30) = P(X > 29.5)$

$$= P\left(Z > \frac{29.5 - 25}{5}\right)$$

$$= P(Z > 0.9)$$

$$= 0.1841 \text{ (From Table A1)}$$

This is indicated in Figure 5.11(b).

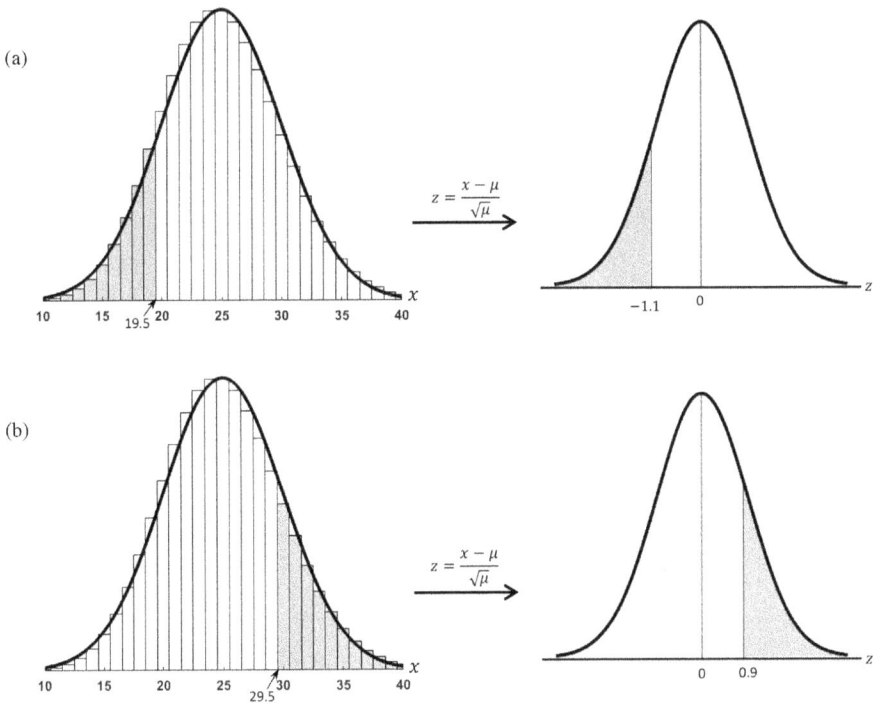

FIGURE 5.11 (a) $P(X < 19.5)$ and corresponding $P(Z < -1.1)$; (b) $P(X > 29.5)$ and corresponding $P(Z > 0.9)$.

5.5   LOGNORMAL DISTRIBUTION

Definition 5.4: If $X \sim N\left(\mu, \sigma^2\right)$ and the random variable $Y = exp(X)$, then the distribution of Y is called the lognormal distribution.

Theorem 5.5: If $X \sim N\left(\mu, \sigma^2\right)$ and $Y = exp(X)$, then the probability density function of Y is given by

$$g(y) = \frac{exp\left(-\frac{1}{2}\left(\frac{\ln y - \mu}{\sigma}\right)^2\right)}{y\sqrt{2\pi}\sigma} \, ; y > 0$$

$$= 0 \ elsewhere$$

Proof: The probability density function of X is,

$$f(x) = \frac{1}{\sigma\sqrt{2\pi}} e^{-\frac{1}{2}\left(\frac{x-\mu}{\sigma}\right)^2} , \quad -\infty < x < \infty$$

We take the transformation $y = \varphi(x) = e^x$, hence $x = \varphi^{-1}(y) = \ln y$. Now, $\ln y$ is well defined since $y = e^x > 0$ for $-\infty < x < \infty$ and it is evident that $0 < y < \infty$. The Jacobian of the transformation is $J = \frac{dx}{dy} = \frac{1}{y}$. Now from Theorem 3.8, the probability density function $g(y)$ of Y can be written as

$$g(y) = f\left(\varphi^{-1}(y)\right)|J|, \ 0 < y < \infty$$

$$= \frac{1}{\sigma\sqrt{2\pi}} e^{-\frac{1}{2}\left(\frac{\ln y - \mu}{\sigma}\right)^2} \frac{1}{y}$$

Therefore, $g(y)$ can be written as,

$$g(y) = \frac{exp\left(-\frac{1}{2}\left(\frac{\ln y - \mu}{\sigma}\right)^2\right)}{y\sqrt{2\pi}\sigma}, \ 0 < y < \infty.$$

The first two moments of the lognormal distribution are $E(Y) = exp\left(\mu + \frac{\sigma^2}{2}\right)$ and $E\left(Y^2\right) = exp\left(2\mu + 2\sigma^2\right)$. From this we have $var(Y) = exp\left(2\mu + \sigma^2\right)\left(exp\left(\sigma^2\right) - 1\right)$. The graph of log-normal probability density for different μ and σ is depicted in Figure 5.12.

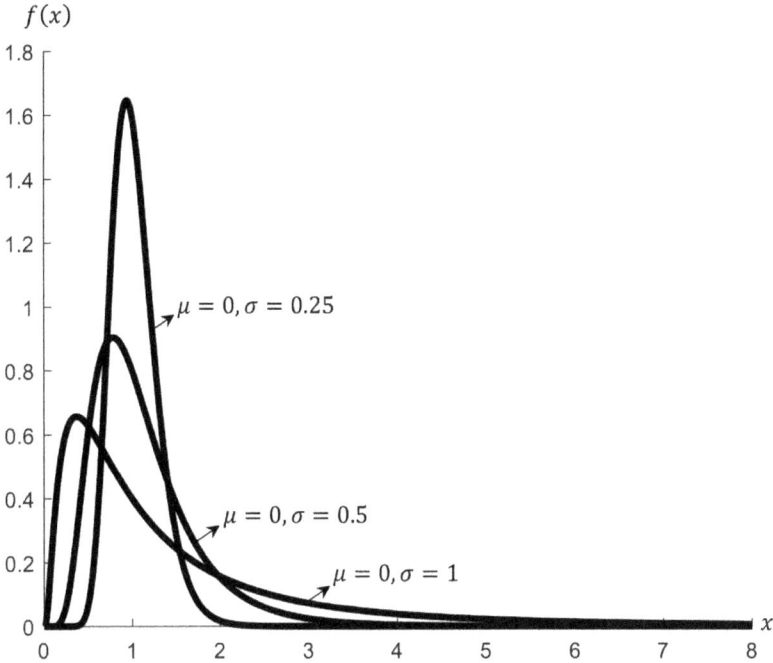

FIGURE 5.12   Log-normal distribution.

Example 5.10:

The concentration of certain pollutant in a textile effluent in parts per million (ppm) has a lognormal distribution with a mean of 4.2 ppm and standard deviation of 1.12 ppm. What is the probability that concentration of the pollutant exceeds 10 ppm?

SOLUTION

Let the random variable X be the concentration of the pollutant. Figure 5.13 shows the lognormal distribution of variable X with $\mu = 4.2$ and $\sigma = 1.12$. The probability that concentration of pollutant exceeds 10 ppm is given by

$$P(X > 10) = P\left(Z > \frac{\ln(10) - 4.2}{1.12}\right)$$

$$= P(Z > -1.69)$$

$$= 1 - P(Z < -1.69)$$

$$= 1 - P(Z > 1.69)$$

$$= 1 - 0.0455 \text{ (From Table A1)}$$

$$= 0.9545$$

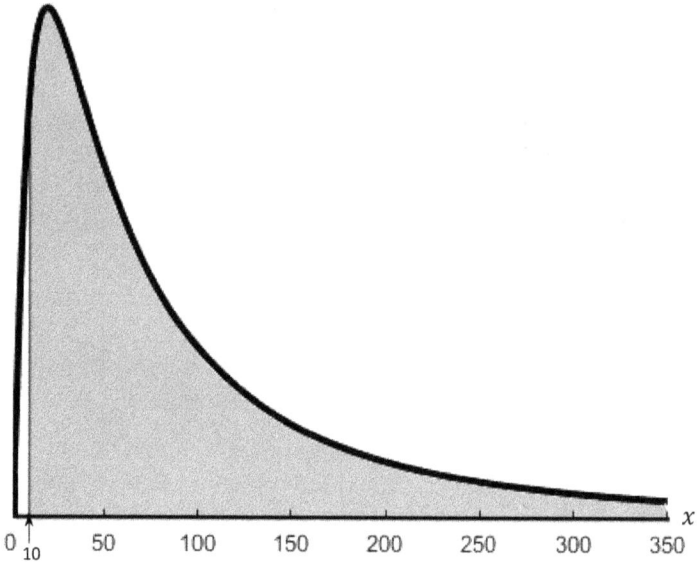

FIGURE 5.13   Lognormal distribution and $P(X > 10)$.

5.6   BIVARIATE NORMAL DISTRIBUTION

Definition 5.5: Two random variables X_1, X_2 are said to follow bivariate normal distribution if the joint density function of X_1, X_2 is given by

$$f(x_1,x_2) = \frac{1}{2\pi\sigma_1\sigma_2\sqrt{1-\rho^2}} exp\left[-\frac{1}{2(1-\rho^2)}\left\{\frac{(x_1-\mu_1)^2}{\sigma_1^2} - 2\rho\frac{(x_1-\mu_1)(x_2-\mu_2)}{\sigma_1\sigma_2} + \frac{(x_2-\mu_2)^2}{\sigma_2^2}\right\}\right]$$

where ρ is the correlation coefficient between X_1 and X_2 (Chapter 9 discusses the correlation coefficient in detail). The mean vector of the bivariate normal distribution is $E\binom{X_1}{X_2} = \binom{\mu_1}{\mu_2}$.

The random variables X_1, X_2 have a variance-covariance structure instead of only variance as in the case of a single random variable. The variance covariance structure is written in a matrix form. For bivariate normal distribution the variance-covariance matrix is given by

$$\Sigma = \begin{pmatrix} \sigma_{11} & \sigma_{12} \\ \sigma_{21} & \sigma_{22} \end{pmatrix}$$

Here the diagonal entries of Σ are the variances of respective random variables and the remaining entries are the covariances between the random variables. Σ is a symmetric matrix that is $\sigma_{12} = \sigma_{21}$. Thus,

$$\sigma_{11} = \sigma_1^2 = E(X_1 - \mu_1)^2$$

$$\sigma_{22} = \sigma_2^2 = E(X_2 - \mu_2)^2$$

$$\sigma_{12} = E\left[(X_1 - \mu_1)(X_2 - \mu_2)\right] = \sigma_1\sigma_2\rho$$

In particular if the random variables X_1 and X_2 are independent, then the correlation coefficient between these two variables is 0, that is $\rho = 0$. In that case, $f(x_1,x_2)$ reduces to

$$f(x_1,x_2) = \frac{1}{2\pi\sigma_1\sigma_2} exp\left[-\frac{1}{2}\left\{\frac{(x_1-\mu_1)^2}{\sigma_1^2} + \frac{(x_2-\mu_2)^2}{\sigma_2^2}\right\}\right]$$

$$= \frac{1}{\sigma_1\sqrt{2\pi}} exp\left[-\frac{1}{2}\frac{(x_1-\mu_1)^2}{\sigma_1^2}\right]\frac{1}{\sigma_2\sqrt{2\pi}} exp\left[-\frac{1}{2}\frac{(x_2-\mu_2)^2}{\sigma_2^2}\right]$$

This shows that the joint density function can be written as the product of the probability density functions of two random variables X_1 and X_2 with $X_1 \sim N(\mu_1,\sigma_1^2)$ and $X_2 \sim N(\mu_2,\sigma_2^2)$, when X_1 and X_2 are uncorrelated or independent.

5.6.1 THE CONDITIONAL DISTRIBUTION OF BIVARIATE NORMAL

Sometimes it becomes necessary to know the distribution of one random variable keeping the other random variable fixed at a particular value. In such cases we may use the conditional distribution of one random variable given the other. The following theorem gives the conditional distribution of one random variable given the other when both the random variables are jointly distributed as normal random variates.

Theorem 5.6: If X_1 and X_2 have joint normal distribution with means μ_1, μ_2 and variances σ_1^2 and σ_2^2 and ρ is the correlation coefficient between X_1 and X_2 with $\rho \neq \pm 1$, then the probability density function of the random variable X_1 given $X_2 = x_2$ is given by

$$g(x_1|x_2) = \frac{1}{\sqrt{2\pi}\sigma_1\sqrt{1-\rho^2}} e^{-\frac{1}{2}\left(\frac{x_1-\mu_1-\rho\frac{\sigma_1}{\sigma_2}(x_2-\mu_2)}{\sigma_1\sqrt{1-\rho^2}}\right)^2}$$

Proof: The conditional distribution of X_1 given $X_2 = x_2$ is given by

$$g(x_1 | x_2) = \frac{f(x_1, x_2)}{f_{X_2}(x_2)} \tag{5.17}$$

where $f_{X_2}(x_2)$ is the marginal density function of the random variable X_2 and hence

$$f_{X_2}(x_2) = \int_{-\infty}^{\infty} f(x_1, x_2) dx_1$$

$$= \int_{-\infty}^{\infty} \frac{1}{2\pi\sigma_1\sigma_2\sqrt{1-\rho^2}} exp\left[-\frac{1}{2(1-\rho^2)}\left\{\frac{(x_1 - \mu_1)^2}{\sigma_1^2} - 2\rho\frac{(x_1 - \mu_1)(x_2 - \mu_2)}{\sigma_1\sigma_2}\right.\right.$$

$$\left.\left. + \frac{(x_2 - \mu_2)^2}{\sigma_2^2}\right\}\right] dx_1$$

$$\tag{5.18}$$

It can be shown that

$$f_{X_2}(x_2) = \frac{1}{\sqrt{2\pi}\sigma_2} e^{-\frac{1}{2}\left(\frac{x_2 - \mu_2}{\sigma_2}\right)^2} \text{ for } -\infty < x_2 < \infty.$$

Thus, the marginal distribution function of X_2 is the normal distribution function with mean μ_2 and variance σ_2^2.

Therefore, from Equation (5.17) we have

$$g(x_1 | x_2) = \frac{f(x_1, x_2)}{f_{X_2}(x_2)} = \frac{1}{\sqrt{2\pi}\sigma_1\sqrt{1-\rho^2}} e^{-\frac{1}{2}\left(\frac{x_1 - \mu_1 - \rho\frac{\sigma_1}{\sigma_2}(x_2 - \mu_2)}{\sigma_1\sqrt{1-\rho^2}}\right)^2}$$

This shows that the conditional distribution of X_1 given $X_2 = x_2$ is normally distributed with mean $\mu_1 + \rho\frac{\sigma_1}{\sigma_2}(x_2 - \mu_2)$ and variance $\sigma_1^2(1-\rho^2)$.

Example 5.11:

The yarn tenacity X_1 and breaking elongation X_2 have a bivariate normal distribution with the following parameters.

$$\mu_1 = 16.3 \text{ cN / tex}, \mu_2 = 4.4\%, \sigma_1 = 1.3 \text{ cN / tex}, \sigma_2 = 0.4\%, \rho = 0.6.$$

a. Find the probability that yarn has tenacity between 18 to 20 cN/tex, given that yarn breaking elongation is 5%.
b. Find the probability that yarn has breaking elongation greater than 5%, given that yarn tenacity is 18.5 cN/tex.

SOLUTION

The bivariate normal distribution of yarn tenacity X_1 and breaking elongation X_2 is shown in Figure 5.14(a).

a. The conditional distribution of X_1, given $X_2 = x_2$, is the normal distribution with mean $\mu_1 + \rho \frac{\sigma_1}{\sigma_2}(x_2 - \mu_2) = 16.3 + 0.6 \times \frac{1.3}{0.4} \times (5 - 4.4) = 17.47$ and standard deviation $\sigma_1 \sqrt{(1 - \rho^2)} = 1.3 \times \sqrt{1 - 0.6^2} = 1.04$ and it is depicted in Figure 5.14(b). Accordingly, given that yarn breaking elongation is 5%, the probability that yarn has tenacity between 18 to 20 cN/tex is given by

$$P\left(18 < X_1 < 20 \mid X_2 = 5\right) = P\left(Z > \frac{18 - 17.47}{1.04}\right) - P\left(Z > \frac{20 - 17.47}{1.04}\right)$$

$$= P\left(Z > 0.51\right) - P\left(Z > 2.43\right)$$

$$= 0.305 - 0.0075 \text{ (From Table A1)}$$

$$= 0.2975$$

Hence, the interval (18, 20) could be thought of as a 29.75% prediction interval for the yarn tenacity in cN/tex, given yarn breaking elongation = 5%.
b. The conditional distribution of X_2, given $X_1 = x_1$, is the normal distribution with mean $\mu_2 + \rho \frac{\sigma_2}{\sigma_1}(x_1 - \mu_1) = 4.4 + 0.6 \times \frac{0.4}{1.3} \times (18.5 - 16.3) = 4.81$ and standard deviation $\sigma_2 \sqrt{(1 - \rho^2)} = 0.4 \times \sqrt{1 - 0.6^2} = 0.32$, and it is depicted in Figure 5.14(c). Accordingly, given that yarn tenacity is 18.5 cN/tex, the probability that yarn has breaking elongation greater than 5% is given by

$$P\left(X_2 > 5 \mid X_1 = 18.5\right) = P\left(Z > \frac{5 - 4.81}{0.32}\right)$$

$$= P\left(Z > 0.59\right)$$

$$= 0.278 \text{ (From Table A1)}$$

Hence, the probability that yarn has breaking elongation greater than 5% is 0.278, given yarn tenacity = 18.5 cN/tex.

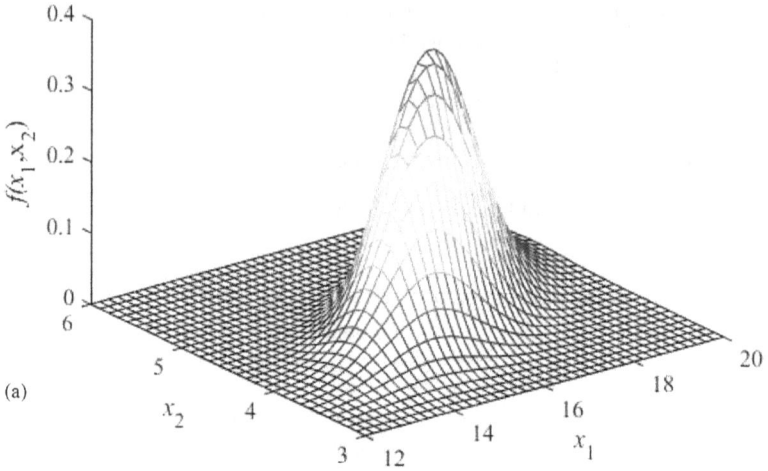

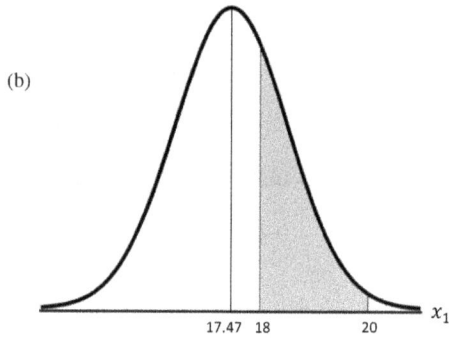

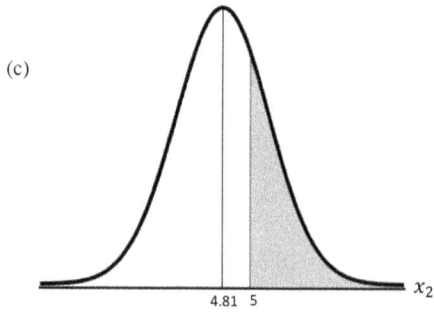

FIGURE 5.14   (a) Bivariate normal distribution of X_1 and X_2; (b) the conditional distribution of X_1, given $X_2 = 5$ showing $P(18 < X_1 < 20|X_2 = 5)$; (c) the conditional distribution of X_2, given $X_1 = 18.5$ showing $P(X_2 > 5|X_1 = 18.5)$.

5.7 MATLAB® CODING

5.7.1 MATLAB® Coding of Example 5.1

```
clc
clear
close all
beta=2.5;
pr_1=expcdf(3,beta)
pr_2=expcdf(6,beta)-expcdf(4,beta)
pr_3=1-expcdf(8,beta)
```

5.7.2 MATLAB® Coding of Example 5.2

```
clc
clear
close all
mu=28.5;
sigma=8;
pr_1=1-normcdf(40,mu,sigma)
pr_2=normcdf(12.5,mu,sigma)
pr_3=normcdf(35,mu,sigma)-normcdf(20,mu,sigma)
n1=pr_1*1000
n2=pr_2*1000
n3=pr_3*1000
```

5.7.3 MATLAB® Coding of Example 5.8

```
clc
clear
close all
n=1000;
p=0.02;
mu=n*p;
sigma=sqrt(n*p*(1-p));
pr_1=normcdf(14.5,mu,sigma)
pr_2=1-normcdf(30.5,mu,sigma)
pr_3=normcdf(30.5,mu,sigma)-normcdf(14.5,mu,sigma)
```

5.7.4 MATLAB® Coding of Example 5.9

```
clc
clear
close all
mu=25;
sigma=sqrt(mu);
pr_1=normcdf(19.5,mu,sigma)
pr_2=1-normcdf(29.5,mu,sigma)
```

5.7.5   MATLAB® CODING OF EXAMPLE 5.10

```
clc
clear
close all
mu=4.2;
sigma=1.12;
pr=1-logncdf(10,mu,sigma)
```

5.7.6   MATLAB® CODING OF EXAMPLE 5.11

```
clc
clear
close all
mu1=16.3;
mu2=4.4;
sigma1=1.3;
sigma2=0.4;
rho=0.6;
mu1_2=mu1+rho*(sigma1/sigma2)*(5-mu2);
sigma1_2=sigma1*sqrt(1-rho^2);
pr1=normcdf(20,mu1_2,sigma1_2)-normcdf(18,mu1_2,sigma1_2)
mu2_1=mu2+rho*(sigma2/sigma1)*(18.5-mu1);
sigma2_1=sigma2*sqrt(1-rho^2);
pr2=1-normcdf(5,mu2_1,sigma2_1)
```

Exercises

5.1 If X is a normally distributed variable with mean $\mu = 40$ and standard deviation $\sigma = 4$ Find out
   i. $P(X < 50)$
   ii. $P(X > 31)$
   iii. $P(40 < X < 45)$

5.2 What is the probability that a random variable having the standard normal distribution will take a value
   i. between 1 and 2?
   ii. greater than 2.2?
   iii. greater than −1.4?
   iv. less than 1.5?
   v. less than −1.3?

5.3 The mean and standard deviation of yarn strength are 256 cN and 25 cN, respectively. If 500 tests are made for yarn strength, how many test results would be expected to have strengths
   i. greater than 270 cN?
   ii. less than 240 cN?
   iii. between 246 and 266 cN?

Assume that the distribution of yarn strength is normal and the strength can be measured to any degree of accuracy.

5.4 A ring frame is expected to spin the yarn of average count 30's Ne with a count CV of 1%. If a ring bobbin is selected for the count, what is the probability that it will show count
   i. more than 30.4 Ne?
   ii. between 29.2 to 30.1 Ne?
   iii. less than 29.5 Ne?
   Assume that the distribution of yarn count is normal and it can be measured to any degree of accuracy.

5.5 Find the probability of yarn count more than 30.4 Ne in Exercise 5.4 if the length is measured to the nearest 0.1 mm.

5.6 In a large batch of hand towels, 15% are found to be less than 62 g in weight and 45% are between 62 and 67 g. Assuming the distribution of weight to be normal, what are the mean weight and standard deviation?

5.7 500 yarn leas are subjected to CSP testing and 24 of the test results have CSP less than 2100 and 276 test results have CSP between 2100 to 2500. What are the mean and standard deviation? Assume that the distribution of CSP is normal.

5.8 The CV% of a certain yarn CSP is 10%. If 1000 leas are tested for CSP, how many leas have CSP greater than 90% of the mean? Assume that the distribution of CSP is normal.

5.9 If 2.5% of the roving bobbins produced by a speed frame are defectives, what is the probability that in a lot of 1000 randomly chosen roving bobbins
   i. less than 30 will be defective?
   ii. more than 30 will be defective?
   iii. exactly 30 will be defective?
   iv. between 20 to 30 inclusive will be defective?

5.10 If 7.4% of the cotton fibres are immatures, what is the probability that out of 500 randomly chosen fibres
   i. at most 50 will be immature?
   ii. 30 or more will be immature?

5.11 Average number of end breaks/1000 spindles/hour of a ring frame is 30. What is the probability that in a given hour there will be
   i. less than 25 breaks/1000 spindles?
   ii. 35 or more breaks/1000 spindles?

5.12 Average number of defects/100 m^2 of a cloth is 18. What is the probability that in a cloth there will be
   i. exactly 20 defects/100 m^2?
   ii. between 15 to 25 inclusive defects/100 m^2?

5.13 The distribution of the diameter of wool fibres is found to follow a lognormal distribution with a mean of 8.25 μm and standard deviation of 2.9 μm. What is the probability that the fibre diameter exceeds 20 μm?

5.14 The pore size (equivalent diameter) distribution of a certain woven fabric is found to follow an exponential distribution with a mean of 15 μm. Find the probability that the pore size is
   i. less than 5 μm.
   ii. between 10 to 20 μm.
   iii. greater than 20 μm.

5.15 The trash and dust size (equivalent diameter) distribution of a cotton bale is found to follow an exponential distribution with a mean of 225 μm. Find the probability that the trash and dust size is
  i. greater than 500 μm.
  ii. less than 200 μm.

5.16 The breaking load X_1 and linear density X_2 of a certain type of cotton fibre have a bivariate normal distribution with the following parameters:

$$\mu_1 = 7 \text{ cN}, \mu_2 = 230 \text{ mtex}, \sigma_1 = 0.75 \text{ cN}, \sigma_2 = 40 \text{ mtex}, \rho = 0.72.$$

  i. Find the probability that fibre breaking load is greater than 8 cN, given that fibre linear density is 220 mtex.
  ii. Find the probability that fibre linear density is less than 200 mtex, given that fibre breaking load is 6 cN.

6 Sampling Distribution and Estimation

6.1 INTRODUCTION

Statistical techniques are mainly meant for predictions and conclusions about the chance factors. It is not always possible to consider the whole population to draw inferences about some properties of it. For example, if we need to judge the quality of yarn produced in a textile industry, we do not have the scope to test the whole lot of the yarn as it will take a considerable amount of time and it is a costly affair. The simplest possible way to judge the quality is to take one or more samples and, based on the information contained in the sample, we arrive at a decision. Very often we apply this idea in our day-to-day life such as when we go to market to buy something, we apply this sampling technique to make a decision before buying it. We start this chapter by defining random sample. The larger set from where a sample is collected is called the population.

Definition 6.1: A set of independent and identically distributed random variables is called a random sample from a population and each random variable follows the same distribution as the population.

For an example, suppose that a person goes to a market to buy some apples. He visits an apple shop and takes a sample of 10 apples. He has a plan to buy apples from the shop if at least 8 out of 10 apples are good. One may wonder how the samples of 10 apples can be thought of as 10 independent and identically distributed random variables. This can be done as follows.

Let $X_i (i = 1, 2, \cdots, 10)$ denotes the random variable taking value 1 if the ith apple chosen is good, otherwise it takes value 0. According to this notation, the person will buy the apple from that particular shop if $\sum_{i=1}^{10} X_i \geq 8$. Note that the person has used an estimator to judge the quality of the apple and that estimator is nothing but $\sum X_i$. Depending on the value of this estimator, the person will judge the quality of the apple. This estimator may also be treated as a statistic. The formal definition of statistic is as follows:

Definition 6.2: A statistic is a function of a set of random variables constituting a random sample.

Two commonly used statistics are sample mean and sample variance and they are defined as follows.

Definition 6.3: If the random variables X_1, X_2, $\cdots$, X_n constitute a random sample of size n, then $\bar{X} = \frac{\sum_{i=1}^{n} X_i}{n}$ is called the sample mean and $S^2 = \frac{\sum_{i=1}^{n}(X_i - \bar{X})^2}{n-1}$ is called the sample variance.

DOI: 10.1201/9781003081234-6

A statistic, being a function of random variables, is itself a random variable and so it has a distribution. In the following section we discuss the sampling distribution of sample mean and sample variance.

6.2   THE DISTRIBUTION OF SAMPLE MEAN

The mean and variance of a random variable carry useful information about the distribution of it. In this section we will explain how to find out the mean and variance of the statistic sample mean.

Theorem 6.1: If X_1, X_2, $\cdots$, X_n constitute a random sample of size n from an infinite population having mean and variance μ and σ^2 respectively, then

$$E(\bar{X}) = \mu \text{ and } var(\bar{X}) = \frac{\sigma^2}{n}$$

Proof: From the definition of $\bar{X}$ we have,

$E(\bar{X}) = E\left(\frac{1}{n}\sum_{i=1}^{n} X_i\right) = \frac{1}{n}\sum_{i=1}^{n} E(X_i) = \frac{1}{n}\sum_{i=1}^{n}\mu = \mu$ [Because each X_i is a member of the same population and so $E(X_i) = \mu \ \forall \ i$]

Now, from Proposition 3.3, we can write

$$var(\bar{X}) = var\left(\frac{1}{n}\sum_{i=1}^{n} X_i\right) = \frac{1}{n^2} var\left(\sum_{i=1}^{n} X_i\right)$$

Again, since X_i's are independent, we have

$$var\left(\sum_{i=1}^{n} X_i\right) = \sum_{i=1}^{n} var(X_i)$$

Therefore, the use of Propositions 3.3 and 3.4 results in,

$$var(\bar{X}) = \frac{1}{n^2} var\left(\sum_{i=1}^{n} X_i\right) = \frac{1}{n^2}\sum_{i=1}^{n} var(X_i) = \frac{\sigma^2 n}{n^2} = \frac{\sigma^2}{n}.$$

We denote the sample mean of a random sample by $\mu_{\bar{x}}$. It is customary to term $\sqrt{var(\bar{X})}$ as standard error of the mean and we denote it by $\sigma_{\bar{x}}$. We can see that $\sigma_{\bar{x}}$ tends to 0 as $n \to \infty$, that is, if the sample size becomes large it carries more information about the population. In that case, we can expect that the sample mean gets closer to the actual mean of the population. For a large sample we can obtain the limiting distribution of the sample mean without knowing the distribution of the population using the following theorem.

Theorem 6.2 (Central Limit Theorem): If X_1, X_2, $\cdots$, X_n constitute a random sample of size n from an infinite population with mean μ, variance σ^2 and moment generating function $M_X(t)$, then $Z = \frac{\bar{X}-\mu}{\sigma/\sqrt{n}}$ follows standard normal distribution as $n \to \infty$.

Proof: Using Theorem 3.5, and Corollary 3.2 we may write

$$M_Z(t) = M_{\frac{\bar{X}-\mu}{\frac{\sigma}{\sqrt{n}}}} = e^{-\sqrt{n}\frac{\mu t}{\sigma}} M_{\bar{X}}\left(\frac{\sqrt{n}t}{\sigma}\right)$$

$$= e^{-\sqrt{n}\frac{\mu t}{\sigma}} M_{n\bar{X}}\left(\frac{t}{\sigma\sqrt{n}}\right)$$

Since $n\bar{X} = \sum_{i=1}^{n} X_i$ and X_i's are independent, it follows from Theorem 3.6 that $M_{n\bar{X}}\left(\frac{t}{\sigma\sqrt{n}}\right) = \prod_{i=1}^{n} M_{X_i}\left(\frac{t}{\sigma\sqrt{n}}\right)$. The random variables X_i's are identically distributed random variables and thus $\prod_{i=1}^{n} M_{X_i}\left(\frac{t}{\sigma\sqrt{n}}\right) = \left[M_X\left(\frac{t}{\sigma\sqrt{n}}\right)\right]^n$. Here, $M_X(t)$ is the common moment generating function (MGF) of each X_i.

Therefore, we can write

$$\ln M_Z(t) = -\frac{\sqrt{n}\mu t}{\sigma} + n \ln M_X\left(\frac{t}{\sigma\sqrt{n}}\right) \tag{6.1}$$

By considering the series expansion of $M_X\left(\frac{t}{\sigma\sqrt{n}}\right)$ in powers of t, we get

$$\ln M_X\left(\frac{t}{\sigma\sqrt{n}}\right) = \ln\left[1 + \frac{\mu'_1}{\sigma\sqrt{n}}t + \frac{\mu'_2}{2\sigma^2 n}t^2 + \frac{\mu'_3}{6\sigma^3 n\sqrt{n}}t^3 + \cdots\right]$$

where μ'_i are the moments of the random variable X about the origin. Using the power series expansion of $\ln(1+x)$, we obtain

$$\ln\left[1 + \frac{\mu'_1}{\sigma\sqrt{n}}t + \frac{\mu'_2}{2\sigma^2 n}t^2 + \frac{\mu'_3}{6\sigma^3 n\sqrt{n}}t^3 + \cdots\right]$$

$$= \left(\frac{\mu'_1}{\sigma\sqrt{n}}t + \frac{\mu'_2}{2\sigma^2 n}t^2 + \frac{\mu'_3}{6\sigma^3 n\sqrt{n}}t^3 + \cdots\right) - \frac{1}{2}\left(\frac{\mu'_1}{\sigma\sqrt{n}}t + \frac{\mu'_2}{2\sigma^2 n}t^2 + \frac{\mu'_3}{6\sigma^3 n\sqrt{n}}t^3 + \cdots\right)^2$$

$$+ \frac{1}{3}\left(\frac{\mu'_1}{\sigma\sqrt{n}}t + \frac{\mu'_2}{2\sigma^2 n}t^2 + \frac{\mu'_3}{6\sigma^3 n\sqrt{n}}t^3 + \cdots\right)^3 - \cdots$$

Therefore, after some simple calculations from Equation (6.1) we get

$$\ln M_z(t) = -\frac{\sqrt{n}\mu t}{\sigma} + n\left[\left(\frac{\mu'_1}{\sigma\sqrt{n}}t + \frac{\mu'_2}{2\sigma^2 n}t^2 + \frac{\mu'_3}{6\sigma^3 n\sqrt{n}}t^3 + \cdots\right)\right.$$

$$- \frac{1}{2}\left(\frac{\mu'_1}{\sigma\sqrt{n}}t + \frac{\mu'_2}{2\sigma^2 n}t^2 + \frac{\mu'_3}{6\sigma^3 n\sqrt{n}}t^3 + \cdots\right)^2$$

$$\left. + \frac{1}{3}\left(\frac{\mu'_1}{\sigma\sqrt{n}}t + \frac{\mu'_2}{2\sigma^2 n}t^2 + \frac{\mu'_3}{6\sigma^3 n\sqrt{n}}t^3 + \cdots\right)^3 - \cdots\right]$$

$$= \left(-\frac{\sqrt{n}\mu}{\sigma} + \frac{\sqrt{n}\mu}{\sigma} \right) t + \left(\frac{\mu_2'}{2\sigma^2} - \frac{\mu_1'^2}{2\sigma^2} \right) t^2 + O\left(\frac{t^3}{\sqrt{n}} \right) \left[\because \mu_1' = \mu \right]$$

$$= \frac{t^2}{2} + O\left(\frac{t^3}{\sqrt{n}} \right) \left[\because \mu_2' - \mu_1'^2 = \sigma^2 \right]$$

Now, for finite t as $n \to \infty$, $\ln M_Z(t) \to \frac{t^2}{2}$ or in other words $M_Z(t) \to e^{\frac{t^2}{2}}$, which is the MGF of a standard normal variate. Since the limiting form of the MGF is that of a standard normal variate, we can conclude that the limiting distribution of Z is standard normal distribution.

When $n \to \infty$, $var(\bar{X}) \to 0$, therefore, it is incorrect to say that the distribution of $\bar{X}$ approaches to normal distribution as $n \to \infty$. However, when n is finite and large, central limit theorem justifies that the distribution of $\bar{X}$ approximately follows normal distribution with mean μ and variance $\frac{\sigma^2}{n}$. In actual practice, this approximation is used when $n \geq 30$ irrespective of the actual shape of the population sampled. However, for $n < 30$, this approximation is doubtful.

Example 6.1:

Twist per meter of a 30 tex cotton yarn has mean of 638 and standard deviation of 52. If a random sample of size 50 is taken for the testing of yarn twist, what is the probability that the sample mean will be between 620 and 650 twist per meter?

SOLUTION

Given data: $\mu = 638$, $\sigma = 52$, $n = 50$.

Let $\bar{X}$ be the mean twist per meter of a random sample of size 50 drawn from a population with mean $\mu = 638$ and standard deviation $\sigma = 52$. According to Theorem 6.1, the distribution of $\bar{X}$ has the mean $\mu_{\bar{x}} = \mu = 638$ and standard deviation $\sigma_{\bar{x}} = \frac{\sigma}{\sqrt{n}} = \frac{52}{\sqrt{50}}$. Again, as $n = 50(> 30)$, it follows from the central limit theorem that $\bar{X}$ is approximately normally distributed with mean $\mu_{\bar{x}} = \mu = 638$ and standard deviation $\sigma_{\bar{x}} = \frac{\sigma}{\sqrt{n}} = \frac{52}{\sqrt{50}}$. The desired probability is given by the area of the shaded region in Figure 6.1. Hence,

$$P(620 < \bar{X} < 650) = P\left(\frac{620 - 638}{52/\sqrt{50}} < Z < \frac{650 - 638}{52/\sqrt{50}} \right)$$

$$= P(-2.45 < Z < 1.63)$$

$$= P(Z > -2.45) - P(Z > 1.63)$$

$$= 1 - P(Z < -2.45) - P(Z > 1.63)$$

$$= 1 - P(Z > 2.45) - P(Z > 1.63)$$

$$= 1 - 0.0071 - 0.0516 \text{ (from Table A1)}$$

$$= 0.9413$$

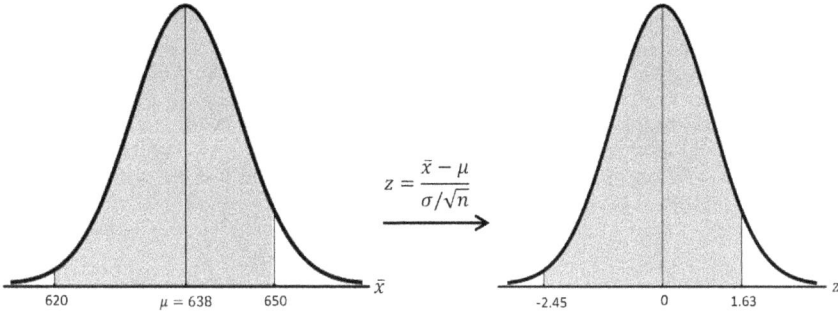

FIGURE 6.1 $P(620 < \bar{X} < 650)$ and corresponding $P(-2.45 < Z < 1.63)$.

Example 6.2:

The mean weight of a full ring frame bobbin is 92 g with standard deviation of 2.5 g. Find the probability that the mean weight of a sample of randomly selected 36 full bobbins will exceed 92.5 g.

SOLUTION

Given data: $\mu = 92$, $\sigma = 2.5$, $n = 36$.

Let $\bar{X}$ be the mean bobbin weight of a random sample of size 36 drawn from a population with mean $\mu = 92$ and standard deviation $\sigma = 2.5$. According to Theorem 6.1, the distribution of $\bar{X}$ has the mean $\mu_{\bar{x}} = \mu = 92$ and standard deviation $\sigma_{\bar{x}} = \frac{\sigma}{\sqrt{n}} = \frac{2.5}{\sqrt{36}}$. Further, as $n = 36 (> 30)$, it follows from the central limit theorem that $\bar{X}$ is approximately normally distributed with mean $\mu_{\bar{x}} = \mu = 92$ and standard deviation $\sigma_{\bar{x}} = \frac{\sigma}{\sqrt{n}} = \frac{2.5}{\sqrt{36}}$. The desired probability is given by the area of the shaded region in Figure 6.2. Hence,

$$P(\bar{X} > 92.5) = P\left(Z > \frac{92.5 - 92}{2.5/\sqrt{36}}\right) = P(Z > 1.2) = 0.1151.$$

In many estimation processes the standard normal variate is not sufficient and in such situations we need to use chi-square distribution, t-distribution and F-distribution.

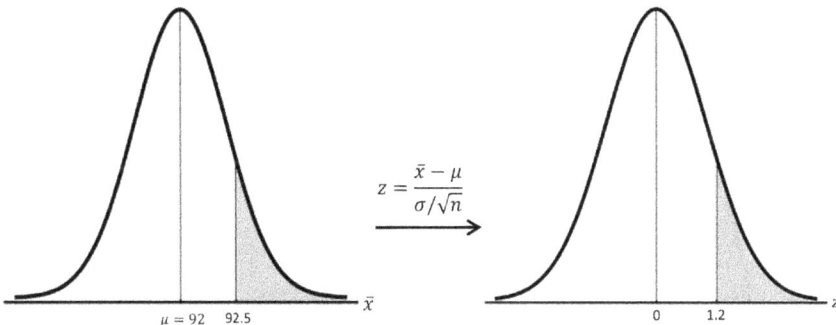

FIGURE 6.2 $P(\bar{X} > 92.5)$ and corresponding $P(Z > 1.2)$.

So, before studying the estimation problem we need to have proper knowledge about these distributions.

6.3 CHI-SQUARE DISTRIBUTION

Definition 6.4: A continuous random variable X is said to follow chi-square distribution if its probability density function is given by

$$f(x) = \frac{e^{-\frac{x}{2}} x^{\frac{v}{2}-1}}{2^{\frac{v}{2}} \Gamma\left(\frac{v}{2}\right)}; \ x > 0, \ v > 0$$

$$= 0; \text{ elsewhere}$$

The parameter v is called the degrees of freedom. In order to denote that a random variable X follows chi-square distribution with v degrees of freedom, we use the notation $X \sim \chi_v^2$. This distribution is useful in case of sampling distribution. Figure 6.3 shows the curves of chi-square distribution for different degrees of freedom. It can be shown that $E\left(\chi_v^2\right) = v$ and $var\left(\chi_v^2\right) = 2v$. The measure of skewness of a χ_v^2 random variate is given by $\beta_1 = \sqrt{\frac{8}{v}}$. Thus, as $v \to \infty$, $\beta_1 \to 0$, which implies that as v becomes large, the chi-square distribution approximates to the normal distribution. If we denote $\alpha = P\left[X \geq \chi_{\alpha,v}^2\right]$, then $\chi_{\alpha,v}^2$ is called the upper $100(1-\alpha)\%$ limit of chi-square distribution. Table A3 in Appendix A shows the values of $\chi_{\alpha,v}^2$ for different values of α and v, where $\chi_{\alpha,v}^2$ is such that the area to its right under the curve of

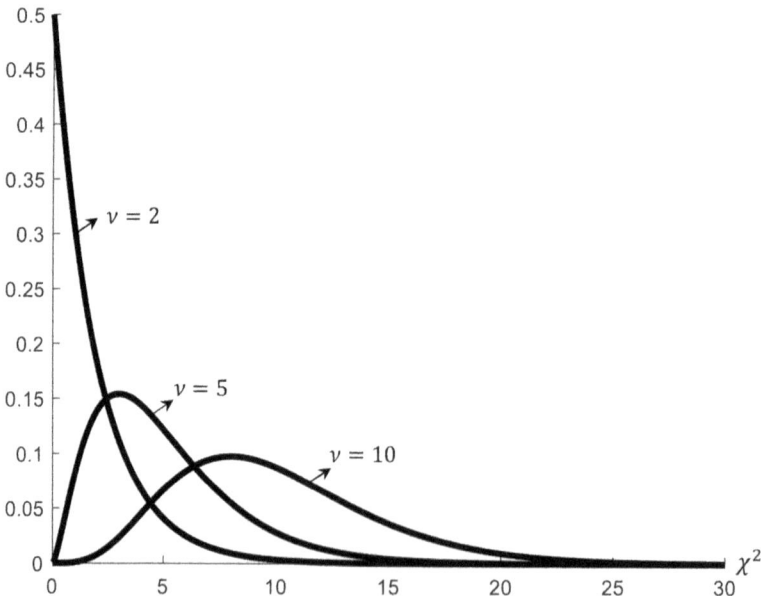

FIGURE 6.3 Chi-square distribution for different values of the degrees of freedom.

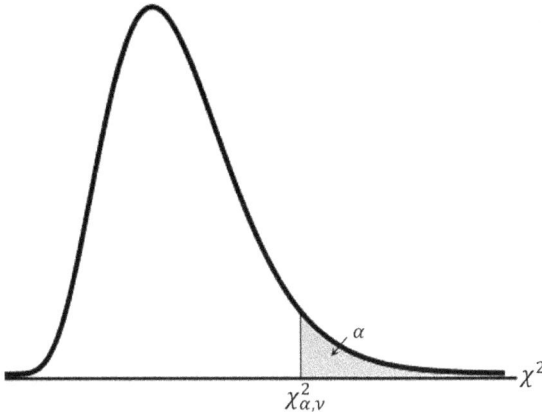

FIGURE 6.4 Chi-square distribution.

the chi-square distribution with v degrees of freedom is equal to α (see Figure 6.4). When v is greater than 30, chi-square distribution can be approximated to normal distribution.

The moment generating function (MGF) of chi-square random variate is useful to establish many important results. In the following theorem we will determine the MGF of chi-square distribution.

Theorem 6.3: The MGF of a chi-square random variate with v degrees of freedom is $(1-2t)^{-\frac{v}{2}}$ for $t < \frac{1}{2}$.

Proof: Suppose X is the random variable which follows chi-square distribution with v degrees of freedom. If $M_X(t)$ denotes the MGF of X, then

$$M_X(t) = E\left(e^{tX}\right) = \int_0^\infty e^{tx} \frac{x^{\frac{v-2}{2}} e^{-\frac{x}{2}}}{2^{\frac{v}{2}} \Gamma\left(\frac{v}{2}\right)} dx$$

$$= \int_0^\infty \frac{x^{\frac{v-2}{2}} e^{-x\left(\frac{1}{2}-t\right)}}{2^{\frac{v}{2}} \Gamma\left(\frac{v}{2}\right)} dx$$

The last integral is convergent if $\left(\frac{1}{2}-t\right) > 0$ or $t < \frac{1}{2}$. On substituting $\left(\frac{1}{2}-t\right)x = z$, it becomes

$$M_X(t) = \frac{(1-2t)^{-\frac{v}{2}}}{\Gamma\left(\frac{v}{2}\right)} \int_0^\infty e^{-z} z^{\frac{v}{2}-1} dz$$

$$= (1-2t)^{-\frac{v}{2}} \text{ for } t < \frac{1}{2} \left[\because \int_0^\infty e^{-z} z^{\frac{v}{2}-1} dz = \Gamma\left(\frac{v}{2}\right)\right]$$

Theorem 6.4: If X_1, X_2, ..., X_k are k independent chi-square random variables with $X_i \sim \chi^2_{v_i}$ then $\sum_{i=1}^{k} X_i \sim \chi^2_{\sum_{i=1}^{k} v_i}$.

Proof: Suppose $M_{X_i}(t)$ denotes the MGF of the random variable X_i following $\chi^2_{v_i}$ distribution. Thus, $M_{X_i}(t) = (1-2t)^{-\frac{v_i}{2}}$. Since the random variables X_i's are independent, the MGF of $X_1 + X_2 + \cdots + X_k$ is given by

$$M_{\sum_{i=1}^{k} X_i}(t) = \prod_{i=1}^{k}(1-2t)^{-\frac{v_i}{2}} = (1-2t)^{-\frac{\sum_{i=1}^{k} v_i}{2}}$$

Now $(1-2t)^{-\frac{\sum_{i=1}^{k} v_i}{2}}$ is the MGF of a random variable having chi-square distribution with $\sum_{i=1}^{k} v_i$ degrees of freedom. So from the uniqueness of MGF, we can infer that $X_1 + X_2 + \cdots + X_k$ is a chi-square random variate with $\sum_{i=1}^{k} v_i$ degrees of freedom.

Theorem 6.5: If X has the standard normal distribution, then X^2 has the χ^2 distribution with $v = 1$ degree of freedom.

Proof: As $X \sim N(0, 1)$, the probability density function of X is $f_X(x) = \frac{1}{\sqrt{2\pi}} e^{-\frac{x^2}{2}}$; $-\infty < x < \infty$. Let us take the transformation $Y = X^2$. Note that the inverse image of the transformation is not one-one. To avoid this problem, we can write the probability density function of X as follows:

$$f_X(x) = \frac{1}{\sqrt{2\pi}} e^{-\frac{x^2}{2}}; 0 < x < \infty.$$

$$= \frac{1}{\sqrt{2\pi}} e^{-\frac{x^2}{2}}; -\infty < x < 0$$

Hence, the above transformation yields $x = \sqrt{y}$ when $x > 0$ and $x = -\sqrt{y}$ when $x \leq 0$ for any $y > 0$. Now for $x = \sqrt{y}$, $\frac{dx}{dy} = \frac{1}{2\sqrt{y}}$ and for $x = -\sqrt{y}$, $\frac{dx}{dy} = -\frac{1}{2\sqrt{y}}$. In both the cases, $|J| = \left|\frac{dx}{dy}\right| = \frac{1}{2\sqrt{y}}$. Therefore, the probability density function of Y is given by

$$f_Y(y) = 2 \cdot \frac{1}{\sqrt{2\pi}} e^{-\frac{y}{2}} \frac{1}{2\sqrt{y}}; y > 0$$

$$= 0; y \leq 0$$

$$f_Y(y) = \frac{e^{-\frac{y}{2}} y^{\frac{1}{2}-1}}{2^{\frac{1}{2}} \Gamma\left(\frac{1}{2}\right)}; y > 0$$

or,

$$= 0; \ y \leq 0$$

Therefore, according to Definition 6.4, $Y \sim \chi^2_1$.

Theorem 6.6: If X_1, X_2, ..., X_n are n independent and identically distributed $N(0, 1)$ random variables, then $Y = \Sigma_i^n X_i^2$ has χ^2 distribution with $\nu = n$ degrees of freedom.

Proof: According to Theorem 6.5 each X_i^2 $(i = 1, 2, \cdots n)$ follows χ^2 distribution with $\nu = 1$ degree of freedom. Now, since X_i's are independent, X_i^2's are also independent. Hence, from Theorem 6.4, we have $\Sigma_{i=1}^n X_i^2 \sim \chi_{\Sigma_{i=1}^n 1}^2 = \chi_n^2$.

Lemma 6.1: If X_1 and X_2 are two independent random variables where X_2 follows chi-square distribution with ν_2 degrees of freedom and $X_1 + X_2$ follows chi-square distribution with $\nu (> \nu_2)$ degrees of freedom then X_1 follows a chi-square distribution with $\nu - \nu_2$ degrees of freedom.

Proof: Suppose $M_{X_1}(t)$ and $M_{X_2}(t)$ are the moment generating functions of the random variables X_1 and X_2, respectively. Now if $M_{X_1+X_2}(t)$ is the MGF of the random variable $X_1 + X_2$, then $M_{X_1+X_2}(t) = M_{X_1}(t) M_{X_2}(t)$, since X_1 and X_2 are independent. Now,

$$M_{X_2}(t) = (1 - 2t)^{-\frac{\nu_2}{2}}$$ since X_2 follows chi-square distribution with ν_2 degrees of freedom. Similarly,

$$M_{X_1+X_2}(t) = (1 - 2t)^{-\frac{\nu}{2}}$$ which implies,

$$M_{X_1}(t) M_{X_2}(t) = (1 - 2t)^{-\frac{\nu}{2}}$$

or,

$$M_{X_1}(t) = \frac{(1 - 2t)^{-\frac{\nu}{2}}}{M_{X_2}(t)} = \frac{(1 - 2t)^{-\frac{\nu}{2}}}{(1 - 2t)^{-\frac{\nu_2}{2}}} = (1 - 2t)^{-\frac{\nu - \nu_2}{2}}.$$

This shows that the MGF of X_1 is same as the MGF of a variate following a chi-square distribution with $\nu - \nu_2$ degrees of freedom. From the uniqueness of MGF we conclude that X_2 is a chi-square variate with $\nu - \nu_2$ degrees of freedom.

Lemma 6.2: If $\bar{X}$ and S^2 are the sample mean and sample variance of a random sample of size n from a normal population with mean μ and variance σ^2, then $\frac{(n-1)S^2}{\sigma^2}$ follows chi-square distribution with $n - 1$ degrees of freedom.

Proof: We have,

$$S^2 = \frac{1}{n-1} \sum_{i=1}^n (X_i - \bar{X})^2$$

or,

$$(n-1)S^2 = \sum_{i=1}^n (X_i - \bar{X})^2$$

$$= \sum_{i=1}^n (X_i - \mu + \mu - \bar{X})^2$$

$$= \sum_{i=1}^{n} \left[(X_i - \mu)^2 + (\bar{X} - \mu)^2 - 2(X_i - \mu)(\bar{X} - \mu) \right]$$

$$= \sum_{i=1}^{n} (X_i - \mu)^2 + n(\bar{X} - \mu)^2 - 2n(\bar{X} - \mu)^2$$

$$= \sum_{i=1}^{n} (X_i - \mu)^2 - n(\bar{X} - \mu)^2$$

Therefore, $\dfrac{(n-1)S^2}{\sigma^2} = \sum_{i=1}^{n} \left(\dfrac{X_i - \mu}{\sigma} \right)^2 - \left(\dfrac{\bar{X} - \mu}{\sigma / \sqrt{n}} \right)^2$

or, $\qquad \sum_{i=1}^{n} \left(\dfrac{X_i - \mu}{\sigma} \right)^2 = \dfrac{(n-1)S^2}{\sigma^2} + \left(\dfrac{\bar{X} - \mu}{\sigma / \sqrt{n}} \right)^2.$

Now since each $X_i \sim N(\mu, \sigma^2)$, $\frac{X_i - \mu}{\sigma} \sim N(0, 1)$ and from Theorems 6.4 and 6.5 we can infer that $\sum_{i=1}^{n} \left(\frac{X_i - \mu}{\sigma} \right)^2 \sim \chi_n^2$. Again, we can show that $\frac{\bar{X} - \mu}{\sigma / \sqrt{n}} \sim N(0, 1)$ and hence $\left(\frac{\bar{X} - \mu}{\sigma / \sqrt{n}} \right)^2 \sim \chi_1^2$ according to Theorem 6.5. Thus, the random variable in the left-hand side of the above equation follows chi-square distribution with n degrees of freedom and the second random variable in the right-hand side follows chi-square distribution with 1 degree of freedom. It thus follows from Lemma 6.1 that $\frac{(n-1)S^2}{\sigma^2}$ has a chi-square distribution with $n - 1$ degrees of freedom.

Example 6.3:

If $X_1 \sim \chi_3^2$, $X_2 \sim \chi_4^2$, $X_3 \sim \chi_5^2$ and they are independent then what is the probability that $X_1 + X_2 + X_3 \geq 21.026$.

SOLUTION

From Theorem 6.4 we know that $X_1 + X_2 + X_3 \sim \chi_{3+4+5}^2$, that is $X_1 + X_2 + X_3 \sim \chi_{12}^2$. Now, from Table A3, it is observed that

$$P\left[\chi_{12}^2 \geq 21.026 \right] = 0.05$$

Hence,

$$P[X_1 + X_2 + X_3 \geq 21.026] = 0.05$$

Example 6.4:

The loop length of a knitted fabric is measured by unravelling a course from the fabric sample and dividing the length by total number of loops unravelled. A

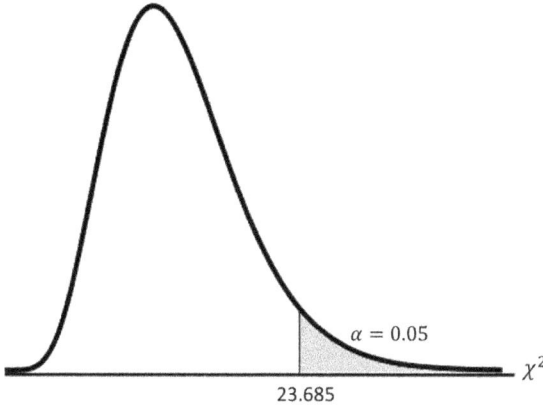

FIGURE 6.5 χ^2-distribution with 24 degrees of freedom.

knitting manufacturer wants to know the variation, as measured by standard deviation, of the loop length for a certain type of single jersey fabrics in relaxed state. From the past experience, it is known that the population standard deviation (σ) of the loop length is 0.5 mm. A technician decides to take 15 randomly chosen courses from the fabric for estimating sample standard deviation (S). Determine the value of b such that $P(S^2 > b) = 0.05$.

SOLUTION

Let s^2 be the variance of a random sample of size n drawn from a normal population with variance σ^2. Therefore, $\frac{(n-1)s^2}{\sigma^2}$ has a chi-square distribution with $n-1$ degrees of freedom. The desired value of b can be obtained by setting the upper tail area equal to 0.05 as shown in Figure 6.5. Hence,

$$P(S^2 > b) = P\left(\frac{(n-1)S^2}{\sigma^2} > \frac{(n-1)b}{\sigma^2}\right) = 0.05$$

Thus,

$$\frac{(n-1)b}{\sigma^2} = \chi^2_{0.05,(n-1)}$$

Substituting $\sigma = 0.5$ and $n = 15$, we have

$$\frac{14b}{0.5^2} = \chi^2_{0.05,14} = 23.685 \text{ (from Table A3)}$$

So, we have $b = \frac{23.685 \times 0.25}{14} = 0.423$.

6.4   THE STUDENT'S *t*-DISTRIBUTION

Definition 6.5: The student's *t*-distribution is defined as

$$T = \frac{Z}{\sqrt{\dfrac{Y}{v}}}$$

where the random variable Z has standard normal distribution and the random variable Y has chi-square distribution with v degrees of freedom.

The premise of independence of Z and Y leads us to write the joint probability density function as

$$f_{Z,Y}(z, y) = \frac{1}{\sqrt{2\pi}} e^{-\frac{1}{2}z^2} \times \frac{1}{\Gamma\left(\dfrac{v}{2}\right) 2^{\frac{v}{2}}} y^{\frac{v}{2}-1} e^{-\frac{y}{2}} \text{ for } -\infty < z < \infty, \ 0 < y < \infty.$$

We take the transformation $t = \frac{z}{\sqrt{\frac{y}{v}}}$ and $w = y$ which gives $z = t\left(\frac{w}{v}\right)^{\frac{1}{2}}$ and the Jacobian of the transformation as $\left(\frac{w}{v}\right)^{\frac{1}{2}}$. We get the joint probability density function of t and w following Theorem 3.7 as

$$f(t, w) = \frac{1}{\sqrt{2\pi}} e^{-\frac{1}{2}t^2\left(\frac{w}{v}\right)} \frac{1}{\Gamma\left(\dfrac{v}{2}\right) 2^{\frac{v}{2}}} w^{\frac{v}{2}-1} e^{-\frac{w}{2}} \left(\frac{w}{v}\right)^{\frac{1}{2}}$$

Therefore, the marginal probability density function of t is given by

$$f_T(t) = \frac{1}{\sqrt{2\pi}} \frac{1}{\Gamma\left(\dfrac{v}{2}\right) 2^{\frac{v}{2}}} \int_0^\infty e^{-\frac{1}{2}t^2 \frac{w}{v}} w^{\frac{w}{2}} w^{\frac{v}{2}} \frac{1}{2} v^{-\frac{1}{2}} \, dw$$

$$= \frac{1}{\sqrt{2\pi}} \frac{1}{\Gamma\left(\dfrac{v}{2}\right) 2^{\frac{v}{2}} v^{\frac{1}{2}}} \int_0^\infty e^{-\frac{w}{2}\left(1+\frac{t^2}{v}\right)} w^{\frac{v+1}{2}-1} \, dw$$

which will eventually give the probability density function of the *t*-distribution as

$$f_T(t) = \frac{\Gamma\left(\dfrac{v+1}{2}\right)}{\Gamma\left(\dfrac{v}{2}\right)} \frac{1}{\sqrt{v\pi}} \left(1+\frac{t^2}{v}\right)^{-\frac{v+1}{2}} \quad -\infty < t < \infty$$

Figure 6.6 depicts the curves of *t*-distribution for different degrees of freedom. In order to denote that a random variable T follows *t*-distribution with v degrees of

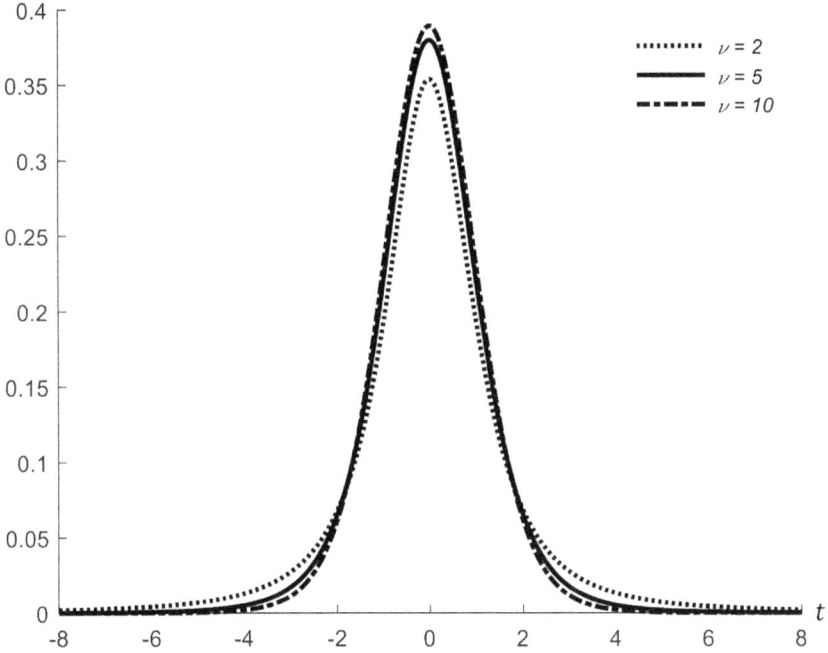

FIGURE 6.6 t-distribution for different degrees of freedom.

freedom, we use the notation $T \sim t_\nu$. If we denote $\alpha = P[T \geq t_{\alpha,\nu}]$, then $t_{\alpha,\nu}$ is called the upper $100(1-\alpha)\%$ limit of a random variable T having a t-distribution with ν degrees of freedom. Table A4 in Appendix A shows the values of $t_{\alpha,\nu}$ for different values of α and ν for t-distribution, where $t_{\alpha,\nu}$ is such that the area to its right under the curve of the t-distribution with ν degrees of freedom is equal to α (see Figure 6.7).

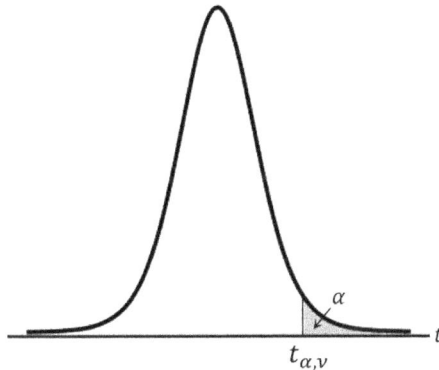

FIGURE 6.7 t-distribution.

6.5   THE *F*-DISTRIBUTION

Definition 6.6: If two random variables Y_1 and Y_2 are independently distributed and $Y_1 \sim \chi^2_{v_1}$ and $Y_2 \sim \chi^2_{v_2}$, then $F = \frac{(Y_1/v_1)}{(Y_2/v_2)} = \frac{v_2}{v_1}\frac{Y_1}{Y_2}$ is said to have an F-distribution with (v_1, v_2) degrees of freedom.

The probability density function of F distribution is obtained as follows.

Since the random variables Y_1 and Y_2 are independent, the joint density of Y_1 and Y_2 is given by

$$f_{Y_1,Y_2}(y_1, y_2) = \frac{e^{-\frac{y_1+y_2}{2}} y_1^{\frac{v_1}{2}-1} y_2^{\frac{v_2}{2}-1}}{2^{\frac{v_1+v_2}{2}} \Gamma\left(\frac{v_1}{2}\right)\Gamma\left(\frac{v_2}{2}\right)}; \ y_1, y_2 > 0$$

$$= 0; \text{ elsewhere}$$

We consider the transformation $U = \frac{v_2}{v_1}\frac{Y_1}{Y_2}$, $V = Y_2$. Therefore, the inverse transformation is $y_1 = \frac{v_1}{v_2}uv$, $y_2 = v$. The Jacobian of the transformation is

$$J = \begin{vmatrix} \frac{v_1}{v_2}v & \frac{v_1}{v_2}u \\ 0 & 1 \end{vmatrix} = \frac{v_1}{v_2}v.$$ Therefore, according to Theorem 3.7 the joint density of

U and V is $f_{U,V}(u, v) = \dfrac{\left(\frac{v_1}{v_2}\right)^{\frac{v_1}{2}} u^{\frac{v_1}{2}-1} v^{\frac{v_1+v_2}{2}-1} e^{-\frac{v}{2}\left(1+\frac{v_1}{v_2}u\right)}}{2^{\frac{v_1+v_2}{2}} \Gamma\left(\frac{v_1}{2}\right)\Gamma\left(\frac{v_2}{2}\right)}$

The marginal density of U will be the required probability density function. Thus,

$$f_U(u) = \int_0^\infty \frac{\left(\frac{v_1}{v_2}\right)^{\frac{v_1}{2}} u^{\frac{v_1}{2}-1} v^{\frac{v_1+v_2}{2}-1} e^{-\frac{v}{2}\left(1+\frac{v_1}{v_2}u\right)}}{2^{\frac{v_1+v_2}{2}} \Gamma\left(\frac{v_1}{2}\right)\Gamma\left(\frac{v_2}{2}\right)} dv$$

or,
$$f_U(u) = \frac{\left(\frac{v_1}{v_2}\right)^{\frac{v_1}{2}} u^{\frac{v_1}{2}-1}}{B\left(\frac{v_1}{2}, \frac{v_2}{2}\right)\left(1+\frac{v_1}{v_2}u\right)^{\frac{v_1+v_2}{2}}}; u > 0 \tag{6.2}$$

$$= 0; \text{ otherwise}$$

where $B(\cdot)$ is the beta function. In Equation (6.2), we have used relation $B\left(\frac{v_1}{2}, \frac{v_2}{2}\right) = \frac{\Gamma\left(\frac{v_1}{2}\right)\Gamma\left(\frac{v_2}{2}\right)}{\Gamma\left(\frac{v_1+v_2}{2}\right)}$.

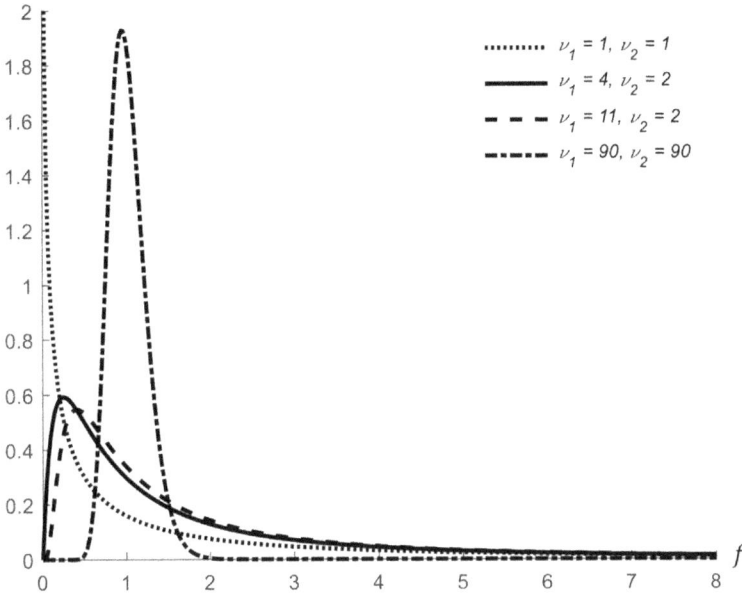

FIGURE 6.8 F-Distribution for different values of the degrees of freedom.

Equation (6.2) gives the probability density function of an F-distributed random variable. Figure 6.8 depicts the curves of F-distribution with different degrees of freedom (v_1, v_2). If we denote $P(F \geq f_{\alpha, v_1, v_2}) = \alpha$, then f_{α, v_1, v_2} is called the upper $100(1-\alpha)\%$ limit of F distribution with v_1 and v_2 degrees of freedom. Table A5 in Appendix A shows the values of f_{α, v_1, v_2} for different values of α and degrees of freedom (v_1, v_2) for F-distribution, where f_{α, v_1, v_2} is such that the area to its right under the curve of the F-distribution with v_1 and v_2 degrees of freedom is equal to α (see Figure 6.9).

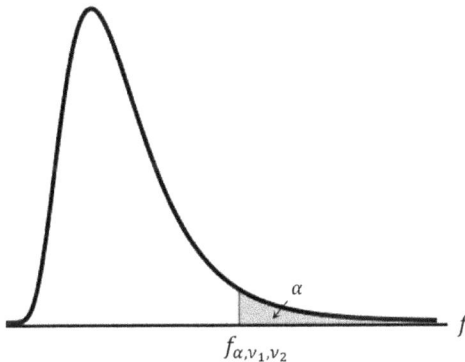

FIGURE 6.9 F-Distribution.

Theorem 6.7: If a random variable X follows F-distribution, that is $X \sim F_{v_1, v_2}$, then $\frac{1}{X} \sim F_{v_2, v_1}$.

Proof: It is given that $X \sim F_{v_1, v_2}$, so the probability density function of X is given by

$$f(x) = \frac{\Gamma\left(\frac{v_1 + v_2}{2}\right)}{\Gamma\left(\frac{v_1}{2}\right)\Gamma\left(\frac{v_2}{2}\right)} \left(\frac{v_1}{v_2}\right)^{\frac{v_1}{2}} x^{\frac{v_1}{2}-1} \left(1 + \frac{v_1}{v_2}x\right)^{-\frac{1}{2}(v_1+v_2)} , x > 0$$

$$= 0 \text{ elsewhere}$$

Now, we take the transformation $y = \frac{1}{x}$. Therefore, the Jacobian of this transformation is written as

$$J = \frac{dx}{dy} = -\frac{1}{y^2}.$$

Thus, according to Theorem 3.8, the probability density function of the random variable Y is given by

$$h(y) = \frac{\Gamma\left(\frac{v_1 + v_2}{2}\right)}{\Gamma\left(\frac{v_1}{2}\right)\Gamma\left(\frac{v_2}{2}\right)} \left(\frac{v_1}{v_2}\right)^{\frac{v_1}{2}} y^{1-\frac{v_1}{2}} \left(1 + \frac{v_1}{v_2}\frac{1}{y}\right)^{-\frac{1}{2}(v_1+v_2)} |J| \left[\text{replacing } x \text{ by } \frac{1}{y} \right]$$

$$= \frac{\Gamma\left(\frac{v_1 + v_2}{2}\right)}{\Gamma\left(\frac{v_1}{2}\right)\Gamma\left(\frac{v_2}{2}\right)} \left(\frac{v_1}{v_2}\right)^{\frac{v_1}{2}} y^{1-\frac{v_1}{2}} \left(1 + \frac{v_1}{v_2}\frac{1}{y}\right)^{-\frac{1}{2}(v_1+v_2)} \frac{1}{y^2}$$

After some simple calculations, $h(y)$ will take the following form

$$h(y) = \frac{\Gamma\left(\frac{v_1 + v_2}{2}\right)}{\Gamma\left(\frac{v_1}{2}\right)\Gamma\left(\frac{v_2}{2}\right)} \left(\frac{v_2}{v_1}\right)^{\frac{v_2}{2}} y^{\frac{v_2}{2}-1} \left(1 + \frac{v_2}{v_1}y\right)^{-\frac{1}{2}(v_1+v_2)} ; y > 0$$

$$= 0; \text{ elsewhere.}$$

This shows that $Y \sim F_{v_2, v_1}$. This theorem will be helpful to prove the following corollary.

Corollary 6.1: If f_{α, v_1, v_2} is the upper $(1-\alpha)100\%$ limit of F-distribution with v_1 and v_2 degrees of freedom, then $\frac{1}{f_{\alpha, v_1, v_2}} = f_{1-\alpha, v_2, v_1}$.

Proof: Suppose X follows F-distribution with v_1, v_2 degrees of freedom. Thus, we have

$$P\left(X \geq f_{\alpha,v_1,v_2}\right) = \alpha \qquad (6.3)$$

We consider the transformation $Y = \frac{1}{X}$. Now according to Theorem 6.7, Y follows F-distribution with v_2, v_1 degrees of freedom. Therefore,

$$P\left(X \geq f_{\alpha,v_1,v_2}\right) = \alpha$$

$$\text{or,} \, P\left(\frac{1}{Y} \geq f_{\alpha,v_1,v_2}\right) = \alpha$$

$$\text{or,} \, P\left(Y \leq \frac{1}{f_{\alpha,v_1,v_2}}\right) = \alpha$$

$$\text{or,} \, 1 - P\left(Y \geq \frac{1}{f_{\alpha,v_1,v_2}}\right) = \alpha$$

$$\text{or,} \, P\left(Y \geq \frac{1}{f_{\alpha,v_1,v_2}}\right) = 1 - \alpha \qquad (6.4)$$

Now if $f_{1-\alpha,v_2,v_1}$ is the upper $100\alpha\%$ limit of the F distribution with v_2, v_1 degrees of freedom, then

$$P\left(Y \geq f_{1-\alpha,v_2,v_1}\right) = 1 - \alpha \qquad (6.5)$$

Comparing Equations (6.4) and (6.5) we have,

$$f_{\alpha,v_1,v_2} = \frac{1}{f_{1-\alpha,v_2,v_1}}.$$

6.6 POINT ESTIMATION

Estimation is one of the major areas in statistics in which the knowledge of statistics is used extensively. Very often, when faced with a real-life problem we need to estimate the unknown parameters of a population. In the case of estimation, the observed value of a statistic is used to estimate the parameter under consideration. For example, if we need to estimate the mean or average of a population, we may use the statistic sample mean $\bar{X}$ as an estimator and the observed value of $\bar{X}$ obtained from a random sample may be used as an estimate of the population mean. We discuss some useful point estimators in the following sub-sections.

6.6.1 UNBIASED ESTIMATOR

Definition 6.7: An estimator Θ is called an unbiased estimator of an unknown population parameter θ if $E(\Theta) = \theta$.

In Theorem 6.1 we have shown that $E(\bar{X}) = \mu$, where $\bar{X}$ is the sample mean and μ is the population mean. Thus, the sample mean of a random sample is an unbiased estimator of the population mean μ. Recall that in order to define sample variance we divide by $n-1$ instead of n. This will be apparent in the following theorem.

Theorem 6.8: The sample variance S^2 is an unbiased estimator of the population variance σ^2 of an infinite population.

Proof: By definition,

$$E\left[\frac{1}{n-1}\sum_{i=1}^{n}(X_i-\bar{X})^2\right] = \frac{1}{n-1}E\left[\sum_{i=1}^{n}(X_i-\mu+\mu-\bar{X})^2\right]$$

$$= \frac{1}{n-1}E\left[\sum_{i=1}^{n}\left\{(X_i-\mu)^2-2(X_i-\mu)(\bar{X}-\mu)+(\bar{X}-\mu)^2\right\}\right]$$

$$= \frac{1}{n-1}E\left[\sum(X_i-\mu)^2-2(\bar{X}-\mu)\sum(X_i-\mu)+\sum(\bar{X}-\mu)^2\right]$$

$$= \frac{1}{n-1}\left[\sum E(X_i-\mu)^2-2nE(\bar{X}-\mu)^2+nE(\bar{X}-\mu)^2\right]$$

$$= \frac{1}{n-1}\left[\sum_{i=1}^{n}var(X_i)-2n\,var(\bar{X})+n\,var(\bar{X})\right]$$

$$= \frac{1}{n-1}\left(n\sigma^2-n\frac{\sigma^2}{n}\right)\quad\left[\because var(\bar{X})=\frac{\sigma^2}{n}\right]$$

$$= \sigma^2$$

This shows that S^2 is an unbiased estimator of σ^2. One should note that S^2 is not an unbiased estimator of the variance σ^2 of a finite population.

6.6.2 CONSISTENT ESTIMATOR

Definition 6.8: An estimator Θ is said to be a consistent estimator of the population parameter θ if for any $\varepsilon > 0$, $\lim_{n\to\infty}P(|\Theta-\theta|<\varepsilon)=1$, where n is the sample size.

We have already mentioned that $\bar{X}$ is an unbiased estimator of the population mean μ. Here we show that $\bar{X}$ is also a consistent estimator of the same. For $\varepsilon > 0$ using Theorem 3.4, we have

$$P(|\bar{X}-E(\bar{X})|<\varepsilon)\geq 1-\frac{var(\bar{X})}{\varepsilon^2}$$

which implies

$$P\left(\left|\bar{X} - \mu\right| < \varepsilon\right) \geq 1 - \frac{\sigma^2}{n\varepsilon^2} \to 1 \text{ as } n \to \infty.$$

This shows that $\bar{X}$ is a consistent estimator of the population mean apart from being unbiased. The following theorem gives an alternate way to verify whether a given estimator is a consistent or not.

Theorem 6.9: If Θ is an unbiased estimator of the parameter θ and $var(\Theta) \to 0$, as $n \to \infty$ then Θ is a consistent estimator of θ.

Proof: It is given that Θ is an unbiased estimator of θ that is $E(\Theta) = \theta$. Now, from Chebyshev's Theorem 3.4 for any real constant k, we have

$$P\left(\left|\Theta - E(\Theta)\right| < k\sigma\right) \geq 1 - \frac{1}{k^2}$$

where $\sigma = \sqrt{var(\Theta)}$. Let us consider any $\varepsilon > 0$, and we put $k = \frac{\varepsilon}{\sigma}$. Thus, we have

$$P\left(\left|\Theta - E(\Theta)\right| < \varepsilon\right) \geq 1 - \frac{\sigma^2}{\varepsilon^2}$$

Taking the sample size n to be large, we get

$$\lim_{n \to \infty} P\left(\left|\Theta - E(\Theta)\right| < \varepsilon\right) \geq 1 - \frac{\sigma^2}{\varepsilon^2} \to 1, \text{ since } \sigma^2 \to 0 \text{ as } n \to \infty.$$

This completes the proof.

As an example, we may consider the sample mean $\bar{X}$. Recall that $\bar{X}$ is an unbiased estimator and $var(\bar{X}) = \frac{\sigma^2}{n}$, n being the sample size. So, $var(\bar{X}) \to 0$ as $n \to \infty$. So, according to Theorem 6.9, $\bar{X}$ is a consistent estimator.

6.6.3 MINIMUM VARIANCE UNBIASED ESTIMATOR

We generally expect the variance of an estimator to be as small as possible because a small variance increases its reliability. In fact, if the variance of an estimator is small, the chance of abrupt deviation from a given observed value of it is small. Thus, among many different unbiased estimators, the estimator with minimum variance is suitable to use. If a given unbiased estimator possesses the smallest possible variance, we call the estimator a minimum variance unbiased estimator (MVUE). We use the help of the Cramer-Rao inequality to check whether a given unbiased estimator has the smallest possible variance or not. The Cramer-Rao inequality is stated below.

Theorem 6.10 (Cramer-Rao inequality): If Θ is an unbiased estimator of θ, then $var(\Theta) \geq \dfrac{1}{n.E\left[\left(\frac{\partial \ln f(X)}{\partial \theta}\right)^2\right]}$, where $f(x)$ is the probability density of the population distribution and n is the sample size.

Definition 6.9: If Θ is an unbiased estimator of θ and $var(\Theta) = \dfrac{1}{n.E\left[\left(\frac{\partial \ln f(X)}{\partial \theta}\right)^2\right]}$, then

Θ is called the minimum variance unbiased estimator (MVUE) of θ.

Example 6.5:

If $X_1, X_2, \cdots, X_n$ constitute a random sample from a Bernoulli population with parameter θ, then show that $\bar{X}$ is a MVUE of the parameter θ.

SOLUTION

Since $\bar{X}$ is the sample mean, $var(\bar{X}) = \frac{\sigma^2}{n}$, where σ^2 is the common variance of each X_i. Now, $E(X_i) = \theta$ and

$$var(X_i) = \theta(1-\theta) \quad \forall \ i = 1, 2, \cdots, n$$

Therefore,

$$var(\bar{X}) = \frac{\theta(1-\theta)}{n}$$

The probability mass function for the Bernoulli distribution is given by

$$f(x) = \theta^x (1-\theta)^{1-x}; \ x = 0, 1$$

Thus,

$$\ln f(x) = x \ln \theta + (1-x)\ln(1-\theta) \text{ and } \frac{\partial}{\partial \theta}\ln f(x) = \frac{x-\theta}{\theta(1-\theta)}.$$

Therefore,

$$E\left[\left(\frac{\partial}{\partial \theta}\ln f(X)\right)^2\right] = E\left[\frac{X-\theta}{\theta(1-\theta)}\right]^2 = \frac{E(X-\theta)^2}{\theta^2(1-\theta)^2} = \frac{var(X)}{\theta^2(1-\theta)^2} = \frac{\theta(1-\theta)}{\theta^2(1-\theta)^2} = \frac{1}{\theta(1-\theta)}$$

Now,

$$var(\bar{X}) = \frac{\theta(1-\theta)}{n} = \frac{1}{n\frac{1}{\theta(1-\theta)}} = \frac{1}{nE\left[\left(\frac{\partial}{\partial \theta}\ln f(X)\right)^2\right]}$$

This shows that $\bar{X}$ is the MVUE.

6.6.4 SUFFICIENCY

Definition 6.10: An estimator Θ is called a sufficient estimator of the population parameter θ if the conditional probability distribution of the random sample $X_1, X_2, \cdots, X_n$ given $\Theta = \theta_0$ is independent of the parameter θ.

For better understanding we consider the following example.

Example 6.6:

Show that the sample mean $\bar{X}$ is a sufficient estimator of the parameter p of the population following binomial distribution with parameter n and p.

SOLUTION

We need to show that the conditional joint probability mass function of the random sample, say $X_1, X_2, \cdots, X_N$ for a given value of $\bar{X}$, is independent of the parameter p. Suppose $f\left(x_1, x_2, \cdots, x_N \mid \bar{X} = \bar{x}\right)$ denotes the said conditional probability mass function.

Therefore,

$$f\left(x_1, x_2, \cdots, x_N \mid \bar{X} = \bar{x}\right) = \frac{P\left(X_1 = x_1 \cap X_2 = x_2 \cdots \cap X_N = x_N \cap \bar{X} = \bar{x}\right)}{P\left(\bar{X} = \bar{x}\right)}$$

$$= \frac{P\left(X_1 = x_1 \cap X_2 = x_2 \cdots \cap X_N = x_N\right)}{P\left(\bar{X} = \bar{x}\right)}$$

$$= \frac{\prod_{i=1}^{N}\binom{n}{x_i} p^{x_i}\left(1-p\right)^{n-x_i}}{P\left(\bar{X} = \bar{x}\right)}$$

$$= \frac{p^{N\bar{x}}\left(1-p\right)^{Nn-N\bar{x}} \prod_{i=1}^{N}\binom{n}{x_i}}{P\left(\bar{X} = \bar{x}\right)} \tag{6.6}$$

Now,

$$P\left(\bar{X} = \bar{x}\right) = P\left(\frac{\sum_{i=1}^{N} x_i}{N} = \bar{x}\right) = P\left(\sum_{i=1}^{N} x_i = N\bar{x}\right)$$

$$= \sum_{\substack{x_1, x_2, \cdots, x_N, \\ \Sigma x_i = N\bar{x}}} \binom{n}{x_1} p^{x_1}\left(1-p\right)^{n-x_1} \binom{n}{x_2} p^{x_2}\left(1-p\right)^{n-x_2} \cdots \binom{n}{x_N} p^{x_N}\left(1-p\right)^{n-x_N}$$

$$= \sum_{\substack{x_1, x_2, \cdots, x_N, \\ \Sigma x_i = N\bar{x}}} \binom{n}{x_1}\binom{n}{x_2} \cdots \binom{n}{x_N} p^{\Sigma x_i}\left(1-p\right)^{Nn-\Sigma x_i}$$

$$= p^{\Sigma x_i}\left(1-p\right)^{Nn-\Sigma x_i} \sum_{\substack{x_1, x_2, \cdots, x_N, \\ \Sigma x_i = N\bar{x}}} \binom{n}{x_1}\binom{n}{x_2} \cdots \binom{n}{x_N}$$

$$= p^{N\bar{x}}\left(1-p\right)^{Nn-N\bar{x}} \sum_{\substack{x_1, x_2, \cdots, x_N, \\ \Sigma x_i = N\bar{x}}} \binom{n}{x_1}\binom{n}{x_2} \cdots \binom{n}{x_N}$$

Therefore, from Equation (6.6) we have

$$f\left(x_1, x_2, \cdots, x_N \mid \bar{X} = \bar{x}\right) = \frac{p^{N\bar{x}}\left(1-p\right)^{Nn-N\bar{x}} \prod_{i=1}^{N}\binom{n}{x_i}}{p^{N\bar{x}}\left(1-p\right)^{Nn-N\bar{x}} \sum_{\substack{x_1,x_2,\cdots,x_N,\\ \sum x_i = N\bar{x}}}\binom{n}{x_1}\binom{n}{x_2}\cdots\binom{n}{x_N}}$$

$$= \frac{\prod_{i=1}^{N}\binom{n}{x_i}}{\sum_{\substack{x_1,x_2,\cdots,x_N,\\ \sum x_i = N\bar{x}}}\binom{n}{x_1}\binom{n}{x_2}\cdots\binom{n}{x_N}}$$

Clearly, the last expression is independent of p. Hence, sample mean is a sufficient estimator of the parameter p.

The direct use of the definition of sufficiency may lead to rigorous calculation. To avoid it the following theorem may be useful.

Theorem 6.11: The statistic Θ is called a sufficient estimator of the population parameter θ if and only if the joint probability distribution or density of the random sample can be written in the form

$$f\left(x_1, x_2, \cdots, x_n; \theta\right) = g\left(\theta, \Theta\right)h\left(x_1, x_2, \cdots, x_n\right)$$

where $g(\theta,\Theta)$ is a function of θ and Θ only and $h(x_1, x_2, \cdots, x_n)$ is independent of θ.

In Example 6.6 we can use this theorem to show that $\bar{X}$ is a sufficient estimator of the parameter p. Note that the joint probability mass function of the sample is given by

$$f\left(x_1, x_2, \cdots, x_N\right) = P\left(X_1 = x_1 \cap X_2 = x_2 \cdots \cap X_N = x_N\right)$$

$$= \prod_{i=1}^{N}\binom{n}{x_i}p^{x_i}\left(1-p\right)^{n-x_i}$$

$$= p^{N\bar{x}}\left(1-p\right)^{Nn-N\bar{x}}\prod_{i=1}^{N}\binom{n}{x_i}$$

$$= g\left(p, \bar{x}\right)h\left(x_1, x_2, \cdots, x_N\right)$$

where $g\left(p, \bar{x}\right) = p^{N\bar{x}}\left(1-p\right)^{Nn-N\bar{x}}$ which is a function of p and $\bar{x}$ only and $h(x_1, x_2, \cdots, x_n)$ is independent of p. This shows that $\bar{X}$ is a sufficient estimator of p.

Example 6.7:

If X_1, X_2 and X_3 constitute a random sample of size 3 from a binomial population with parameters n and p, show that $Y = X_1 + 2X_2 + X_3$ is not a sufficient estimator of the probability of success p.

SOLUTION

We consider a particular case $x_1 = 1$, $x_2 = 0$ and $x_3 = 1$ with $y = 2$.
Now,

$$P\left(x_1 = 1,\ x_2 = 0,\ x_3 = 1 \mid y = 2\right)$$

$$= \frac{P\left(x_1 = 1,\ x_2 = 0,\ x_3 = 1, y = 2\right)}{P\left(y = 2\right)}$$

$$= \frac{P\left(x_1 = 1,\ x_2 = 0,\ x_3 = 1\right)}{P\left(y = 2\right)}$$

$$= \frac{\left[\binom{n}{1} p(1-p)^{n-1}\right]^2 (1-p)^n}{P\left(x_1 = 1,\ x_2 = 0,\ x_3 = 1\right) + P\left(x_1 = 0,\ x_2 = 1,\ x_3 = 0\right) + P\left(x_1 = 2,\ x_2 = 0,\ x_3 = 0\right) + P\left(x_1 = 0,\ x_2 = 0,\ x_3 = 2\right)}$$

$$= \frac{\binom{n}{1}^2 p^2 (1-p)^{3n-2}}{\binom{n}{1}^2 p^2 (1-p)^{3n-2} + (1-p)^{2n}\binom{n}{1} p(1-p)^{n-1} + 2\binom{n}{2} p^2 (1-p)^{3n-2}}$$

$$= \frac{\binom{n}{1}^2}{\binom{n}{1}^2 + 2\binom{n}{2} - \binom{n}{1} + \frac{1}{p}\binom{n}{1}}$$

Thus, $P\left(X_1 = 1,\ X_2 = 0,\ X_3 = 1 \mid Y = 2\right)$ is not independent of p. Hence, Y is not a sufficient estimator of p.

Example 6.8:

Show that $\bar{X}$ is a sufficient estimator of the mean μ of a normal population with known variance σ^2.

SOLUTION

The joint probability density function of the random sample X_1, X_2, $\cdots$, X_n is given by

$$f\left(x_1,\ x_2, \cdots, x_n; \mu\right) = \prod_{i=1}^{n} \frac{1}{\sigma\sqrt{2\pi}} e^{-\frac{1}{2}\left(\frac{x_i - \mu}{\sigma}\right)^2} = \left(\frac{1}{\sigma\sqrt{2\pi}}\right)^n e^{-\frac{1}{2}\sum_{i=1}^{n}\left(\frac{x_i - \mu}{\sigma}\right)^2} \tag{6.7}$$

Now, $\displaystyle\sum_{i=1}^{n}\left(\frac{x_i - \mu}{\sigma}\right)^2 = \left(\frac{1}{\sigma^2}\right)\sum_{i=1}^{n}(x_i - \mu)^2$

$$= \left(\frac{1}{\sigma^2}\right)\sum_{i=1}^{n}\left[(x_i - \bar{x}) - (\mu - \bar{x})\right]^2$$

$$= \left(\frac{1}{\sigma^2}\right)\sum_{i=1}^{n}\left[(x_i - \bar{x})^2 + (\mu - \bar{x})^2 - 2(\mu - \bar{x})(x_i - \bar{x})\right]$$

$$= \left(\frac{1}{\sigma^2}\right)\left[\sum_{i=1}^{n}(x_i - \bar{x})^2 + n(\mu - \bar{x})^2 - 2(\mu - \bar{x})\sum_{i=1}^{n}(x_i - \bar{x})\right]$$

$$= \left(\frac{1}{\sigma^2}\right)\left[\sum_{i=1}^{n}(x_i - \bar{x})^2 + n(\bar{x} - \mu)^2 - 2(\mu - \bar{x})\left(\sum_{i=1}^{n}x_i - n\bar{x}\right)\right]$$

$$= \left(\frac{1}{\sigma^2}\right)\left[\sum_{i=1}^{n}(x_i - \bar{x})^2 + n(\bar{x} - \mu)^2 - 2(\mu - \bar{x})(n\bar{x} - n\bar{x})\right]$$

$$= \left(\frac{1}{\sigma^2}\right)\left[\sum_{i=1}^{n}(x_i - \bar{x})^2 + n(\bar{x} - \mu)^2\right]$$

Therefore, from Equation (6.7) we get

$$f(x_1, x_2, \cdots, x_n; \mu) = \left[e^{-\frac{1}{2}\left(\frac{\bar{x} - \mu}{\sigma/\sqrt{n}}\right)^2}\right]\left[\left(\frac{1}{\sigma\sqrt{2\pi}}\right)^n e^{-\frac{1}{2}\sum_{i=1}^{n}\left(\frac{x_i - \bar{x}}{\sigma}\right)^2}\right]$$

It is evident that the first factor in the above expression depends only on the parameter μ and the estimator $\bar{x}$. The second factor is independent of the parameter μ. Thus, from Theorem 6.11, it follows that $\bar{X}$ is a sufficient estimator of the mean μ of a normal population with known variance σ^2.

6.6.5 Maximum Likelihood Estimator

Definition 6.11: If $X_1, X_2, \cdots, X_n$ constitutes a random sample of size n from a population with parameter vector θ, then the function $L(\theta) = f(x_1, x_2, \cdots, x_n; \theta)$ is called the likelihood function. Here, $f(x_1, x_2, \cdots, x_n; \theta)$ is the value of the joint probability distribution or joint probability density function of the given random sample $X_1, X_2, \cdots, X_n$ at the observed value $(x_1, x_2, \cdots, x_n)$.

The method of maximum likelihood estimation involves maximizing $L(\theta)$ with respect to θ as follows.

Suppose $X_1, X_2, \cdots, X_n$ is a random sample of size n and $x_1, x_2, \cdots, x_n$ is the observed value of the sample. The likelihood function of this observed value is given by,

$$L(x_1, x_2, \cdots, x_n; \theta) = \prod_{i=1}^{n} f(x_i, \theta)$$

where $f(x, \theta)$ is the probability density of the population random variable. This likelihood function is maximized with respect to the vector parameter θ. To illustrate it we consider the following example.

Example 6.9:

For a random sample of size n from a normal population with mean μ and variance σ^2, find the simultaneous maximum likelihood estimators of the parameters μ and σ^2.

SOLUTION

The likelihood function of a random sample of size n drawn from a normal population with mean μ and variance σ^2 is given as

$$L(\mu, \sigma^2) = \left(\frac{1}{\sigma\sqrt{2\pi}} \right)^n \exp\left[-\frac{1}{2} \sum_{i=1}^{n} \left(\frac{x_i - \mu}{\sigma} \right)^2 \right]$$

Therefore,

$$\ln L = -\frac{n}{2}\ln 2\pi - n\ln\sigma - \frac{1}{2}\sum_{i=1}^{n}\left(\frac{x_i - \mu}{\sigma}\right)^2$$

By differentiating $\ln L$ partially with respect to μ we have

$$\frac{\partial}{\partial\mu}\ln L = -\frac{1}{2\sigma^2}\sum_{i=1}^{n} 2(x_i - \mu) = \frac{n}{\sigma^2}(\mu - \bar{x}) \qquad (6.8)$$

where $\bar{x}$ is the sample mean of the given sample. Again, one can verify that

$$\ln L = -\frac{n}{2}\ln 2\pi - \frac{n}{2}\ln\sigma^2 - \frac{1}{2}\sum_{i=1}^{n}\left(\frac{x_i - \mu}{\sigma}\right)^2$$

By differentiating both sides partially with respect to σ^2 we have

$$\frac{\partial}{\partial\sigma^2}\ln L = -\frac{n}{2\sigma^2} + \frac{1}{2\sigma^4}\sum_{i=1}^{n}(x_i - \mu)^2 \qquad (6.9)$$

Equating (6.8) to 0, we get $\mu = \bar{x}$ and equating (6.9) to 0 after substituting $\mu = \bar{x}$ we get $\sigma^2 = \frac{1}{n}\sum_{i=1}^{n}(x_i - \bar{x})^2$. It can be shown that L is maximum for these two values of μ and σ^2. Thus, the maximum likelihood estimators of μ and σ^2 are given by

$$\hat{\mu} = \bar{x} \text{ and } \widehat{\sigma^2} = \frac{1}{n}\sum_{i=1}^{n}(x_i - \bar{x})^2.$$

Note that, the maximum likelihood estimate of σ can be obtained by taking the square root of $\widehat{\sigma^2}$. Thus, if $\hat{\sigma}$ is the estimate of σ, then $\hat{\sigma} = \sqrt{\frac{1}{n}\sum_{i=1}^{n}(x_i - \bar{x})^2}$. In general, if $\hat{\theta}$ is a maximum likelihood estimate of the parameter θ, then for any continuous function $h(\theta)$, $h(\hat{\theta})$ is also the maximum likelihood estimate of $h(\theta)$.

6.7   INTERVAL ESTIMATION

The point estimation of a population parameter does not tell anything about the error involved in the future prediction of the parameter. More precisely, in case of the point estimation, we cannot say with which probability the obtained point estimate will be exactly equal to the population parameter under consideration. In such cases, determining an interval in which the parameter has a chance to lie with some probability can be used to estimate the unknown parameter.

We say that an interval $\left(\hat{\theta}_1, \hat{\theta}_2\right)$ is an $(1-\alpha)100\%$ confidence interval for the parameter θ if $P\left(\hat{\theta}_1 < \theta < \hat{\theta}_2\right) = 1 - \alpha$ for a given $0 < \alpha < 1$. Or in other words, the unknown parameter has $(1-\alpha)100\%$ chance to lie inside the said interval. Here $\hat{\theta}_1$ and $\hat{\theta}_2$ are called the confidence limits and $(1-\alpha)$ is the specified probability which is also called the degree of confidence. For instance, if $\alpha = 0.05$, then the degree of confidence is 0.95 and this corresponds to 95% confidence interval.

6.7.1   THE ESTIMATION OF MEAN OF NORMAL POPULATION WITH KNOWN POPULATION VARIANCE

Suppose the sample mean $\bar{X}$ of a random sample of size n from a normal population with unknown mean μ and known variance σ^2 is to be used to estimate the mean of the population. Now, it can be shown that $\bar{X}$ also follows the normal distribution with mean μ and variance $\frac{\sigma^2}{n}$. Thus, according to Theorem 5.3 the random variable $Z = \frac{\bar{X} - \mu}{\sigma/\sqrt{n}}$ follows the standard normal distribution.

Suppose for a given $\alpha \in (0, 1)$, $z_{\alpha/2}$ is such a point that $P(Z > z_{\alpha/2}) = \frac{\alpha}{2}$ or $P(Z < -z_{\alpha/2}) = \frac{\alpha}{2}$ due to symmetry of the standard normal distribution.

This assures us to write

$$P(|Z| < z_{\alpha/2}) = 1 - \alpha$$

or,

$$P\left(\left|\bar{X} - \mu\right| < z_{\alpha/2} \cdot \frac{\sigma}{\sqrt{n}}\right) = 1 - \alpha$$

Hence,

$$P\left(\bar{X} - z_{\alpha/2} \cdot \frac{\sigma}{\sqrt{n}} < \mu < \bar{X} + z_{\alpha/2} \cdot \frac{\sigma}{\sqrt{n}}\right) = 1 - \alpha$$

This leads to the following theorem.

Theorem 6.12: If a random sample of size n is chosen from a normal population with mean μ and variance σ^2, then the $(1-\alpha)100\%$ confidence interval for the mean with known variance σ^2 is given by $\bar{x} - z_{\frac{\alpha}{2}} \cdot \frac{\sigma}{\sqrt{n}} < \mu < \bar{x} + z_{\frac{\alpha}{2}} \cdot \frac{\sigma}{\sqrt{n}}$, where $\bar{x}$ is the value of the sample mean.

In order to calculate the confidence interval for mean, it is necessary that σ is known. However, when $n \geq 30$, but σ is unknown, we substitute s for σ.

Example 6.10:

50 pieces of a 20 tex cotton yarn were tested for strength. The mean and standard deviation of the test results were 18.3 cN/tex and 1.7 cN/tex, respectively. Calculate 95% confidence limits for the mean yarn strength of the population. If we want to be 95% certain that our estimate of the population mean of yarn strength correct to within ±0.25 cN/tex, how many tests should be required?

SOLUTION

Given data: $n = 50$, $\bar{x} = 18.3$, $s = 1.7$. As $n > 30$, we can assume that $\sigma = s$. For 95% confidence limits, $(1 - \alpha) = 0.95$, which leads to $\alpha/2 = 0.025$ and $z_{\alpha/2} = 1.96$. Hence, 95% confidence limits for population mean μ are

$$\bar{x} \pm z_{\alpha/2} \frac{\sigma}{\sqrt{n}} = 18.3 \pm 1.96 \times \frac{1.7}{\sqrt{50}}$$

$$= 18.3 \pm 0.47$$

$$= (17.83, 18.77)$$

Therefore, we can assert with 95% confidence that the population mean of yarn strength lies between 17.83 and 18.77 cN/tex.

Let N be the number of tests required to make the 95% confidence limits for population mean μ equal to 18.3 ± 0.25 cN/tex. Hence,

$$\pm 1.96 \times \frac{\sigma}{\sqrt{N}} = \pm 0.25$$

or,

$$N = \left(\frac{1.96 \times 1.7}{0.25}\right)^2 \cong 178.$$

So approximately 178 tests should be required.

Example 6.11:

Cotton samples of 100 specimens were tested for trash content. The average value of the test results was 4.1% and their standard deviation was 0.24%.

- a. Calculate 95% and 99% confidence limits for the mean trash content of the population.
- b. If it was required to estimate the mean trash content correct to within ±0.05% (with 99% chance of being correct), how many tests would be needed?

SOLUTION

Given data: $n = 100$, $\bar{x} = 4.1$, $s = 0.24$. As $n > 30$, we can assume that $\sigma = s$.

- a. For 95% confidence limits, $(1 - \alpha) = 0.95$, which leads to $\alpha/2 = 0.025$ and $z_{\alpha/2} = 1.96$. Hence, 95% confidence limits for population mean μ are

$$4.1 \pm 1.96 \times \frac{0.24}{\sqrt{100}}$$

$$= 4.1 \pm 0.047$$

$$= (4.053,\ 4.147)$$

Therefore, we can assert with 95% confidence that the population mean of trash content lies between 4.053% and 4.147%.

For 99% confidence limits, $(1 - \alpha) = 0.99$, which leads to $\alpha/2 = 0.005$ and $z_{\alpha/2} = 2.58$. Hence, 99% confidence limits for population mean μ are

$$4.1 \pm 2.58 \times \frac{0.24}{\sqrt{100}}$$

$$= 4.1 \pm 0.062$$

$$= (4.038,\ 4.162)$$

Therefore, we can assert with 99% confidence that the population mean of trash content lies between 4.038% and 4.162%.

- b. Let N be the number of tests required to make the 99% confidence limits for population mean μ equal to 4.1 ± 0.05 cN/tex. Hence,

$$\pm 2.58 \times \frac{\sigma}{\sqrt{N}} = \pm 0.05$$

or,
$$N = \left(\frac{2.58 \times 0.24}{0.05} \right)^2 \cong 153$$

So approximately 153 tests should be required.

Example 6.12:

The strength values of single fibre have 20% CV. How many tests are required to get the population mean strength with an accuracy of 2% at 95% confidence limits?

SOLUTION

Let $\bar{x}$ be the observed mean of a random sample of size n. The 95% confidence limits for population mean μ based on a sample size n are

$$\bar{x} \pm 1.96 \frac{\sigma}{\sqrt{n}}$$

If E is the tolerance, the 95% confidence limits can be expressed as

$$\bar{x} \pm E$$

Hence,

$$1.96 \frac{\sigma}{\sqrt{n}} = E$$

or,

$$\text{or, } n = \left(\frac{1.96\sigma}{E}\right)^2$$

By dividing $\frac{100}{\mu}$ in both numerator and denominator, we get

$$n = \left(\frac{1.96 \frac{\sigma}{\mu} \times 100}{\frac{E}{\mu} \times 100}\right)^2$$

As per definition, $CV\% = \frac{\sigma}{\mu} \times 100$ and Accuracy $\% = \frac{E}{\mu} \times 100$.

Therefore, sample size n to be chosen is expressed as

$$n = \left(\frac{1.96 \times CV\%}{\text{Accuracy}\%}\right)^2$$

Substituting $CV\% = 20$ and Accuracy$\% = 2$, we have

$$n = \left(\frac{1.96 \times 20}{2}\right)^2 \cong 384$$

So approximately 384 tests should be required.

6.7.2 ESTIMATION OF MEAN OF A NORMAL POPULATION WHEN VARIANCE σ^2 IS UNKNOWN

Suppose a random sample of small size $(n < 30)$ is chosen from a normal population whose variance (σ^2) is unknown and mean is to be estimated. In this case, Theorem 6.12 can not be used and we can not substitute s for σ. Instead, we use the following facts.

We know that for a sample of size n from a normal population, $Z = \frac{\bar{X} - \mu}{\sigma/\sqrt{n}} \sim N(0, 1)$ and from Lemma 6.2 shows that $\frac{(n-1)s^2}{\sigma^2} \sim \chi^2_{n-1}$ where χ^2_{n-1} is a random variable having chi-square distribution with $n-1$ degrees of freedom. In Section 6.4 we observe that $\frac{Z}{\sqrt{\frac{\chi^2_{n-1}}{n-1}}}$ follows t-distribution with $n-1$ degrees of freedom, i.e., $\frac{Z}{\sqrt{\frac{\chi^2_{n-1}}{n-1}}} = \frac{\bar{X} - \mu}{S/\sqrt{n}} \sim t_{n-1}$.

Thus, for a given level of significance α we can write $P\left(\left|\frac{\bar{X}-\mu}{S/\sqrt{n}}\right| < t_{\alpha/2,n-1}\right) = 1 - \alpha$ and this leads to the following theorem.

Theorem 6.13: If $\bar{x}$ and s are the observed values of the sample mean and the standard error of the random sample of size n from a normal population then the $(1-\alpha)100\%$ confidence interval of population mean μ is given by

$$\bar{x} - t_{\alpha/2,n-1} \cdot \frac{s}{\sqrt{n}} < \mu < \bar{x} + t_{\alpha/2,n-1} \cdot \frac{s}{\sqrt{n}}$$

Remark 6.1: Theorem 6.13 gives a small sample $(n < 30)$ interval estimate of μ. For a large sample $(n \geq 30)$ we can safely replace σ^2 by the sample variance s^2 in Theorem 6.12 since for large sample the distribution of t-statistics can be approximated by standard normal distribution.

Example 6.13:

Count values in English system (Ne) were tested for 12 randomly chosen cops from a ring frame with the following results:

39.8, 40.3, 43.1, 39.6, 42.5, 41.0, 39.9, 42.1, 40.7, 41.6, 42.1, 40.8

a. Calculate 95% confidence limits for the mean count.
b. If it was required to estimate the mean count correct to within ±0.5 Ne (with 95% chance of being correct), how many tests would be needed?

SOLUTION

From the given data we have: $n = 12$, $\bar{x} = \frac{\sum_{i=1}^{12} x_i}{12} = 41.125$, $s = \sqrt{\frac{\sum_{i=1}^{12}(x_i - \bar{x})^2}{12-1}} = 1.147$, and the degrees of freedom $= n - 1 = 11$. As $n < 30$, the population standard deviation σ remains unknown; hence, in this case t-distribution is appropriate in order to find 95% confidence limits of population mean μ.

a. For 95% confidence limits, $(1-\alpha) = 0.95$, so that $\alpha/2 = 0.025$. From Table A4 we find $t_{0.025,11} = 2.2$. Hence, 95% confidence limits for population mean μ are

$$\bar{x} \pm t_{\frac{\alpha}{2}, n-1} \frac{s}{\sqrt{n}} = 41.125 \pm 2.2 \times \frac{1.147}{\sqrt{12}}$$

$$= 41.125 \pm 0.728$$

$$= \left(40.397, \ 41.853\right)$$

Therefore, we can assert with 95% confidence that the population mean of yarn count lies between 40.397 and 41.853 Ne.

b. Let N be the number of tests required to make the 95% confidence limits for population mean μ equal to 41.125 ± 0.5 Ne. Hence,

$$\pm 2.2 \times \frac{s}{\sqrt{N}} = \pm 0.5$$

or,

$$N = \left(\frac{2.2 \times 1.147}{0.5}\right)^2 \cong 25.$$

So approximately 25 tests should be required.

6.7.3 THE ESTIMATION OF DIFFERENCE BETWEEN TWO MEANS

6.7.3.1 When Variances of the Two Populations Are Known

We can find the interval estimation of the difference between two means when the variances of the two populations from which the samples are taken are known using the following theorem.

Theorem 6.14: Suppose $\bar{x}_1$ and $\bar{x}_2$ are observed sample means of two independent random samples of sizes n_1 and n_2 drawn from two normal populations having unknown means μ_1 and μ_2 and known variances σ_1^2 and σ_2^2, respectively. Then,

$$(\bar{x}_1 - \bar{x}_2) - z_{\frac{\alpha}{2}} \sqrt{\frac{\sigma_1^2}{n_1} + \frac{\sigma_2^2}{n_2}} < \mu_1 - \mu_2 < (\bar{x}_1 - \bar{x}_2) + z_{\frac{\alpha}{2}} \sqrt{\frac{\sigma_1^2}{n_1} + \frac{\sigma_2^2}{n_2}}$$

is a $(1-\alpha)100\%$ confidence interval for the difference between the two means.

Proof: Suppose the samples are $X_{11}, X_{12}, \cdots, X_{1n_1}$ and $X_{21}, X_{22}, \cdots, X_{2n_2}$ where $X_{1i} \sim N\left(\mu_1, \sigma_1^2\right)$ for all $i = 1, 2, \cdots, n_1$ and $X_{2j} \sim N\left(\mu_2, \sigma_2^2\right)$ for all $j = 1, 2, \cdots, n_2$. Therefore,

$$\bar{X}_1 = \frac{1}{n_1} \sum_{i=1}^{n_1} X_{1i} \sim N\left(\mu_1, \frac{\sigma_1^2}{n_1}\right) \text{ and}$$

$$\bar{X}_2 = \frac{1}{n_2} \sum_{i=1}^{n_2} X_{2i} \sim N\left(\mu_2, \frac{\sigma_2^2}{n_2}\right)$$

It can be shown that $\bar{X}_1 - \bar{X}_2$ also follows normal distribution. Now we have

$$E\left(\bar{X}_1 - \bar{X}_2\right) = \mu_1 - \mu_2$$

Since the random samples are independent, $\bar{X}_1$ and $\bar{X}_2$ are also so. This leads to the following:

$$var\left(\bar{X}_1 - \bar{X}_2\right) = var\left(\bar{X}_1\right) + var\left(\bar{X}_2\right) = \frac{\sigma_1^2}{n_1} + \frac{\sigma_2^2}{n_2}$$

Hence, the distribution of $\bar{X}_1 - \bar{X}_2$ is given by

$$\bar{X}_1 - \bar{X}_2 \sim N\left(\mu_1 - \mu_2, \frac{\sigma_1^2}{n_1} + \frac{\sigma_2^2}{n_2}\right)$$

Therefore, we can write

$$Z = \frac{\left(\bar{X}_1 - \bar{X}_2\right) - E\left(\bar{X}_1 - \bar{X}_2\right)}{\sqrt{var\left(\bar{X}_1 - \bar{X}_2\right)}} = \frac{\left(\bar{X}_1 - \bar{X}_2\right) - \left(\mu_1 - \mu_2\right)}{\sqrt{\frac{\sigma_1^2}{n_1} + \frac{\sigma_2^2}{n_2}}} \sim N(0,1).$$

Thus,

$$P\left(-z_{\frac{\alpha}{2}} < Z < z_{\frac{\alpha}{2}}\right) = 1 - \alpha$$

or, $$P\left(-z_{\frac{\alpha}{2}} < \frac{\left(\bar{X}_1 - \bar{X}_2\right) - \left(\mu_1 - \mu_2\right)}{\sqrt{\frac{\sigma_1^2}{n_1} + \frac{\sigma_2^2}{n_2}}} < z_{\frac{\alpha}{2}}\right) = 1 - \alpha$$

or, $$P\left(\left(\bar{X}_1 - \bar{X}_2\right) - z_{\frac{\alpha}{2}}\sqrt{\frac{\sigma_1^2}{n_1} + \frac{\sigma_2^2}{n_2}} < \mu_1 - \mu_2 < \left(\bar{X}_1 - \bar{X}_2\right) + z_{\frac{\alpha}{2}}\sqrt{\frac{\sigma_1^2}{n_1} + \frac{\sigma_2^2}{n_2}}\right) = 1 - \alpha$$

This indicates that,

$$\left(\bar{x}_1 - \bar{x}_2\right) - z_{\frac{\alpha}{2}}\sqrt{\frac{\sigma_1^2}{n_1} + \frac{\sigma_2^2}{n_2}} < \mu_1 - \mu_2 < \left(\bar{x}_1 - \bar{x}_2\right) + z_{\frac{\alpha}{2}}\sqrt{\frac{\sigma_1^2}{n_1} + \frac{\sigma_2^2}{n_2}}$$

is the $(1-\alpha)100\%$ confidence interval for the difference between the means.

Remark 6.2: The confidence interval of Theorem 6.14 can be used even if the samples are drawn from two populations which do not follow normal distribution if the sample sizes n_1 and n_2 are large by virtue of the central limit theorem.

6.7.3.2   When the Variances Are Unknown

There is no straightforward way to find the interval estimation of the difference between the two means when the variances of the populations are unknown. So, for simplicity we assume that the variances of both the populations from where the samples are drawn are same i.e., $\sigma_1 = \sigma_2 = \sigma$ (say). We use the pooled estimator for the common variance to find the interval estimation of the difference between two means in the following theorem.

Theorem 6.15: If $\bar{x}_1$ and $\bar{x}_2$ are observed sample means of two independent random samples of sizes n_1 and n_2 drawn from two normal populations having unknown means μ_1 and μ_2, respectively and common unknown variance, then

$$(\bar{x}_1 - \bar{x}_2) \pm t_{\frac{\alpha}{2},(n_1+n_2-2)} s_p \sqrt{\frac{1}{n_1} + \frac{1}{n_2}}$$

is a $(1-\alpha)100\%$ confidence interval for the difference between the two means where the pooled estimator s_p^2 is given by $s_p^2 = \frac{(n_1-1)s_1^2+(n_2-1)s_2^2}{n_1+n_2-2}$ and s_1^2, s_2^2 are the sample variances of the two random samples.

Proof: The sample variances of two random samples $X_{1i}(i=1,2,\cdots,n_1)$ and $X_{2j}(j=1,2,\cdots,n_2)$ of sizes n_1 and n_2, respectively, can be expressed as

$$S_1^2 = \frac{1}{n_1-1}\sum_{i=1}^{n_1}(X_{1i}-\bar{X}_1)^2$$

and

$$S_2^2 = \frac{1}{n_2-1}\sum_{i=1}^{n_2}(X_{2i}-\bar{X}_2)^2$$

Thus, we can write

$$(n_1-1)S_1^2 + (n_2-1)S_2^2$$

$$= \sum_{i=1}^{n_1}(X_{1i}-\bar{X}_1)^2 + \sum_{i=1}^{n_2}(X_{2i}-\bar{X}_2)^2$$

$$= \sum_{i=1}^{n_1}(X_{1i}-\mu_1+\mu_1-\bar{X}_1)^2 + \sum_{i=1}^{n_2}(X_{2i}-\mu_2+\mu_2-\bar{X}_2)^2$$

$$= \sum_{i=1}^{n_1}(X_{1i}-\mu_1)^2 - n_1(\bar{X}_1-\mu_1)^2 + \sum_{i=1}^{n_2}(X_{2i}-\mu_2)^2 - n_2(\bar{X}_2-\mu_2)^2$$

Therefore,

$$\frac{(n_1-1)S_1^2+(n_2-1)S_2^2}{\sigma^2} = \sum_{i=1}^{n_1}\left(\frac{X_{1i}-\mu_1}{\sigma}\right)^2 - \left(\frac{\bar{X}_1-\mu_1}{\sigma/\sqrt{n_1}}\right)^2$$

$$+ \sum_{i=1}^{n_2}\left(\frac{X_{2i}-\mu_2}{\sigma}\right)^2 - \left(\frac{\bar{X}_2-\mu_2}{\sigma/\sqrt{n_2}}\right)^2$$

Now from the given condition $\frac{X_{1i}-\mu_1}{\sigma} \sim N(0,1)$ and hence each $\left(\frac{X_{1i}-\mu_1}{\sigma}\right)^2 \sim \chi_1^2$. Therefore, from Theorems 6.4 and 6.5 $\sum_{i=1}^{n_1}\left(\frac{X_{1i}-\mu_1}{\sigma}\right)^2 \sim \chi_{n_1}$

Similarly, $\sum_{i=1}^{n_2}\left(\frac{X_{2i}-\mu_2}{\sigma}\right)^2 \sim \chi_{n_2}$

Moreover, $\left(\frac{\bar{X}_1-\mu_1}{\sigma/\sqrt{n_1}}\right)^2$ and $\left(\frac{\bar{X}_2-\mu_2}{\sigma/\sqrt{n_2}}\right)^2 \sim \chi_1^2$

Hence, from Lemma 6.2 we can write

$$\frac{(n_1-1)S_1^2+(n_2-1)S_2^2}{\sigma^2} \sim \chi_{n_1+n_2-2}^2$$

Further, we know that

$$\frac{\left(\bar{X}_1-\bar{X}_2\right)-\left(\mu_1-\mu_2\right)}{\sigma\sqrt{\dfrac{1}{n_1}+\dfrac{1}{n_2}}} \sim N(0,1)$$

It thus follows that

$$\frac{\dfrac{\left(\bar{X}_1-\bar{X}_2\right)-\left(\mu_1-\mu_2\right)}{\sigma\sqrt{\dfrac{1}{n_1}+\dfrac{1}{n_2}}}}{\sqrt{\dfrac{(n_1-1)S_1^2+(n_2-1)S_2^2}{\sigma^2(n_1+n_2-2)}}} = \frac{\left(\bar{X}_1-\bar{X}_2\right)-\left(\mu_1-\mu_2\right)}{\sqrt{\dfrac{1}{n_1}+\dfrac{1}{n_2}}\sqrt{\dfrac{(n_1-1)S_1^2+(n_2-1)S_2^2}{(n_1+n_2-2)}}} \sim t_{n_1+n_2-2}$$

Now, if $t_{\frac{\alpha}{2},n_1+n_2-2}$ is the lower $\frac{\alpha}{2}$ limit of the t-distribution with n_1+n_2-2 degrees of freedom, then

$$P\left(\left|\frac{\left(\bar{X}_1-\bar{X}_2\right)-\left(\mu_1-\mu_2\right)}{\sqrt{\dfrac{1}{n_1}+\dfrac{1}{n_2}}\sqrt{\dfrac{(n_1-1)S_1^2+(n_2-1)S_2^2}{(n_1+n_2-2)}}}\right| \ge t_{\frac{\alpha}{2},n_1+n_2-2}\right) = \alpha$$

From the symmetry of t-distribution, we have

$$
P\left(\left| \frac{\left(\bar{X}_1 - \bar{X}_2\right) - \left(\mu_1 - \mu_2\right)}{\sqrt{\frac{1}{n_1} + \frac{1}{n_2}}\sqrt{\frac{(n_1 - 1)S_1^2 + (n_2 - 1)S_2^2}{(n_1 + n_2 - 2)}}} \right| \leq t_{\frac{\alpha}{2}, n_1 + n_2 - 2} \right) = 1 - \alpha
$$

As $S_p^2 = \frac{(n_1 - 1)S_1^2 + (n_2 - 1)S_2^2}{n_1 + n_2 - 2}$, from the above equations we can write

$$
P\left(-t_{\frac{\alpha}{2}, n_1 + n_2 - 2} \leq \frac{\left(\bar{X}_1 - \bar{X}_2\right) - \left(\mu_1 - \mu_2\right)}{S_p\sqrt{\frac{1}{n_1} + \frac{1}{n_2}}} \leq t_{\frac{\alpha}{2}, n_1 + n_2 - 2} \right) = 1 - \alpha.
$$

This indicates that

$$
\left(\bar{X}_1 - \bar{X}_2\right) - t_{\frac{\alpha}{2}, n_1 + n_2 - 2}\left(S_p\sqrt{\frac{1}{n_1} + \frac{1}{n_2}} \right) < \mu_1 - \mu_2 < \left(\bar{X}_1 - \bar{X}_2\right) + t_{\frac{\alpha}{2}, n_1 + n_2 - 2}\left(S_p\sqrt{\frac{1}{n_1} + \frac{1}{n_2}} \right)
$$

is the $(1 - \alpha)100\%$ confidence interval for the difference between the means.

Remark 6.3: The confidence interval of Theorem 6.15 is mainly used when n_1 and n_2 are small, less than 30, that's why we refer it as a small-sample confidence interval for $\mu_1 - \mu_2$. For large samples $(n_1 \geq 30, n_2 \geq 30)$ Theorem 6.14 can be used to find the interval of estimation by replacing σ_1^2 and σ_2^2 by the respective sample variances s_1^2 and s_2^2.

Example 6.14:

The following data refer to the test results of bursting strength for two knitted fabrics A and B.

	Fabric A	Fabric B
Number of tests	50	50
Mean bursting strength (KPa)	490.6	450.8
Standard deviation (KPa)	29.2	22.3

Calculate 95% confidence limits for the difference between the mean bursting strengths of two fabrics.

SOLUTION

Given data: $n_1 = 50$, $\bar{x}_1 = 490.6$, $s_1 = 29.2$, $n_2 = 50$, $\bar{x}_2 = 450.8$, $s_2 = 22.3$. Because n_1 and n_2 are both greater than 30, we can assume that $\sigma_1 = s_1$ and $\sigma_2 = s_2$. For 95%

confidence limits, $z_{\alpha/2} = 1.96$. Hence, 95% confidence limits for the difference between two population means $\mu_1 - \mu_2$ are

$$(\bar{x}_1 - \bar{x}_2) \pm z_{\alpha/2}\sqrt{\frac{\sigma_1^2}{n_1} + \frac{\sigma_2^2}{n_2}}$$

$$= (490.6 - 450.8) \pm 1.96 \times \sqrt{\frac{29.2^2}{50} + \frac{22.3^2}{50}}$$

$$= 39.8 \pm 10.18$$

$$= (29.62,\ 49.98)$$

Therefore, we can assert with 95% confidence that the difference between the mean bursting strengths of two fabrics lies between 29.62 and 49.98 KPa.

Example 6.15:

A modification in the draw-ratio was made to a manmade fibre production process with the purpose of increasing the fibre modulus. The results of the modulus tests in GPa on the original and modified filaments were as follows:

Modified filament: 19.7, 18.8, 20.4, 19.9, 20.0
Unmodified filament: 15.3, 14.7, 15.9, 16.1, 15.5

Calculate 95% confidence limits for the mean increase in yarn modulus caused by the modification.

SOLUTION

From the given data we have: $n_1 = n_2 = 5$, $\bar{x}_1 = \frac{\sum_{i=1}^5 x_{1i}}{5} = 19.76$, $s_1 = \sqrt{\frac{\sum_{i=1}^5 (x_{1i} - \bar{x}_1)^2}{5-1}} =$ 0.594, $\bar{x}_2 = \frac{\sum_{i=1}^5 x_{2i}}{5} = 15.5$, $s_2 = \sqrt{\frac{\sum_{i=1}^5 (x_{2i} - \bar{x}_2)^2}{5-1}} = 0.548$, and the degrees of freedom = $n_1 + n_2 - 2 = 8$.

Thus, $s_p = \sqrt{\frac{(n_1-1)s_1^2 + (n_2-1)s_2^2}{n_1+n_2-2}} = \sqrt{\frac{(5-1)0.594^2 + (5-1)0.548^2}{5+5-2}} = 0.571$.

For 95% confidence limits, $\alpha/2 = 0.025$. From Table A4 we find $t_{0.025,\,8} = 2.31$. Hence, 95% confidence limits for the difference between two population means $\mu_1 - \mu_2$ are

$$(\bar{x}_1 - \bar{x}_2) \pm t_{\frac{\alpha}{2},(n_1+n_2-2)} s_p \sqrt{\frac{1}{n_1} + \frac{1}{n_2}}$$

$$= (19.76 - 15.5) \pm 2.31 \times 0.571 \times \sqrt{\frac{1}{5} + \frac{1}{5}}$$

$$= 4.26 \pm 0.834$$

$$= (3.426,\ 5.094)$$

Therefore, we can assert with 95% confidence that the mean increase in yarn modulus caused by the modification lies between 3.426 and 5.094 GPa.

6.7.4 THE ESTIMATION OF PROPORTION

An important application of interval estimation is related to finding confidence interval of proportion or the success probability of a binomial distribution. The following theorem gives the $(1-\alpha)100\%$ confidence interval for the proportion.

Theorem 6.16: If x is the number of successes in n number of Bernoulli trials, then an $(1-\alpha)100\%$ confidence interval of the success probability p is given by

$$\frac{x}{n} - z_{\alpha/2}\sqrt{\frac{\left(\frac{x}{n}\right)\left(1-\frac{x}{n}\right)}{n}} \leq p \leq \frac{x}{n} + z_{\alpha/2}\sqrt{\frac{\left(\frac{x}{n}\right)\left(1-\frac{x}{n}\right)}{n}}$$

when n is large.

Proof: In Section 5.4.1 it has been shown that if a random variable X follows a binomial distribution with parameters n and p, then X will follow normal distribution with mean np and variance $np(1-p)$ when n is sufficiently large. Thus, $Z = \frac{X-np}{\sqrt{np(1-p)}} \sim N(0, 1)$ or $\frac{\frac{X}{n}-p}{\sqrt{\frac{p(1-p)}{n}}} \sim N(0, 1)$. This ensures that

$$P\left[-z_{\alpha/2} \leq \frac{\frac{X}{n}-p}{\sqrt{\frac{p(1-p)}{n}}} \leq z_{\alpha/2}\right] = 1-\alpha$$

for a given significance level α and

$$P\left[\frac{X}{n} - z_{\alpha/2}\sqrt{\frac{p(1-p)}{n}} \leq p \leq \frac{X}{n} + z_{\alpha/2}\sqrt{\frac{p(1-p)}{n}}\right] = 1-\alpha$$

Now as an estimator of p inside the square root of the above expression we use $\hat{P} = \frac{X}{n}$.

This shows that the $(1-\alpha)100\%$ confidence interval for p is

$$\frac{x}{n} - z_{\alpha/2}\sqrt{\frac{\left(\frac{x}{n}\right)\left(1-\frac{x}{n}\right)}{n}} \leq p \leq \frac{x}{n} + z_{\alpha/2}\sqrt{\frac{\left(\frac{x}{n}\right)\left(1-\frac{x}{n}\right)}{n}}$$

Example 6.16:

Snap readings were taken in order to estimate the proportion of time a machine operative is working or not. An operative was observed to be working 80 occasions out of 110 observations.

    a. Calculate 95% confidence limits for the proportion of time he works.
    b. If it is required to estimate the proportion within ±0.03 (with 95% chance of being correct), how many observations should be made?

SOLUTION

Given data: $n = 110$ and $x = 80$, thus, the observed value of proportion $\hat{p} = \frac{x}{n} = \frac{80}{110} = 0.727$.

    a. As n is large, a normal approximation to the binomial distribution can be made. For 95% confidence limits, $z_{\alpha/2} = 1.96$. Hence, 95% confidence limits for the proportion p are

$$\hat{p} \pm z_{\alpha/2}\sqrt{\frac{\hat{p}(1-\hat{p})}{n}} = 0.727 \pm 1.96\sqrt{\frac{0.727(1-0.727)}{111}}$$

$$= 0.727 \pm 0.083$$

$$= (0.644,\ 0.81)$$

    b. Let N be the number of observations required to make the 95% confidence limits for proportion p equal to 0.727 ± 0.03. Hence,

$$\pm 1.96 \times \sqrt{\frac{\hat{p}(1-\hat{p})}{N}} = \pm 0.03$$

or, $N = \left(\frac{1.96 \times \sqrt{0.727(1-0.727)}}{0.03}\right)^2 \cong 847$

So approximately 847 observations should be required.

6.7.5 THE ESTIMATION OF DIFFERENCES BETWEEN PROPORTIONS

In many practical problems we need to estimate the difference between the proportions for two different independent binomial distributions. The $(1-\alpha)100\%$ confidence interval for the estimate of the difference between proportions is given in the following theorem.

Theorem 6. 17: Suppose X_1 and X_2 are independent binomially distributed random variables having parameters n_1, p_1 and n_2, p_2, respectively where n_1 and n_2 are large. If

x_1 and x_2 are observed values of the random variables X_1, X_2 respectively and $\hat{p}_1 = \frac{x_1}{n_1}$ and $\hat{p}_2 = \frac{x_2}{n_2}$, then

$$\left(\hat{p}_1 - \hat{p}_2\right) - z_{\frac{\alpha}{2}}\sqrt{\frac{\hat{p}_1\left(1 - \hat{p}_1\right)}{n_1} + \frac{\hat{p}_2\left(1 - \hat{p}_2\right)}{n_2}} < p_1 - p_2 < \left(\hat{p}_1 - \hat{p}_2\right)$$

$$+ z_{\frac{\alpha}{2}}\sqrt{\frac{\hat{p}_1\left(1 - \hat{p}_1\right)}{n_1} + \frac{\hat{p}_2\left(1 - \hat{p}_2\right)}{n_2}}$$

is an approximate $(1 - \alpha)100\%$ confidence interval for $p_1 - p_2$.

Proof: It is known that X_1 and X_2 follow binomial distribution with parameters n_1, p_1 and n_2, p_2 respectively. The random variable representing the sample proportions are $\widehat{P}_1 = \frac{X_1}{n_1}$ and $\widehat{P}_2 = \frac{X_2}{n_2}$. Now,

$$E\left(\frac{X_1}{n_1} - \frac{X_2}{n_2}\right) = \frac{E(X_1)}{n_1} - \frac{E(X_2)}{n_2} = \frac{n_1 p_1}{n_1} - \frac{n_2 p_2}{n_2} = p_1 - p_2$$

$$var\left(\frac{X_1}{n_1} - \frac{X_2}{n_2}\right) = \frac{var(X_1)}{n_1^2} + \frac{var(X_2)}{n_2^2}$$

$$= \frac{n_1 p_1\left(1 - p_1\right)}{n_1^2} + \frac{n_2 p_2\left(1 - p_2\right)}{n_2^2}$$

$$= \frac{p_1\left(1 - p_1\right)}{n_1} + \frac{p_2\left(1 - p_2\right)}{n_2}$$

When n_1 and n_2 are large, it can be approximated that

$$var\left(\frac{X_1}{n_1} - \frac{X_2}{n_2}\right) = \frac{\widehat{P}_1\left(1 - \widehat{P}_1\right)}{n_1} + \frac{\widehat{P}_2\left(1 - \widehat{P}_2\right)}{n_2}$$

and

$$Z = \frac{\left(\frac{X_1}{n_1} - \frac{X_2}{n_2}\right) - E\left(\frac{X_1}{n_1} - \frac{X_2}{n_2}\right)}{\sqrt{var\left(\frac{X_1}{n_1} - \frac{X_2}{n_2}\right)}} = \frac{\left(\frac{X_1}{n_1} - \frac{X_2}{n_2}\right) - \left(p_1 - p_2\right)}{\sqrt{\frac{\widehat{P}_1\left(1 - \widehat{P}_1\right)}{n_1} + \frac{\widehat{P}_2\left(1 - \widehat{P}_2\right)}{n_2}}} \sim N(0,1)$$

Thus, we can write

$$P\left(-z_{\frac{\alpha}{2}} < Z < z_{\frac{\alpha}{2}}\right) = 1 - \alpha$$

The substitution of the value of Z by $\dfrac{\left(\frac{X_1}{n_1} - \frac{X_2}{n_2}\right) - (p_1 - p_2)}{\sqrt{\frac{\hat{P}_1\left(1-\hat{P}_1\right)}{n_1} + \frac{\hat{P}_2\left(1-\hat{P}_2\right)}{n_2}}}$ and the estimators in the

denominator by their respective observed values will complete the proof.

Example 6.17:

A fabric manufacturing company conducts a market survey to assess the attitudes toward the denim jeans among college students. If 264 of 400 male students and 180 of 320 female students favour to wear jeans, find the 99% confidence interval for the difference between the actual proportions of male and female students who favour to wear jeans.

SOLUTION

Given data: $n_1 = 400$, $\hat{p}_1 = \frac{264}{400} = 0.66$, $n_2 = 320$, $\hat{p}_2 = \frac{180}{320} = 0.562$.

As both n_1 and n_2 are large, a normal approximation to the binomial distribution can be made. For 99% confidence limits, $z_{\alpha/2} = 2.58$. Hence, 99% confidence limits for the difference between the actual proportions of male and female students who favour to wear jeans are

$$(\hat{p}_1 - \hat{p}_2) \pm z_{\alpha/2}\sqrt{\frac{\hat{p}_1\left(1-\hat{p}_1\right)}{n_1} + \frac{\hat{p}_2\left(1-\hat{p}_2\right)}{n_2}}$$

$$= (0.66 - 0.562) \pm 2.58\sqrt{\frac{0.66(1-0.66)}{400} + \frac{0.562(1-0.562)}{320}}$$

$$= 0.098 \pm 2.58\sqrt{0.000561 + 0.000769}$$

$$= 0.098 \pm 0.094$$

$$= (0.004,\ 0.192)$$

6.7.6 THE ESTIMATION OF VARIANCE

Theorem 6.18: If s^2 is the observed value of the sample variance of a random sample of size n from a normal population, then the $(1-\alpha)100\%$ confidence interval of σ^2 is given by

$$\frac{(n-1)s^2}{\chi^2_{\alpha/2,n-1}} < \sigma^2 < \frac{(n-1)s^2}{\chi^2_{1-\alpha/2,n-1}}$$

As in the case of point estimation, the $(1-\alpha)100\%$ confidence limits of σ can be obtained by taking the square root of the confidence limits of σ^2.

Proof: From Lemma 6.2, we have $\frac{(n-1)S^2}{\sigma^2} = \chi^2_{n-1}$ is a chi-square random variate with $n-1$ degrees of freedom. Now for a given significance level α, we can write

$$P\left(\chi^2_{n-1} \geq \chi^2_{\alpha/2,n-1}\right) = \frac{\alpha}{2} \text{ and } P\left(\chi^2_{n-1} \geq \chi^2_{1-\alpha/2,n-1}\right) = 1 - \frac{\alpha}{2}$$

Therefore, from the above two equations, we have

$$P\left(\chi^2_{n-1} \geq \chi^2_{1-\alpha/2,n-1}\right) - P\left(\chi^2_{n-1} \geq \chi^2_{\alpha/2,n-1}\right) = 1 - \alpha$$

It thus follows that

$$P\left(\chi^2_{1-\alpha/2,n-1} \leq \chi^2_{n-1} \leq \chi^2_{\alpha/2,n-1}\right) = 1 - \alpha \text{ (see Figure 6.10)}$$

Therefore,

$$P\left(\chi^2_{1-\alpha/2,n-1} < \frac{(n-1)S^2}{\sigma^2} < \chi^2_{\alpha/2,n-1}\right) = 1 - \alpha$$

$$\text{or, } P\left(\frac{(n-1)S^2}{\chi^2_{\alpha/2,n-1}} < \sigma^2 < \frac{(n-1)S^2}{\chi^2_{1-\alpha/2,n-1}}\right) = 1 - \alpha$$

Thus, if s^2 is the observed value of S^2, we can write the $(1-\alpha)100\%$ confidence interval of σ^2 as $\frac{(n-1)s^2}{\chi^2_{\alpha/2,n-1}} < \sigma^2 < \frac{(n-1)s^2}{\chi^2_{1-\alpha/2,n-1}}$.

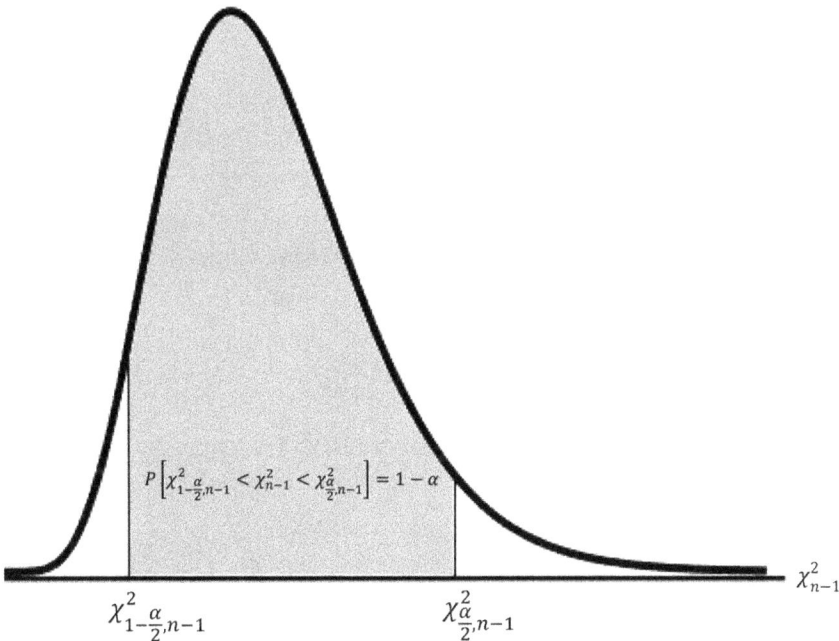

$$P\left[\chi^2_{1-\frac{\alpha}{2},n-1} < \chi^2_{n-1} < \chi^2_{\frac{\alpha}{2},n-1}\right] = 1 - \alpha$$

χ^2_{n-1}

$\chi^2_{1-\frac{\alpha}{2},n-1}$ $\chi^2_{\frac{\alpha}{2},n-1}$

FIGURE 6.10 Chi-square distribution.

Example 6.18:

The following are the areal density (g/m²) of 10 fabrics:

191.7, 186.9, 195.2, 194.4, 190.2, 192.6, 193.8, 189.2, 188.3 and 187.5.

Construct a 95% confidence interval for variance which measures the true variability of the fabric areal density.

SOLUTION

From the given data we have: $n = 10$, $\bar{x} = \frac{\sum_{i=1}^{10} x_i}{10} = 190.98$, $s^2 = \frac{\sum_{i=1}^{10}(x_i-\bar{x})^2}{10-1} = 8.924$, and the degrees of freedom $= n-1 = 9$.

For 95% confidence limits, $\alpha/2 = 0.025$ and $\left(1-\frac{\alpha}{2}\right) = 0.975$. Using Table A3 with 9 degrees of freedom, we find $\chi^2_{0.025,9} = 19.02$ and $\chi^2_{0.975,9} = 2.7$. Therefore, a 95% confidence interval for σ^2 is

$$\frac{(n-1)}{\chi^2_{0.025,9}}s^2 < \sigma^2 < \frac{(n-1)}{\chi^2_{0.975,9}}s^2$$

Substituting n, s^2, $\chi^2_{0.025,9}$ and $\chi^2_{0.075,9}$, we have

$$\frac{9 \times 8.924}{19.02} < \sigma^2 < \frac{9 \times 8.924}{2.7}$$

or, $4.22 < \sigma^2 < 29.75$.

6.7.7   THE ESTIMATION OF THE RATIO OF TWO VARIANCES

Before we determine the interval estimation for the ratio of two random variables, we state the following theorem.

Theorem 6.19: If S_1^2 and S_2^2 are sample variances of two independent random samples of sizes n_1 and n_2 from two normal populations with the variances σ_1^2 and σ_2^2, respectively, then the ratio $\frac{\sigma_1^2 S_2^2}{\sigma_2^2 S_1^2}$ follows F-distribution having $n_2 - 1$ and $n_1 - 1$ degrees of freedom.

Proof: According to Lemma 6.2, $\frac{(n_1-1)S_1^2}{\sigma_1^2}$ and $\frac{(n_2-1)S_2^2}{\sigma_2^2}$ follow chi-square distribution with $n_1 - 1$ and $n_2 - 1$ degrees of freedom, respectively. Thus, from the Definition 6.6 of F-distribution,

$$\frac{\left(\frac{(n_2-1)S_2^2}{\sigma_2^2(n_2-1)}\right)}{\left(\frac{(n_1-1)S_1^2}{\sigma_1^2(n_1-1)}\right)} = \frac{\sigma_1^2 S_2^2}{\sigma_2^2 S_1^2}$$ follows F-distribution with $n_2 - 1$ and $n_1 - 1$ degrees of freedom.

Theorem 6.20: If s_1^2 and s_2^2 are the observed values of the sample variances of two independent random samples of size n_1 and n_2 from two normal populations with the

variances σ_1^2 and σ_2^2, respectively, then the $(1-\alpha)100\%$ confidence interval of the ratio $\frac{\sigma_1^2}{\sigma_2^2}$ is given by,

$$\frac{s_1^2}{s_2^2} \cdot \frac{1}{f_{\frac{\alpha}{2},n_1-1,\,n_2-1}} < \frac{\sigma_1^2}{\sigma_2^2} < \frac{s_1^2}{s_2^2} \cdot f_{\frac{\alpha}{2},n_2-1,\,n_1-1}$$

Proof: From Definition 6.6 we know that $\frac{\sigma_1^2 S_2^2}{\sigma_2^2 S_1^2}$ follows F-distribution with n_2-1 and n_1-1 degrees of freedom. Thus if $f_{\frac{\alpha}{2},v_1,v_2}$ is the upper $\frac{\alpha}{2}$ limit of F-distribution with v_1,v_2 degrees of freedom, then

$$P\left(\frac{\sigma_1^2 S_2^2}{\sigma_2^2 S_1^2} \geq f_{\frac{\alpha}{2},v_1,v_2}\right) = \frac{\alpha}{2} \tag{6.10}$$

Further if $f'_{\frac{\alpha}{2},v_1,v_2}$ is the lower $\frac{\alpha}{2}$ point of F-distribution with v_1,v_2 degrees of freedom, then

$$P\left(\frac{\sigma_1^2 S_2^2}{\sigma_2^2 S_1^2} \leq f'_{\frac{\alpha}{2},v_1,v_2}\right) = \frac{\alpha}{2} \tag{6.11}$$

or,

$$1 - P\left(\frac{\sigma_1^2 S_2^2}{\sigma_2^2 S_1^2} \geq f'_{\frac{\alpha}{2},v_1,v_2}\right) = \frac{\alpha}{2}$$

or,

$$P\left(\frac{\sigma_1^2 S_2^2}{\sigma_2^2 S_1^2} \geq f'_{\frac{\alpha}{2},v_1,v_2}\right) = 1 - \frac{\alpha}{2}$$

This shows that $f'_{\frac{\alpha}{2},v_1,v_2}$ is the upper $1-\frac{\alpha}{2}$ limit of the F-distribution with v_1,v_2 degrees of freedom. This enables us to write, $f'_{\frac{\alpha}{2},v_1,v_2} = f_{1-\frac{\alpha}{2},v_1,v_2}$. Now from Equations (6.10) and (6.11), we can write

$$P\left(\frac{\sigma_1^2 S_2^2}{\sigma_2^2 S_1^2} \leq f'_{\frac{\alpha}{2},v_1,v_2} \cup \frac{\sigma_1^2 S_2^2}{\sigma_2^2 S_1^2} \geq f_{\frac{\alpha}{2},v_1,v_2}\right) = \alpha$$

or,

$$1 - P\left(f'_{\frac{\alpha}{2},v_1,v_2} \leq \frac{\sigma_1^2 S_2^2}{\sigma_2^2 S_1^2} \leq f_{\frac{\alpha}{2},v_1,v_2}\right) = \alpha \;\left[\text{Using De'Morgan's rule}\right]$$

or,

$$P\left(f'_{\frac{\alpha}{2},v_1,v_2} \leq \frac{\sigma_1^2 S_2^2}{\sigma_2^2 S_1^2} \leq f_{\frac{\alpha}{2},v_1,v_2}\right) = 1 - \alpha$$

Therefore, the $(1-\alpha)100\%$ confidence interval for $\frac{\sigma_1^2}{\sigma_2^2}$ is given by,

$$f'_{\frac{\alpha}{2},v_1,v_2} \leq \frac{\sigma_1^2 s_2^2}{\sigma_2^2 s_1^2} \leq f_{\frac{\alpha}{2},v_1,v_2}$$

or,
$$\frac{s_1^2}{s_2^2} f'_{\frac{\alpha}{2},v_1,v_2} \leq \frac{\sigma_1^2}{\sigma_2^2} \leq \frac{s_1^2}{s_2^2} f_{\frac{\alpha}{2},v_1,v_2}$$

or,
$$\frac{s_1^2}{s_2^2} f_{1-\frac{\alpha}{2},v_1,v_2} \leq \frac{\sigma_1^2}{\sigma_2^2} \leq \frac{s_1^2}{s_2^2} f_{\frac{\alpha}{2},v_1,v_2}$$

or,
$$\frac{s_1^2}{s_2^2} \frac{1}{f_{\frac{\alpha}{2},v_2,v_1}} \leq \frac{\sigma_1^2}{\sigma_2^2} \leq \frac{s_1^2}{s_2^2} f_{\frac{\alpha}{2},v_1,v_2} \quad [\text{Using Corollary 6.1}]$$

Here, s_1^2 and s_2^2 are respectively the observed sample variances of the two samples. Substitution of $v_1 = n_2 - 1$ and $v_2 = n_1 - 1$ will complete the proof.

Example 6.19:

From the past experience it is known that the population variance of carded yarn strength is two times the population variance of combed yarn strength. Random samples of sizes 10 and 8 are taken for measuring the variances of carded and combed yarn strengths, respectively. If S_1^2 and S_2^2 denote the sample variances of carded and combed yarn strengths, respectively determine the value of b such that $P\left(\frac{S_1^2}{S_2^2} > b\right) = 0.05$.

SOLUTION

Let S_1^2 and S_2^2 be the sample variances of the two independent random samples of sizes n_1 and n_2, respectively which are drawn from two normal populations with variances σ_1^2 and σ_2^2.

Hence, $\frac{S_1^2/\sigma_1^2}{S_2^2/\sigma_2^2}$ has F-distribution with $(n_1 - 1)$ and $(n_2 - 1)$ degrees of freedom. The desired value of b can be obtained by setting the upper tail area equal to 0.05 as shown in Figure 6.11. It is given that $\sigma_1^2 = 2\sigma_2^2$, therefore, $\frac{S_1^2/\sigma_1^2}{S_2^2/\sigma_2^2} = \frac{S_1^2}{2S_2^2}$. Substituting $n_1 = 10$ and $n_2 = 8$, we have

$$P\left(\frac{S_1^2}{2S_2^2} > f_{0.05,9,7}\right) = 0.05$$

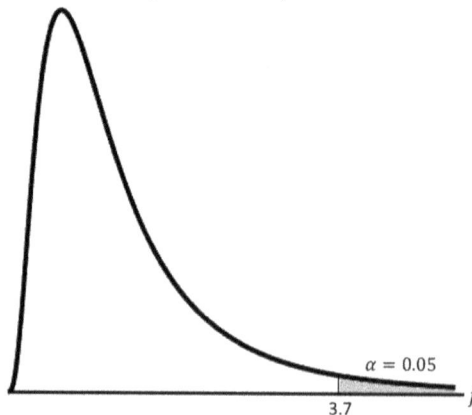

$\alpha = 0.05$

3.7

f

FIGURE 6.11 F-Distribution with 9 and 7 degrees of freedom.

Thus,

$\frac{s_1^2}{2s_2^2} > 3.7$ (from Table A5)

or, $\frac{s_1^2}{s_2^2} > 7.4$

So, $b = 7.4$.

Example 6.20:

The tear strengths of two fabrics in warp direction have been compared. Ten samples of first fabric have an average tear strength of 54.7 N with a standard deviation of 6.1 N, while eight samples of second fabric have an average tear strength of 40.4 N with a standard deviation of 4.7 N. Find 95% confidence interval for the ratio of population variances of tear strengths.

SOLUTION

From the given data we have: $n_1 = 10$, $s_1^2 = 6.1^2$, $n_2 = 8$, $s_2^2 = 4.7^2$.

For 95% confidence limits, $\alpha/2 = 0.025$. Using Table A5, we find $f_{0.025,9,7} = 4.8$ and $f_{0.025,7,9} = 4.2$. Therefore 95% confidence interval for $\frac{\sigma_1^2}{\sigma_2^2}$ is

$$\frac{s_1^2}{s_2^2} \frac{1}{f_{\alpha/2,n_1-1,n_2-1}} < \frac{\sigma_1^2}{\sigma_2^2} < \frac{s_1^2}{s_2^2} f_{\alpha/2,n_2-1,n_1-1}$$

Substituting n_1, s_1^2, n_2, s_2^2, $f_{0.025,9,7}$ and $f_{0.025,7,9}$, we have

$$\frac{6.1^2}{4.7^2 \times 4.8} < \frac{\sigma_1^2}{\sigma_2^2} < \frac{6.1^2 \times 4.2}{4.7^2}$$

or, $0.3492 < \frac{\sigma_1^2}{\sigma_2^2} < 7.0748$

6.8 MATLAB® CODING

6.8.1 MATLAB® CODING OF EXAMPLE 6.1

```
clc
clear
close all
mu=638;
n=50;
sigma=52;
se=sigma/sqrt(n);
pr=normcdf(650,mu,se)-normcdf(620,mu,se)
```

6.8.2 MATLAB® CODING OF EXAMPLE 6.2

```
clc
clear
close all
mu=92;
n=36;
sigma=2.5;
se=sigma/sqrt(n);
pr=1-normcdf(92.5,mu,se)
```

6.8.3 MATLAB® CODING OF EXAMPLE 6.4

```
clc
clear
close all
n=15;
sigma=0.5;
chi2=chi2inv(0.95,n-1);
b=chi2*sigma^2/(n-1)
```

6.8.4 MATLAB® CODING OF EXAMPLE 6.10

```
clc
clear
close all
x_bar=18.3;
n=50;
s=1.7;
se=s/sqrt(n);
z=norminv([0.025  0.975],0,1)
ci=x_bar+z*se
```

6.8.5 MATLAB® CODING OF EXAMPLE 6.11

```
clc
clear
close all
x_bar=4.1;
n=100;
s=0.24;
se=s/sqrt(n);
z1=norminv([0.025  0.975],0,1);
z2=norminv([0.005  0.995],0,1);
ci_95=x_bar+z1*se
ci_99=x_bar+z2*se
```

6.8.6 MATLAB® CODING OF EXAMPLE 6.13

```
clc
clear
```

```
close all
x=[39.8 40.3 43.1 39.6 42.5 41.0 39.9 42.1 40.7 41.6 42.1
40.8];
x_bar=mean(x);
se=std(x)/sqrt(length(x));
t=tinv([0.025  0.975],length(x)-1);
ci=x_bar+t*se
```

6.8.7   MATLAB® Coding of Example 6.14

```
clc
clear
close all
n1=50;
n2=50;
x1_bar=490.6;
x2_bar=450.8;
sigma1=29.2;
sigma2=22.3;
se=sqrt((sigma1^2/n1)+(sigma2^2/n2));
z=norminv([0.025  0.975],0,1)
ci=(x1_bar-x2_bar)+z*se
```

6.8.8   MATLAB® Coding of Example 6.15

```
clc
clear
close all
x1=[19.7 18.8 20.4 19.9 20.0];
n1=length(x1);
x1_bar=mean(x1);
s1=std(x1);
x2=[15.3 14.7 15.9 16.1 15.5];
n2=length(x2);
x2_bar=mean(x2);
s2=std(x2);
sp=sqrt((((n1-1)*s1^2+(n2-1)*s2^2))/(n1+n2-2));
t=tinv([0.025  0.975],n1+n2-2);
ci=(x1_bar-x2_bar)+t*sp*sqrt((1/n1)+(1/n2))
```

6.8.9   MATLAB® Coding of Example 6.16

```
clc
clear
close all
n=110;
p_cap=80/110;
z=norminv([0.025  0.975],0,1);
ci= p_cap +z*sqrt((p_cap *(1- p_cap)/n))
```

6.8.10 MATLAB® CODING OF EXAMPLE 6.17

```
clc
clear
close all
n1=400;
p1_cap=264/400;
n2=320;
p2_cap=180/320;
z=norminv([0.005  0.995],0,1);
ci=(p1_cap-p2_cap)+z*sqrt((p1_cap*(1-p1_cap)/n1)+(p2_
cap*(1-p2_cap)/n2))
```

6.8.11 MATLAB® CODING OF EXAMPLE 6.18

```
clc
clear
close all
x=[191.7 186.9 195.2 194.4 190.2 192.6 193.8 189.2 188.3
187.5];
n=length(x);
x_bar=mean(x);
s=std(x);
chi2=chi2inv([0.975  0.025],n-1);
ci=(n-1)*s^2./chi2
```

6.8.12 MATLAB® CODING OF EXAMPLE 6.19

```
clc
clear
close all
n1=10;
n2=8;
f=finv(0.95,n1-1,n2-1)
b=2*f
```

6.8.13 MATLAB® CODING OF EXAMPLE 6.20

```
clc
clear
close all
n1=10;
v1=6.1^2;
n2=8;
v2=4.7^2;
f=finv([0.025  0.975],n2-1,n1-1)
ci= (v1/v2)*f
```

Exercises

6.1 The mean weight of a certain garment is 211 g with standard deviation of 5.8 g. If a random sample of 100 garments is taken, what is the probability that the sample mean will be between 212 and 215 g?

6.2 The mean length of a certain garment is 52.5 cm with standard deviation of 1.75 cm. Find the probability that the mean length of a sample of randomly selected 50 garments will exceed 53 cm.

6.3 The mean chest girth of the male students in a college is 95 cm with standard deviation of 8.25 cm. Find the probability that the mean chest girth of 40 randomly selected male students will be less than 94 cm.

6.4 Given a random sample of size n form a population that has the known mean μ and the finite variance σ^2, show that $\frac{1}{n}\sum_{i=1}^{n}(X_i - \mu)^2$ is an unbiased estimator of σ^2.

6.5 Show that $\frac{X+1}{n+1}$ is a biased estimator of the parameter θ of a Binomial distribution.

6.6 Show that $\bar{X}$ is a minimum variance unbiased estimator of the parameter p of a Bernoulli distribution.

6.7 If $X_1 + X_2 + X_3$ is a chi-square random variate with $X_1 \sim \chi_2^2$ and $X_2 \sim \chi_4^2$, and $P(X_1 + X_2 + X_3 \geq 19.023) = 0.025$, then what is the degree of freedom of X_3?

6.8 The 95% confidence limits of mean yarn tenacity (cN/tex) based on 64 test samples is 20 ± 1.5. Find out the number of test samples required to obtain 95% confidence limits of 20 ± 1.

6.9 If the 99% confidence range of the mean based on 100 test samples is ± 5, determine the number of tests to be conducted to obtain 99% confidence range of ± 3.

6.10 Thirty-six tests were done for yarn count. The mean and standard deviation of the test results were 19.6 tex and 1.4 tex, respectively. Calculate 95% confidence limits for the mean yarn strength of the population. If we want to be 95% certain that our estimate of the population means of yarn strength correct to within ± 0.2 tex, how many tests should be required?

6.11 Cotton samples of 40 specimens were tested for 2.5% span length. The average value of the test results was 25.6 mm and their standard deviation was 1.75 mm. Calculate 95% and 99% confidence limits for the population mean of 2.5% span length. If it was required to estimate the population mean of 2.5% span length correct to within ± 0.5 mm (with 99% chance of being correct), how many tests would be needed?

6.12 The CV of wool diameter is 25%. How many tests are required to get the population mean diameter with an accuracy of 3% at 95% confidence limits?

6.13 CSP tests were carried out on 12 cops of woollen yarns with the following results:

2490, 2670, 2130, 2400, 2550, 2310, 2520, 2250, 2400, 2700, 2280, 2100.

Calculate 95% and 99% confidence limits for the mean yarn CSP. If it is required to estimate the average yarn CSP to a precision such that 95% confidence limits are ±100, how many tests are needed?

6.14 Thickness values (mm) of 10 randomly chosen fabric samples are as follows. 0.41, 0.48, 0.43, 0.44, 0.49, 0.42, 0.43, 0.47, 0.40, 0.45.

Calculate 95% and 99% confidence limits for the mean fabric thickness. If it is required to estimate the average fabric thickness to a precision such that 99% confidence limits are ±0.025 mm, how many tests are needed?

6.15 In a random sample of 1525 fibres chosen from a large batch of mixed white and black fibres. It was observed that there were 777 black fibres. Use these data to calculate 95% and 99% confidence limits for the proportion of black fibres in the batch.

6.16 A random sample of 750 cocoons was chosen and 80 of them were found to be defective. Calculate 95% confidence limits for the proportion of defective cocoons.

6.17 The following data refer to the test results of fabric crimp (%) in weft and warp directions.

	Weft Crimp	Warp Crimp
Number of tests	36	36
Mean (%)	9.5	5.2
CV (%)	7.5	6.1

Calculate 95% confidence limits for the difference between the mean crimp (%) in weft and warp directions.

6.18 A modification in the finishes was made in the production process of a flame-retardant fabric to increase its flame retardancy. The results of the limiting oxygen index (LOI) tests on the original and modified flame-retardant fabrics were as follows:

Modified fabrics: 29.9, 30.4, 30.7, 29.5, 30.1
Unmodified fabrics: 25.6, 24.8, 25.9, 25.1, 25.3

Calculate 95% confidence limits for the mean increase in LOI caused by the modification.

6.19 Two varieties of cotton fibre were tested for maturity. If 315 of 500 fibres of the first variety and 218 of 400 fibres of the second variety were fully matured, find 95% confidence interval for the difference between the actual proportions of fully matured fibres in two varieties of cottons.

6.20 The population standard deviation (σ) of yarn hairiness index is 1.15. Fourteen tests are made for estimating sample standard deviation (S) of yarn hairiness. Determine the value of b such that $P(S^2 > b) = 0.025$.

6.21 The following are the results of twists per meter (TPM) of 8 yarn samples having nominal count of 30's Ne.
882, 846, 861, 813, 834, 845, 831, and 872.

Construct a 95% confidence interval for variance which measures the true variability of the yarn twist.

6.22 From past experience, it is known that the population variance of mass/unit length of open-end friction spun yarn is twice than that of ring spun yarn. Random samples of sizes 10 from each yarn are taken for measuring the variances of mass/unit length. If S_1^2 and S_2^2 denote the sample variances of open-end friction spun yarn and ring spun yarn, respectively determine the value of b such that $P\left(\frac{S_1^2}{S_2^2} > b\right) = 0.025$.

6.23 The CV% of the strengths of ring and rotor spun yarns each having 20's Ne nominal count were found to be 7.8% and 6.2%, respectively based on 10 tests on each yarn. Find 95% confidence interval for the ratio of population variances of the strengths of two yarns.

7 Test of Significance

7.1 INTRODUCTION

We begin with the formulation of the practical problem in terms of two hypotheses, namely null hypothesis, H_0 and alternative hypothesis, H_1. The null hypothesis signifies the status quo, i.e. there is no difference between the processes being tested. Whereas, a positive test on alternative hypothesis indicates that there is significant difference between the processes and we should take action of some kind. An analogy of a jury trial may be drawn to illustrate the null and alternative hypotheses. The null hypothesis that the defendant is innocent cannot be rejected unless the alternative hypothesis that the defendant is guilty is supported by evidence beyond a reasonable doubt. However, accepting H_0 not necessarily implies innocence, but it merely indicates that the evidence was insufficient to convict. So if it is not supported by evidence, the jury does not necessarily accept H_0, but fails to reject H_0.

In order to illustrate the basic conceptions involved in deciding whether a statement is true or false, suppose that a consumer of nylon fibre wants to test the fibre manufacturer's claim that the average initial modulus of nylon fibre is 30 cN/denier. The consumer has instructed a member of the quality control staff to take 36 tests with the intention of rejecting the claim if the mean initial modulus falls short of 29.4 cN/denier; else it will accept the claim. Because the decision is based on the sample, there will be some probability that the sample mean is less than 29.4 cN/denier when the true mean is $\mu = 30$ cN/denier. In addition, there will be some probability that the sample mean is greater than 29.4 cN/denier when the true mean is, say, $\mu = 29$ cN/denier. Therefore, before setting a criterion, it is better to investigate the probability that the criterion may lead to a wrong decision.

Assume that the population standard deviation of the nylon fibre modulus to be $\sigma = 2$ cN/denier. If $\bar{X}$ denotes the sample mean, from the central limit theorem, we know that $\bar{X}$ is approximately normally distributed with standard deviation $\sigma_{\bar{X}} = \frac{\sigma}{\sqrt{n}}$. Hence, $\sigma_{\bar{X}} = \frac{2}{\sqrt{36}} = \frac{1}{3}$. The probability that the sample mean is less than 29.4 cN/denier when the true mean is $\mu = 30$ cN/denier is estimated as

$$P\left(\bar{X} < 29.4 \text{ when } \mu = 30\right) = P\left(Z < \frac{29.4 - 30}{1/3}\right) = P(Z < -1.8) = 0.036$$
$$\text{(from Table A1).}$$

Therefore, the probability of erroneously rejecting the $H_0: \mu = 30$ cN/denier when it is actually true is 0.036 as depicted by the area of the shaded region in Figure 7.1. This type of error is termed as type I error and the probability of committing this error is termed as level of significance which is designated by α. Thus, the level of significance is the risk we are ready to take in rejecting H_0 when it is in fact true.

DOI: 10.1201/9781003081234-7

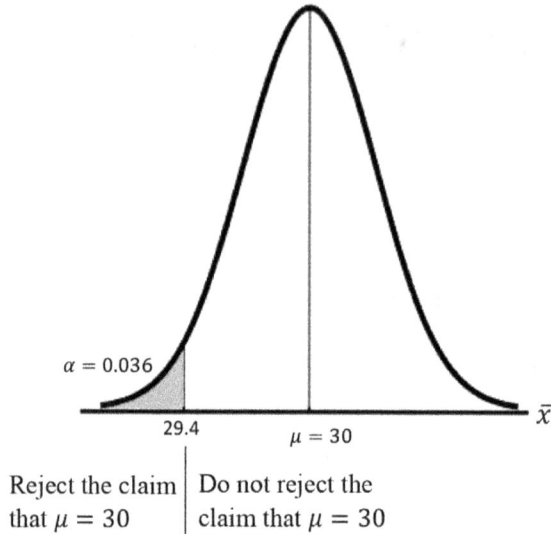

FIGURE 7.1   Probability of type I error.

The probability that the sample mean is greater than 29.4 cN/denier when the true mean is $\mu = 29$ cN/denier is estimated as

$$P(\bar{X} > 29.4 \text{ when } \mu = 29) = P\left(Z > \frac{29.4 - 29}{1/3}\right) = P(Z > 1.2) = 0.115$$
$$\text{(from Table A1).}$$

Therefore, the probability of erroneously not rejecting H_0: $\mu = 30$ cN/denier when it is actually false is 0.115 as depicted by the area of the shaded region in Figure 7.2. This type of error is termed as type II error and the probability of committing this error is designated by β. In this example we have arbitrarily chosen the alternative hypothesis that $\mu = 29$ cN/denier for computing β. Thus, it is impossible to compute β unless we have a specific alternative hypothesis.

The situation described in the aforesaid example may be summarized as follows:

	H_0 is Not Rejected	H_0 is Rejected
H_0 is true	Correct decision	Type I error
H_0 is false	Type II error	Correct decision

It should be noted that both errors cannot be minimized simultaneously. In reality, if we try to minimize one error, the other type of error will be increased.

The null hypothesis is assumed to be true until it appears to be inconsistent with the sample data. The decision of whether the H_0 is rejected or not is determined on the basis of test statistics. The observed value of a test statistic is calculated from the sample data. The probability distribution of a test statistic should be calculable under the assumption that H_0 is true.

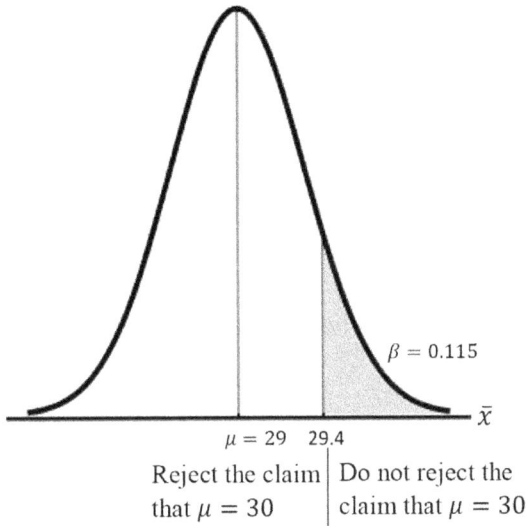

$\beta = 0.115$

$\bar{x}$

$\mu = 29$ 29.4

| Reject the claim | Do not reject the |
| that $\mu = 30$ | claim that $\mu = 30$ |

FIGURE 7.2 Probability of type II error.

Next we decide a critical region. In order to test the null hypothesis, the critical region C is defined as the set such that if the observed value of the test statistic falls inside this set, the null hypothesis is rejected. Thus, we can define the level of significance α and the probability of committing the type II error β as follows:

$$\alpha = P\left(\hat{t} \in C \mid H_0 \text{ is true}\right)$$

and

$$\beta = P\left(\hat{t} \in C^c \mid H_1 \text{ is true}\right)$$

where $\hat{t}$ is the observed value of the test statistic used and C^c is the complement of the critical region C.

Hypothesis testing can be three types: right-tail test, so that we reject H_0 if the observed value of test statistic is greater than a critical value in the right side; left-tail test, so that we reject H_0 if the observed value of test statistic is less than a critical value in the left side; and double-tail test, so that we reject H_0 if the observed value of test statistic is either greater than a critical value in the right side or less than a critical value in the left side.

In any hypothesis testing problem the following step-by-step procedures should be performed.

1. State the null hypothesis (H_0), alternative hypothesis (H_1) and specify the level of significance (α).
2. Determine the critical region.
3. Calculate the observed value of test statistic.
4. Apply decision rules and draw conclusion.

7.2   TESTS CONCERNING SINGLE POPULATION MEAN: LARGE SAMPLE AVAILABLE (z-Test)

1. Suppose that a random sample of large size n $(n \geq 30)$ has calculated sample mean $\bar{x}$. If the sample is drawn from a population with assumed mean μ_0 and known variance σ^2, we will test the null hypothesis that the true population mean μ is equal to μ_0 as expressed below:

$$H_0: \mu = \mu_0$$

The alternative hypothesis is that μ is either greater or less or different than μ_0 as expressed below:

$$H_1: \begin{array}{l} \mu > \mu_0 \text{ or} \\ \mu < \mu_0 \text{ or} \\ \mu \neq \mu_0 \end{array}$$

If $\mu > \mu_0$, the right-tail test is performed, but if $\mu < \mu_0$ the left-tail test is to be done. Whereas if $\mu \neq \mu_0$, the test becomes a double-tailed. Then we decide upon the level of significance α. Usually α is set to 0.05 or 0.01. This test is often called the z-test.

   The value of population variance σ^2 is known from the past experience. If σ^2 is unknown, sample variance s^2 provides a good estimate of σ^2, when the sample size is large enough. Hence, σ^2 can be replaced by s^2, when it is unknown for large sample size.

2. If the random variable $\bar{X}$ has a normal distribution with the mean μ_0 and the variance σ^2/n, then the standardized variable $Z = \frac{\bar{X} - \mu_0}{\sigma/\sqrt{n}}$ has the standard normal distribution. For a double-tail test, the total area under standard normal curve to the right of the critical value $z_{\alpha/2}$ and left of the critical value $-z_{\alpha/2}$ is equal to α, hence the critical regions are $z > z_{\alpha/2}$ and $z < -z_{\alpha/2}$ as depicted in Figure 7.3. In case of a right-tail test, the area under standard

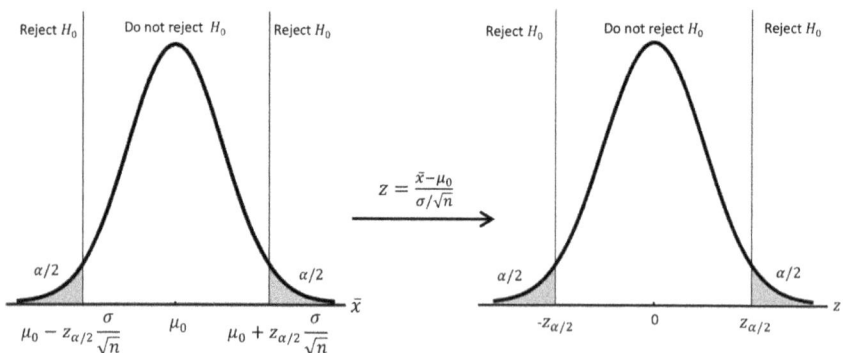

FIGURE 7.3   Critical regions for testing $H_1: \mu \neq \mu_0$.

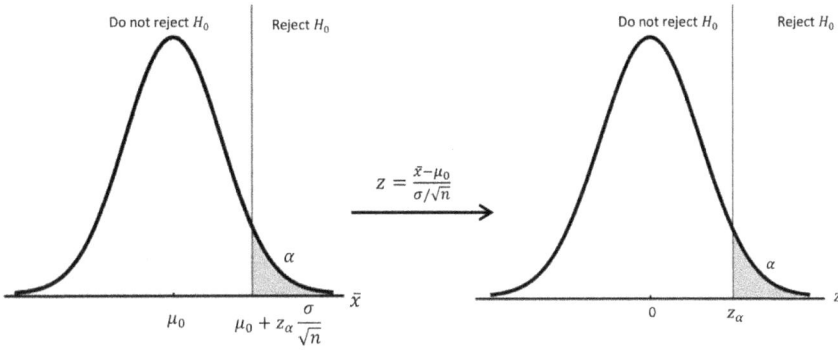

FIGURE 7.4 Critical region for testing H_1: $\mu > \mu_0$.

normal curve to the right of the critical value z_α is equal to α, therefore, the critical region is $z > z_\alpha$ as shown in Figure 7.4. Similarly, for a left-tail test, the area under standard normal curve to the left of the critical value $-z_\alpha$ is equal to α, thus the critical region is $z < -z_\alpha$ as shown in Figure 7.5. In Figures 7.3–7.5, the dividing lines of the critical regions require the value $z_{\alpha/2}$ or z_α which can be readily obtained from the table of standard normal distribution for given α (from Table A1).

3. Substituting the values of $\bar{x}$, μ_0, σ and n, the observed value of test statistic z is calculated from the following equation:

$$z = \frac{\bar{x} - \mu_0}{\sigma/\sqrt{n}}$$

4. The decision rules for rejecting H_0 are shown in the Table 7.1.

By comparing the observed value of test statistic with the critical value, finally the conclusion on rejecting or not rejecting H_0 is drawn.

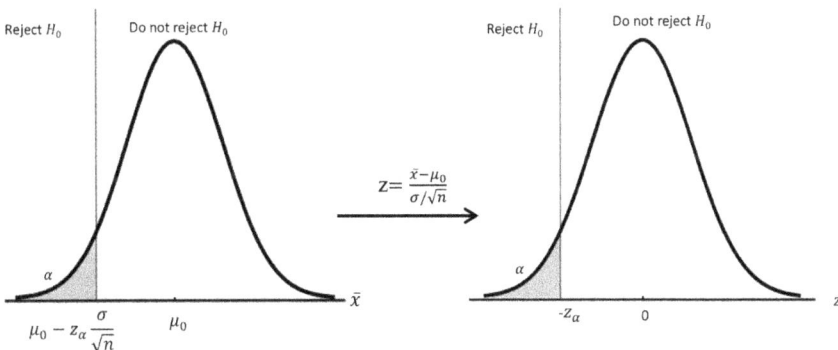

FIGURE 7.5 Critical region for testing H_1: $\mu < \mu_0$.

TABLE 7.1

Decision Rules for Rejecting H_0: $\mu = \mu_0$ (Large Sample)

Statement	Decision Rule
$H_1: \mu > \mu_0$	Do not reject H_0 if $z < z_\alpha$, else reject H_0
$H_1: \mu < \mu_0$	Do not reject H_0 if $z > -z_\alpha$, else reject H_0
$H_1: \mu \neq \mu_0$	Do not reject H_0 if $-z_{\alpha/2} < z < z_{\alpha/2}$, else reject H_0

Example 7.1:

A ring frame is nominally spinning 40 tex yarn. One-hundred-fifty ring frame bobbins are tested for count. The mean and standard deviation are found to be 40.25 tex and 1.1 tex, respectively. Justify whether the mean count of the delivery is different from the nominal at the 0.05 level of significance.

SOLUTION

Given data: $\mu_0 = 40$, $n = 150$, $\bar{x} = 40.25$, $s = 1.1$. Because $n > 30$, we can assume that $\sigma = s$.

a. $H_0: \mu = 40$
   $H_1: \mu \neq 40$ (double-tail test)
   $\alpha = 0.05$.
b. The critical values are $z_{0.025} = 1.96$ and $-z_{0.025} = -1.96$ (from Table A1).
c. The observed value of test statistic is $z = \frac{\bar{x} - \mu_0}{\sigma/\sqrt{n}} = \frac{40.25 - 40}{1.1/\sqrt{150}} = 2.78$.
d. As $2.78 > 1.96$, the null hypothesis is rejected. Therefore, there is sufficient evidence to conclude that the mean count of the delivery is significantly different from the nominal count. Figure 7.6 depicts the pictorial representation of this hypothesis testing.

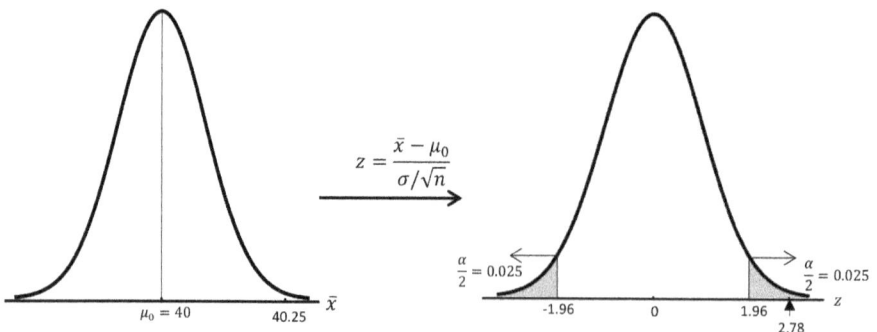

FIGURE 7.6 z-Test of H_0: $\mu = 40$ against H_1: $\mu \neq 40$.

Example 7.2:

A spinning mill produces a yarn that has mean strength of 260 cN and standard deviation of 20 cN. A change in the spinning process parameters claims to produces a stronger yarn. Determine whether the claim is correct at the 0.01 level of significance, if the new process produces a yarn that has mean strength of 266 cN out of 50 tests.

SOLUTION

Given data: $\mu_0 = 260$, $n = 50$, $\bar{x} = 266$, $\sigma = 20$.

a. H_0: $\mu = 260$
   H_1: $\mu > 260$ (right-tail test)
   $\alpha = 0.01$.
b. The critical value is $z_{0.01} = 2.326$ (from Table A1).
c. The observed value of test statistic is $z = \frac{\bar{x} - \mu_0}{\sigma/\sqrt{n}} = \frac{266 - 260}{20/\sqrt{50}} = 2.12$.
d. As $2.12 < 2.326$, the null hypothesis cannot be rejected. Therefore, there is no sufficient evidence to conclude that the change in the spinning process parameters produces a stronger yarn. This hypothesis testing is schematically shown in Figure 7.7.

Example 7.3:

A fabric manufacturer placed an order to a yarn manufacturer for a large batch of yarn that has strength of 20 cN/tex. Upon delivery, the fabric manufacturer found that the average strength of the yarn is 19.65 cN/tex with 1.2 standard deviation based on 40 tests. Carry out a statistical test at 0.05 level of confidence to justify whether the fabric manufacturer would ask the yarn manufacturer for replacement of yarn batch.

SOLUTION

Given data: $\mu_0 = 20$, $n = 40$, $\bar{x} = 19.65$, $s = 1.2$. Because $n > 30$, we can assume that $\sigma = s$.

a. H_0: $\mu = 20$
   H_1: $\mu < 20$ (left-tail test)
   $\alpha = 0.05$.

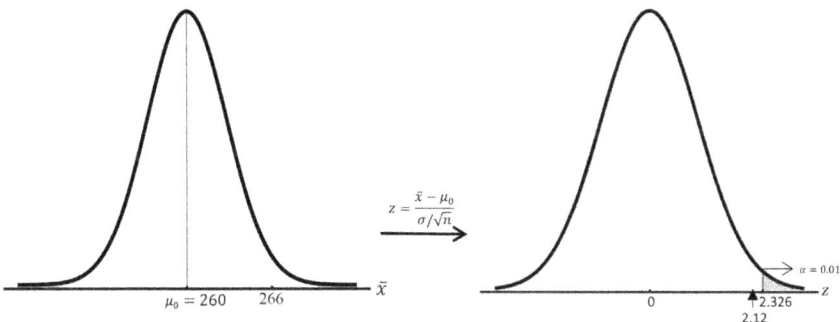

FIGURE 7.7 z-Test of H_0: $\mu = 260$ against H_1: $\mu > 260$.

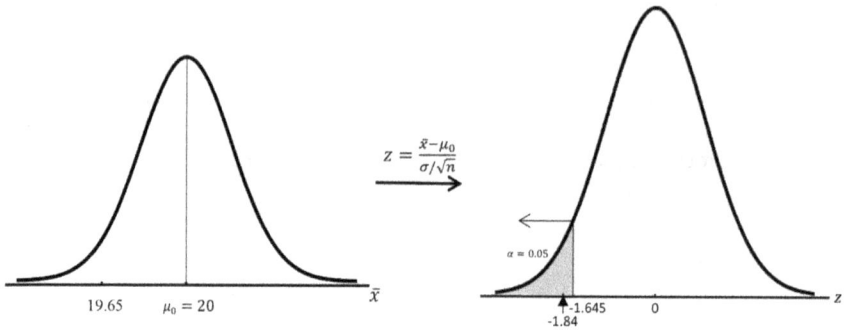

FIGURE 7.8 z-Test of H_0: $\mu = 20$ against H_1: $\mu < 20$.

b. The critical value is $-z_{0.05} = -1.645$ (from Table A1).

c. The observed value of test statistic is $z = \frac{\bar{x}-\mu_0}{\sigma/\sqrt{n}} = \frac{19.65-20}{1.2/\sqrt{40}} = -1.84$.

d. As $-1.84 < -1.645$, the null hypothesis is rejected. Therefore, there is sufficient evidence to conclude that the fabric manufacturer would ask for replacement of yarn batch. Figure 7.8 shows the schematic diagram of this hypothesis testing.

7.3 TESTS CONCERNING SINGLE POPULATION MEAN: SMALL SAMPLE AVAILABLE (*t*-Test)

1. Suppose that a random sample of small size n $(n < 30)$ has calculated sample mean $\bar{x}$ and sample variance s^2. If the sample is drawn from a population with assumed mean μ_0 and unknown variance, we will test the null hypothesis that

$$H_0: \mu = \mu_0$$

against the alternative hypothesis

$$\mu > \mu_0 \text{ or}$$
$$H_1: \quad \mu < \mu_0 \text{ or}$$
$$\mu \neq \mu_0$$

where μ is the true population mean. We then decide on the level of significance α. For small sample size with unknown population variance, we need to perform t-test.

2. If $\bar{X}$ and S^2 are the mean and the variance of a random sample of small size n drawn from a normal population with mean μ_0 and unknown variance then the random variable $T = \frac{\bar{X}-\mu_0}{S/\sqrt{n}}$ has the t-distribution with $n-1$ degrees of freedom. For a double-tail test, the critical regions are $t > t_{\alpha/2,n-1}$ and $t < -t_{\alpha/2,n-1}$ as depicted in Figure 7.9. In case of a right-tail test, the critical

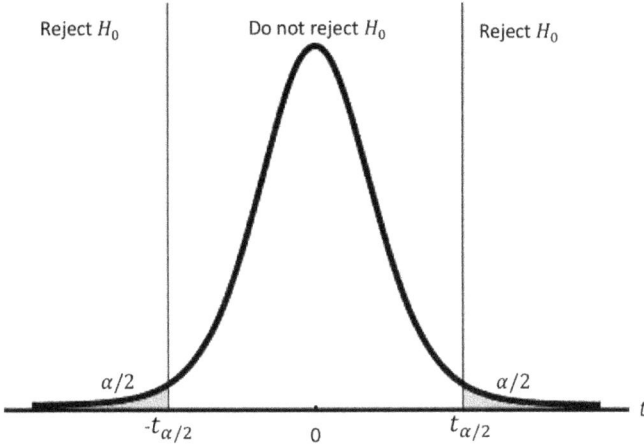

FIGURE 7.9   Critical regions for a double-tail t-test.

region is $t > t_{\alpha,n-1}$ as shown in Figure 7.10. Similarly for a left-tail test, the critical region is $t < -t_{\alpha,n-1}$ as shown in Figure 7.11. The values of $t_{\alpha/2,n-1}$ or $t_{\alpha,n-1}$ can be obtained from the table of t-distribution for given degrees of freedom and α (from Table A4).

3. Substituting the values of $\bar{x}$, μ_0, s and n, the observed value of test statistic t is calculated from the following equation:

$$t = \frac{\bar{x} - \mu_0}{s / \sqrt{n}}$$

4. The decision rules for rejecting H_0 are shown in the Table 7.2.

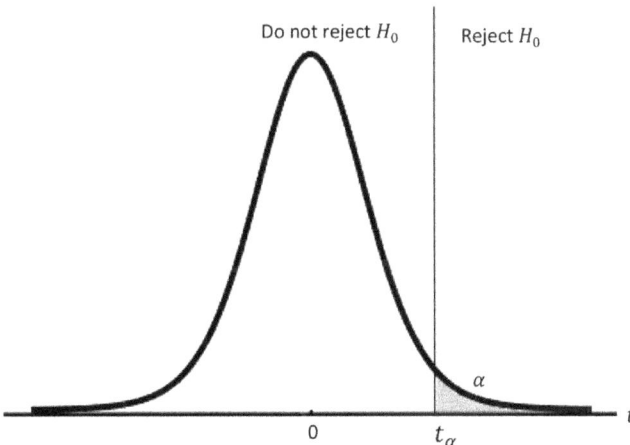

FIGURE 7.10   Critical region for a right-tail t-test.

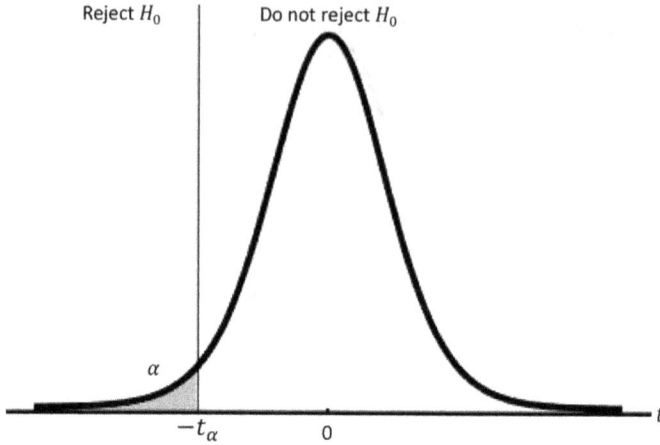

FIGURE 7.11   Critical region for a left-tail t-test.

TABLE 7.2

Decision Rules for Rejecting H_0: $\mu = \mu_0$ (Small Sample)

Statement	Decision Rule
$H_1: \mu > \mu_0$	Do not reject H_0 if $t < t_{\alpha,n-1}$, else reject H_0
$H_1: \mu < \mu_0$	Do not reject H_0 if $t > -t_{\alpha,n-1}$, else reject H_0
$H_1: \mu \neq \mu_0$	Do not reject H_0 if $-t_{\alpha/2,n-1} < t < t_{\alpha/2,n-1}$, else reject H_0

Example 7.4:

The nominal count of a ring frame is 16^s Ne. 12 ring frame bobbins are tested for English count. The mean and standard deviation are found to be 16.8^s Ne and 0.8^s Ne, respectively. Test whether the mean count of the delivery is different from the nominal at the 0.01 level of significance.

SOLUTION

Given data: $\mu_0 = 16$, $n = 12$, $\bar{x} = 16.8$, $s = 0.8$.

a. $H_0: \mu = 16$
   $H_1: \mu \neq 16$ (double-tail test)
   $\alpha = 0.01$.
b. The critical values are $t_{0.005,11} = 3.11$ and $-t_{0.005,11} = -3.11$ (from Table A4).
c. The observed value of test statistic is $t = \frac{\bar{x} - \mu_0}{s/\sqrt{n}} = \frac{16.8 - 16}{0.8/\sqrt{12}} = 3.46$.
d. As $3.46 > 3.11$, the null hypothesis is rejected. Therefore, there is sufficient evidence to conclude that the mean count of the delivery is different from the nominal count. Figure 7.12 depicts the schematic diagram of this hypothesis testing.

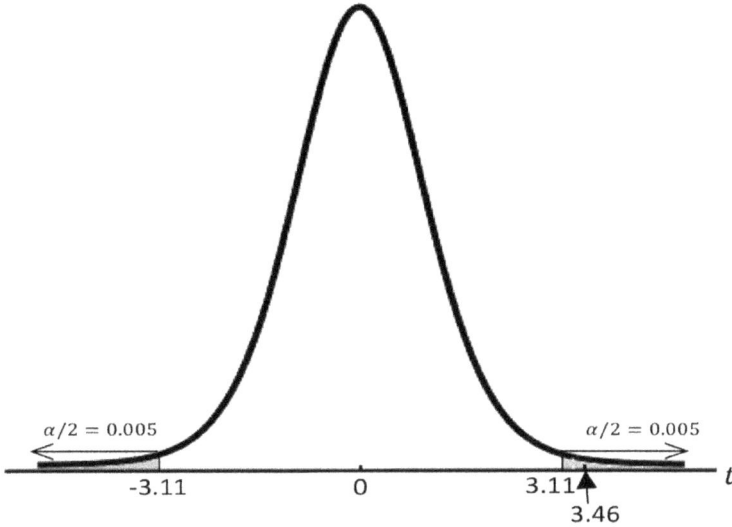

FIGURE 7.12 t-Test of H_0: $\mu = 16$ against H_1: $\mu \neq 16$.

Example 7.5:

It is specified that the trash content in cotton bales should not exceed 4%. To test this, eight cotton bales were subjected for the determination of percentage trash content, with the following results:

$$4.2, 4.4, 4.1, 4.3, 3.9, 4.3, 4.0, 3.8.$$

Do these results tend to confirm that cotton bales contain significantly more trash than the specified value at 0.05 level of significance?

SOLUTION

From the given data we have: $\mu_0 = 4$, $n = 8$, $\bar{X} = \frac{\sum_{i=1}^{8} x_i}{8} = 4.125$ and $s = \sqrt{\frac{\sum_{i=1}^{8}(x_i - \bar{x})^2}{8-1}} = 0.212$.

a. H_0: $\mu = 4$
   H_1: $\mu > 4$ (right-tail test)
   $\alpha = 0.05$.
b. The critical value is $t_{0.05,7} = 1.9$ (from Table A4).
c. The observed value of test statistic is $t = \frac{\bar{x} - \mu_0}{s/\sqrt{n}} = \frac{4.125 - 4}{0.212/\sqrt{8}} = 1.67$.
d. As $1.67 < 1.9$, the null hypothesis cannot be rejected. Therefore, there is no sufficient evidence to conclude that the cotton bales contain significantly more trash than the specified value. Figure 7.13 shows the pictorial representation of this hypothesis testing.

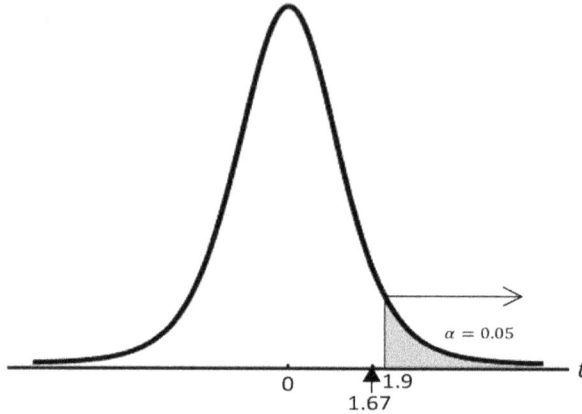

FIGURE 7.13 t-Test of H_0: $\mu = 4$ against H_1: $\mu > 4$.

7.4 TESTS CONCERNING THE DIFFERENCE BETWEEN TWO POPULATION MEANS: TWO LARGE SAMPLES AVAILABLE (z-Test)

1. Suppose that a random sample of large size n_1 $(n_1 > 30)$ with calculated sample mean $\bar{x}_1$ is drawn from a population with mean μ_1 and known variance σ_1^2, and also suppose that a random sample of large size n_2 $(n_2 > 30)$ with calculated sample mean $\bar{x}_2$ is drawn from a population with mean μ_2 and known variance σ_2^2. We will test the null hypothesis that

$$H_0: \mu_1 - \mu_2 = \delta$$

against the alternative hypothesis

$$H_1: \begin{array}{l} \mu_1 - \mu_2 > \delta \text{ or} \\ \mu_1 - \mu_2 < \delta \text{ or} \\ \mu_1 - \mu_2 \neq \delta \end{array}$$

where δ is a constant. Next, we specify the level of significance α.

If σ_1^2 and σ_2^2 are unknown, then σ_1^2 and σ_2^2 can be replaced by sample variances s_1^2 and s_2^2, respectively, when n_1 and n_2 are both greater than 30.

2. If the random variables $\bar{X}_1$ and $\bar{X}_2$ are normally and independently distributed with the means μ_1 and μ_2 and the variances σ_1^2/n_1 and σ_2^2/n_2, respectively, then $\bar{X}_1 - \bar{X}_2$ is normally distributed with the mean $\mu_1 - \mu_2$ and the variance $\frac{\sigma_1^2}{n_1} + \frac{\sigma_2^2}{n_2}$, and the standardized variable $Z = \frac{(\bar{X}_1 - \bar{X}_2) - (\mu_1 - \mu_2)}{\sqrt{\frac{\sigma_1^2}{n_1} + \frac{\sigma_2^2}{n_2}}}$ has the

standard normal distribution. For a double-tail test, the critical regions are $z > z_{\alpha/2}$ and $z < -z_{\alpha/2}$ as depicted in Figure 7.14. In case of a right-tail test, the critical region is $z > z_\alpha$ as shown in Figure 7.15. For a left-tail test, the critical region is $z < -z_\alpha$ as shown in Figure 7.16.

Reject H_0 Do not reject H_0 Reject H_0

$$z = \frac{(\bar{x}_1 - \bar{x}_2) - (\mu_1 - \mu_2)}{\sqrt{\dfrac{\sigma_1^2}{n_1} + \dfrac{\sigma_2^2}{n_2}}}$$

Reject H_0 Do not reject H_0 Reject H_0

$\alpha/2$ $\alpha/2$ $\alpha/2$ $\alpha/2$

$\mu_1 - \mu_2$

$\bar{x}_1 - \bar{x}_2$

$\mu_1 - \mu_2 - z_{\alpha/2}\dfrac{\sigma}{\sqrt{n}}$ $\mu_1 - \mu_2 + z_{\alpha/2}\dfrac{\sigma}{\sqrt{n}}$

$-z_{\alpha/2}$ 0 $z_{\alpha/2}$ z

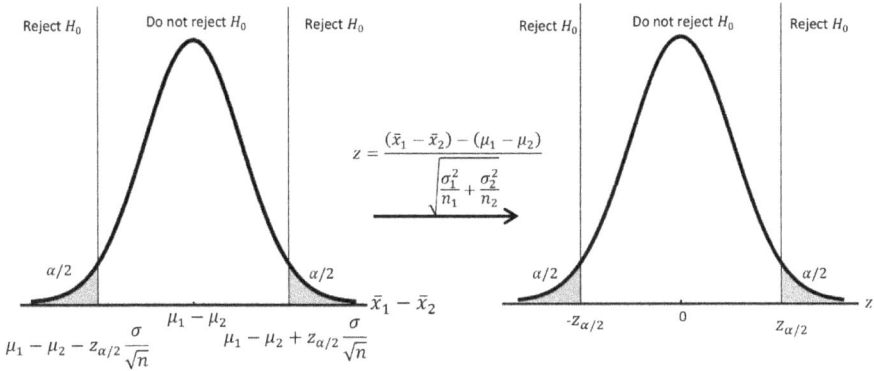

FIGURE 7.14   Critical regions for testing H_1: $\mu_1 - \mu_2 \neq \delta$.

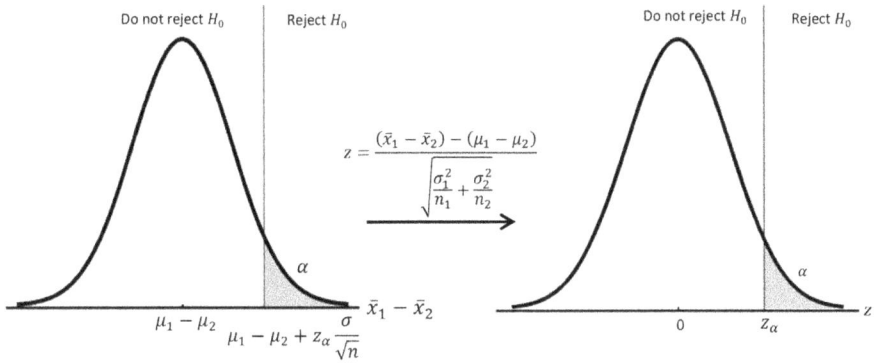

Do not reject H_0 Reject H_0

$$z = \frac{(\bar{x}_1 - \bar{x}_2) - (\mu_1 - \mu_2)}{\sqrt{\dfrac{\sigma_1^2}{n_1} + \dfrac{\sigma_2^2}{n_2}}}$$

Do not reject H_0 Reject H_0

α α

$\mu_1 - \mu_2$

$\bar{x}_1 - \bar{x}_2$

$\mu_1 - \mu_2 + z_{\alpha}\dfrac{\sigma}{\sqrt{n}}$

0 z_{α} z

FIGURE 7.15   Critical region for testing H_1: $\mu_1 - \mu_2 > \delta$.

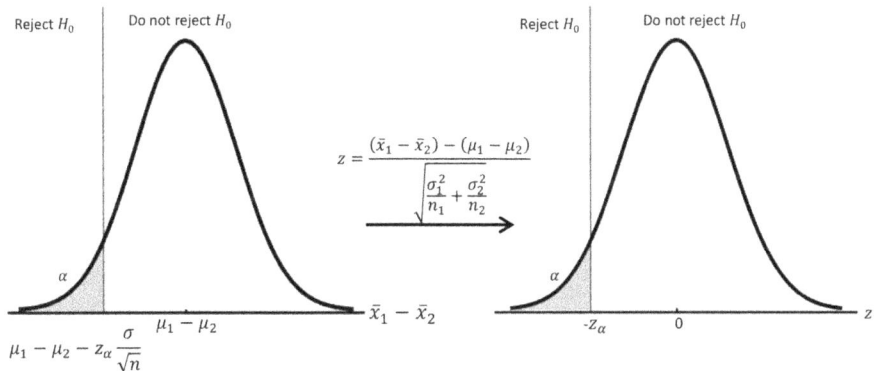

Reject H_0 Do not reject H_0

$$z = \frac{(\bar{x}_1 - \bar{x}_2) - (\mu_1 - \mu_2)}{\sqrt{\dfrac{\sigma_1^2}{n_1} + \dfrac{\sigma_2^2}{n_2}}}$$

Reject H_0 Do not reject H_0

α α

$\mu_1 - \mu_2$

$\bar{x}_1 - \bar{x}_2$

$\mu_1 - \mu_2 - z_{\alpha}\dfrac{\sigma}{\sqrt{n}}$

$-z_{\alpha}$ 0 z

FIGURE 7.16   Critical region for testing H_1: $\mu_1 - \mu_2 < \delta$.

TABLE 7.3

Decision Rules for Rejecting H_0: $\mu_1 - \mu_2 = \delta$ (Two Large Samples)

Statement	Decision Rule
$H_1: \mu_1 - \mu_2 > \delta$	Do not reject H_0 if $z < z_\alpha$, else reject H_0
$H_1: \mu_1 - \mu_2 < \delta$	Do not reject H_0 if $z > -z_\alpha$, else reject H_0
$H_1: \mu_1 - \mu_2 \neq \delta$	Do not reject H_0 if $-z_{\alpha/2} < z < z_{\alpha/2}$, else reject H_0

3. Substituting the values of $\bar{x}_1$, $\bar{x}_2$, $\mu_1 - \mu_2$, σ_1, σ_2, n_1 and n_2, the observed value of test statistic z is calculated from the following equation:

$$z = \frac{(\bar{x}_1 - \bar{x}_2) - (\mu_1 - \mu_2)}{\sqrt{\dfrac{\sigma_1^2}{n_1} + \dfrac{\sigma_2^2}{n_2}}}$$

4. The decision rules for rejecting H_0 are shown in the Table 7.3.

Example 7.6:

A mill produces 80^s Ne polyester/cotton blended yarn on ring frames manufactured by two different makes, A and B. While testing 40 yarn samples from both the frames for yarn hairiness, the mean hairiness indices were found to be 6.1 and 6.6, respectively for frame-A and frame-B. The CV% of yarn hairiness indices were 12% and 18% for frame-A and frame-B, respectively. Justify whether frame-A produces less hairy yarn than frame-B at the 0.05 level of significance.

SOLUTION

Given data: $n_1 = n_2 = 40$, $\bar{x}_1 = 6.1$, $\bar{x}_2 = 6.6$, $CV_1\% = 12$, $CV_2\% = 18$.

Thus, $s_1 = \frac{\bar{x}_1 \times CV_1\%}{100} = \frac{6.1 \times 12}{100} = 0.732$ and $s_2 = \frac{\bar{x}_2 \times CV_2\%}{100} = \frac{6.6 \times 18}{100} = 1.188$. Because n_1 and n_2 are both greater than 30, we can assume that $\sigma_1 = s_1$ and $\sigma_2 = s_2$.

a. $H_0: \mu_1 - \mu_2 = 0$
   $H_1: \mu_1 - \mu_2 < 0$ (left-tail test)
   $\alpha = 0.05$.
b. The critical value is $-z_{0.05} = -1.645$ (from Table A1).
c. The observed value of test statistic is $z = \frac{(\bar{x}_1 - \bar{x}_2) - (\mu_1 - \mu_2)}{\sqrt{\frac{\sigma_1^2}{n_1} + \frac{\sigma_2^2}{n_2}}} = \frac{(6.1 - 6.6) - 0}{\sqrt{\frac{0.732^2}{40} + \frac{1.188^2}{40}}} = -2.27$.
d. As $-2.27 < -1.645$, the null hypothesis is rejected. Therefore, there is sufficient evidence to conclude that the frame-A produces less hairy yarn than frame-B. A pictorial diagram of this hypothesis testing is depicted in Figure 7.17.

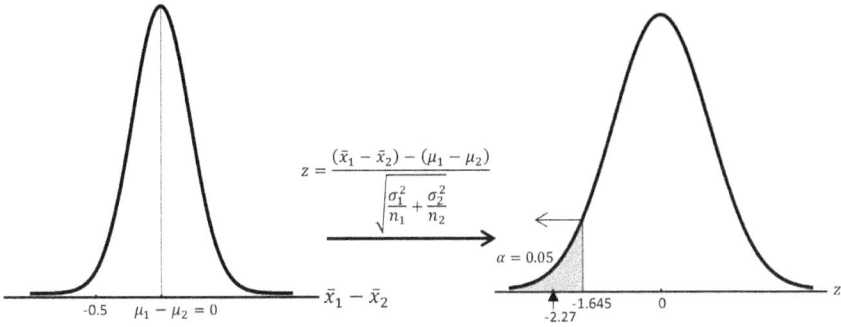

FIGURE 7.17 z-Test of H_0: $\mu_1 - \mu_2 = 0$ against H_1: $\mu_1 - \mu_2 < 0$.

Example 7.7:

An experiment is performed in a Vibrodyn instrument to determine whether the average denier of one kind of polyester staple fibre exceeds than that of another kind by 0.25 denier. If 50 tests on first kind of fibre had an average denier of 1.61 with a standard deviation of 0.06 and 40 tests on second kind of fibre had an average denier of 1.34 with a standard deviation of 0.07, justify whether the difference between the average denier of two kinds of fibres really exceeds by 0.25 denier at the 0.05 level of significance.

SOLUTION

Given data: $n_1 = 50$, $n_2 = 40$, $\bar{x}_1 = 1.61$, $\bar{x}_2 = 1.34$, $s_1 = 0.06$, $s_2 = 0.07$. Because n_1 and n_2 are both greater than 30, we can assume that $\sigma_1 = s_1$ and $\sigma_2 = s_2$.

a. $H_0 : \mu_1 - \mu_2 = 0.25$
$H_1 : \mu_1 - \mu_2 > 0.25$ (right-tail test)
$\alpha = 0.05$.

b. The critical value is $z_{0.05} = 1.645$ (from Table A1).

c. The observed value of test statistic is $z = \dfrac{(\bar{x}_1 - \bar{x}_2) - (\mu_1 - \mu_2)}{\sqrt{\dfrac{\sigma_1^2}{n_1} + \dfrac{\sigma_2^2}{n_2}}} = \dfrac{(1.61 - 1.34) - 0.25}{\sqrt{\dfrac{0.06^2}{50} + \dfrac{0.07^2}{40}}} = 1.43$.

d. As $1.43 < 1.645$, the null hypothesis cannot be rejected. Therefore, there is no sufficient evidence to conclude that the difference between the average denier of two kinds of fibres exceeds by 0.25 denier. Figure 7.18 shows the schematic representation of this hypothesis testing.

Example 7.8:

A rotor spun yarn produced from a cotton mixing is claimed to give lower strength than the ring spun yarn produced from the same mixing. The experimental result based on 100 tests shows that the average strength of ring spun yarn is 18 cN/tex. What would be the average strength of the rotor spun yarn at same number of test to say that this system produces a weaker yarn than the ring spinning at the 0.05 level of significance? The standard deviations of strength of ring and rotor spun yarns are 1.8 cN and 1.5 cN, respectively.

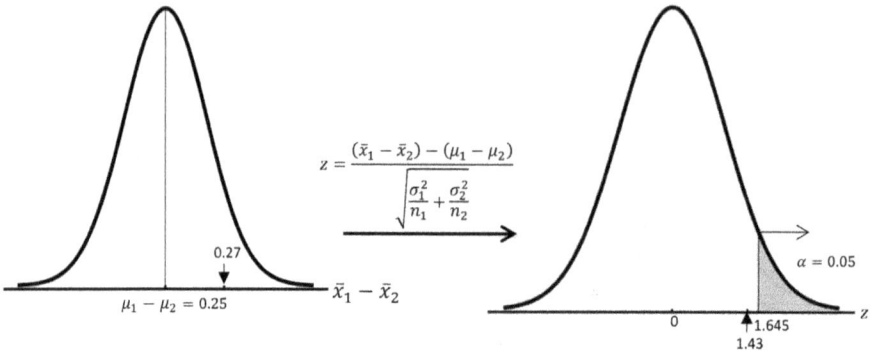

$$z = \frac{(\bar{x}_1 - \bar{x}_2) - (\mu_1 - \mu_2)}{\sqrt{\frac{\sigma_1^2}{n_1} + \frac{\sigma_2^2}{n_2}}}$$

FIGURE 7.18 z-Test of H_0: $\mu_1 - \mu_2 = 0.25$ against H_1: $\mu_1 - \mu_2 > 0.25$.

SOLUTION

Given data: $n_1 = n_2 = 100$, $\bar{x}_1 = 18$, $\sigma_1 = 1.8$, $\sigma_2 = 1.5$.

a. H_0: $\mu_1 - \mu_2 = 0$
   H_1: $\mu_1 - \mu_2 > 0$ (right-tail test)
   $\alpha = 0.05$.
b. The critical value is $z_{0.05} = 1.645$ (from Table A1).
c. The observed value of test statistic is $z = \frac{(\bar{x}_1 - \bar{x}_2) - (\mu_1 - \mu_2)}{\sqrt{\frac{\sigma_1^2}{n_1} + \frac{\sigma_2^2}{n_2}}} = \frac{(18 - \bar{x}_2) - 0}{\sqrt{\frac{1.8^2}{100} + \frac{1.5^2}{100}}} = \frac{18 - \bar{x}_2}{0.2343}$.

d. The null hypothesis will be rejected if $\frac{18 - \bar{x}_2}{0.2343} > 1.645$, or $\bar{x}_2 < 17.61$. Therefore, in order to say that the rotor spun yarn is weaker than the ring spun yarn the average strength of the rotor spun yarn has to be lower than 17.61 cN/tex. This hypothesis testing is depicted in Figure 7.19.

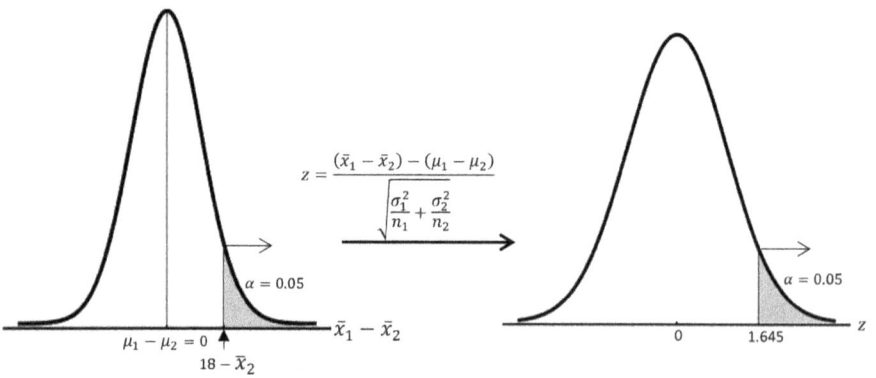

$$z = \frac{(\bar{x}_1 - \bar{x}_2) - (\mu_1 - \mu_2)}{\sqrt{\frac{\sigma_1^2}{n_1} + \frac{\sigma_2^2}{n_2}}}$$

FIGURE 7.19 z-Test of H_0: $\mu_1 - \mu_2 = 0$ against H_1: $\mu_1 - \mu_2 > 0$.

7.5 TESTS CONCERNING THE DIFFERENCE BETWEEN TWO POPULATION MEANS: TWO SMALL SAMPLES AVAILABLE (*t*-Test)

1. Suppose that a random sample of small size n_1 $(n_1 < 30)$ with calculated sample mean $\bar{x}_1$ is drawn from a normal population with mean μ_1 and unknown variance σ_1^2, and also suppose that a random sample of small size n_2 $(n_2 < 30)$ with calculated sample mean $\bar{x}_2$ is drawn from a normal population with mean μ_2 and unknown variance σ_2^2. We will test the null hypothesis that

$$H_0: \mu_1 - \mu_2 = \delta$$

against the alternative hypothesis

$$H_1: \begin{array}{l} \mu_1 - \mu_2 > \delta \text{ or} \\ \mu_1 - \mu_2 < \delta \text{ or} \\ \mu_1 - \mu_2 \neq \delta \end{array}$$

where δ is a constant. Next, we specify the level of significance α. When n_1, n_2 are both less than 30 and σ_1^2, σ_2^2 are unknown, we use *t*-test.

2. If $\bar{X}_1$ and S_1^2 are the mean and the variance of a random sample of small size n_1 drawn from a normal population with mean μ_1 and unknown variance σ_1^2, and $\bar{X}_2$ and S_2^2 are the mean and the variance of a random sample of small size n_2 drawn from a normal population with mean μ_2 and unknown variance σ_2^2, then in order to test the null hypothesis $\mu_1 - \mu_2 = \delta$, it is reasonable to assume that the two populations have a common variance $\sigma_1^2 = \sigma_2^2 = \sigma^2$. Under this condition the random variable $T = \frac{(\bar{X}_1 - \bar{X}_2) - (\mu_1 - \mu_2)}{\hat{\sigma}_{\bar{x}_1 - \bar{x}_2}}$

has the *t*-distribution with $n_1 + n_2 - 2$ degrees of freedom, where $\hat{\sigma}_{\bar{x}_1 - \bar{x}_2}$ is the square root of an estimate of the variance $\sigma_{\bar{x}_1 - \bar{x}_2}^2 = \sigma^2 \left(\frac{1}{n_1} + \frac{1}{n_2} \right)$. We estimate σ^2 by pooling the sum of square deviations from the respective sample means as follows:

$$S_p^2 = \frac{\sum \left(X_1 - \bar{X}_1 \right)^2 + \sum \left(X_2 - \bar{X}_2 \right)^2}{n_1 + n_2 - 2} \tag{7.1}$$

where S_p^2 is the pooled sample variance, $\sum \left(X_1 - \bar{X}_1 \right)^2$ and $\sum \left(X_2 - \bar{X}_2 \right)^2$ are the sum of square deviations from the corresponding sample means for the first and second samples, respectively. As there are $n_1 - 1$ and $n_2 - 1$

independent deviations from the corresponding sample means for the first and second samples, respectively, therefore in order to estimate the population variance we divide total sum of square deviations from the respective means of both the samples by $n_1 + n_2 - 2$. The sample variances S_1^2 and S_2^2 are expressed as follows:

$$S_1^2 = \frac{\sum (X_1 - \bar{X}_1)^2}{n_1 - 1} \tag{7.2}$$

$$S_2^2 = \frac{\sum (X_2 - \bar{X}_2)^2}{n_2 - 1} \tag{7.3}$$

From Equations (7.2) and (7.3) we get
$\sum (X_1 - \bar{X}_1)^2 = (n_1 - 1)S_1^2$ and $\sum (X_2 - \bar{X}_2)^2 = (n_2 - 1)S_2^2$, thus by replacing these expressions into Equation (7.1) we can write

$$S_p^2 = \frac{(n_1 - 1)S_1^2 + (n_2 - 1)S_2^2}{n_1 + n_2 - 2} \tag{7.4}$$

Now by substituting the estimate of σ^2 into the expression for $\sigma_{\bar{X}_1 - \bar{X}_2}^2$ and then by taking the square root we get

$$\hat{\sigma}_{\bar{X}_1 - \bar{X}_2} = S_p \sqrt{\frac{1}{n_1} + \frac{1}{n_2}} \tag{7.5}$$

Thus, we obtain the formula of T as

$$T = \frac{(\bar{X}_1 - \bar{X}_2) - (\mu_1 - \mu_2)}{S_p \sqrt{\frac{1}{n_1} + \frac{1}{n_2}}} \tag{7.6}$$

where

$$S_p = \sqrt{\frac{(n_1 - 1)S_1^2 + (n_2 - 1)S_2^2}{n_1 + n_2 - 2}} \tag{7.7}$$

For a double-tail test, the critical regions are $t > t_{\alpha/2, n_1 + n_2 - 2}$ and $t < -t_{\alpha/2, n_1 + n_2 - 2}$ as depicted in Figure 7.9. In case of a right-tail test, the critical region is $t > t_{\alpha, n_1 + n_2 - 2}$ as shown in Figure 7.10. Similarly for a left-tail test, the critical region is $t < -t_{\alpha, n_1 + n_2 - 2}$ as shown in Figure 7.11.

TABLE 7.4
Decision Rules for Rejecting H_0: $\mu_1 - \mu_2 = \delta$ (Two Small Samples)

Statement	Decision Rule
$H_1: \mu_1 - \mu_2 > \delta$	Do not reject H_0 if $t < t_{\alpha, n_1+n_2-2}$, else reject H_0
$H_1: \mu_1 - \mu_2 < \delta$	Do not reject H_0 if $t > -t_{\alpha, n_1+n_2-2}$, else reject H_0
$H_1: \mu_1 - \mu_2 \neq \delta$	Do not reject H_0 if $-t_{\alpha/2, n_1+n_2-2} < t < t_{\alpha/2, n_1+n_2-2}$, else reject H_0

3. Substituting the values of $\bar{x}_1$, $\bar{x}_2$, $\mu_1 - \mu_2$, s_p, n_1 and n_2, the observed value of test statistic t is calculated from the following equation:

$$t = \frac{(\bar{x}_1 - \bar{x}_2) - (\mu_1 - \mu_2)}{s_p \sqrt{\dfrac{1}{n_1} + \dfrac{1}{n_2}}}$$

4. The decision rules for rejecting H_0 are shown in the Table 7.4.

Example 7.9:

Two cotton mixings are tested for 2.5% span length (mm) in an HVI tester. Five samples were chosen at random from each mixing and the results are as follows:

Mixing I: 30.0, 30.3, 31.0, 30.8, 30.5
Mixing II: 30.8, 31.3, 31.2, 31, 30.5

On this evidence, are the two mixings different in 2.5% span lengths at 0.05 level of significance?

SOLUTION

From the given data we have: $n_1 = n_2 = 5$, $\bar{x}_1 = \frac{\sum_{i=1}^{5} x_{1i}}{5} = 30.52$, $s_1 = \sqrt{\frac{\sum_{i=1}^{5}(x_{1i}-\bar{x}_1)^2}{5-1}} = 0.396$,

$\bar{x}_2 = \frac{\sum_{i=1}^{5} x_{2i}}{5} = 30.96$ and $s_2 = \sqrt{\frac{\sum_{i=1}^{5}(x_{2i}-\bar{x}_2)^2}{5-1}} = 0.321$.

Thus, $s_p = \sqrt{\frac{(n_1-1)s_1^2 + (n_2-1)s_2^2}{n_1+n_2-2}} = \sqrt{\frac{(5-1)0.396^2 + (5-1)0.321^2}{5+5-2}} = 0.36$.

a. H_0: $\mu_1 - \mu_2 = 0$
   H_1: $\mu_1 - \mu_2 \neq 0$ (double-tail test)
   $\alpha = 0.05$.
b. The critical values are $t_{0.025,8} = 2.31$ and $-t_{0.025,8} = -2.31$ (from Table A4).
c. The observed value of test statistic is $t = \frac{(\bar{x}_1 - \bar{x}_2) - (\mu_1 - \mu_2)}{s_p\sqrt{\frac{1}{n_1}+\frac{1}{n_2}}} = \frac{(30.52-30.96)-0}{0.36\sqrt{\frac{1}{5}+\frac{1}{5}}} = -1.93$.
d. As $-2.31 < -1.93 < 2.31$, the null hypothesis cannot be rejected. Therefore, there is no sufficient evidence to conclude that the difference between the 2.5% span lengths of two mixings is significant. Figure 7.20 shows the schematic diagram of this hypothesis testing.

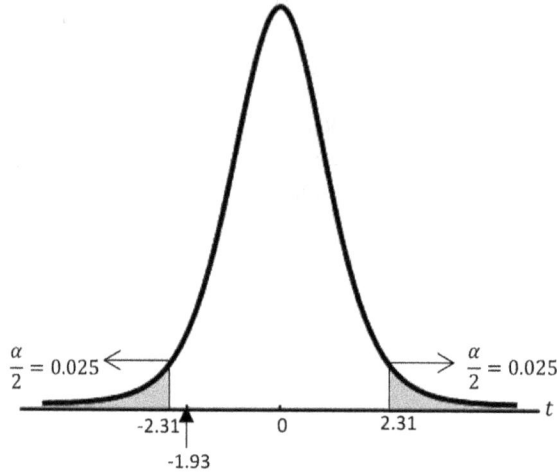

FIGURE 7.20 t-Test of H_0: $\mu_1 - \mu_2 = 0$ against H_1: $\mu_1 - \mu_2 \neq 0$.

Example 7.10:

A modification was made to a filament production process with the objective that the average breaking strain exceeds by at least 2% than the original filament. The breaking strain (%) tests on the modified and original filaments showed the following results:

Modified filament: 17.2, 17.7, 17.5, 17.9, 18.1.
Original filament: 14.1, 14.7, 15.1, 14.3, 15.6, 14.8.

Test whether the difference between the average breaking strains of the modified and original filaments really exceeds by 2% at the 0.05 level of significance.

SOLUTION

From the given data we have: $n_1 = 5$, $n_2 = 6$, $\bar{x}_1 = \frac{\sum_{i=1}^{5} x_{1i}}{5} = 17.68$, $s_1 = \sqrt{\frac{\sum_{i=1}^{5}(x_{1i} - \bar{x}_1)^2}{5-1}} = $

0.349, $\bar{x}_2 = \frac{\sum_{i=1}^{6} x_{2i}}{6} = 14.77$ and $s_2 = \sqrt{\frac{\sum_{i=1}^{6}(x_{2i} - \bar{x}_2)^2}{6-1}} = 0.543$, $s_p = \sqrt{\frac{(n_1-1)s_1^2 + (n_2-1)s_2^2}{n_1+n_2-2}} = $

$\sqrt{\frac{(5-1)0.349^2 + (6-1)0.543^2}{5+6-2}} = 0.467$.

a. H_0: $\mu_1 - \mu_2 = 2$
   H_1: $\mu_1 - \mu_2 > 2$ (right-tail test)
   $\alpha = 0.05$.
b. The critical value is $t_{0.05,9} = 1.83$ (from Table A4).
c. The observed value of test statistic is $t = \frac{(\bar{x}_1 - \bar{x}_2) - (\mu_1 - \mu_2)}{s_p \sqrt{\frac{1}{n_1} + \frac{1}{n_2}}} = \frac{(17.68 - 14.77) - 2}{0.467\sqrt{\frac{1}{5} + \frac{1}{6}}} = 3.23$.
d. As $3.23 > 1.83$, the null hypothesis is rejected. Therefore, there is sufficient evidence to conclude that the difference between the average breaking strains of the modified and original filaments exceeds by 2%. Figure 7.21 depicts the pictorial diagram of this hypothesis testing.

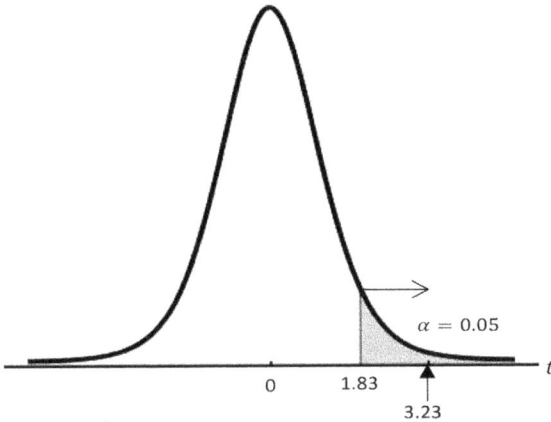

FIGURE 7.21 t-Test of H_0: $\mu_1 - \mu_2 = 2$ against H_1: $\mu_1 - \mu_2 > 2$.

7.6 TESTS CONCERNING SINGLE PROPORTION (z-Test)

1. Suppose that x members of n $(n > 30)$ elements are found to possess some characteristic so that the sample proportion is $\hat{p} = x / n$. If the sample is drawn from a population with assumed mean proportion p_0, we will test the null hypothesis that

H_0: $p = p_0$

against the alternative hypothesis

$$H_1: \begin{array}{l} p > p_0 \ \text{or} \\ p < p_0 \ \text{or} \\ p \neq p_0 \end{array}$$

where p is the true mean proportion of population. We then decide on the level of significance α.

2. If X follows a binomial distribution with parameters n and p_0, then for large n the binomial distribution of X can be approximated with a normal distribution, that is the standardized variable $Z = \frac{X - np_0}{\sqrt{np_0(1-p_0)}}$ will approximately follow a standard normal distribution. Dividing both numerator and denominator by n, we get the expression Z as

$$Z = \frac{\dfrac{X}{n} - p_0}{\sqrt{\dfrac{p_0(1-p_0)}{n}}}$$

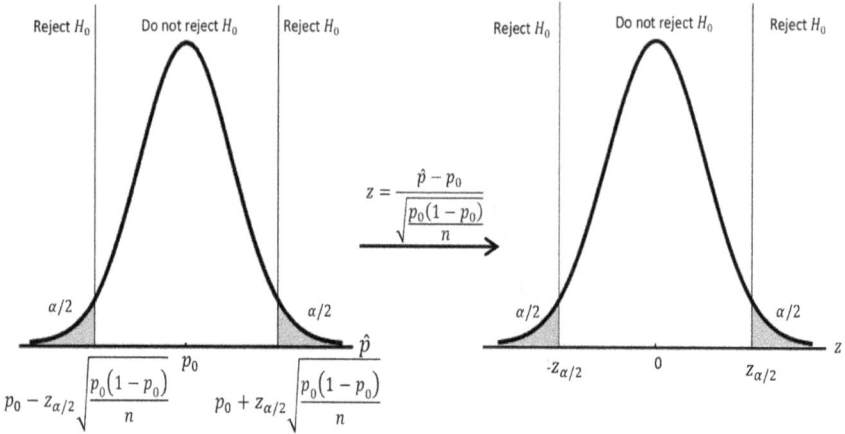

FIGURE 7.22   Critical regions for testing $H_1: p \neq p_0$.

For a double-tail test, the critical regions are $z > z_{\alpha/2}$ and $z < -z_{\alpha/2}$ as depicted in Figure 7.22. In case of a right-tail test, the critical region is $z > z_\alpha$ as shown in Figure 7.23. Similarly for a left-tail test, the critical region is $z < -z_\alpha$ as shown in Figure 7.24.

3. Substituting the values of $\hat{p}$, p_0 and n, the observed value of test statistic z is calculated from the following equation:

$$z = \frac{\hat{p} - p_0}{\sqrt{\dfrac{p_0\left(1 - p_0\right)}{n}}}$$

4. The decision rules for rejecting H_0 are shown in the Table 7.5.

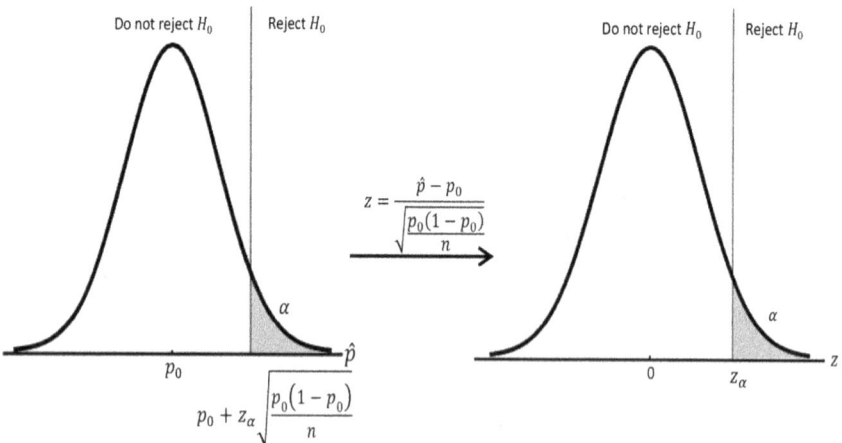

FIGURE 7.23   Critical region for testing $H_1: p > p_0$.

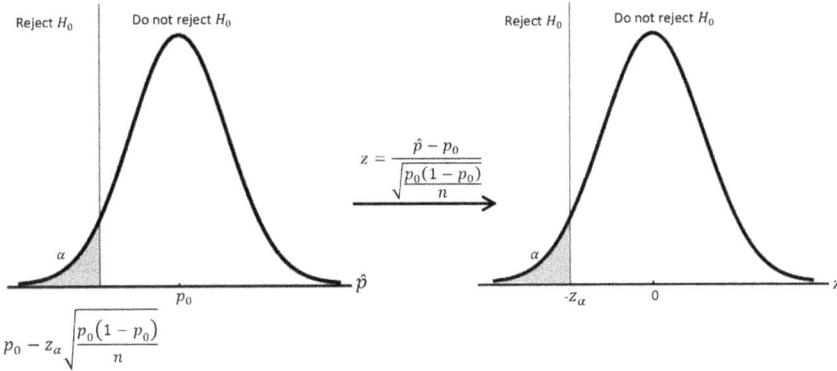

FIGURE 7.24 Critical region for testing $H_1: p < p_0$.

TABLE 7.5
Decision Rules for Rejecting $H_0: p = p_0$

Statement	Decision Rule
$H_1: p > p_0$	Do not reject H_0 if $z < z_\alpha$, else reject H_0
$H_1: p < p_0$	Do not reject H_0 if $z > -z_\alpha$, else reject H_0
$H_1: p \neq p_0$	Do not reject H_0 if $-z_{\alpha/2} < z < z_{\alpha/2}$, else reject H_0

Example 7.11:

In a study designated to investigate whether the piecing workers in a ring frame meet the requirement that at least 90% of yarn piecing works will be perfect at the first attempt. It is observed that 261 out of 300 piecing works are done perfectly at the first attempt. From this result, test whether the piecing workers fail to meet the required standard at the 0.05 level of significance.

SOLUTION

Given data: $p_0 = 0.9$, $n = 300$, $\hat{p} = \frac{261}{300} = 0.87$. As n is large, a normal approximation to the binomial distribution can be made.

    a. $H_0: p = 0.9$
       $H_1: p < 0.9$ (left-tail test)
       $\alpha = 0.05$.
    b. The critical value is $-z_{0.05} = -1.645$ (from Table A1).
    c. The observed value of test statistic is $z = \dfrac{\hat{p} - p_0}{\sqrt{\frac{p_0(1-p_0)}{n}}} = \dfrac{0.87 - 0.9}{\sqrt{\frac{0.9(1-0.9)}{300}}} = -1.732$.

    d. As $-1.732 < -1.645$, the null hypothesis is rejected. Therefore, there is sufficient evidence to conclude that the piecing workers fail to meet the required standard. A schematic representation of this hypothesis testing is shown in Figure 7.25.

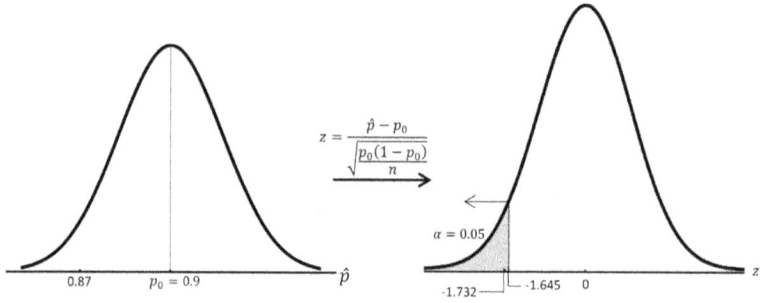

FIGURE 7.25 z-Test of H_0: $p = 0.9$ against H_1: $p < 0.9$.

Example 7.12:

A batch of cotton fibres is expected to contain mature and immature fibres in the ratio of 3:1. An experiment was conducted on 250 fibres from this batch and it was observed that 180 fibres were mature. From this result, test whether the batch of cotton is meeting the required ratio of mature and immature fibres at the 0.01 level of significance.

SOLUTION

Given data: $p_0 = \frac{3}{4} = 0.75$, $n = 250$, $\hat{p} = \frac{180}{250} = 0.72$. As n is large, a normal approximation to the binomial distribution can be made.

a. H_0: $p = 0.75$
   H_1: $p \neq 0.75$ (double-tail test)
   $\alpha = 0.01$.
b. The critical values are $z_{0.005} = 2.58$ and $-z_{0.005} = -2.58$ (from Table A1).
c. The observed value of test statistic is $z = \dfrac{\hat{p} - p_0}{\sqrt{\frac{p_0(1-p_0)}{n}}} = \dfrac{0.72 - 0.75}{\sqrt{\frac{0.75(1-0.75)}{250}}} = -1.095$.
d. As $-2.58 < -1.095 < 2.58$, the null hypothesis cannot be rejected. Therefore, there is no sufficient evidence to conclude that the batch of cotton does not meeting the required ratio of mature and immature fibres. Figure 7.26 shows the pictorial diagram of this hypothesis testing.

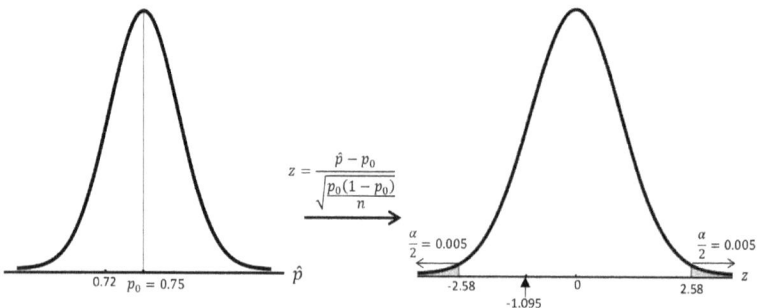

FIGURE 7.26 z-Test of H_0: $p = 0.75$ against H_1: $p \neq 0.75$.

7.7   TESTS CONCERNING DIFFERENCE BETWEEN TWO PROPORTIONS (z-Test)

1. Suppose that n_1 and n_2 are independent random samples of large sizes $(n_1, n_2 > 30)$ drawn from two populations and x_1 and x_2 represent the number elements that possess some characteristic so that the sample proportions are $\hat{p}_1 = x_1/n_1$ and $\hat{p}_2 = x_2/n_2$. Let p_1 and p_2 denote, respectively, the mean proportions of the two populations from which the two samples are drawn. We will test the null hypothesis that

$$H_0: p_1 - p_2 = 0$$

against the alternative hypothesis

$$
\begin{aligned}
& p_1 - p_2 > 0 \text{ or} \\
H_1:\ & p_1 - p_2 < 0 \text{ or} \\
& p_1 - p_2 \neq 0
\end{aligned}
$$

We then decide on the level of significance α.

2. Suppose that X_1 follows a binomial distribution with parameters n_1 and p_1, and X_2 follows another binomial distribution with parameters n_2 and p_2. When n_1 and n_2 are large, then the difference between proportions $\frac{X_1}{n_1} - \frac{X_2}{n_2}$ will be approximately normally distributed with mean $p_1 - p_2$ and variance $\frac{p_1(1-p_1)}{n_1} + \frac{p_2(1-p_2)}{n_2}$. As a result, the standardized variable $Z = \dfrac{\left(\frac{X_1}{n_1} - \frac{X_2}{n_2}\right) - (p_1 - p_2)}{\sqrt{\frac{p_1(1-p_1)}{n_1} + \frac{p_2(1-p_2)}{n_2}}}$ will approximately follow a standard normal distribution. Under the null hypothesis that the mean proportions are equal, let p denotes their common mean proportion, that is $p_1 = p_2 = p$. So, the standardized variable Z becomes

$$
Z = \frac{\dfrac{X_1}{n_1} - \dfrac{X_2}{n_2}}{\sqrt{\overline{p}(1-\overline{p})\left(\dfrac{1}{n_1} + \dfrac{1}{n_2}\right)}} \tag{7.8}
$$

where $\overline{p}$ is the pooled estimator of p. If it is observed that $x_1 + x_2$ number of elements possess the characteristic of interest when we combine the two samples of sizes n_1 and n_2. Thus, $\overline{p}$ can be expressed as

$$
\overline{p} = \frac{x_1 + x_2}{n_1 + n_2} \tag{7.9}
$$

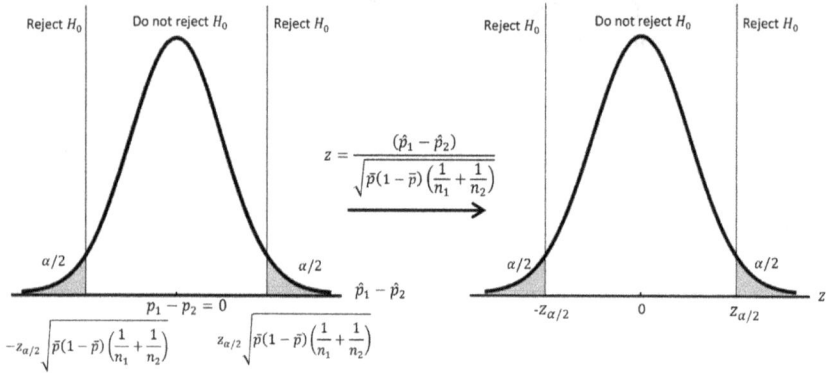

FIGURE 7.27 Critical regions for testing $H_1: p_1 - p_2 \neq 0$.

Replacing x_1 by $n_1 \hat{p}_1$ and x_2 by $n_2 \hat{p}_2$ into Equation (7.9), we get

$$\bar{p} = \frac{n_1 \hat{p}_1 + n_2 \hat{p}_2}{n_1 + n_2} \tag{7.10}$$

For a double-tail test, the critical regions are $z > z_{\alpha/2}$ and $z < -z_{\alpha/2}$ as depicted in Figure 7.27. In case of a right-tail test, the critical region is $z > z_\alpha$ as shown in Figure 7.28. Similarly for a left-tail test, the critical region is $z < -z_\alpha$ as shown in Figure 7.29.

3. Substituting the values of $\hat{p}_1$, $\hat{p}_2$, $\bar{p}$, n_1 and n_2, the observed value of test statistic z is calculated from the following equation:

$$z = \frac{\left(\hat{p}_1 - \hat{p}_2 \right)}{\sqrt{\bar{p}\left(1-\bar{p}\right)\left(\dfrac{1}{n_1} + \dfrac{1}{n_2} \right)}}$$

4. The decision rules for rejecting H_0 are shown in the Table 7.6.

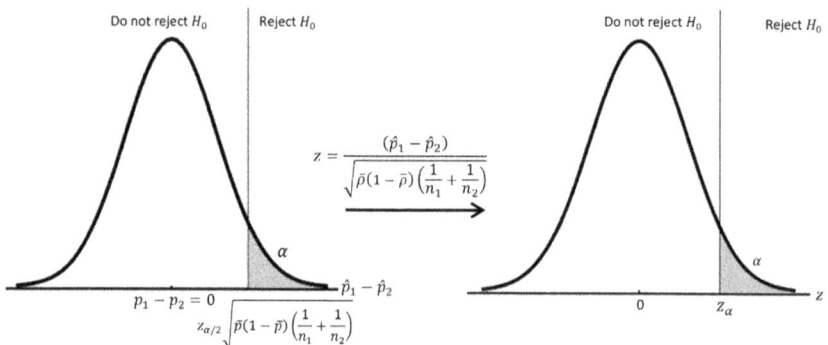

FIGURE 7.28 Critical region for testing $H_1: p_1 - p_2 > 0$.

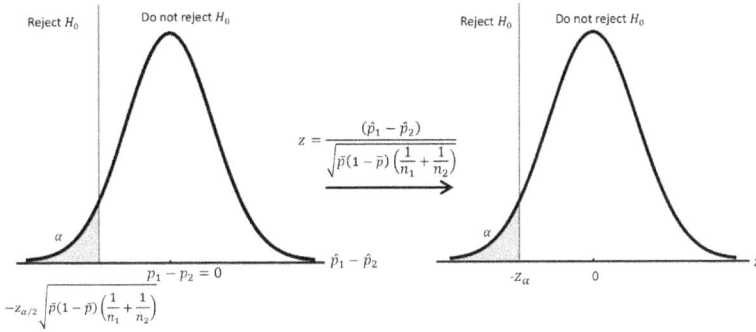

FIGURE 7.29   Critical region for testing $H_1: p_1 - p_2 < 0$.

TABLE 7.6
Decision Rules for Rejecting H_0: $p_1 - p_2 = 0$

Statement	Decision Rule
$H_1: p_1 - p_2 > 0$	Do not reject H_0 if $z < z_\alpha$, else reject H_0
$H_1: p_1 - p_2 < 0$	Do not reject H_0 if $z > -z_\alpha$, else reject H_0
$H_1: p_1 - p_2 \neq 0$	Do not reject H_0 if $-z_{\alpha/2} < z < z_{\alpha/2}$, else reject H_0

Example 7.13:

Two batches of cotton fibres contain the same mix of black and grey fibres. In an experiment, 300 and 325 fibres were selected from the first and second batches, respectively, and it was observed that first batch contains 210 black fibres and second batch contains 220 black fibres. From this result, test whether the two batches of cotton fibres maintain the same ratio of black and grey fibres at the 0.05 level of significance.

SOLUTION

Given data: $n_1 = 300$, $\hat{p}_1 = \frac{90}{300} = 0.3$, $n_2 = 325$, $\hat{p}_2 = \frac{105}{325} = 0.323$, so that the pooled estimator $\bar{p} = \frac{n_1 \hat{p}_1 + n_2 \hat{p}_2}{n_1 + n_2} = \frac{300 \times 0.3 + 325 \times 0.323}{300 + 325} = 0.312$. As both n_1 and n_2 are large, a normal approximation to the binomial distribution can be made.

a. $H_0: p_1 - p_2 = 0$
   $H_1: p_1 - p_2 \neq 0$ (double-tail test)
   $\alpha = 0.05$.
b. The critical values are $z_{0.025} = 1.96$ and $-z_{0.025} = -1.96$ (from Table A1).
c. The observed value of test statistic is $z = \dfrac{(\hat{p}_1 - \hat{p}_2)}{\sqrt{\bar{p}(1-\bar{p})\left(\frac{1}{n_1}+\frac{1}{n_2}\right)}} = \dfrac{0.3 - 0.323}{\sqrt{0.312(1-0.312)\left(\frac{1}{300}+\frac{1}{325}\right)}} =$
   -0.622.
d. As $-1.96 < -0.622 < 1.96$, the null hypothesis cannot be rejected. Therefore, there is no sufficient evidence to conclude that the two batches of cotton fibres do not maintain the same ratio of black and grey fibres. A schematic representation of this hypothesis testing is shown in Figure 7.30.

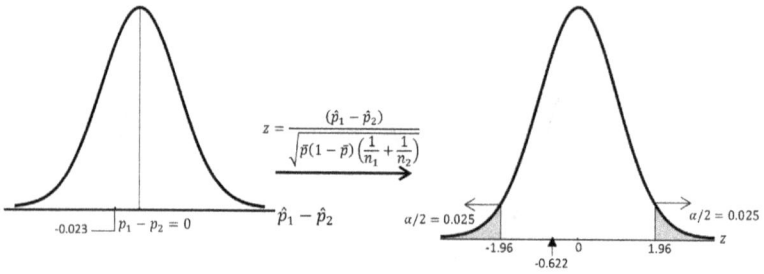

FIGURE 7.30 z-Test of H_0: $p_1 - p_2 = 0$ against H_1: $p_1 - p_2 \neq 0$.

Example 7.14:

Two random samples are taken from two populations of readymade garments which are sold as two different brands. The first sample of size 950 yielded 104 manufacturing defects and the second sample of size 1160 yielded 151 manufacturing defects. From this result, test whether the first brand yields less manufacturing defects than the second brand at the 0.05 level of significance.

SOLUTION

Given data: $n_1 = 950$, $\hat{p}_1 = \frac{104}{950} = 0.1095$, $n_2 = 1160$, $\hat{p}_2 = \frac{151}{1160} = 0.1302$, so that the pooled estimator $\bar{p} = \frac{n_1\hat{p}_1 + n_2\hat{p}_2}{n_1 + n_2} = \frac{950 \times 0.1095 + 1160 \times 0.1302}{950 + 1160} = 0.1209$. As both n_1 and n_2 are large, a normal approximation to the binomial distribution can be made.

a. H_0: $p_1 - p_2 = 0$
   H_1: $p_1 - p_2 < 0$ (left-tail test)
   $\alpha = 0.05$.
b. The critical value is $-z_{0.005} = -1.645$ (from Table A1).
c. The observed value of test statistic is $z = \frac{(\hat{p}_1 - \hat{p}_2)}{\sqrt{\bar{p}(1-\bar{p})\left(\frac{1}{n_1} + \frac{1}{n_2}\right)}} = \frac{0.1095 - 0.1302}{\sqrt{0.1209(1-0.1209)\left(\frac{1}{950} + \frac{1}{1160}\right)}} = -1.451$.

d. As $-1.451 > -1.645$, the null hypothesis cannot be rejected. Therefore, there is no sufficient evidence to conclude that the first brand yields less manufacturing defects than the second brand. Figure 7.31 shows a pictorial representation of this hypothesis testing.

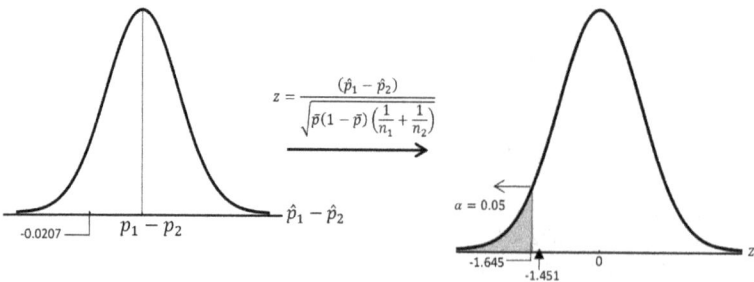

FIGURE 7.31 z-Test of H_0: $p_1 - p_2 = 0$ against H_1: $p_1 - p_2 < 0$.

7.8 TESTS CONCERNING ONE VARIANCE (χ^2-Test)

1. Suppose that s^2 is the value of variance of a random sample of size n drawn from a normal population with assumed variance σ_0^2, we will test the null hypothesis that

$$H_0: \sigma^2 = \sigma_0^2$$

against the alternative hypothesis

$$H_1: \begin{array}{l} \sigma^2 > \sigma_0^2 \text{ or} \\ \sigma^2 < \sigma_0^2 \text{ or} \\ \sigma^2 \neq \sigma_0^2 \end{array}$$

where σ^2 is the true population variance. We then decide on the level of significance α.

2. If $\bar{X}$ and S^2 are the sample mean and sample variance of a random sample of size n, then the random variable $\frac{(n-1)S^2}{\sigma^2}$ follows chi-square distribution with $n-1$ degrees of freedom. For a double-tail test, the critical regions are $\chi^2 > \chi_{\alpha/2,n-1}^2$ and $\chi^2 < \chi_{1-\alpha/2,n-1}^2$ as depicted in Figure 7.32. In case of a right-tail test, the critical region is $\chi^2 > \chi_{\alpha,n-1}^2$ as shown in Figure 7.33. Similarly for a left-tail test, the critical region is $\chi^2 < \chi_{1-\alpha,n-1}^2$ as shown in Figure 7.34. The values of $\chi_{\alpha/2,n-1}^2$ or $\chi_{\alpha,n-1}^2$ can be obtained from the table of chi-square distribution for given degrees of freedom and α (from Table A3).

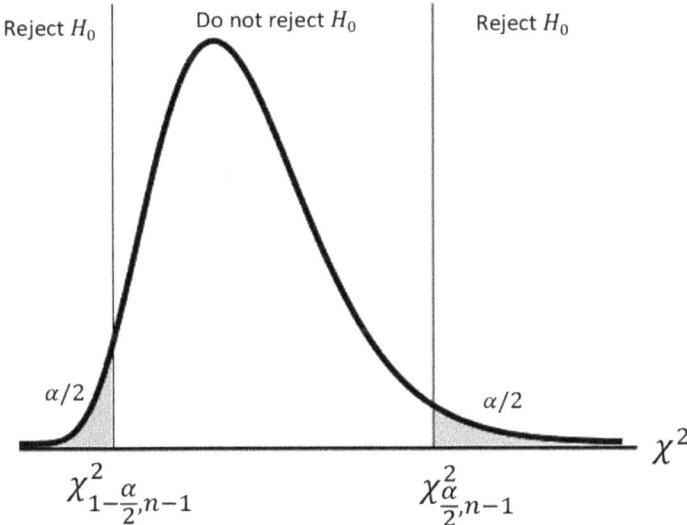

FIGURE 7.32 Critical regions for testing $H_1: \sigma^2 \neq \sigma_0^2$.

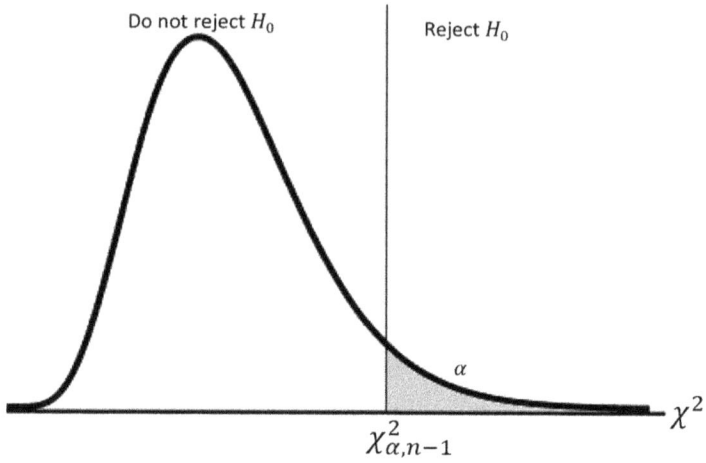

FIGURE 7.33 Critical region for testing $H_1: \sigma^2 > \sigma_0^2$.

3. Substituting the values of s, σ_0 and n, the observed value of test statistic χ^2 is calculated from the following equation:

$$\chi^2 = \frac{(n-1)s^2}{\sigma_0^2}$$

4. The decision rules for rejecting H_0 are shown in the Table 7.7.

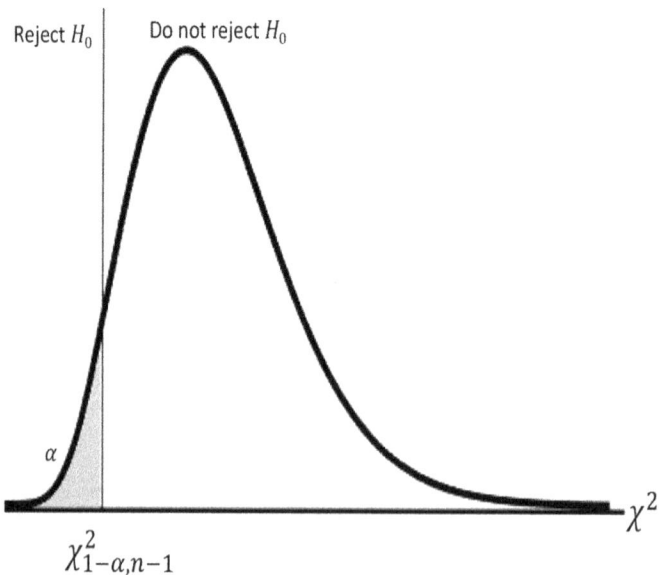

FIGURE 7.34 Critical region for testing $H_1: \sigma^2 < \sigma_0^2$.

TABLE 7.7
Decision Rules for Rejecting H_0: $\sigma^2 = \sigma_0^2$

Statement	Decision Rule
H_1: $\sigma^2 > \sigma_0^2$	Do not reject H_0 if $\chi^2 < \chi^2_{\alpha,n-1}$, else reject H_0
H_1: $\sigma^2 < \sigma_0^2$	Do not reject H_0 if $\chi^2 > \chi^2_{1-\alpha,n-1}$, else reject H_0
H_1: $\sigma^2 \neq \sigma_0^2$	Do not reject H_0 if $\chi^2_{1-\alpha/2,n-1} < \chi^2 < \chi^2_{\alpha/2,n-1}$, else reject H_0

Example 7.15:

The nominal count of the yarn is 20 tex with the standard deviation of 0.8 tex. A sample of 16 leas tested has shown average count of 20.6 tex with standard deviation of 0.92 tex. From the sample results can we say that the yarn has a different count variation than the nominal at the 0.05 level of significance?

SOLUTION

Given data: $\mu_0 = 20$, $\sigma_0 = 0.8$, $n = 16$, $\bar{x} = 20.6$, $s = 0.92$.

a. H_0: $\sigma^2 = 0.64$
   H_1: $\sigma^2 \neq 0.64$ (double-tail test)
   $\alpha = 0.05$.
b. The critical values are $\chi^2_{0.025,15} = 27.49$ and $\chi^2_{0.975,15} = 6.26$ (from Table A3).
c. The observed value of test statistic is $\chi^2 = \frac{(n-1)s^2}{\sigma_0^2} = \frac{(16-1)0.92^2}{0.8^2} = 19.84$.
d. As $6.26 < 19.84 < 27.49$, the null hypothesis cannot be rejected. Therefore, there is no sufficient evidence to conclude that, the yarn count variation is different from the nominal count variation. This hypothesis testing is schematically depicted in Figure 7.35.

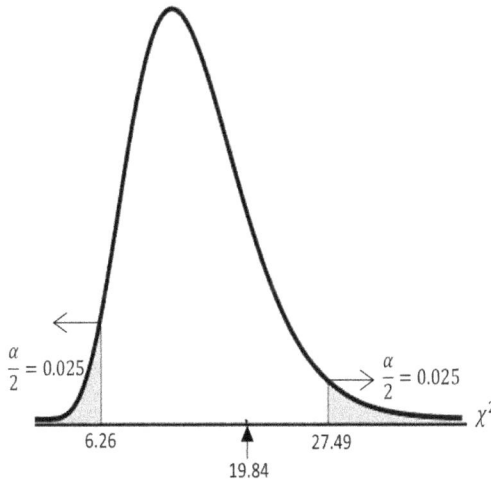

FIGURE 7.35 χ^2-Test of H_0: $\sigma^2 = 0.64$ against H_1: $\sigma^2 \neq 0.64$.

Example 7.16:

In a certain geo-textile application, polypropylene geo-synthetics are acceptable only if the standard deviation of their thicknesses is at most 0.1 mm. If the thicknesses of the 15 samples of a lot of polypropylene geo-synthetics have a standard deviation of 0.135 mm, test whether this lot is accepted at the 0.05 level of significance.

SOLUTION

Given data: $\sigma_0 = 0.1$, $n = 15$, $s = 0.135$.

   a. H_0: $\sigma^2 = 0.01$
      H_1: $\sigma^2 > 0.01$ (right-tail test)
      $\alpha = 0.05$.
   b. The critical value is $\chi^2_{0.05,14} = 23.685$ (from Table A3).
   c. The observed value of test statistic is $\chi^2 = \frac{(n-1)s^2}{\sigma_0^2} = \frac{(15-1)0.135^2}{0.1^2} = 25.52$.
   d. As $25.52 > 23.685$, the null hypothesis is rejected. Therefore, there is sufficient evidence to conclude that the geo-synthetics lot would be rejected. Figure 7.36 shows the pictorial diagram of this hypothesis testing.

Example 7.17:

From the data of Example 7.5, can it be concluded that variance of percentage trash content is significantly lower than 0.05 at the 0.05 level of significance?

SOLUTION

Given data: $n = 8$, $s = 0.212$, $\sigma_0^2 = 0.05$.

   a. H_0: $\sigma^2 = 0.05$
      H_1: $\sigma^2 < 0.05$ (left-tail test)
      $\alpha = 0.05$.

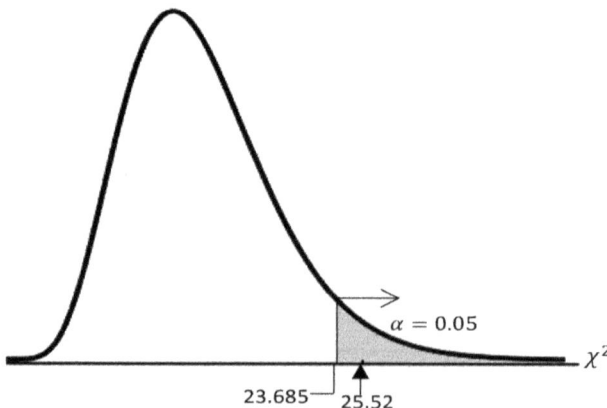

FIGURE 7.36 χ^2-Test of H_0: $\sigma^2 = 0.01$ against H_1: $\sigma^2 > 0.01$.

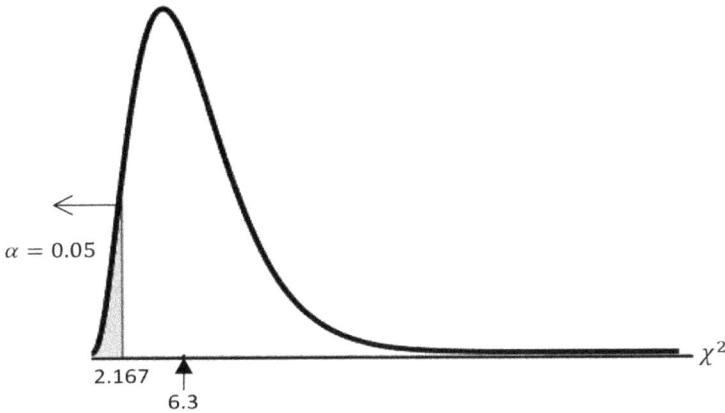

FIGURE 7.37 χ^2-Test of H_0: $\sigma^2 = 0.05$ against H_1: $\sigma^2 < 0.05$.

b. The critical value is $\chi^2_{0.95,7} = 2.167$ (from Table A3).

c. The observed value of test statistic is $\chi^2 = \frac{(n-1)s^2}{\sigma_0^2} = \frac{(8-1)0.212^2}{0.05} = 6.3$.

d. As $6.3 > 2.167$, the null hypothesis cannot be rejected. Therefore, there is no sufficient evidence to conclude that the variance of percentage trash content is significantly lower than 0.05. Figure 7.37 shows the schematic representation of this hypothesis testing.

7.9   TESTS CONCERNING TWO VARIANCES (f-Test)

1. Suppose that two independent random samples of sizes n_1 and n_2 have the values of sample variances s_1^2 and s_2^2, respectively, and the samples are drawn from two normal populations with variances σ_1^2 and σ_2^2, respectively, we will test the null hypothesis that

H_0: $\sigma_1^2 = \sigma_2^2$

against the alternative hypothesis

$$H_1: \begin{array}{l} \sigma_1^2 > \sigma_2^2 \text{ or} \\ \sigma_1^2 < \sigma_2^2 \text{ or} \\ \sigma_1^2 \neq \sigma_2^2 \end{array}$$

We then specify the level of significance α.

2. If S_1^2 and S_2^2 are sample variances of two independent random samples of size n_1 and n_2 from two normal populations with the variances σ_1^2 and σ_2^2, respectively, then $\chi_1^2 = \frac{(n_1-1)S_1^2}{\sigma_1^2}$ and $\chi_2^2 = \frac{(n_2-1)S_2^2}{\sigma_2^2}$ are the two independent random variables having chi-square distributions with (n_1-1) and (n_2-1) degrees of freedom, respectively. In addition, the random variable

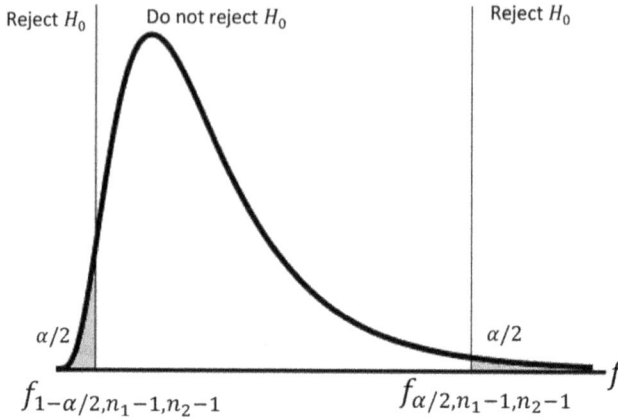

FIGURE 7.38 Critical regions for testing H_1: $\sigma_1^2 \neq \sigma_2^2$.

$F = \dfrac{\chi_1^2/(n_1-1)}{\chi_2^2/(n_2-1)} = \dfrac{S_1^2 \sigma_2^2}{S_2^2 \sigma_1^2}$ follows F-distribution with $(n_1 - 1)$ and $(n_2 - 1)$ degrees of freedom. Under the null hypothesis that $\sigma_1^2 = \sigma_2^2$, $F = \dfrac{S_1^2}{S_2^2}$ is a random variable having F-distribution with $(n_1 - 1)$ and $(n_2 - 1)$ degrees of freedom. For a double-tail test, the critical regions are $f > f_{\alpha/2,n_1-1,n_2-1}$ and $f < f_{1-\alpha/2,n_1-1,n_2-1}$ as depicted in Figure 7.38. In case of a right-tail test, the critical region is $f > f_{\alpha,n_1-1,n_2-1}$ as shown in Figure 7.39. Similarly, for a left-tail test, the critical region is $f < f_{1-\alpha,n_1-1,n_2-1}$ as shown in Figure 7.40. In order to find $f_{1-\alpha,n_1-1,n_2-}$, we simply use the identity $f_{1-\alpha,n_1-1,n_2-1} = 1/f_{\alpha,n_2-1,n_1-1}$. The values of $f_{\alpha/2,n_1-1,n_2-1}$ or f_{α,n_1-1,n_2-1} can be obtained from the table of F-distribution for given degrees of freedom and α (from Table A5).

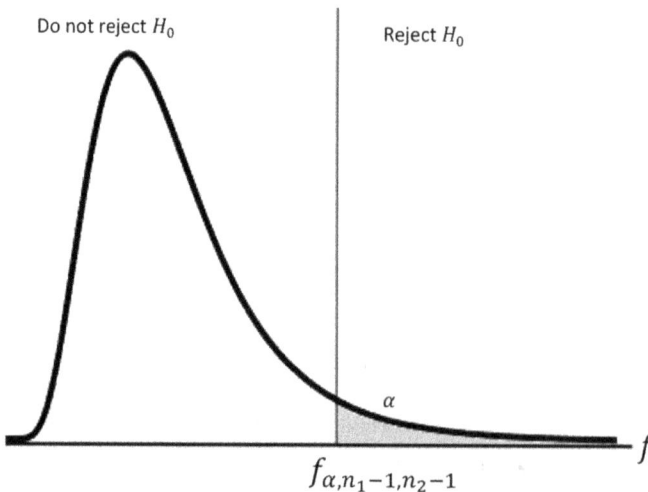

FIGURE 7.39 Critical region for testing H_1: $\sigma_1^2 > \sigma_2^2$.

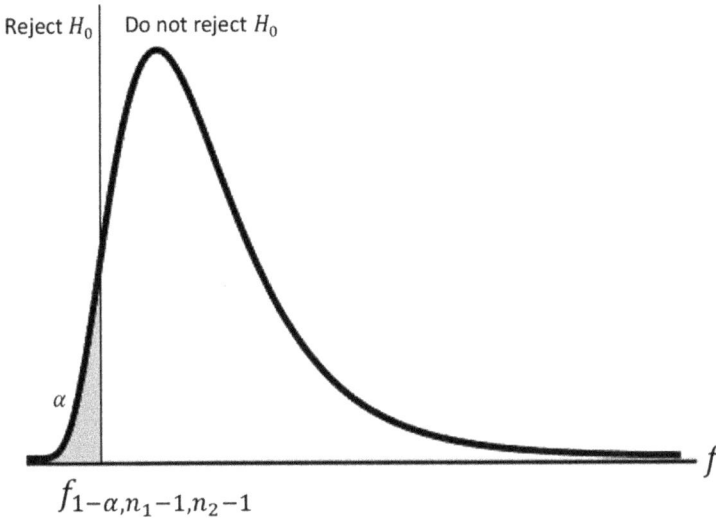

FIGURE 7.40   Critical region for testing $H_1: \sigma_1^2 < \sigma_2^2$.

TABLE 7.8
Decision Rules for Rejecting $H_0: \sigma_1^2 = \sigma_2^2$

Statement	Decision Rule
$H_1: \sigma_1^2 > \sigma_2^2$	Do not reject H_0 if $f < f_{\alpha, n_1-1, n_2-1}$, else reject H_0
$H_1: \sigma_1^2 < \sigma_2^2$	Do not reject H_0 if $f > f_{1-\alpha, n_1-1, n_2-1}$, else reject H_0
$H_1: \sigma_1^2 \neq \sigma_2^2$	Do not reject H_0 if $f_{1-\alpha/2, n_1-1, n_2-1} < f < f_{\alpha/2, n_1-1, n_2-1}$, else reject H_0

---

3. Substituting the values of s_1^2 and s_2^2, the observed value of test statistic f is calculated from the following equation:

$$f = \frac{s_1^2}{s_2^2}$$

4. The decision rules for rejecting H_0 are shown in the Table 7.8.

Example 7.18:

On testing two yarn samples from two different spindles for twist variation, it was observed that the standard deviations of yarn twist for spindle-1 and spindle-2 were 2.85 and 1.78, respectively based on 20 tests for each sample. Can it be concluded that the spindle-1 generates more twist variation than that of spindle-2 at the 0.05 level of significance?

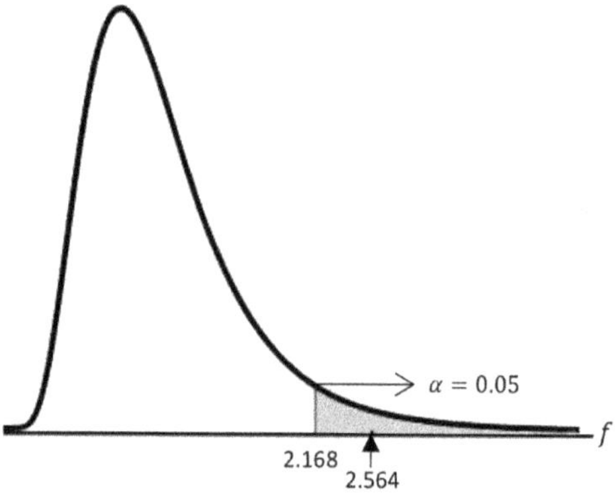

FIGURE 7.41 f-Test of H_0: $\sigma_1^2 = \sigma_2^2$ against H_1: $\sigma_1^2 > \sigma_2^2$.

SOLUTION

Given data: $n_1 = n_2 = 20$, $s_1 = 2.85$, $s_2 = 1.78$.

a. H_0: $\sigma_1^2 = \sigma_2^2$
   H_1: $\sigma_1^2 > \sigma_2^2$ (right-tail test)
   $\alpha = 0.05$.
b. The critical value is $f_{0.05,19,19} = 2.168$ (from Table A5).
c. The observed value of test statistic is $f = \frac{s_1^2}{s_2^2} = \frac{2.85^2}{1.78^2} = 2.564$.
d. As $2.564 > 2.168$, the null hypothesis is rejected. Therefore, there is sufficient evidence to conclude that spindle-1 generates more twist variation than that of spindle-2. This hypothesis testing is schematically shown in Figure 7.41.

Example 7.19:

The strength and its variation of yarns A and B were studied and the following results were obtained:

	Sample Size	Mean Strength (cN/tex)	CV% of Strength
Yarn-A	12	17.8	8.7
Yarn-B	15	20.2	9.8

Can it be concluded that two yarns differ significantly in terms of strength variation at the 0.05 level of significance?

SOLUTION

Given data: $n_1 = 12$, $n_2 = 15$, $\bar{x}_1 = 17.8$, $\bar{x}_2 = 20.2$, $CV_1\% = 8.7$, $CV_2\% = 9.8$.

Thus, $s_1 = \frac{\bar{x}_1 \times CV_1\%}{100} = \frac{17.8 \times 8.7}{100} = 1.549$ and $s_2 = \frac{\bar{x}_2 \times CV_2\%}{100} = \frac{20.2 \times 9.8}{100} = 1.98$.

a. H_0: $\sigma_1^2 = \sigma_2^2$
   H_1: $\sigma_1^2 \neq \sigma_2^2$ (double-tail test)
   $\alpha = 0.05$.

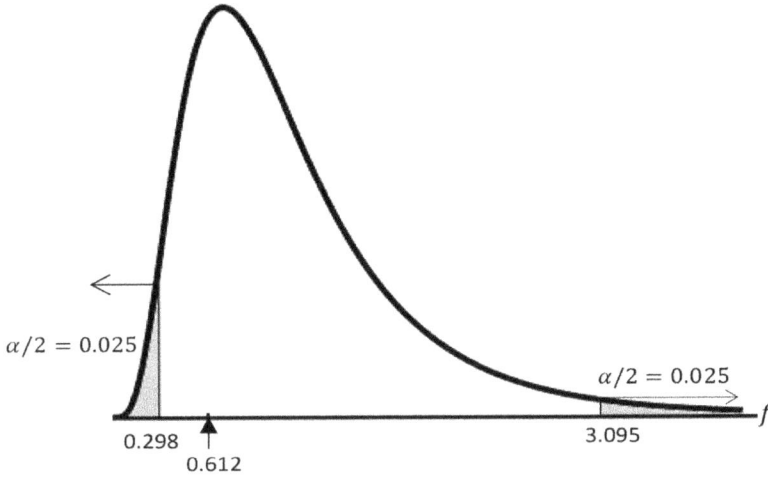

FIGURE 7.42 f-Test of H_0: $\sigma_1^2 = \sigma_2^2$ against H_1: $\sigma_1^2 \neq \sigma_2^2$.

b. The critical values are $f_{0.025,11,14} = 3.095$ and $f_{0.975,11,14} = 0.298$ (from Table A5).

c. The observed value of test statistic is $f = \frac{s_1^2}{s_2^2} = \frac{1.549^2}{1.98^2} = 0.612$.

d. As $0.298 < 0.612 < 3.095$, the null hypothesis cannot be rejected. Therefore, there is no sufficient evidence to conclude that two yarns differ significantly in terms of strength variation. This hypothesis testing is schematically depicted in Figure 7.42.

Example 7.20:

From the data of Example 7.9, can it be concluded that first mixing has higher span length variation than that of second mixing at the 0.05 level of significance?

SOLUTION

Given data: $n_1 = n_2 = 5$, $\bar{x}_1 = 30.52$, $s_1 = 0.396$, $\bar{x}_2 = 30.96$ and $s_2 = 0.321$.

a. H_0: $\sigma_1^2 = \sigma_2^2$
   H_1: $\sigma_1^2 > \sigma_2^2$ (right-tail test)
   $\alpha = 0.05$.

b. The critical value is $f_{0.05,4,4} = 6.39$ (from Table A5).

c. The observed value of test statistic is $f = \frac{s_1^2}{s_2^2} = \frac{0.396^2}{0.321^2} = 1.53$.

d. As $1.53 < 6.39$, the null hypothesis cannot be rejected. Therefore, there is no sufficient evident to conclude that first mixing has higher span length variation than that of second mixing. This hypothesis testing is represented in Figure 7.43.

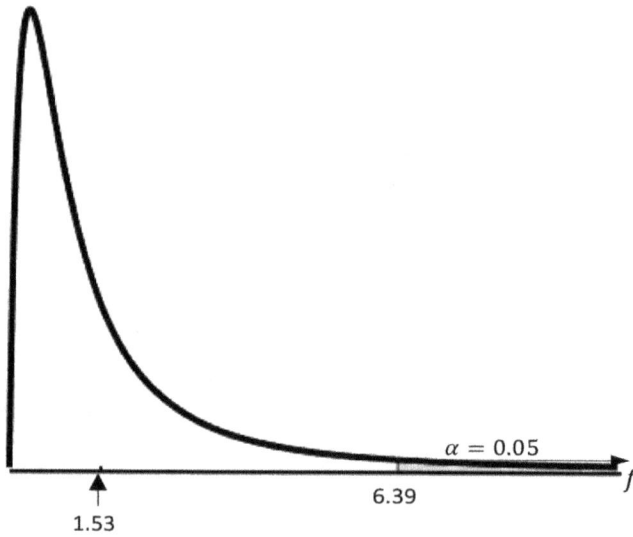

FIGURE 7.43 f-Test of H_0: $\sigma_1^2 = \sigma_2^2$ against H_1: $\sigma_1^2 > \sigma_2^2$.

7.10 TESTS CONCERNING EXPECTED AND OBSERVED FREQUENCIES (χ^2-Test)

1. Suppose that we have k mutually exclusive classes and the values of observed and expected frequencies of ith class are o_i and e_i, respectively, we will test the null hypothesis that
   H_0: There are no differences in observed frequencies and expected frequencies against the alternative hypothesis
   H_1: There are differences in observed frequencies and expected frequencies.
   We then decide on the level of significance α.

2. The observed frequency of ith class O_i is a random variable having a Poisson distribution with mean and variance both are equal to E_i. If the mean is large enough, the Poisson distribution approaches to normal distribution and under this condition the standardized variable $Z_i = \frac{O_i - E_i}{\sqrt{E_i}}$ has the standard normal distribution. Therefore, the random variable $\sum_{i=1}^{k} Z_i^2 = \sum_{i=1}^{k} \frac{(O_i - E_i)^2}{E_i}$ has a chi-square distribution. In this case, the values of Z_i are not all independent, because $\sum_{i=1}^{k} (O_i - E_i) = 0$, thus if $(k-1)$ values of Z_i are known, the last one can be determined. Consequently, the random variable $\sum_{i=1}^{k} \frac{(O_i - E_i)^2}{E_i}$ has $(k-1)$ degrees of freedom. A small value of χ^2 indicates that the observed frequencies are close to the corresponding expected frequencies and it leads to the acceptance of H_0. On the other hand, if observed frequencies differ substantially from the expected frequencies, χ^2 value would be large and it will lead to the rejection of H_0. Therefore, the critical region falls in the right tail of the chi-square distribution; hence, it is always a right-tail test. In this case the critical region is $\chi^2 > \chi^2_{\alpha, k-1}$ as shown in Figure 7.44.

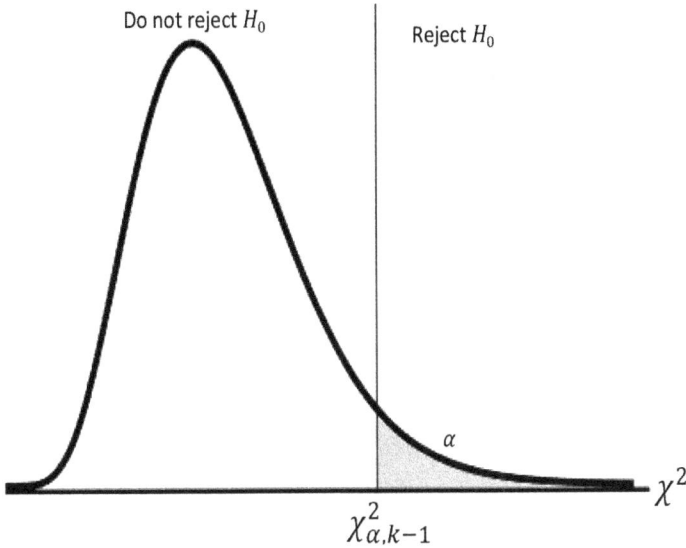

FIGURE 7.44 Critical region for right-tail χ^2-test.

3. Substituting o_i, e_i and k, the observed value of test statistic χ^2 is calculated from the following equation:

$$\chi^2 = \sum_{i=1}^{k} \frac{(o_i - e_i)^2}{e_i}$$

4. The decision rule is do not reject H_0 if $\chi^2 < \chi^2_{\alpha,k-1}$, else reject H_0.

Example 7.21:

The number of neps per Km length of similar yarn spun on five different ring frames was counted, with the following results.

Ring Frame Number	1	2	3	4	5
Number of Neps/Km	104	78	96	85	112

Do these data suggest that there were significant differences among the machines at the 0.05 level of significance?

SOLUTION

Given data: $k = 5$, $o_1 = 104$, $o_2 = 78$, $o_3 = 96$, $o_4 = 85$ and $o_5 = 112$.

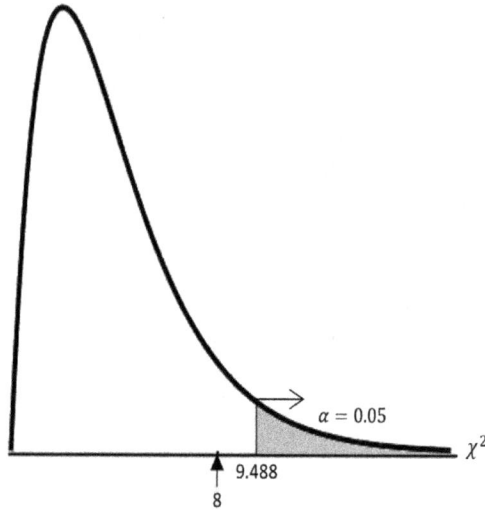

FIGURE 7.45 χ^2-Test of verifying significance differences in observed and expected frequencies.

The expected frequency of each frame can be estimated as the average of these observed frequencies. Hence the estimate of the expected frequency of each frame is $e_i = \frac{104+78+96+85+112}{5} = 95$.

a. H_0: There are no differences in observed frequencies and expected frequencies.
H_1: There are differences in observed frequencies and expected frequencies.
$\alpha = 0.05$.
b. The critical value is $\chi^2_{0.05,4} = 9.488$ (from Table A3).
c. The observed value of test statistic is $\chi^2 = \sum_{i=1}^{5} \frac{(o_i - e_i)^2}{e_i} = \frac{(104-95)^2}{95} + \frac{(78-95)^2}{95} + \frac{(96-95)^2}{95} + \frac{(85-95)^2}{95} + \frac{(112-95)^2}{95} = 8$.
d. As $8 < 9.488$, the null hypothesis cannot be rejected. Therefore, there is no sufficient evidence to conclude that there were significant differences among the machines in terms of generating neps. Figure 7.45 shows the schematic diagram of this hypothesis testing.

Example 7.22:

Magenta, violet and green fibres are nominally mixed in the proportions 5:6:9 to form a certain mélange blend. A sample of 500 fibres chosen at random from a large batch of the mixture was examined and the number of fibres of each colour counted as given below:

Colour	Magenta	Violet	Green
Number of Fibres	138	142	220

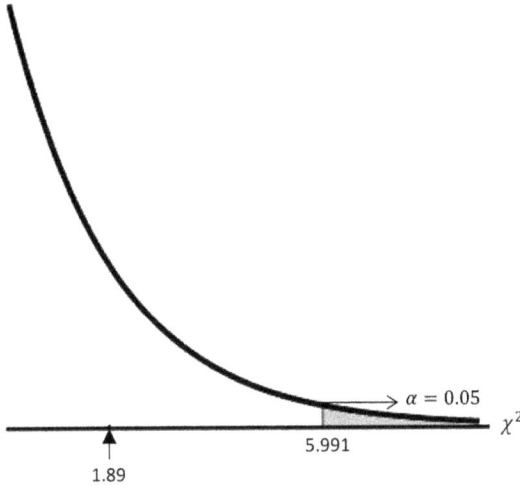

FIGURE 7.46 χ^2-Test of verifying significance differences in observed and expected frequencies.

Did the proportion differ significantly from the nominal at the 0.05 level of significance?

SOLUTION

Given data: $k = 3$, $o_1 = 138$, $o_2 = 142$, $o_3 = 220$, $e_1 = \frac{5}{20} \times 500 = 125$, $e_2 = \frac{6}{20} \times 500 = 150$, $e_3 = \frac{9}{20} \times 500 = 225$.

a. H_0: There are no differences in observed frequencies and expected frequencies.
   H_1: There are differences in observed frequencies and expected frequencies.
   $\alpha = 0.05$.
b. The critical value is $\chi^2_{0.05,2} = 5.991$ (from Table A3).
c. The observed value of the test statistic is $\chi^2 = \sum_{i=1}^{3} \frac{(o_i - e_i)^2}{e_i} = \frac{(138-125)^2}{125} +$
   $\frac{(142-150)^2}{150} + \frac{(220-225)^2}{225} = 1.89$.
d. As $1.89 < 5.991$, the null hypothesis cannot be rejected. Therefore, there is no sufficient evidence to conclude that there were significant differences among the proportions of three fibres. A pictorial representation of this hypothesis testing is depicted in Figure 7.46.

7.11 TESTS CONCERNING THE INDEPENDENCE OF CATEGORICAL DATA IN A CONTINGENCY TABLE (χ^2-Test)

7.11.1 $r \times c$ CONTINGENCY TABLE

Suppose that the large sample of size n is categorized into c classes by first criterion and r classes by second criterion as shown in Table 7.9. Let us denote the observed frequency of the cell belonging to ith row and jth column by f_{ij}, sum of the

TABLE 7.9

$r \times c$ **Contingency Table**

		First Criterion						
		1	2	...	j	...	c	Total
Second Criterion	1	f_{11}	f_{12}	...	f_{1j}	...	f_{1c}	$f_{1.}$
	2	f_{21}	f_{22}	...	f_{2j}	...	f_{2c}	$f_{2.}$
	$\vdots$	$\vdots$	$\vdots$		$\vdots$		$\vdots$	$\vdots$
	i	f_{i1}	f_{i2}	...	f_{ij}	...	f_{ic}	$f_{i.}$
	$\vdots$	$\vdots$	$\vdots$		$\vdots$		$\vdots$	$\vdots$
	r	f_{r1}	f_{r2}	...	f_{rj}	...	f_{rc}	$f_{r.}$
	Total	$f_{.1}$	$f_{.2}$	...	$f_{.j}$	...	$f_{.c}$	n

frequencies of ith row by $f_{i.}$, sum of the frequencies of the jth column by $f_{.j}$ and the sum of all cell frequencies by n. Our objective is to test the dependency or contingency between two criteria of classification. A contingency table with r rows and c columns is referred to as $r \times c$ table.

1. Suppose that p_{ij} is the probability that an item occupies the cell belonging to ith row and jth column, $p_{i.}$ is the probability that an item falls into ith row and $p_{.j}$ is the probability that an item falls into jth column, we will test the null hypothesis that two criteria of classification are independent, that is

H_0: $p_{ij} = p_{i.} \cdot p_{.j}$

against the alternative hypothesis that two criteria of classification are dependent, that is

H_1: $p_{ij} \neq p_{i.} \cdot p_{.j}$

We then decide on the level of significance α.
2. The probabilities $p_{i.}$ and $p_{.j}$ are estimated as follows

$$\hat{p}_{i.} = \frac{f_{i.}}{n} \text{ and } \hat{p}_{.j} = \frac{f_{.j}}{n}$$

where $\hat{p}_{i.}$ and $\hat{p}_{.j}$ are the estimates of $p_{i.}$ and $p_{.j}$, respectively. If e_{ij} denotes the expected frequency of ith row and jth column, under the null hypothesis of independence we can obtain e_{ij} as follows

$$e_{ij} = \hat{p}_{i.} \cdot \hat{p}_{.j} \cdot n = \frac{f_{i.}}{n} \cdot \frac{f_{.j}}{n} \cdot n = \frac{f_{i.} \cdot f_{.j}}{n}$$

Hence, e_{ij} is obtained as follows

$$e_{ij} = \frac{(i\text{th row total}) \times (j\text{th column total})}{\text{grand total}}$$

The total of observed frequencies and the total of expected frequencies are same for each row and each column, therefore, $(r-1)(c-1)$ of the e_{ij} have to be calculated directly, while the others can be calculated by subtraction from appropriate row or column.

3. The observed value of the test statistic can be calculated from the following equation

$$\chi^2 = \sum_{i=1}^{r} \sum_{j=1}^{c} \frac{\left(f_{ij} - e_{ij}\right)^2}{e_{ij}}$$

where χ^2 is a value of a random variable having approximately chi-square distribution with $(r-1)(c-1)$ degrees of freedom under the condition that none of the e_{ij} is less than 5.

4. The decision rule is do not reject H_0 if $\chi^2 < \chi^2_{\alpha,(r-1)(c-1)}$, else reject H_0.

Example 7.23:

Yarn samples of three ring-frame machines are tested for the numbers of thick places, thin places and neps per Km length. The following are the test results:

	Machine-1	Machine-2	Machine-3
Thick places	88	80	74
Thin places	10	13	8
Neps	108	98	89

Test at the 0.05 level of significance whether the thick places, thin places and neps are independent of the machines.

SOLUTION

From the given data: $f_{11} = 88$, $f_{12} = 80$, $f_{13} = 74$, $f_{21} = 10$, $f_{22} = 13$, $f_{23} = 8$, $f_{31} = 108$, $f_{32} = 98$, $f_{33} = 89$, $f_{1.} = 242$, $f_{2.} = 31$, $f_{3.} = 295$, $f_{.1} = 206$, $f_{.2} = 191$, $f_{.3} = 171$, $n = 568$.

Thus the expected frequencies are $e_{11} = \frac{242 \times 206}{568} = 87.77$, $e_{12} = \frac{242 \times 191}{568} = 81.38$, $e_{13} = \frac{242 \times 171}{568} = 72.86$, $e_{21} = \frac{31 \times 206}{568} = 11.24$, $e_{22} = \frac{31 \times 191}{568} = 10.42$, $e_{23} = \frac{31 \times 171}{568} = 9.33$, $e_{31} = \frac{295 \times 206}{568} = 106.99$, $e_{32} = \frac{295 \times 191}{568} = 99.2$, $e_{33} = \frac{295 \times 171}{568} = 88.81$.

1. H_0: The thick places, thin places and neps are independent of the machines.
   H_1: The thick places, thin places and neps are dependent of the machines.
   $\alpha = 0.05$.
2. The critical value is $\chi^2_{0.05,4} = 9.488$ (from Table A3).

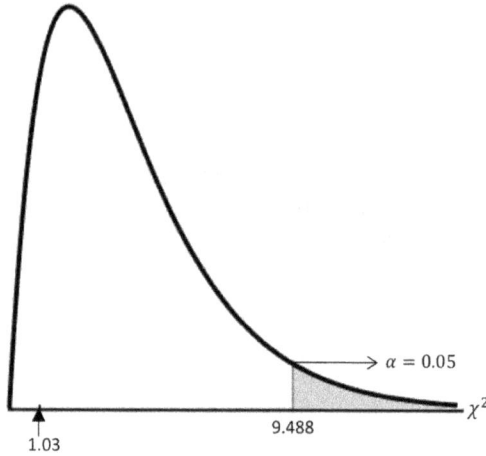

FIGURE 7.47 χ^2-Test of verifying the dependency between two criteria of classification in a $r \times c$ contingency table.

3. The observed value of test statistic is $\chi^2 = \sum_{i=1}^{3}\sum_{j=1}^{3}\frac{(f_{ij}-e_{ij})^2}{e_{ij}} = \frac{(88-87.77)^2}{87.77} +$

$\frac{(80-81.38)^2}{81.38} + \frac{(74-72.86)^2}{72.86} + \frac{(10-11.24)^2}{11.24} + \frac{(13-10.42)^2}{10.42} + \frac{(8-9.33)^2}{9.33} + \frac{(108-106.99)^2}{106.99} + \frac{(98-99.2)^2}{99.2} +$

$\frac{(89-88.81)^2}{88.81} = 1.03.$

4. As $1.03 < 9.488$, the null hypothesis cannot be rejected. Therefore, there is no sufficient evidence to conclude that the thick places, thin places and neps are dependent of the machines. Figure 7.47 shows a pictorial representation of this hypothesis testing.

7.11.2   2 × 2 CONTINGENCY TABLE

A contingency table with 2 rows and 2 columns is referred to as 2×2 contingency table as shown in Table 7.10, where a sample of size n has been classified according to two criteria each having two classes. It is a special case of $r \times c$ table where both r and c are equal to 2.

Table 7.10 shows the sample frequencies and their totals in a 2×2 contingency table. Hence, the observed value of test statistic χ^2 can be calculated as follows:

$$\chi^2 = \frac{\left\{a-\dfrac{(a+b)(a+c)}{n}\right\}^2}{\dfrac{(a+b)(a+c)}{n}} + \frac{\left\{b-\dfrac{(a+b)(b+d)}{n}\right\}^2}{\dfrac{(a+b)(b+d)}{n}} + \frac{\left\{c-\dfrac{(c+d)(a+c)}{n}\right\}^2}{\dfrac{(c+d)(a+c)}{n}}$$

$$+ \frac{\left\{d-\dfrac{(c+d)(b+d)}{n}\right\}^2}{\dfrac{(c+d)(b+d)}{n}}$$

TABLE 7.10

2 × 2 Contingency Table

		First Criterion		
		1	2	Total
Second	1	a	b	$a+b$
Criterion	2	c	d	$c+d$
	Total	$a+c$	$b+d$	$n = a+b+c+d$

It can be shown that square of the difference between any observed frequency and the corresponding expected frequency is same for all four cells in a 2×2 contingency table as expressed below.

$$\left\{ a - \frac{(a+b)(a+c)}{n} \right\}^2 = \left\{ b - \frac{(a+b)(b+d)}{n} \right\}^2 = \left\{ c - \frac{(c+d)(a+c)}{n} \right\}^2$$

$$= \left\{ d - \frac{(c+d)(b+d)}{n} \right\}^2$$

Therefore, the observed value of test statistic becomes

$$\chi^2 = \left\{ a - \frac{(a+b)(a+c)}{n} \right\}^2 \left\{ \frac{1}{\frac{(a+b)(a+c)}{n}} + \frac{1}{\frac{(a+b)(b+d)}{n}} + \frac{1}{\frac{(c+d)(a+c)}{n}} + \frac{1}{\frac{(c+d)(b+d)}{n}} \right\}$$

$$= \frac{1}{n} \left\{ na - (a+b)(a+c) \right\}^2 \left\{ \frac{1}{(a+b)(a+c)} + \frac{1}{(a+b)(b+d)} + \frac{1}{(c+d)(a+c)} + \frac{1}{(c+d)(b+d)} \right\}$$

$$= \frac{1}{n} \left\{ na - (a^2 + ab + ac + bc) \right\}^2 \left\{ \frac{(b+d)(c+d) + (a+c)(c+d) + (a+b)(b+d) + (a+b)(a+c)}{(a+b)(a+c)(b+d)(c+d)} \right\}$$

$$= \frac{1}{n} \left\{ na - a(a+b+c) - bc \right\}^2 \left\{ \frac{(c+d)(a+b+c+d) + (a+b)(a+b+c+d)}{(a+b)(a+c)(b+d)(c+d)} \right\}$$

$$= \frac{1}{n} \left\{ na - a(n-d) - bc \right\}^2 \left\{ \frac{(a+b+c+d)(a+b+c+d)}{(a+b)(a+c)(b+d)(c+d)} \right\}$$

$$= \frac{1}{n} (ad - bc)^2 \left\{ \frac{n^2}{(a+b)(a+c)(b+d)(c+d)} \right\}$$

$$= \frac{n(ad - bc)^2}{(a+b)(a+c)(b+d)(c+d)}$$

This test statistic is compared with a value obtained from the table of chi-square distribution with 1 degree of freedom (from Table A3). However, if any of the expected frequencies is less than 5, the Yates' correction is applied and the observed value of test statistic becomes

$$\chi^2 = \frac{n(|ad - bc| - 0.5n)^2}{(a+b)(a+c)(b+d)(c+d)}.$$

Example 7.24:

In a spinning mill two production lines are producing the same cotton yarns. Random samples of cones were examined from each unit and the numbers of defective and non-defective cones were counted with the results shown in the following table.

	Defective	Non-defective
Line I	3	24
Line II	7	40

Test at the 0.05 level of significance whether the numbers of defective cones are dependent on the production lines.

SOLUTION

Given data: $a = 3$, $b = 24$, $c = 7$, $d = 40$, $n = 3 + 24 + 7 + 40 = 74$.

1. H_0: Production lines and defectives/non-defectives patterns are independent.
   H_1: Production lines and defectives/non-defectives patterns are dependent.
   $\alpha = 0.05$.
2. The critical value is $\chi^2_{0.05,1} = 3.841$ (from Table A3).
3. As one of the expected frequencies is less than 5, the corrected observed value of test statistic is $\chi^2 = \frac{n(|ad-bc|-0.5n)^2}{(a+b)(a+c)(b+d)(c+d)} = \frac{74(|3\times40-24\times7|-0.5\times74)^2}{(3+24)(3+7)(24+40)(7+40)} = 0.011$.
4. As $0.011 < 3.841$, the null hypothesis cannot be rejected. Therefore, there is no sufficient evidence to conclude that numbers of defective cones are dependent on the production lines. Figure 7.48 shows a schematic diagram of this hypothesis testing.

7.12 MATLAB® CODING

7.12.1 MATLAB® CODING OF EXAMPLE 7.1

```
clc
clear
close all
n=150;
```

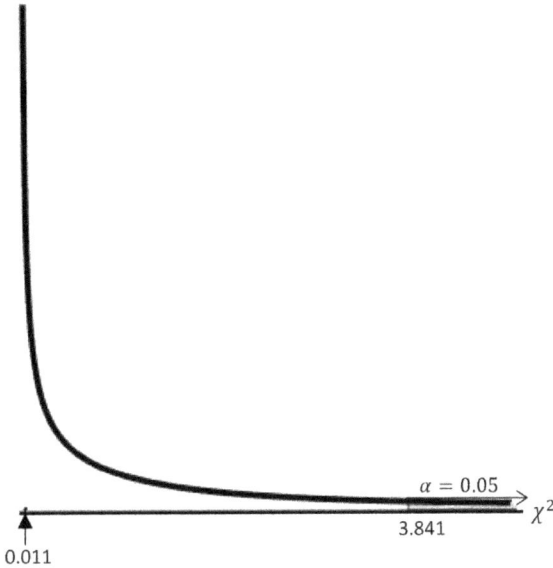

FIGURE 7.48 χ^2-Test of verifying the dependency between two criteria of classification in a 2×2 contingency table.

```
mu=40;
sigma=1.1;
sd=sigma/sqrt(n);
x_bar=40.25;
[h,p,ci,zval] = ztest(x_bar,mu,sd,'Tail','both','Alpha',0.05)
```

7.12.2   MATLAB® CODING OF EXAMPLE 7.2

```
clc
clear
close all
n=50;
mu=260;
sigma=20;
sd=sigma/sqrt(n);
x_bar=266;
[h,p,ci,zval] = ztest(x_bar,mu,sd,'Tail','right','Alpha',0.01)
```

7.12.3   MATLAB® CODING OF EXAMPLE 7.3

```
clc
clear
close all
n=40;
mu=20;
sigma=1.2;
```

```
sd=sigma/sqrt(n);
x_bar=19.65;
[h,p,ci,zval] = ztest(x_bar,mu,sd,'Tail','left','Alpha',0.05)
```

7.12.4   MATLAB® CODING OF EXAMPLE 7.5

```
clc
clear
close all
x=[4.2 4.4 4.1 4.3 3.9 4.3 4.0 3.8];
mu=4;
[h,p,ci,zval] = ttest(x,mu,'Tail','right','Alpha',0.05)
```

7.12.5   MATLAB® CODING OF EXAMPLE 7.6

```
clc
clear
close all
n1=40;
n2=40;
mu=0;
sigma1=0.732;
sigma2=1.188;
sd=sqrt((sigma1^2/n1)+(sigma2^2/n2));
x1_bar=6.1;
x2_bar=6.6;
[h,p,ci,zval] =
ztest(x1_bar-x2_bar,mu,sd,'Tail','left','Alpha',0.05)
```

7.12.6   MATLAB® CODING OF EXAMPLE 7.9

```
clc
clear
close all
x1=[30 30.3 31 30.8 30.5];
x2=[30.8 31.3 31.2 31 30.5];
[h,p,ci,zval] = ttest2(x1,x2,'Tail','both','Alpha',0.05)
```

7.12.7   MATLAB® CODING OF EXAMPLE 7.11

```
clc
clear
close all
n=300;
p0=0.9;
sd=sqrt((p0*(1-p0)/n));
p_cap=261/300;
[h,p,ci,zval] = ztest(p_cap,p0,sd,'Tail','left','Alpha',0.05)
```

7.12.8   MATLAB® CODING OF EXAMPLE 7.13

```
clc
clear
close all
n1=300;
n2=325;
p0=0;
p1_cap=90/300;
p2_cap=105/325;
p_bar=(n1*p1_cap+n2*p2_cap)/(n1+n2);
sd=sqrt(p_bar*(1-p_bar)*((1/n1)+(1/n2)));
[h,p,ci,zval] =
ztest(p1_cap-p2_cap,p0,sd,'Tail','both','Alpha',0.05)
```

7.12.9   MATLAB® CODING OF EXAMPLE 7.17

```
clc
clear
close all
x=[4.2 4.4 4.1 4.3 3.9 4.3 4.0 3.8];
v=0.05;
[h,p,ci,zval] = vartest(x,v,'Tail','left','Alpha',0.05)
```

7.12.10   MATLAB® CODING OF EXAMPLE 7.20

```
clc
clear
close all
x1=[30 30.3 31 30.8 30.5];
x2=[30.8 31.3 31.2 31 30.5];
[h,p,ci,zval] = vartest2(x1,x2,'Tail','right','Alpha',0.05)
```

7.12.11   MATLAB® CODING OF EXAMPLE 7.23

```
clc
clear
close all
x=[88    80   74
   10    13   8
   108   98   89];
[m,n]=size(x);
c_s=sum(x);
r_s=sum(x');
k=1;
for i=1:n
    for j=1:m
        fo(k)=x(j,i);
```

```
        fe(k)=r_s(j)*c_s(i)/sum(r_s);
        k=k+1;
    end
end
frequencies=[fo' fe']
chi_sq=sum((fo-fe).^2./fe)
```

7.12.12   MATLAB® CODING OF EXAMPLE 7.24

```
clc
clear
close all
x=[3     24
   7     40];
a=x(1,1);
b=x(1,2);
c=x(2,1);
d=x(2,2);
n=a+b+c+d;
c_sq_m=n*(abs(a*d-b*c)-0.5*n)^2/((a+b)*(c+d)*(a+c)*(b+d))
```

Exercises

7.1 A comber is nominally producing sliver having linear density of 3.4 ktex. 36 tests are made for sliver linear density. The mean and standard deviation are found to be 3.45 ktex and 0.052 ktex, respectively. Justify whether the mean linear density of the combed sliver is different from the nominal at the 0.05 level of significance.

7.2 A garment manufacturer produces a certain type of garment that should have an average mass of 200 g. Forty randomly selected garments were tested and the mean and standard deviation of the masses were found to be 202 g and 6 g, respectively. Justify whether the mean mass of the garment is more than the expected at the 0.01 level of significance.

7.3 The strength of ring yarn produced from given cotton mixing is 200 gf. Air-jet spun yarn of same count produced from that mixing is claimed to give lower strength. What would be the average strength of the air-jet spun yarn based on 36 tests to say that this system produces a weaker yarn than the ring spinning at the 0.05 level of significance? Assume the standard deviation of strength for both yarns is 40 gf.

7.4 The strength of a 100% cotton yarn is 220 gf. Cotton-polyester blended yarn of the same count is claimed to give higher strength. What would be the average strength of the cotton-polyester blended yarn based on 64 tests to say that it is stronger than the 100% cotton yarn at the 0.05 level of significance? Assume the standard deviation of strength for both yarns is 35 gf.

7.5 Test the hypothesis at the 0.01 level of significance that the random samples 12.1, 12.3, 11.8, 11.9, 12.8 and 12.4 come from a normal population with mean 12.

7.6 A ring frame is nominally spinning 30^s Ne yarn. 15 ring frame bobbins are tested for count. The mean and standard deviation are found to be 30.9 Ne and 2.1 Ne, respectively. Justify whether the mean count of the delivery is too fine at the 0.05 level of significance.

7.7 It is specified that the yarn breaking elongation should not be less than 4%. To test this, 6 yarn samples were subjected for the determination of breaking elongation percentage with the following results: 4.2, 4.1, 3.6, 3.4, 4.2, 3.9.

Do these results tend to confirm that yarn breaking elongation is significantly lower than the specified value at 0.05 level of significance?

7.8 Lea strength tests on 2/60^s Ne (English count) cotton yarns spun from different fibre mixes produces the following results:

Mix	Strength (N)	Standard Deviation (N)	Number of Leas Tested
Mix I	340	47	105
Mix II	328	42	72

Is the difference in yarn strength produced from two mixes significant at 0.01 level of significance?

7.9 The average twist multiplier (TM) of warp yarn exceeds than that of hosiery yarn by 1. If 36 tests on warp yarn had an average TM of 4.45 with a standard deviation of 0.31 and 48 tests on hosiery yarn had an average TM of 3.30 with a standard deviation of 0.24, justify whether the difference between the average TM of two kinds of yarns really exceeds by 1 at the 0.05 level of significance.

7.10 Two cotton mixings are tested for micronaire value in HVI tester. Five samples were chosen at random from each mixing and the results are as follows:
Mixing I: 3.5, 3.7, 3.4, 3.6, 3.4
Mixing II: 3.9, 3.8, 3.7, 3.9, 3.8
On this evidence, are the two mixings different in micronaire values at 0.01 level of significance?

7.11 The breaking length (Km) of air-jet spun yarn is claimed to be higher than equivalent rotor spun yarn by at least 1.5 Km. The breaking length tests on randomly chosen seven samples from air-jet spun yarn and six samples from rotor spun yarn showed the following results:
Air-jet spun yarn: 16.4, 17.3, 16.6, 17.0, 16.5, 16.6, 17.4
Rotor spun yarn: 15.2, 15.7, 15.1, 15.3, 15.6, 15.8.
Test whether the difference between the average breaking lengths of the air-jet and rotor spun yarns really exceeds by 1.5 Km at the 0.05 level of significance.

7.12 Black-dyed cotton and undyed polyester fibres were blended in the ratio 2:1 by number. A yarn cross-section was studied and 115 cotton fibres were found out of 150 fibres. From this result, test whether the yarn is meeting the required ratio of cotton and polyester fibres at the 0.05 level of significance.

7.13 Two batches of cotton fibres were studied for immature fibre content. In an experiment 500 and 525 fibres were selected from the first and second batches, respectively and it was observed that first batch contains 38 immature fibres and second batch contains 45 immature fibres. From this result, test whether the two batches of cotton fibres maintain the same proportion of immature fibres at the 0.05 level of significance.

7.14 The nominal count and its CV of a roving are 0.85^s Ne and 1.5%, respectively. A sample of 12 rovings tested has shown average count of 0.83^s Ne with count CV of 2%. From the sample results can we say that the roving has a higher count variation than the nominal at the 0.01 level of significance?

7.15 In a cotton fibre grading system, the cotton is classified as good quality if the standard deviation of fibre bundle strength is less than 5 cN/tex. If the bundle strengths of the 10 samples of a cotton lot have a standard deviation of 4.7 cN/tex, test whether this lot is accepted as a good quality at the 0.05 level of significance.

7.16 Cleaning efficiencies (%) of two blow room lines were measured over seven days with the following results:

Blow room line I: 40.7, 38.3, 41.9, 39.4, 38.2, 40.1, 41.2
Blow room line II: 37.8, 35.2, 42.7, 43.1, 36.5, 40.6, 35.6

Do these data suggests that there was significant difference between the variation of cleaning efficiencies in two blow room lines at the 0.01 level of significance?

7.17 The waste (%) and its variation of two combers, A and B, were studied and the following results were obtained:

	Number of Readings	Mean Waste (%)	CV% of Waste
Comber-A	20	18.1	6.2
Comber-B	20	16.6	5.3

Can it be concluded that comber-A has more waste variation than comber-B at the 0.05 level of significance?

7.18 The numbers of Imperfections per Km length of similar yarns produced on four different spinning mills were measured with the following results.

Spinning Mill:	1	2	3	4
Imperfections/Km:	119	136	99	108

Do these data suggest that there were significant differences among the spinning mills at the 0.05 level of significance?

7.19 A mill processes cotton through carding machines A and B procured from two different machine manufactures. Neps/100 square inch in card webs were found to be 34 and 53, respectively, for card A and card B. Does card B generate more neps than card A at 0.05 level of significance?

7.20 In a knitting factory, the inspections of fabrics were done to find out the occurrences of different types of fabric defects for four consecutive weeks with the following results:

	1st Week	2nd Week	3rd Week	4th Week
Holes	79	112	86	95
Needle breakages	20	13	8	33
Oil spots	130	166	87	102
Yarn faults	311	435	551	388

Test at the 0.05 level of significance whether the different types of fabric defects are independent of the weeks.

7.21 The following table shows the number of defective and non-defective ready-made denim samples taken before and after the stone wash finish treatment.

	Defective	Non-defective
Before treatment	4	402
After treatment	10	416

Test at the 0.01 level of significance whether the stone wash finish treatment results in a different number of defective items.

7.22 An experiment was carried out to compare the effect of two shrink-resist treatments A and B on wool fibres. 100 fibres from each treated sample were examined. The following table shows the number of damaged and undamaged fibres.

	A	B
Damaged fibres	52	31
Undamaged fibres	48	69

Is the difference between the treatments significant at the 0.05 level of significance?

8 Analysis of Variance

8.1 INTRODUCTION

Analysis of variance (ANOVA) is a statistical methodology for comparing multiple population means across different groups while the number of groups are more than two. It is important to note that analysis of variance is not dealing about the analyzing of population variance, rather it is useful in analyzing population means by identifying the variation present in the observations. In this procedure the total variation is split into individual components that can be ascribed to recognizable sources of variation. These individual components are used for testing the pertinent hypotheses. The causes of variation present in the observation are classified as assignable causes and chance causes. The assignable causes are identifiable and preventable. But the chance causes are neither detectable nor they can be totally eliminated. The variation in the observation due to the chance causes is called the error variation. The hypothesis testing is done by comparing the variation due to assignable causes with the variation due to the chance causes.

There are two classifications of analysis of variation, namely one-way and two-way. One-way analysis of variance compares three or more categorical groups of one factor, whereas two-way analysis of variance compares multiple groups of two factors. Two-way analysis of variance can be with or without replication.

8.2 ONE-WAY ANALYSIS OF VARIANCE

Suppose that there are k number of treatments or groups or manufacturing processes, etc. and an experimenter has the available data of independent random samples of size n from each of the k populations. The experimenter wants to test the hypothesis that the means of these k populations are all equal. Let y_{ij} denote the jth observation from the ith treatment and the schema for one-way of variance is shown in Table 8.1. We use the notation T_i for the total of all observations in the sample from the ith treatment and $\bar{y}_i$ for the mean of all observations in the sample from the ith treatment. Thus, T_i and $\bar{y}_i$ can be expressed as

$$T_i = \sum_{j=1}^{n} y_{ij}$$

$$\bar{y}_i = \frac{T_i}{n}.$$

DOI: 10.1201/9781003081234-8

TABLE 8.1
General Layout of the Experimental Plan for One-Way Analysis of Variance

	Observations					Totals	Means	Sum of Square
1	y_{11}	y_{12}	$\cdots$	y_{1j}	$\cdots$	y_{1n}	T_1	$\bar{y}_1$	$\sum_{j=1}^{n}(y_{1j}-\bar{y}_1)^2$
2	y_{21}	y_{22}	$\cdots$	y_{2j}	$\cdots$	y_{2n}	T_2	$\bar{y}_2$	$\sum_{j=1}^{n}(y_{2j}-\bar{y}_2)^2$
$\vdots$	$\vdots$	$\vdots$		$\vdots$		$\vdots$	$\vdots$	$\vdots$	$\vdots$
i	y_{i1}	y_{i2}	$\cdots$	y_{ij}	$\cdots$	y_{in}	T_i	$\bar{y}_i$	$\sum_{j=1}^{n}(y_{ij}-\bar{y}_i)^2$
$\vdots$	$\vdots$	$\vdots$		$\vdots$		$\vdots$	$\vdots$	$\vdots$	$\vdots$
k	y_{k1}	y_{k2}	$\cdots$	y_{kj}	$\cdots$	y_{kn}	T_k	$\bar{y}_k$	$\sum_{j=1}^{n}(y_{kj}-\bar{y}_k)^2$

(Row label: Treatments)

The grand total T and grand mean $\bar{y}$ of all kn observations are expressed as follows.

$$T = \sum_{i=1}^{k}\sum_{j=1}^{n}y_{ij} = \sum_{i=1}^{k}T_i$$

$$\bar{y} = \frac{T}{kn}.$$

Figure 8.1 depicts a schematic representation of the observations and sample means for k treatments. For ith treatment, the deviation of an observation y_{ij} from the grand mean $\bar{y}$ can be written as

$$y_{ij} - \bar{y} = (\bar{y}_i - \bar{y}) + (y_{ij} - \bar{y}_i) \tag{8.1}$$

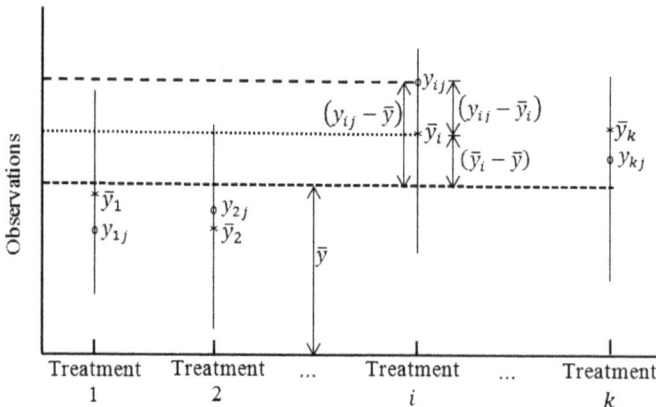

FIGURE 8.1   Schematic representation of the observations and sample means for k treatments.

where $(\bar{y}_i - \bar{y})$ is the effect of ith treatment and $(y_{ij} - \bar{y}_i)$ is the random error associated with the observation y_{ij}. By squaring both sides of Equation (8.1) and summing on i and j, we get

$$\sum_{i=1}^{k}\sum_{j=1}^{n}(y_{ij} - \bar{y})^2 = \sum_{i=1}^{k}\sum_{j=1}^{n}\left\{(\bar{y}_i - \bar{y}) + (y_{ij} - \bar{y}_i)\right\}^2$$

$$= \sum_{i=1}^{k}\sum_{j=1}^{n}(\bar{y}_i - \bar{y})^2 + \sum_{i=1}^{k}\sum_{j=1}^{n}(y_{ij} - \bar{y}_i)^2$$

$$+ 2\sum_{i=1}^{k}\sum_{j=1}^{n}(\bar{y}_i - \bar{y})(y_{ij} - \bar{y}_i)$$

$$= n\sum_{i=1}^{k}(\bar{y}_i - \bar{y})^2 + \sum_{i=1}^{k}\sum_{j=1}^{n}(y_{ij} - \bar{y}_i)^2$$

$$+ 2\sum_{i=1}^{k}(\bar{y}_i - \bar{y})\sum_{j=1}^{n}(y_{ij} - \bar{y}_i).$$

Since $\bar{y}_i$ is the mean of the ith treatment, $\sum_{j=1}^{n}(y_{ij} - \bar{y}_i) = 0$. Thus, it becomes

$$\sum_{i=1}^{k}\sum_{j=1}^{n}(y_{ij} - \bar{y})^2 = n\sum_{i=1}^{k}(\bar{y}_i - \bar{y})^2 + \sum_{i=1}^{k}\sum_{j=1}^{n}(y_{ij} - \bar{y}_i)^2 \qquad (8.2)$$

In Equation (8.2) $\sum_{i=1}^{k}\sum_{j=1}^{n}(y_{ij} - \bar{y})^2$ is the measure of the total variability in the data and it is commonly termed as 'total sum of square' (SST), $n\sum_{i=1}^{k}(\bar{y}_i - \bar{y})^2$ is the 'between treatment sum of square' (SST_r) and $\sum_{i=1}^{k}\sum_{j=1}^{n}(y_{ij} - \bar{y}_i)^2$ is the 'within treatment sum of square' or simply 'error sum of square' (SSE). Symbolically, the sum of square identity can be expressed as

$$SST = SST_r + SSE \qquad (8.3)$$

From Equation (8.1) we can write

$$y_{ij} = \bar{y} + (\bar{y}_i - \bar{y}) + (y_{ij} - \bar{y}_i)$$

$$= \bar{y} + t_i + e_{ij}$$

where t_i denotes the effect of ith treatment and e_{ij} denotes the random error associated with the observation y_{ij}.

Suppose that for each value of i, y_{ij} are the values of a random sample of size n from a normal population with the common variance σ^2. The treatment means $\bar{y}_i$ are the sample estimates of population treatment means μ_i, and the grand mean $\bar{y}$ is the estimate of the overall population mean μ. Therefore, the assumed model of the random variable Y_{ij} is given by

$$Y_{ij} = \mu + (\mu_i - \mu) + (Y_{ij} - \mu_i)$$

$$= \mu + \alpha_i + \varepsilon_{ij}$$

where α_i are the population treatment effects for $i = 1, 2, \cdots, k$ and ε_{ij} are the error terms. The random errors ε_{ij} are assumed to be independent, normally distributed random variables, each with zero mean and variance σ^2. We will test the null hypothesis that population means are all equal; that is

$H_0: \mu_1 = \mu_2 = \cdots = \mu_k$

or, equivalently,

$H_0: \alpha_i = 0$, for $i = 1, 2, \cdots, k$

against the alternative hypothesis that population means are not equal; that is

$H_1: \alpha_i \neq 0$, for at least one value of i.

To test the null hypothesis that the k population means are all equal, the variance σ^2 can be estimated in two independent ways as follows.

The first estimate of σ^2 is based on the variation among the treatment means and it is termed as 'variance between treatments'. If H_0 is true, $\bar{y}_i$ are the values of the independent random variables having identical normal distributions with the mean μ and variance $\frac{\sigma^2}{n}$. Hence, the sample variance $\frac{\sum_{i=1}^{k}(\bar{y}_i - \bar{y})^2}{(k-1)}$ is an estimate of population variance $\frac{\sigma^2}{n}$. Hence, the first estimate of σ^2 is expressed as

$$MST_r = \frac{n \sum_{i=1}^{k}(\bar{y}_i - \bar{y})^2}{(k-1)} = \frac{SST_r}{(k-1)}$$

where MST_r is the 'treatment mean square' and it is also called the 'variance between treatments'.

The second estimate of σ^2 is based on the variation within each treatment and it is termed as 'variance within treatments'. The individual observations

$y_{i1}, y_{i2}, \cdots, y_{ij}, \cdots, y_{in}$ of the ith treatment vary about the mean $\bar{y}_i$ and their sample variance s_i^2 is given by

$$s_i^2 = \frac{\sum_{j=1}^{n}\left(y_{ij} - \bar{y}_i\right)^2}{(n-1)}$$

Thus, for $i = 1, 2, \cdots, k$; we have k number of separate sample variances. Each of these sample variances is an estimate of σ^2 which is assumed constant for all treatments. Thus, the second estimate of σ^2 is obtained by pooling the s_i^2 as follows

$$MSE = \frac{\sum_{i=1}^{k} s_i^2}{k} = \frac{\sum_{i=1}^{k}\sum_{j=1}^{n}\left(y_{ij} - \bar{y}_i\right)^2}{k(n-1)} = \frac{SSE}{k(n-1)}$$

where MSE is the 'error mean square' and it is also called the 'variance within treatments'.

If H_0 is true, $\frac{SST_r}{\sigma^2}$ and $\frac{SSE}{\sigma^2}$ are the values of independent random variables having chi-square distributions with $(k-1)$ and $k(n-1)$ degrees of freedom and the ratio $f = \frac{SST_r/\sigma^2(k-1)}{SSE/\sigma^2 k(n-1)} = \frac{MST_r}{MSE}$ is a value of a random variable having F-distribution with $(k-1)$ and $k(n-1)$ degrees of freedom. However, when H_0 is false, MST_r is expected to exceed MSE. We reject H_0 if the observed value of f exceeds $f_{\alpha,(k-1),k(n-1)}$, where α is the level of significance. Figure 8.2 shows the critical region for rejecting H_0: $\alpha_i = 0$, for $i = 1, 2, \cdots, k$.

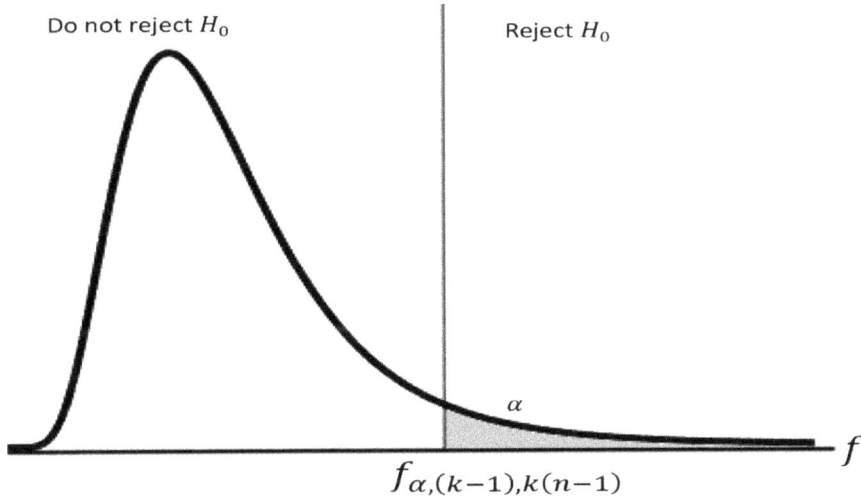

FIGURE 8.2 Critical region for rejecting H_0: $\alpha_i = 0$, for $i = 1, 2, \cdots, k$.

It is required to simplify the sum of squares for the ease of calculation. The short-cut ways of expressing the sum of squares are given below.

$$SST = \sum_{i=1}^{k}\sum_{j=1}^{n}\left(y_{ij} - \bar{y}\right)^2$$

$$= \sum_{i=1}^{k}\sum_{j=1}^{n}y_{ij}^2 + \sum_{i=1}^{k}\sum_{j=1}^{n}\bar{y}^2 - 2\sum_{i=1}^{k}\sum_{j=1}^{n}y_{ij}\bar{y}$$

$$= \sum_{i=1}^{k}\sum_{j=1}^{n}y_{ij}^2 + kn\bar{y}^2 - 2\bar{y}\sum_{i=1}^{k}\sum_{j=1}^{n}y_{ij}$$

$$= \sum_{i=1}^{k}\sum_{j=1}^{n}y_{ij}^2 + kn\left(\frac{T}{kn}\right)^2 - 2\left(\frac{T}{kn}\right)T$$

$$= \sum_{i=1}^{k}\sum_{j=1}^{n}y_{ij}^2 - \frac{T^2}{kn}$$

$$SST_r = n\sum_{i=1}^{k}\left(\bar{y}_i - \bar{y}\right)^2$$

$$= n\left(\sum_{i=1}^{k}\bar{y}_i^2 - k\bar{y}^2\right)$$

$$= n\left(\sum_{i=1}^{k}\frac{T_i^2}{n^2} - \frac{kT^2}{k^2n^2}\right)$$

$$= \frac{1}{n}\sum_{i=1}^{k}T_i^2 - \frac{T^2}{kn}$$

From Equation (8.3) we can write

$$SSE = SST - SST_r$$

$$= \sum_{i=1}^{k}\sum_{j=1}^{n}y_{ij}^2 - \frac{T^2}{kn} - \frac{1}{n}\sum_{i=1}^{k}T_i^2 + \frac{T^2}{kn}$$

$$= \sum_{i=1}^{k}\sum_{j=1}^{n}y_{ij}^2 - \frac{1}{n}\sum_{i=1}^{k}T_i^2$$

TABLE 8.2
One-Way Analysis of Variance Table

Source of Variation	Sum of Square	Degrees of Freedom	Mean Square	f
Treatments	$SST_r = \dfrac{1}{n}\sum_{i=1}^{k} T_i^2 - \dfrac{T^2}{kn}$	$(k-1)$	$MST_r = SST_r / (k-1)$	$\dfrac{MST_r}{MSE}$
Error	$SSE = \sum_{i=1}^{k}\sum_{j=1}^{n} y_{ij}^2 - \dfrac{1}{n}\sum_{i=1}^{k} T_i^2$	$k(n-1)$	$MSE = SSE / k(n-1)$	
Total	$SST = \sum_{i=1}^{k}\sum_{j=1}^{n} y_{ij}^2 - \dfrac{T^2}{kn}$	$kn-1$		

The procedure we have described in this section is called one-way analysis of variance and it is summarized in Table 8.2.

Example 8.1:

The following are the results of whiteness readings of 12 white cloths which were first soiled and then washed with three detergents.

Detergent-A	82	75	74	84
Detergent-B	90	84	87	81
Detergent-C	61	72	64	76

Test at the 0.05 level of significance whether there are significant differences among the detergents.

SOLUTION

From the given data we have: $k = 3$, $n = 4$, $T_1 = 315$, $T_2 = 342$, $T_3 = 273$, $T = 930$, $\sum_{i=1}^{k}\sum_{j=1}^{n} y_{ij}^2 = 72944$.

a. $H_0: \alpha_i = 0$, for $i = 1,2,3$
   $H_1: \alpha_i \neq 0$, for at least one value of i
   $\alpha = 0.05$.
b. The critical value is $f_{0.05,(3-1),3\times(4-1)} = f_{0.05,2,\,9} = 4.3$ (Table A5).
c. SST_r and SSE are calculated as

$$SST = \sum_{i=1}^{k}\sum_{j=1}^{n} y_{ij}^2 - \frac{T^2}{kn} = 72944 - \frac{930^2}{3\times 4} = 869$$

$$SST_r = \frac{1}{n}\sum_{i=1}^{k} T_i^2 - \frac{T^2}{kn} = \frac{1}{4}\left(315^2 + 342^2 + 273^2\right) - \frac{930^2}{3\times 4} = 604.5$$

By subtraction we obtain, $SSE = SST - SST_r = 869 - 604.5 = 264.5$
Therefore,

$$MST_r = \frac{SST_r}{(k-1)} = \frac{604.5}{3-1} = 302.25$$

and

$$MSE = \frac{SSE}{k(n-1)} = \frac{264.5}{3 \times (4-1)} = 29.39$$

Hence, the calculated value of $f = \frac{MST_r}{MSE} = \frac{302.25}{29.39} = 10.28$.
Thus we get the following analysis of variance table.

Source of Variation	Sum of Square	Degrees of Freedom	Mean Square	f
Treatments	604.5	2	302.25	10.28
Error	264.5	9	29.39	
Total	869	11		

d. As $10.28 > 4.3$, the null hypothesis is rejected. Therefore, we do have statistical evidence that there are significant differences among the detergents.

8.2.1 Multiple Comparisons of Treatment Means

The analysis of variance determines whether the differences among the treatment means are statistically significant, however it cannot able to determine which treatment means are different from which others. For this purpose we need to perform multiple comparisons of treatment means. There are several standard methods for making multiple comparisons tests. Tukey's procedure based on studentized range distribution is one of the widely used methods which compares between all pairs of treatment means. In this method the formation of simultaneous $100(1-\alpha)\%$ confidence intervals for all paired comparisons are made. The method of paired comparison are comprised of finding significant difference between treatment means if

$$\left|\bar{y}_h - \bar{y}_i\right| > q_{\alpha,k,v}\sqrt{\frac{MSE}{n}} \qquad (8.4)$$

where h and i are the treatment numbers being compared such that $h \neq i$, k is the total number of treatments, v is the degrees of freedom associated with MSE, n is the sample size of each treatment, α is the level of significance and $q_{\alpha,k,v}$ denotes the studentized range whose value is given in Table A6 (Appendix A). In Equation (8.4) $q_{\alpha,k,v}\sqrt{\frac{MSE}{n}}$ is called the least significant difference.

Example 8.2:

The following are the strength readings (cN/tex) of yarns made from four different spinning technologies.

Ring yarn	20.5	20.7	20.9	20.2	20.6	20.4	21.0
Rotor yarn	19.4	18.6	18.9	19.2	18.8	19.7	18.9
Air-jet yarn	18.9	19.5	20.2	19.8	19.6	20.1	19.2
Friction yarn	17.6	16.9	16.7	17.4	17.5	17.2	16.8

Test at the 0.05 level of significance whether there are significant differences among the strengths of different spun yarns. Also make multiple comparisons of sample means.

SOLUTION

From the given data we have: $k = 4$, $n = 7$, $T_1 = 144.3$, $T_2 = 133.5$, $T_3 = 137.3$, $T_4 = 120.1$, $T = 535.2$, $\sum_{i=1}^{k}\sum_{j=1}^{n} y_{ij}^2 = 10277.72$.

a. $H_0: \alpha_i = 0$, for $i = 1,2,3,4$
   $H_1: \alpha_i \neq 0$, for at least one value of i
   $\alpha = 0.05$.
b. The critical value is $f_{0.05,(4-1),4\times(7-1)} = f_{0.05,3,\ 24} = 3.01$ (Table A5).
c. SST_r and SSE are calculated as

$$SST = \sum_{i=1}^{k}\sum_{j=1}^{n} y_{ij}^2 - \frac{T^2}{kn} = 10277.72 - \frac{535.2^2}{4\times 7} = 47.76$$

$$SST_r = \frac{1}{n}\sum_{i=1}^{k} T_i^2 - \frac{T^2}{kn} = \frac{1}{7}\left(144.3^2 + 133.5^2 + 137.3^2 + 120.1^2\right) - \frac{535.2^2}{4\times 7} = 44.33$$

By subtraction we obtain, $SSE = SST - SST_r = 47.76 - 44.33 = 3.43$
   Therefore,

$$MST_r = \frac{SST_r}{(k-1)} = \frac{44.33}{4-1} = 14.78$$

and

$$MSE = \frac{SSE}{k(n-1)} = \frac{3.43}{4\times(7-1)} = 0.143$$

Hence, the calculated value of $f = \frac{MST_r}{MSE} = \frac{14.78}{0.143} = 103.36$.

Thus we get the following analysis of variance table.

Source of Variation	Sum of Square	Degrees of Freedom	Mean Square	f
Treatments	44.33	3	14.78	103.36
Error	3.43	24	0.143	
Total	47.76	27		

d. As $103.36 > 3.01$, the null hypothesis is rejected. Therefore, there is sufficient statistical evidence that the differences among the strengths of different spun yarns are significant.

For multiple comparisons of sample means, we first calculate the least significant difference. With $\alpha = 0.05$, $k = 4$ and $v = 24$, the value of the studentized range is $q_{0.05,4,24} = 3.9$ (Table A6). Therefore, the least significant difference is $q_{\alpha,k,v}\sqrt{\frac{MSE}{n}} = 3.9\sqrt{\frac{0.143}{7}} = 0.557$. Thus, any absolute difference between the sample means greater than this value is significant at the 0.05 level of significance. Using Tukey's method, all pairs of sample means are compared as shown in the following table.

h	i	$\bar{y}_h$	$\bar{y}_i$	$\lvert\bar{y}_h - \bar{y}_i\rvert$	Is the Difference Significant?
1	2	20.61	19.07	1.54	Yes
1	3	20.61	19.61	1	Yes
1	4	20.61	17.16	3.45	Yes
2	3	19.07	19.61	0.54	No
2	4	19.07	17.16	1.91	Yes
3	4	19.61	17.16	2.46	Yes

The sample means are arranged in descending order as follows.

$\bar{y}_1$	$\bar{y}_3$	$\bar{y}_2$	$\bar{y}_4$
20.61	19.61	19.07	17.16

As far as the strength is concerned, we can conclude from the results that the ring yarn is significantly superior to any of the others, whereas the friction spun yarn is inferior to any of the others. However, there is no significant difference between the strengths of air-jet and rotor yarns.

8.3   TWO-WAY ANALYSIS OF VARIANCE

Two-way analysis of variance is used when we compare the treatment means in the presence of an extraneous variable. For instance, an experiment needs to be carried out to compare the yield of three makes of machines which are running in three

TABLE 8.3
General Layout of the Experimental Plan for Two-Way Analysis of Variance

	Blocks							
	1	2	$\cdots$	j	$\cdots$	b	Totals	Means
Treatments 1	y_{11}	y_{12}	$\cdots$	y_{1j}	$\cdots$	y_{1b}	$T_{1.}$	$\bar{y}_{1.}$
2	y_{21}	y_{22}	$\cdots$	y_{2j}	$\cdots$	y_{2b}	$T_{2.}$	$\bar{y}_{2.}$
$\vdots$	$\vdots$	$\vdots$		$\vdots$		$\vdots$	$\vdots$	$\vdots$
i	y_{i1}	y_{i2}	$\cdots$	y_{ij}	$\cdots$	y_{ib}	$T_{i.}$	$\bar{y}_{i.}$
$\vdots$	$\vdots$	$\vdots$		$\vdots$		$\vdots$	$\vdots$	$\vdots$
a	y_{a1}	y_{a2}	$\cdots$	y_{aj}	$\cdots$	y_{ab}	$T_{a.}$	$\bar{y}_{a.}$
Totals	$T_{.1}$	$T_{.2}$	$\cdots$	$T_{.j}$	$\cdots$	$T_{.b}$	Grand total $= T$	
Means	$\bar{y}_{.1}$	$\bar{y}_{.2}$	$\cdots$	$\bar{y}_{.j}$	$\cdots$	$\bar{y}_{.b}$	Grand mean $= \bar{y}$	

shifts per day. It is known a priori that a machine has different yield in different shifts. We run each machine randomly on each shift to ensure that the comparisons between the machines do not get affected by any difference between shifts. Thus, in this experiment, each shift is referred to as a block.

Suppose that there are a number of treatments which are randomly allocated to b number of blocks. Let y_{ij} denote the observation pertaining to the ith treatment and the jth block and the data is summarized in the $a \times b$ rectangular array as shown in Table 8.3. As the data are classified based on two criteria, namely treatments and blocks, this kind of arrangement is termed as 'two-way classification' or 'two-way analysis of variance'. There are twofold objectives of two-way analysis of variance. The first objective is to test the null hypothesis that the treatment means are all equal and the second objective is to test another null hypothesis that the block means are all equal. For this purpose we use the notation $T_{i.}$ and $\bar{y}_{i.}$, respectively for the total and mean of b observations in the ith treatment, and $T_{.j}$ and $\bar{y}_{.j}$, respectively for the total and mean of a observations in the jth block. Thus, $T_{i.}$, $\bar{y}_{i.}$, $T_{.j}$ and $\bar{y}_{.j}$ can be expressed as

$$T_{i.} = \sum_{j=1}^{b} y_{ij}$$

$$\bar{y}_{i.} = \frac{T_{i.}}{b}$$

$$T_{.j} = \sum_{i=1}^{a} y_{ij}$$

$$\bar{y}_{.j} = \frac{T_{.j}}{a}$$

The grand total T and grand mean $\bar{y}$ of all ab observations are expressed as

$$T = \sum_{i=1}^{a}\sum_{j=1}^{b} y_{ij} = \sum_{i=1}^{a} T_{i.} = \sum_{j=1}^{b} T_{.j}$$

$$\bar{y} = \frac{T}{ab}$$

For ith treatment, the deviation of an observation y_{ij} from the grand mean $\bar{y}$ can be written as

$$y_{ij} - \bar{y} = \left(\bar{y}_{i.} - \bar{y}\right) + \left(\bar{y}_{.j} - \bar{y}\right) + \left(y_{ij} - \bar{y}_{i.} - \bar{y}_{.j} + \bar{y}\right) \tag{8.5}$$

where $\left(\bar{y}_{i.} - \bar{y}\right)$ is the effect of ith treatment, $\left(\bar{y}_{.j} - \bar{y}\right)$ is the effect of jth block, and $\left(y_{ij} - \bar{y}_{i.} - \bar{y}_{.j} + \bar{y}\right)$ is the random error associated with the observation y_{ij}. By squaring both sides of Equation (8.5) and summing on i and j, and following the same argument as in the proof of Equation (8.2) it can be shown that

$$\sum_{i=1}^{a}\sum_{j=1}^{b}\left(y_{ij} - \bar{y}\right)^2 = b\sum_{i=1}^{a}\left(\bar{y}_{i.} - \bar{y}\right)^2 + a\sum_{j=1}^{b}\left(\bar{y}_{.j} - \bar{y}\right)^2$$

$$+ \sum_{i=1}^{a}\sum_{j=1}^{b}\left(y_{ij} - \bar{y}_{i.} - \bar{y}_{.j} + \bar{y}\right)^2 \tag{8.6}$$

In the above equation, $\sum_{i=1}^{a}\sum_{j=1}^{b}\left(y_{ij} - \bar{y}\right)^2$ is the 'total sum of square' (SST), $b\sum_{i=1}^{a}\left(\bar{y}_{i.} - \bar{y}\right)^2$ is the 'between treatment sum of square' (SST$_r$), $a\sum_{j=1}^{b}\left(\bar{y}_{.j} - \bar{y}\right)^2$ is the 'between block sum of square'(SSB) and $\sum_{i=1}^{a}\sum_{j=1}^{b}\left(y_{ij} - \bar{y}_{i.} - \bar{y}_{.j} + \bar{y}\right)^2$ is the 'error sum of square' (SSE). Symbolically, the sum of square identity can be expressed as

$$SST = SST_r + SSB + SSE \tag{8.7}$$

As in the one-way analysis of variance, the shortcut ways of expressing the sum of squares in this case are given below.

$$SST = \sum_{i=1}^{a}\sum_{j=1}^{b}\left(y_{ij} - \bar{y}\right)^2$$

$$= \sum_{i=1}^{a}\sum_{j=1}^{b} y_{ij}^2 - \frac{T^2}{ab}$$

$$SST_r = b \sum_{i=1}^{a} (\bar{y}_{i.} - \bar{y})^2$$

$$= \frac{1}{b} \sum_{i=1}^{a} T_{i.}^2 - \frac{T^2}{ab}$$

$$SSB = a \sum_{j=1}^{b} (\bar{y}_{.j} - \bar{y})^2$$

$$= \frac{1}{a} \sum_{j=1}^{b} T_{.j}^2 - \frac{T^2}{ab}$$

From Equation (8.7) we have

$$SSE = SST - SST_r - SSB$$

$$= \sum_{i=1}^{a} \sum_{j=1}^{b} y_{ij}^2 - \frac{T^2}{ab} - \frac{1}{b} \sum_{i=1}^{a} T_{i.}^2 + \frac{T^2}{ab} - \frac{1}{a} \sum_{j=1}^{b} T_{.j}^2 + \frac{T^2}{ab}$$

$$= \sum_{i=1}^{a} \sum_{j=1}^{b} y_{ij}^2 - \frac{1}{b} \sum_{i=1}^{a} T_{i.}^2 - \frac{1}{a} \sum_{j=1}^{b} T_{.j}^2 + \frac{T^2}{ab}$$

Now each observation y_{ij} may be written in the form

$$y_{ij} = \bar{y} + (\bar{y}_{i.} - \bar{y}) + (\bar{y}_{.j} - \bar{y}) + (y_{ij} - \bar{y}_{i.} - \bar{y}_{.j} + \bar{y})$$

$$= \bar{y} + t_i + b_j + e_{ij}$$

where t_i is the effect of ith treatment, b_j is the effect of jth block and e_{ij} is the random error associated with the observation y_{ij}.

We assume that Y_{ij}, for $i = 1, 2, \cdots, a$ and $j = 1, 2, \cdots, b$ are the independent random variables having normal distributions with the common variance σ^2. The treatment means $\bar{y}_{i.}$ are the sample estimates of population treatment means $\mu_{i.}$, The block means $\bar{y}_{.j}$ are the sample estimates of population block means $\mu_{.j}$, and the grand mean $\bar{y}$ is the estimate of the overall population mean μ. Therefore, the assumed model of Y_{ij} is given by

$$Y_{ij} = \mu + (\mu_{i.} - \mu) + (\mu_{.j} - \mu) + (Y_{ij} - \mu_{i.} - \mu_{.j} + \mu)$$

$$= \mu + \alpha_i + \beta_j + \varepsilon_{ij}$$

where α_i are the population treatment effects for $i = 1,2,\cdots,a$; β_j are the population block effects for $j = 1,2,\cdots,b$; and ε_{ij} are the error terms. The random errors ε_{ij} are the values of independent, normally distributed random variables, each with zero mean and variance σ^2. We will test two null hypotheses that the treatment effects are all equal to zero and the block effects are all equal to zero; that is

$$H_0: \alpha_i = 0, \text{ for } i = 1,2,\cdots,a$$

and

$$H_0': \beta_j = 0, \text{ for } j = 1,2,\cdots,b$$

against the alternative hypotheses that

$$H_1: \alpha_i \neq 0, \text{ for at least one value of } i$$

and

$$H_1': \beta_j \neq 0, \text{ for at least one value of } j$$

As there are three sources of variation, the variance σ^2 can be estimated in three independent ways as follows.

The first estimate of σ^2 is based on the variation among the treatment means. If H_0 is true, $\bar{y}_{i.}$ are the values of the independent random variables having identical normal distributions with the mean μ and variance $\frac{\sigma^2}{b}$. Therefore, $\frac{\sum_{i=1}^{a}(\bar{y}_i - \bar{y})^2}{(a-1)}$ is an estimate of $\frac{\sigma^2}{b}$. Hence, the first estimate of σ^2 is expressed as

$$MST_r = \frac{b\sum_{i=1}^{a}\left(\bar{y}_{i.} - \bar{y}\right)^2}{(a-1)} = \frac{SST_r}{(a-1)}$$

where MST_r is the 'treatment mean square'.

The second estimate of σ^2 is based on the variation among the block means. If H_0' is true, $\bar{y}_{.j}$ are the values of the independent random variables having identical normal distributions with the mean μ and variance $\frac{\sigma^2}{a}$. Therefore, $\frac{\sum_{i=1}^{b}(\bar{y}_{.j}-\bar{y})^2}{(b-1)}$ is an estimate of $\frac{\sigma^2}{a}$. Hence, the second estimate of σ^2 is expressed as

$$MSB = \frac{a\sum_{j=1}^{b}\left(\bar{y}_{.j} - \bar{y}\right)^2}{(b-1)} = \frac{SSB}{(b-1)}$$

where MSB is the 'block mean square'.

The third estimate of σ^2 is based on the random error which is expressed as

$$MSE = \frac{\sum_{i=1}^{a} \sum_{j=1}^{b} \left(y_{ij} - \bar{y}_{i.} - \bar{y}_{.j} + \bar{y}\right)^2}{(a-1)(b-1)} = \frac{SSE}{(a-1)(b-1)}$$

where MSE is the 'error mean square'.

If H_0 is true, $\frac{SST_r}{\sigma^2}$ and $\frac{SSE}{\sigma^2}$ are the values of independent random variables having chi-square distributions with $(a-1)$ and $(a-1)(b-1)$ degrees of freedom, respectively and the ratio $f_{T_r} = \frac{SST_r/\sigma^2(a-1)}{SSE/\sigma^2(a-1)(b-1)} = \frac{MST_r}{MSE}$ is a value of the random variable having F-distribution with $(a-1)$ and $(a-1)(b-1)$ degrees of freedom. However, when H_0 is false, MST_r is expected to exceed MSE. We reject H_0 if the value of f_{T_r} exceeds $f_{\alpha,(a-1),(a-1)(b-1)}$, where α is the level of significance.

Similarly, if H_0' is true, $\frac{SSB}{\sigma^2}$ and $\frac{SSE}{\sigma^2}$ are the values of independent random variables having chi-square distributions with $(b-1)$ and $(a-1)(b-1)$ degrees of freedom, respectively and the ratio $f_B = \frac{SSB/\sigma^2(b-1)}{SSE/\sigma^2(a-1)(b-1)} = \frac{MSB}{MSE}$ is a value of the random variable having F-distribution with $(b-1)$ and $(a-1)(b-1)$ degrees of freedom. However, when H_0' is false, MSB is expected to exceed MSE. We reject H_0' if the value of f_B exceeds $f_{\alpha,(b-1),(a-1)(b-1)}$.

The procedure we have described in this section is called two-way analysis of variance and it is summarized in Table 8.4.

TABLE 8.4
Two-Way Analysis of Variance Table

Source of Variation	Sum of Square	Degrees of Freedom	Mean Square	f
Treatments	$SST_r = \frac{1}{b}\sum_{i=1}^{a} T_{i.}^2 - \frac{T^2}{ab}$	$(a-1)$	$MST_r = \frac{SST_r}{(a-1)}$	$f_{T_r} = \frac{MST_r}{MSE}$
Blocks	$SSB = \frac{1}{a}\sum_{j=1}^{b} T_{.j}^2 - \frac{T^2}{ab}$	$(b-1)$	$MSB = \frac{SSB}{(b-1)}$	$f_B = \frac{MSB}{MSE}$
Error	$SSE = \sum_{i=1}^{a}\sum_{j=1}^{b} y_{ij}^2 - \frac{1}{b}\sum_{i=1}^{a} T_{i.}^2 - \frac{1}{a}\sum_{j=1}^{b} T_{.j}^2 + \frac{T^2}{ab}$	$(a-1)(b-1)$	$MSE = \frac{SSE}{(a-1)(b-1)}$	
Total	$SST = \sum_{i=1}^{a}\sum_{j=1}^{b} y_{ij}^2 - \frac{T^2}{ab}$	$ab-1$		

Example 8.3:

Table below shows the production in terms of front roller delivery in Km per machine per shift of three different makes of ring frame machines in three shifts each of 8 hours. Test at the 0.05 level of significance 1) whether there are significant differences in the production due to the makes of machines, and 2) whether there are significant differences in the production due to the shifts.

	Shift I	Shift II	Shift III
Machine A	10.7	12.8	14.4
Machine B	16.2	15.8	16.6
Machine C	11.8	13.6	11.4

SOLUTION

From the given data we have: $a = 3, b = 3, T_{1.} = 37.9, T_{2.} = 48.6, T_{3.} = 36.8, T_{.1} = 38.7, T_{.2} = 42.2, T_{.3} = 42.4, T = 123.3, \sum_{i=1}^{a}\sum_{j=1}^{b} y_{ij}^2 = 1727.49$.

a. $H_0: \alpha_i = 0$, for $i = 1,2,3$
   $H_0': \beta_j = 0$, for $j = 1,2,3$
   $H_1: \alpha_i \neq 0$, for at least one value of i
   $H_1': \beta_j \neq 0$, for at least one value of j
   $\alpha = 0.05$.

b. For treatments, the critical value is $f_{\alpha,(a-1),(a-1)(b-1)} = f_{0.05,(3-1),(3-1)\times(3-1)} = f_{0.05,2,4} = 6.9$ (Table A5). For blocks, the critical value is also equal to 6.9, because $a = b$.

c. SST_r, SSB and SSE are calculated as

$$SST_r = \frac{1}{b}\sum_{i=1}^{a} T_{i.}^2 - \frac{T^2}{ab} = \frac{1}{3}\left(37.9^2 + 48.6^2 + 36.8^2\right) - \frac{123.3^2}{3\times 3} = 28.33$$

$$SSB = \frac{1}{a}\sum_{i=1}^{b} T_{.j}^2 - \frac{T^2}{ab} = \frac{1}{3}\left(38.7^2 + 42.2^2 + 42.4^2\right) - \frac{123.3^2}{3\times 3} = 2.89$$

$$SST = \sum_{i=1}^{a}\sum_{j=1}^{b} y_{ij}^2 - \frac{T^2}{ab} = 1727.49 - \frac{123.3^2}{3\times 3} = 38.28$$

By subtraction we obtain, $SSE = SST - SST_r - SSB = 38.28 - 28.33 - 2.89 = 7.06$

Therefore,

$$MST_r = \frac{SST_r}{(a-1)} = \frac{28.33}{3-1} = 14.16$$

$$MSB = \frac{SSB}{(b-1)} = \frac{2.89}{3-1} = 1.44$$

and

$$MSE = \frac{SSE}{(a-1)(b-1)} = \frac{7.06}{(3-1)\times(3-1)} = 1.76$$

Hence, $f_{T_r} = \frac{MST_r}{MSE} = \frac{14.16}{1.76} = 8.04$
and

$$f_B = \frac{MSB}{MSE} = \frac{1.44}{1.76} = 0.82$$

Thus, we get the following analysis of variance table where treatments are represented by the makes of the machines and blocks are represented by the shifts.

Source of Variation	Sum of Square	Degrees of Freedom	Mean Square	f
Treatments	28.33	2	14.16	8.04
Blocks	2.89	2	1.44	0.82
Error	7.06	4	1.76	
Total	38.28	8		

d. For treatments, as $8.04 > 6.9$, the null hypothesis is rejected. Therefore, we do have statistical evidence that there are significant differences in the production due to the makes of machines. For blocks, as $0.82 < 6.9$, the null hypothesis cannot be rejected. Therefore, there is no sufficient statistical evidence to conclude that the differences in the production due to the shifts are significant.

8.3.1 MULTIPLE COMPARISONS OF TREATMENT MEANS AND BLOCK MEANS

Tukey's method as discussed in Section 8.2.1 can also be applied for multiple comparisons of treatment means as well as block means. In case of multiple comparisons of treatment means, the difference is significant if

$$\left| \bar{y}_{h.} - \bar{y}_{i.} \right| > q_{\alpha,a,v} \sqrt{\frac{MSE}{b}} \tag{8.8}$$

where h and i are the treatment numbers being compared such that $h \neq i$. For multiple comparisons of block means, the difference is significant if

$$\left| \bar{y}_{.g} - \bar{y}_{.j} \right| > q_{\alpha,b,v} \sqrt{\frac{MSE}{a}} \tag{8.9}$$

where g and j are the block numbers being compared such that $g \neq j$. Here v is the degrees of freedom associated with MSE.

Example 8.4:

With reference to Example 8.3, make a multiple comparisons of average production rates among the makes of machine.

SOLUTION

From the given data of Example 8.3 we have: $a = b = 3$, $\bar{y}_{1.} = 12.63$, $\bar{y}_{2.} = 16.2$, $\bar{y}_{3.} = 12.27$ and $MSE = 1.76$. With $\alpha = 0.05$, $a = 3$ and $v = 4$, the value of the studentized range is $q_{0.05,3,4} = 5.04$ (Table A6). Therefore, the least significant difference is $q_{\alpha,a,v}\sqrt{\frac{MSE}{b}} = 5.04\sqrt{\frac{1.76}{3}} = 3.86$. Thus, any absolute difference between the sample means greater than this value is significant at the 0.05 level of significance. Using Tukey's method, all pairs of sample means are compared as shown in the following table.

h	i	$\bar{y}_{h.}$	$\bar{y}_{i.}$	$\left|\bar{y}_{h.} - \bar{y}_{i.}\right|$	Is the Difference Significant?
1	2	12.63	16.2	3.57	No
1	3	12.63	12.27	0.36	No
2	3	16.2	12.27	3.93	Yes

The sample means are arranged in the descending order as follows.

$\bar{y}_{2.}$	$\bar{y}_{1.}$	$\bar{y}_{3.}$
16.2	12.63	12.27

As far as the production rate is concerned, we can conclude from the results that machine B is superior followed by machine A and machine C, and except the difference between machine B and machine C, the other differences are not significant.

8.4   TWO-WAY ANALYSIS OF VARIANCE WITH REPLICATION

In the case of two-way analysis of variance as discussed in the last section, the data were classified on the basis of treatment and block. A generalized representation of the two-way analysis of variance is done by replacing the block with another treatment and considering n replications at each treatment combination. In the context of the experimental design, the term 'treatment' can be used in a broad sense. Sometimes the term 'factor' is also used to represent 'treatment'. Table 8.5 shows the general layout of the data for two-way analysis of variance with replications, where there are a levels of treatment A and b levels of treatment B. We denote y_{ijk} as the kth observation taken at the ith level of treatment A and jth level of treatment B. Each treatment combination constitutes a cell containing n observations and there are altogether ab cells. In Table 8.5, the notations $T_{i..}$ and $\bar{y}_{i..}$ are the total and mean of bn observations for the ith level of treatment A,

TABLE 8.5
General Layout of the Experimental Plan for Two-Way Analysis of Variance with Replications

		1	2	...	j	...	b	Totals	Means
					Levels of Treatment B				
Levels of Treatment A	1	y_{111}	y_{121}	...	y_{1j1}	...	y_{1b1}	$T_{1..}$	$\bar{y}_{1..}$
		y_{112}	y_{122}	...	y_{1j2}	...	y_{1b2}		
		$\vdots$	$\vdots$		$\vdots$		$\vdots$		
		y_{11n}	y_{12n}	...	y_{1jn}	...	y_{1bn}		
	2	y_{211}	y_{221}	...	y_{2j1}	...	y_{2b1}	$T_{2..}$	$\bar{y}_{2..}$
		y_{211}	y_{222}	...	y_{2j2}	...	y_{2b2}		
		$\vdots$	$\vdots$		$\vdots$		$\vdots$		
		y_{21n}	y_{22n}	...	y_{2jn}	...	y_{2bn}		
	$\vdots$	$\vdots$	$\vdots$		$\vdots$		$\vdots$	$\vdots$	$\vdots$
	i	y_{i11}	y_{i21}	...	y_{ij1}	...	y_{ib1}	$T_{i..}$	$\bar{y}_{i..}$
		y_{i12}	y_{i22}	...	y_{ij2}	...	y_{ib2}		
		$\vdots$	$\vdots$		$\vdots$		$\vdots$		
		y_{i1n}	y_{i2n}	...	y_{ijn}	...	y_{ibn}		
	$\vdots$	$\vdots$	$\vdots$		$\vdots$		$\vdots$	$\vdots$	$\vdots$
	a	y_{a11}	y_{a21}	...	y_{aj1}	...	y_{ab1}	$T_{a..}$	$\bar{y}_{a..}$
		y_{a12}	y_{a22}	...	y_{aj2}	...	y_{ab2}		
		$\vdots$	$\vdots$		$\vdots$		$\vdots$		
		y_{a1n}	y_{a2n}	...	y_{ajn}	...	y_{abn}		
Totals		$T_{.1.}$	$T_{.2.}$	...	$T_{.j.}$	...	$T_{.b.}$	Grand total $= T$	
Means		$\bar{y}_{.1.}$	$\bar{y}_{.2.}$	...	$\bar{y}_{.j.}$	...	$\bar{y}_{.b.}$	Grand mean $= \bar{y}$	

respectively; $T_{.j.}$ and $\bar{y}_{.j.}$ are the total and mean of an observations for the jth level of treatment B, respectively.

Thus, $T_{i..}$, $\bar{y}_{i..}$, $T_{.j.}$ and $\bar{y}_{.j.}$ can be expressed as

$$T_{i..} = \sum_{j=1}^{b}\sum_{k=1}^{n} y_{ijk}$$

$$\bar{y}_{i..} = \frac{T_{i..}}{bn}$$

$$T_{.j.} = \sum_{i=1}^{a}\sum_{k=1}^{n} y_{ijk}$$

$$\bar{y}_{.j.} = \frac{T_{.j.}}{an}$$

The grand total T and grand mean $\bar{y}$ of all abn observations are expressed as

$$T = \sum_{i=1}^{a}\sum_{j=1}^{b}\sum_{k=1}^{n} y_{ijk}$$

$$\bar{y} = \frac{T}{abn}$$

We assume that the observations in the (ij)th cell constitute a random sample of size n from a normally distributed population with variance σ^2. Further, all ab populations are assumed to have common variance σ^2. The deviation of an observation y_{ijk} from the grand mean $\bar{y}$ can be written as

$$y_{ijk} - \bar{y} = \left(\bar{y}_{i..} - \bar{y}\right) + \left(\bar{y}_{.j.} - \bar{y}\right) + \left(\bar{y}_{ij.} - \bar{y}_{i..} - \bar{y}_{.j.} + \bar{y}\right) + \left(y_{ijk} - \bar{y}_{ij.}\right) \qquad (8.10)$$

where $\left(\bar{y}_{i..} - \bar{y}\right)$ is the effect of ith level of treatment A, $\left(\bar{y}_{.j.} - \bar{y}\right)$ is the effect of jth level of treatment B, $\left(\bar{y}_{ij.} - \bar{y}_{i..} - \bar{y}_{.j.} + \bar{y}\right)$ is the interaction effect of the ith level of treatment A and jth level of treatment B, $\bar{y}_{ij.}$ is the mean of observations in the (ij)th cell, and $\left(y_{ijk} - \bar{y}_{ij.}\right)$ is the random error. By squaring both sides of Equation (8.10) and summing on i, j and k it can be shown that

$$\sum_{i=1}^{a}\sum_{j=1}^{b}\sum_{k=1}^{n}\left(y_{ij} - \bar{y}\right)^2 = bn\sum_{i=1}^{a}\left(\bar{y}_{i..} - \bar{y}\right)^2 + an\sum_{j=1}^{b}\left(\bar{y}_{.j.} - \bar{y}\right)^2$$

$$+ n\sum_{i=1}^{a}\sum_{j=1}^{b}\left(\bar{y}_{ij.} - \bar{y}_{i..} - \bar{y}_{.j.} + \bar{y}\right)^2$$

$$+ \sum_{i=1}^{a}\sum_{j=1}^{b}\sum_{k=1}^{n}\left(y_{ijk} - \bar{y}_{ij.}\right)^2 \qquad (8.11)$$

In the above equation, $\sum_{i=1}^{a}\sum_{j=1}^{b}\sum_{k=1}^{n}\left(y_{ij} - \bar{y}\right)^2$ is the 'total sum of square' (SST), $bn\sum_{i=1}^{a}\left(\bar{y}_{i..} - \bar{y}\right)^2$ is the 'treatment A sum of square' (SS_A), $an\sum_{j=1}^{b}\left(\bar{y}_{.j.} - \bar{y}\right)^2$ is the 'treatment B sum of square' (SS_B), $n\sum_{i=1}^{a}\sum_{j=1}^{b}\left(\bar{y}_{ij.} - \bar{y}_{i..} - \bar{y}_{.j.} + \bar{y}\right)^2$ is the interaction sum of square $\left(SS_{AB}\right)$ and $\sum_{i=1}^{a}\sum_{j=1}^{b}\sum_{k=1}^{n}\left(y_{ijk} - \bar{y}_{ij.}\right)^2$ is the 'error sum of square' (SSE). Symbolically, the sum of square identity can be expressed as

$$SST = SS_A + SS_B + SS_{AB} + SSE \qquad (8.12)$$

The shortcut ways of expressing the sum of squares in this case are given below.

$$SST = \sum_{i=1}^{a} \sum_{j=1}^{b} \sum_{k=1}^{n} \left(y_{ij} - \bar{y} \right)^2$$

$$= \sum_{i=1}^{a} \sum_{j=1}^{b} \sum_{k=1}^{n} y_{ijk}^2 - \frac{T^2}{abn}$$

$$SS_A = bn \sum_{i=1}^{a} \left(\bar{y}_{i..} - \bar{y} \right)^2$$

$$= \frac{1}{bn} \sum_{i=1}^{a} T_{i..}^2 - \frac{T^2}{abn}$$

$$SS_B = an \sum_{j=1}^{b} \left(\bar{y}_{.j.} - \bar{y} \right)^2$$

$$= \frac{1}{an} \sum_{j=1}^{b} T_{.j.}^2 - \frac{T^2}{abn}$$

$$SS_{AB} = n \sum_{i=1}^{a} \sum_{j=1}^{b} \left(\bar{y}_{ij.} - \bar{y}_{i..} - \bar{y}_{.j.} + \bar{y} \right)^2$$

$$= \frac{1}{n} \sum_{i=1}^{a} \sum_{j=1}^{b} T_{ij.}^2 - \frac{1}{bn} \sum_{i=1}^{a} T_{i..}^2 - \frac{1}{an} \sum_{j=1}^{b} T_{.j.}^2 + \frac{T^2}{abn}$$

where $T_{ij.}$ is the sum of observations in the (ij)th cell. From Equation (8.12) we have

$$SSE = SST - SS_A - SS_B - SS_{AB}$$

$$= \sum_{i=1}^{a} \sum_{j=1}^{b} \sum_{k=1}^{n} y_{ijk}^2 - \frac{1}{n} \sum_{i=1}^{a} \sum_{j=1}^{b} T_{ij.}^2$$

Now, from Equation (8.10) we have

$$y_{ijk} = \bar{y} + \left(\bar{y}_{i..} - \bar{y} \right) + \left(\bar{y}_{.j.} - \bar{y} \right) + \left(\bar{y}_{ij.} - \bar{y}_{i..} - \bar{y}_{.j.} + \bar{y} \right) + \left(y_{ijk} - \bar{y}_{ij.} \right)$$

Thus, the assumed model becomes

$$Y_{ijk} = \mu + \alpha_i + \beta_j + \left(\alpha\beta \right)_{ij} + \varepsilon_{ij}$$

where α_i are the effects of the treatment A for the levels $i = 1, 2, \cdots, a$; β_j are the effects of treatment B for the levels $j = 1, 2, \cdots, b$; $\left(\alpha\beta \right)_{ij}$ are the effects of the interaction between level i of treatment A and level j of treatment B; and ε_{ij} are the error terms. The random errors ε_{ij} are values of independent, normally distributed random variables, each with zero mean and variance σ^2. We will test the null hypotheses that

H_0: $\alpha_i = 0$, for $i = 1, 2, \cdots, a$
H_0': $\beta_j = 0$, for $j = 1, 2, \cdots, b$
H_0'': $(\alpha\beta)_{ij} = 0$, for $i = 1, 2, \cdots, a$ and $j = 1, 2, \cdots, b$

against the alternative hypotheses that

H_1: $\alpha_i \neq 0$, for at least one i
H_1': $\beta_j \neq 0$, for at least one j
H_1'': $(\alpha\beta)_{ij} \neq 0$, for at least one (i, j)

The independent estimates of σ^2 under the conditions that there are no effects of α_i, β_j and $(\alpha\beta)_{ij}$ are given by the following mean squares

$$MS_A = \frac{SS_A}{(a-1)}$$

$$MS_B = \frac{SS_B}{(b-1)}$$

$$MS_{AB} = \frac{SS_{AB}}{(a-1)(b-1)}$$

$$MSE = \frac{SSE}{ab(n-1)}$$

where MS_A is the 'treatment A mean square', MS_B is the 'treatment B mean square', MS_{AB} is the 'interaction mean square' and MSE is the 'error mean square'. It can be shown that if H_0, H_0' and H_0'' are true, the ratios

$$f_A = \frac{SS_A / \sigma^2(a-1)}{SSE / \sigma^2 ab(n-1)} = \frac{MS_A}{MSE}$$

$$f_B = \frac{SS_B / \sigma^2(b-1)}{SSE / \sigma^2 ab(n-1)} = \frac{MS_B}{MSE}$$

$$f_{AB} = \frac{SS_A B / \sigma^2(a-1)(b-1)}{SSE / \sigma^2 ab(n-1)} = \frac{MS_{AB}}{MSE}$$

all have the values of the random variables having F-distributions with respectively, $(a-1)$, $(b-1)$ and $(a-1)(b-1)$ degrees of freedom in the numerator and $ab(n-1)$ degrees of freedom in the denominator. However, we reject a null hypothesis at the level of significance α if the calculated value of f exceeds f_{α, v_1, v_2}, as obtained from Table A5, with the corresponding numerator and denominator degrees of freedom.

The procedure we have described in this section is called two-way analysis of variance with replication and it is summarized in Table 8.6.

TABLE 8.6
Two-Way Analysis of Variance with Replication Table

Source of Variation	Sum of Square	Degrees of Freedom	Mean Square	f
Treatment A	$SS_A = \dfrac{1}{bn}\sum_{i=1}^{a} T_{i..}^2 - \dfrac{T^2}{abn}$	$(a-1)$	$MS_A = \dfrac{SS_A}{(a-1)}$	$f_A = \dfrac{MS_A}{MSE}$
Treatment B	$SS_B = \dfrac{1}{an}\sum_{j=1}^{b} T_{.j.}^2 - \dfrac{T^2}{abn}$	$(b-1)$	$MS_B = \dfrac{SS_B}{(b-1)}$	$f_B = \dfrac{MS_B}{MSE}$
Interaction $(A \times B)$	$SS_{AB} = \dfrac{1}{n}\sum_{i=1}^{a}\sum_{j=1}^{b} T_{ij.}^2$ $-\dfrac{1}{bn}\sum_{i=1}^{a} T_{i..}^2 - \dfrac{1}{an}\sum_{j=1}^{b} T_{.j.}^2 + \dfrac{T^2}{abn}$	$(a-1)(b-1)$	$MS_{AB} = \dfrac{SS_{AB}}{(a-1)(b-1)}$	$f_{AB} = \dfrac{MS_{AB}}{MSE}$
Error	$SSE = \sum_{i=1}^{a}\sum_{j=1}^{b}\sum_{k=1}^{n} y_{ijk}^2$ $-\dfrac{1}{n}\sum_{i=1}^{a}\sum_{j=1}^{b} T_{ij.}^2$	$ab(n-1)$	$MSE = \dfrac{SSE}{ab(n-1)}$	
Total	$SST = \sum_{i=1}^{a}\sum_{j=1}^{b}\sum_{k=1}^{n} y_{ijk}^2 - \dfrac{T^2}{abn}$	$abn-1$		

Example 8.5:

An experiment was conducted to measure the breaking strength of cotton yarns having four different counts viz., 16, 24, 32 and 40 tex using three different tensile testing instruments, namely I_1, I_2 and I_3 and the results of yarn strength in cN/tex are as follows.

	I_1	I_2	I_3
16 tex yarn	19.6	18.7	18.8
	20.4	19.4	19.2
	19.9	20.2	19.7
24 tex yarn	21.4	20.7	19.9
	20.9	21.6	20.6
	21.2	21.1	20.8
32 tex yarn	22.4	23.3	21.9
	23.1	22.3	22.4
	22.7	23.0	22.2
40 tex yarn	24.8	23.7	22.9
	23.5	24.2	23.3
	24.4	25.1	23.6

Make an analysis of variance of the above data and conclude.

SOLUTION

From the given data we have: $a = 4$, $b = 3$, $n = 3$, $\sum_{i=1}^{a}\sum_{j=1}^{b}\sum_{k=1}^{n} y_{ijk}^2 = 17136.38$. The required sums and sum of squares can be calculated by forming the following two-way table giving the sums $T_{ij.}$.

	I_1	I_2	I_3	Totals
16 tex yarn	59.9	58.3	57.7	175.9
24 tex yarn	63.5	63.4	61.3	188.2
32 tex yarn	68.2	68.6	66.5	203.3
40 tex yarn	72.7	73.0	69.8	215.5
Totals	264.3	263.3	255.3	782.9

Thus we obtain, $T_{1..} = 175.9$, $T_{2..} = 188.2$, $T_{3..} = 203.3$, $T_{4..} = 215.5$, $T_{.1.} = 264.3$, $T_{.2.} = 263.3$, $T_{.3.} = 255.3$, $T = 782.9$ and $\sum_{i=1}^{a}\sum_{j=1}^{b} T_{ij.}^2 = 51391.47$.

a. H_0: $\alpha_i = 0$, for $i = 1, 2, 3, 4$
H_0': $\beta_j = 0$, for $j = 1, 2, 3$
H_0'': $(\alpha\beta)_{ij} = 0$, for $i = 1, 2, 3, 4$ and $j = 1, 2, 3$
H_1: $\alpha_i \neq 0$, for at least one i
H_1': $\beta_j \neq 0$, for at least one j
H_1'': $(\alpha\beta)_{ij} \neq 0$, for at least one (i, j)
$\alpha = 0.05$.
b. For treatment A, the critical value is $f_{\alpha,(a-1),ab(n-1)} = f_{0.05,3,24} = 3.01$. For treatment B, the critical value is $f_{\alpha,(b-1),ab(n-1)} = f_{0.05,2,24} = 3.4$. For Interaction $A \times B$, the critical value is $f_{\alpha,(a-1)(b-1),ab(n-1)} = f_{0.05,6,24} = 2.51$ (Table A5).
c. SS_A, SS_B, SS_{AB} and SSE are calculated as

$$SS_A = \frac{1}{bn}\sum_{i=1}^{a} T_{i..}^2 - \frac{T^2}{abn} = \frac{1}{3 \times 3}\left(175.9^2 + 188.2^2 + 203.3^2 + 215.5^2\right) - \frac{782.9^2}{4 \times 3 \times 3}$$

$$= 17125.69 - 17025.90 = 99.79$$

$$SS_B = \frac{1}{an}\sum_{j=1}^{b} T_{.j.}^2 - \frac{T^2}{abn} = \frac{1}{4 \times 3}\left(264.3^2 + 263.3^2 + 255.3^2\right) - \frac{782.9^2}{4 \times 3 \times 3}$$

$$= 17029.96 - 17025.90 = 4.06$$

$$SS_{AB} = \frac{1}{n}\sum_{i=1}^{a}\sum_{j=1}^{b} T_{ij.}^2 - \frac{1}{bn}\sum_{i=1}^{a} T_{i..}^2 - \frac{1}{an}\sum_{j=1}^{b} T_{.j.}^2 + \frac{T^2}{abn}$$

$$= \frac{51391.47}{3} - 17125.69 - 17029.96 + 17025.90 = 0.75$$

$$SST = \sum_{i=1}^{a}\sum_{j=1}^{b}\sum_{k=1}^{n} y_{ijk}^2 - \frac{T^2}{abn} = 17136.38 - 17025.90 = 110.48$$

By subtraction we obtain $SSE = SST - SS_A - SS_B - SS_{AB}$

$$= 110.48 - 99.79 - 4.06 - 0.75 = 5.88$$

Therefore,

$$MS_A = \frac{SS_A}{(a-1)} = \frac{99.79}{4-1} = 33.26$$

$$MS_B = \frac{SS_B}{(b-1)} = \frac{4.06}{3-1} = 2.03$$

$$MS_{AB} = \frac{SS_{AB}}{(a-1)(b-1)} = \frac{0.75}{(4-1)\times(3-1)} = 0.125$$

and

$$MSE = \frac{SSE}{ab(n-1)} = \frac{5.88}{4\times3\times(3-1)} = 0.245$$

Hence,

$$f_A = \frac{MS_A}{MSE} = \frac{33.26}{0.245} = 135.76$$

$$f_B = \frac{MS_B}{MSE} = \frac{2.03}{0.245} = 8.29$$

and

$$f_{AB} = \frac{MS_{AB}}{MSE} = \frac{0.125}{0.245} = 0.51$$

Thus, we get the following analysis of variance table, where treatment A is represented by the yarn count in tex and treatment B is represented by the tensile testing instrument.

Source of Variation	Sum of Square	Degrees of Freedom	Mean Square	f
Treatment A	99.79	3	33.26	135.76
Treatment B	4.06	2	2.03	8.29
Interaction $(A\times B)$	0.75	6	0.125	0.51
Error	5.88	24	0.245	
Total	110.48	35		

d. For treatment *A*, as 135.76 > 3.01, the null hypothesis is rejected. Therefore, we do have statistical evidence that there are significant differences in the yarn strengths due to the yarn count. For treatment *B*, as 8.29 > 3.4, the null hypothesis is rejected. Therefore, there is sufficient statistical evidence to conclude that the differences in the yarn strengths due to the instruments are significant. For interaction $A \times B$, as 0.51 < 2.51, the null hypothesis cannot be rejected. Therefore, there is no sufficient statistical evidence to conclude that the interaction between yarn count and instrument has influence on the yarn strength.

8.5   MATLAB® CODING

8.5.1   MATLAB® CODING OF EXAMPLE 8.1

```
clc
clear
close all
x=[82    75   74   84
   90    84   87   81
   61    72   64   76];
y=x';
[p,tbl,stats]=anova1(y)
```

8.5.2   MATLAB® CODING OF EXAMPLE 8.2

```
clc
clear
close all
x=[20.5  20.7   20.9   20.2   20.6   20.4   21.0
   19.4  18.6   18.9   19.2   18.8   19.7   18.9
   18.9  19.5   20.2   19.8   19.6   20.1   19.2
   17.6  16.9   16.7   17.4   17.5   17.2   16.8];
y=x';
[p,tbl,stats]=anova1(y)
results = multcompare(stats,'CType','tukey-kramer')
```

8.5.3   MATLAB® CODING OF EXAMPLE 8.3

```
clc
clear
close all
y=[10.7  12.8    14.4
   16.2  15.8    16.6
   11.8  13.6    11.4];
[p,tbl,stats]=anova2(y)
```

8.5.4   MATLAB® CODING OF EXAMPLE 8.4

```
clc
clear
```

```
close all
format bank
y=[10.7    12.8    14.4
    16.2    15.8    16.6
    11.8    13.6    11.4];
[p,tbl,stats]=anova2(y');
[results,means] = multcompare(stats,'CType','tukey-kramer')
```

8.5.5 MATLAB® CODING OF EXAMPLE 8.5

```
clc
clear
close all
y=[19.6    18.7    18.8
    20.4    19.4    19.2
    19.9    20.2    19.7
    21.4    20.7    19.9
    20.9    21.6    20.6
    21.2    21.1    20.8
    22.4    23.3    21.9
    23.1    22.3    22.4
    22.7    23.0    22.2
    24.8    23.7    22.9
    23.5    24.2    23.3
    24.4    25.1    23.6];
[p,tbl,stats]=anova2(y,3)
```

Exercises

8.1 The following are the strength readings (cN/tex) of a spun yarn at three different gauge lengths.

50 mm gauge length	15.16	14.95	15.23	14.78	14.99
200 mm gauge length	14.47	14.64	14.39	14.54	14.75
500 mm gauge length	13.78	13.94	13.67	13.59	14.05

Test at the 0.05 level of significance whether there are significant differences of yarn strengths at different gauge lengths.

8.2 The following are the strength readings (cN/tex) of a spun yarn at three different strain rates.

0.1/min strain rate	14.61	13.85	14.39	14.84	13.68
1/min strain rate	16.70	15.94	16.34	15.87	16.24
10/min strain rate	17.56	18.21	18.06	17.79	17.64

Test at the 0.05 level of significance whether there are significant differences of yarn strengths at different strain rates.

8.3 With reference to Exercises 8.1 and 8.2, make multiple comparisons of variables means for both the cases.

8.4 Table below shows the average count (Ne) of yarns produced in four different ring frames at three shifts in a day. Test at the 0.01 level of significance i) whether there are significant differences in the yarn count due to the machines, and ii) whether there are significant differences in the yarn count due to the shifts.

	Shift I	Shift II	Shift III
Machine A	28.72	30.12	29.74
Machine B	29.05	28.94	30.22
Machine C	30.08	31.14	29.75
Machine D	31.23	30.40	30.79

8.5 An experiment was conducted to compare the effect of three shrink-resist treatments on fabric of a large roll. For this purpose, five blocks of fabric were prepared from the large roll and each block was randomly subjected to three shrink-resist treatments. Table below shows the results of percentage area shrinkage of all test specimens after washing and drying. Test at the 0.05 level of significance i) whether there are significant differences among the treatments, and ii) whether there are significant differences among the blocks.

	Block I	Block II	Block III	Block IV	Block V
Treatment A	5.09	5.16	5.00	5.39	4.98
Treatment B	4.75	5.12	4.93	4.88	5.00
Treatment C	5.71	5.98	5.87	5.39	5.42

8.6 With reference to Exercise 8.5, make multiple comparisons of treatment means and block means.

8.7 An experiment was conducted to measure the effect of detachment settings (mm) and batt weights (ktex) on the noil (%) of a combing machine. The results of the noil (%) at three detachment settings viz., 17, 19 and 21 mm and three batt weights, viz., 60, 70 and 80 Ktex are as follows.

		Batt Weight	
	60 Ktex	70 Ktex	80 Ktex
17 mm	13.5	14.1	13.2
	13.9	14.3	13.4
19 mm	14.7	15.1	14.5
	14.6	14.9	14.2
21 mm	15.3	16.0	15.4
	15.8	16.1	15.7

Detachment Setting

Make an analysis of variance of the above data and write conclusions.

8.8 An experiment was conducted to study the effect of polyester-cotton blend ratios and yarn counts on the vortex yarn strength. For this purpose, the yarn was spun with three blend ratios at three different counts. The results of yarn strengths (cN/tex) are as follows.

		Yarn Count		
		10 tex	16 tex	24 tex
Blend Ratio (polyester: cotton)	67:33	18.6	18.3	19.2
		17.9	18.7	18.8
		18.3	18.9	18.4
	50:50	16.8	17.2	17.5
		15.9	16.6	17.8
		16.7	16.8	17.3
	33:67	15.4	15.8	16.4
		14.8	16.0	16.1
		14.6	15.5	16.7

Carry out an analysis of variance of the above data and write the conclusions.

9 Regression and Correlation

9.1 INTRODUCTION

An important objective of statistical investigations is to establish the relationship between a dependent variable and one or more independent variables in order to create a model that can be used for the purpose of prediction. The random variable to be predicted is called the dependent variable. The independent variable is the variable the experimenter changes or controls and is assumed to have a direct effect on the dependent variable. When we are looking for some kind of relationship between variables we are trying to see if the independent variables cause some kind of change in the dependent variables. The process of finding a mathematical equation that best fits the data using the least square method is called the regression analysis. The regression analysis is used to study the dependence of one variable with respect to another or multiple variables. There are different forms of regressions, for example linear, nonlinear, multiple etc. The linear regression is the most common form of regression analysis. The simple linear regression is a linear regression model with a single independent variable.

In statistics, the data related to two different variables, where each value of one of the variables is paired with a value of the other variable is called the bivariate data. When we study bivariate data, the correlation commonly refers to the degree to which a pair of variables is linearly related.

9.2 SIMPLE LINEAR REGRESSION

Example 9.1:

The following data were obtained in an experiment to investigate the relation between relative humidity (%) and moisture content (%) of the certain variety of cotton fibre.

Relative Humidity (%)	20	30	40	50	60	70	80
Moisture Content (%)	2.5	3.6	5.7	6.8	7.0	7.8	9.9

In Example 9.1 there are two types of variables, namely independent and dependent. The variable such as relative humidity (%) is under control of the experimenter, since the values of this variable can be varied deliberately. Therefore, it is termed as independent variable and denoted by X. On the other hand, the moisture content (%) depends on the values of relative humidity (%), therefore it is termed as dependent variable which is denoted by Y. We denote the values of variables X and Y by x and y, respectively. In this context, Y is a random variable whose distribution depends on x. The purpose of the

DOI: 10.1201/9781003081234-9

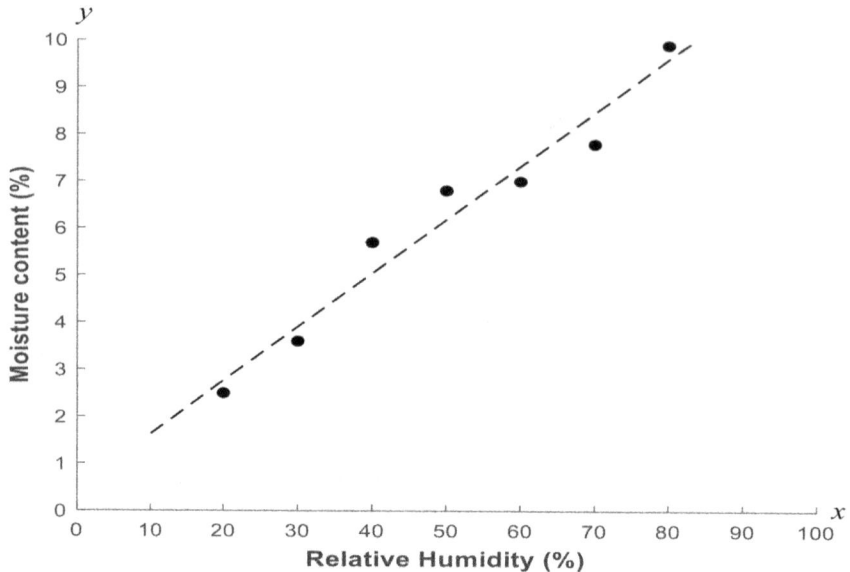

FIGURE 9.1 Plot of the data on relative humidity and moisture content.

regression analysis is to find the equation of the line or curve that gives the best prediction to the mean value of Y for given value of $X = x$. Plotting the data of Example 9.1, it
is apparent from Figure 9.1 that a straight line provides a reasonably good fit; hence, the
average moisture content of the cotton fibre for any given value of relative humidity may
well be related to the relative humidity by means of the following straight-line equation

$$\mu_{Y|x} = mx + c \qquad (9.1)$$

where $\mu_{Y|x}$ is the mean of Y for the given value of $X = x$. In Equation (9.1), m and c are
the regression coefficients, which are also termed as slope and intercept coefficients,
respectively. In general, Y will differ from this mean which we denote by ε, hence the
regression equation of Y on X is expressed as follows

$$Y = mx + c + \varepsilon \qquad (9.2)$$

In Equation (9.2), ε is the random error. The regression coefficients m and c are the
population parameters. Now from the given set of n number of paired sample data
$\{(x_i, y_i); \text{ for } i = 1, 2, \ldots, n\}$, we can only make an estimate of the coefficients m and c
which are denoted as $\hat{m}$ and $\hat{c}$, respectively. Thus, the estimated regression line which
provides the best fit to the given data has the following equation

$$\hat{y}_i = \hat{m}x_i + \hat{c} \qquad (9.3)$$

where $\hat{y}_i$ is the estimated value of the dependent variable and x_i is the given value
of the independent variable, for $i = 1, 2, \cdots, n$. Our objective is to determine $\hat{m}$ and $\hat{c}$

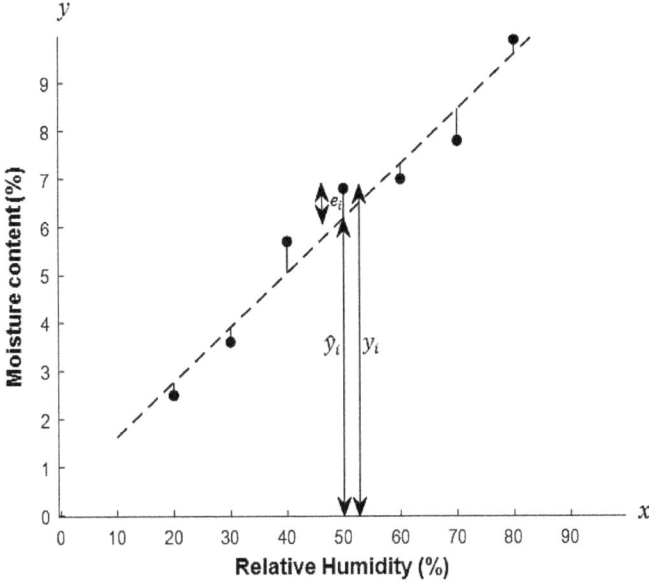

FIGURE 9.2 Least square method.

such that the estimated regression line of Equation (9.3) provides the best possible fit to the given data.

Denoting the actual value of the dependent variable by y_i and the error or the vertical distance from a point to the line by e_i, for $i = 1, 2, ..., n$, as depicted in Figure 9.2, the least square method requires that we find a minimum of the sum of square of these vertical distances as given in the following expression:

$$\xi = \sum_{i=1}^{n} e_i^2 \tag{9.4}$$

As $e_i = y_i - \hat{y}_i$, we can write

$$\xi = \sum_{i=1}^{n} \left(y_i - \hat{y}_i \right)^2$$

$$= \sum_{i=1}^{n} \left(y_i - \hat{m}x_i - \hat{c} \right)^2 \tag{9.5}$$

ξ is also termed as error sum of square which is minimized by setting the partial derivatives $\frac{\partial \xi}{\partial \hat{c}}$ and $\frac{\partial \xi}{\partial \hat{m}}$ equal to zero. Taking the partial derivative of ξ with respect to $\hat{c}$ and setting it to zero we get

$$\frac{\partial \xi}{\partial \hat{c}} = -2 \sum_{i=1}^{n} \left(y_i - \hat{m}x_i - \hat{c} \right) = 0$$

or,

$$\sum_{i=1}^{n} y_i = \sum_{i=1}^{n} \hat{m} x_i + n\hat{c} \tag{9.6}$$

By dividing both the sides by n we have,

$$\bar{y} = \hat{m}\bar{x} + \hat{c} \tag{9.7}$$

Thus,

$$\hat{c} = \bar{y} - \hat{m}\bar{x} \tag{9.8}$$

By replacing $\hat{c}$ from Equation (9.8) into Equation (9.5) we get

$$\xi = \sum_{i=1}^{n} \left(y_i - \hat{m} x_i - \bar{y} + \hat{m}\bar{x} \right)^2$$

$$= \sum_{i=1}^{n} \left\{ \left(y_i - \bar{y} \right) - \hat{m}\left(x_i - \bar{x} \right) \right\}^2$$

Now taking the partial derivative of ξ with respect to $\hat{m}$ and setting it to zero we get

$$\frac{\partial \xi}{\partial \hat{m}} = -2\sum_{i=1}^{n} \left\{ \left(y_i - \bar{y} \right) - \hat{m}\left(x_i - \bar{x} \right) \right\}\left(x_i - \bar{x} \right) = 0$$

or,

$$\sum_{i=1}^{n} \left(x_i - \bar{x} \right)\left(y_i - \bar{y} \right) = \hat{m}\sum_{i=1}^{n} \left(x_i - \bar{x} \right)^2 \tag{9.9}$$

Thus,

$$\hat{m} = \frac{\sum_{i=1}^{n} \left(x_i - \bar{x} \right)\left(y_i - \bar{y} \right)}{\sum_{i=1}^{n} \left(x_i - \bar{x} \right)^2} \tag{9.10}$$

In least square regression, it is assumed that the random variables Y_i are independently normally distributed having means $mx_i + c$ and common variance σ^2, for $i = 1, 2, \cdots, n$. Consequently, we can write that

$$Y_i = mx_i + c + \varepsilon_i \tag{9.11}$$

where random variables ε_i are independently normally distributed with means zero and common variance σ^2, for $i = 1, 2, \cdots, n$. Figure 9.3 gives a schematic representation of the linear regression model where the normal distribution of Y_i about the regression line $mx + c$ is depicted for four selected values of x_i.

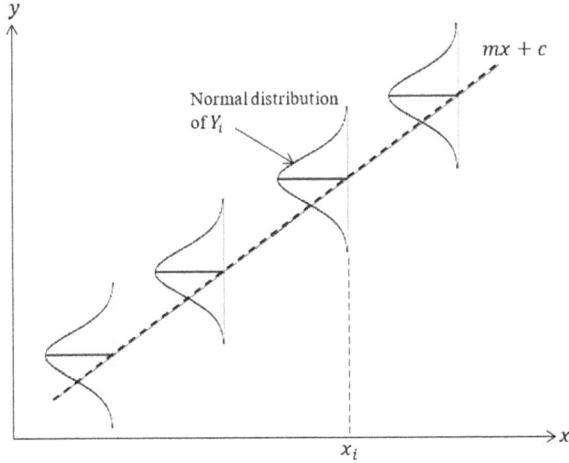

FIGURE 9.3 The assumptions of a regression model.

SOLUTION OF PROBLEM 9.1

From the data of Example 9.1, we obtain the following table.

x	y	x^2	xy
20	2.5	400	50
30	3.6	900	108
40	5.7	1600	228
50	6.8	2500	340
60	7.0	3600	420
70	7.8	4900	546
80	9.9	6400	792
Totals 350	43.30	20,300	2484

The values of $\hat{m}$ and $\hat{c}$ are computed as follows:

$$n = 7$$

$$\bar{x} = \frac{350}{7} = 50$$

$$\bar{y} = \frac{43.30}{7} = 6.19$$

$$\sum (x - \bar{x})^2 = \sum x^2 - \frac{\left(\sum x\right)^2}{n} = 20,300 - \frac{350^2}{7} = 2800$$

$$\sum (x - \bar{x})(y - \bar{y}) = \sum xy - \frac{\sum x \sum y}{n} = 2484 - \frac{350 \times 43.30}{7} = 319$$

$$\hat{m} = \frac{\sum (x - \bar{x})(y - \bar{y})}{\sum (x - \bar{x})^2} = \frac{319}{2800} = 0.11$$

$$\hat{c} = \bar{y} - \hat{m}\bar{x} = 6.19 - 0.11 \times 50 = 0.69$$

So, the equation of the regression line is estimated as

$$\hat{y} = 0.11x + 0.69$$

This line is drawn in Figure 9.1.

9.3 CORRELATION ANALYSIS

From Figure 9.4, geometrically, for ith observation the error e_i can be decomposed as

$$(y_i - \hat{y}_i) = (y_i - \bar{y}) - (\hat{y}_i - \bar{y}) \tag{9.12}$$

where $(y_i - \bar{y})$ is the total variation from the mean value $\bar{y}$ and $(\hat{y}_i - \bar{y})$ is the variation due to the regression. By taking square on both the sides of Equation (9.12), we can write

$$(y_i - \hat{y}_i)^2 = (y_i - \bar{y})^2 + (\hat{y}_i - \bar{y})^2 - 2(y_i - \bar{y})(\hat{y}_i - \bar{y})$$

Taking summation on both the sides, it becomes

$$\sum_{i=1}^{n}(y_i - \hat{y}_i)^2 = \sum_{i=1}^{n}(y_i - \bar{y})^2 + \sum_{i=1}^{n}(\hat{y}_i - \bar{y})^2 - 2\sum_{i=1}^{n}(y_i - \bar{y})(\hat{y}_i - \bar{y}) \tag{9.13}$$

By subtracting Equation (9.7) from Equation (9.3) we get

$$(\hat{y}_i - \bar{y}) = \hat{m}(x_i - \bar{x}) \tag{9.14}$$

Taking square and summation on both the sides, it becomes

$$\sum_{i=1}^{n}(\hat{y}_i - \bar{y})^2 = \hat{m}^2 \sum_{i=1}^{n}(x_i - \bar{x})^2 \tag{9.15}$$

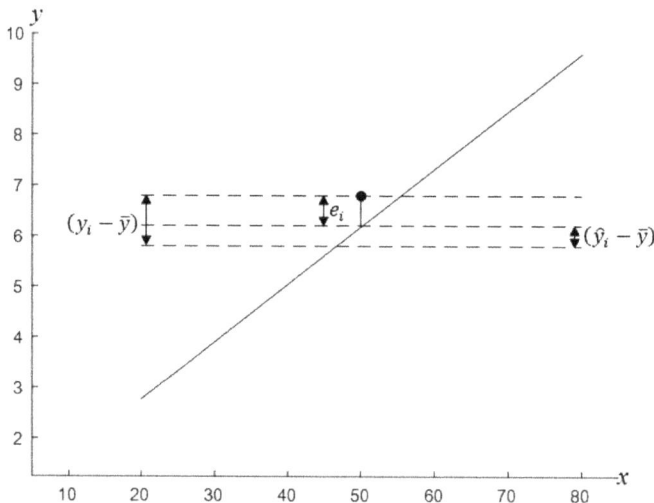

FIGURE 9.4 Breakdown of the error into two components.

Using Equation (9.14) we can write

$$\sum_{i=1}^{n}(y_i - \bar{y})(\hat{y}_i - \bar{y}) = \hat{m}\sum_{i=1}^{n}(x_i - \bar{x})(y_i - \bar{y}) \tag{9.16}$$

By replacing the expression of $\sum_{i=1}^{n}(x_i - \bar{x})(y_i - \bar{y})$ from Equation (9.9) into Equation (9.16) we get

$$\sum_{i=1}^{n}(y_i - \bar{y})(\hat{y}_i - \bar{y}) = \hat{m}^2\sum_{i=1}^{n}(x_i - \bar{x})^2 \tag{9.17}$$

Thus using Equation (9.15), we can write

$$\sum_{i=1}^{n}(y_i - \bar{y})(\hat{y}_i - \bar{y}) = \sum_{i=1}^{n}(\hat{y}_i - \bar{y})^2 \tag{9.18}$$

By replacing the expression of $\sum_{i=1}^{n}(y_i - \bar{y})(\hat{y}_i - \bar{y})$ from Equation (9.18) into Equation (9.13), we get

$$\sum_{i=1}^{n}(y_i - \hat{y}_i)^2 = \sum_{i=1}^{n}(y_i - \bar{y})^2 - \sum_{i=1}^{n}(\hat{y}_i - \bar{y})^2 \tag{9.19}$$

By rearranging the above equation becomes

$$\sum_{i=1}^{n}(y_i - \bar{y})^2 = \sum_{i=1}^{n}(\hat{y}_i - \bar{y})^2 + \sum_{i=1}^{n}(y_i - \hat{y}_i)^2 \tag{9.20}$$

In the above equation, $\sum_{i=1}^{n}(y_i - \bar{y})^2$ is the 'total sum of square variation', $\sum_{i=1}^{n}(\hat{y}_i - \bar{y})^2$ is the 'sum of square variation due to regression' or simply 'explained variation by regression' and $\sum_{i=1}^{n}(y_i - \hat{y}_i)^2$ is the 'error sum of square' or simply 'unexplained variation by regression'. Therefore, the total sum of square variation is the sum total of explained and unexplained variations by regression. The coefficient of determination (r^2) is defined as follows

$$r^2 = \frac{\text{Explained variation by regression}}{\text{Total sum of square variation}} \tag{9.21}$$

Thus,

$$r^2 = \frac{\sum_{i=1}^{n}(\hat{y}_i - \bar{y})^2}{\sum_{i=1}^{n}(y_i - \bar{y})^2} \tag{9.22}$$

The coefficient of determination measures the proportion or percentage of the total variation that can be explained by the regression model. Its limits are $0 \le r^2 \le 1$. An r^2 of 1 indicates a perfect fit that the regression model is able to explain 100% of the total variation, whereas r^2 of 0 means that there is no relationship between dependent and independent variables.

Now considering the linear regression, if we replace the expression of explained variation due to regression from Equation (9.15) into Equation (9.22), we get

$$r^2 = \frac{\hat{m}^2 \sum_{i=1}^{n} (x_i - \bar{x})^2}{\sum_{i=1}^{n} (y_i - \bar{y})^2} \tag{9.23}$$

By replacing the expression of $\hat{m}$ from (9.10) into (9.23) we have

$$r^2 = \frac{\left[\sum_{i=1}^{n} (x_i - \bar{x})(y_i - \bar{y}) \right]^2 \sum_{i=1}^{n} (x_i - \bar{x})^2}{\left[\sum_{i=1}^{n} (x_i - \bar{x})^2 \right]^2 \sum_{i=1}^{n} (y_i - \bar{y})^2} \tag{9.24}$$

or,

$$r^2 = \frac{\left[\sum_{i=1}^{n} (x_i - \bar{x})(y_i - \bar{y}) \right]^2}{\sum_{i=1}^{n} (x_i - \bar{x})^2 \sum_{i=1}^{n} (y_i - \bar{y})^2} \tag{9.25}$$

By taking square root, we can write

$$r = \frac{\sum_{i=1}^{n} (x_i - \bar{x})(y_i - \bar{y})}{\sqrt{\sum_{i=1}^{n} (x_i - \bar{x})^2} \sqrt{\sum_{i=1}^{n} (y_i - \bar{y})^2}} \tag{9.26}$$

By dividing both the numerator and denominator of the above equation with $(n-1)$, we have

$$r = \frac{\dfrac{\sum_{i=1}^{n} (x_i - \bar{x})(y_i - \bar{y})}{n-1}}{\sqrt{\dfrac{\sum_{i=1}^{n} (x_i - \bar{x})^2}{n-1}} \sqrt{\dfrac{\sum_{i=1}^{n} (y_i - \bar{y})^2}{n-1}}} \tag{9.27}$$

In the above equation, $\frac{\sum_{i=1}^{n}(x_i - \bar{x})(y_i - \bar{y})}{n-1}$ is the covariance of x and y and denoted as $cov(x,y)$, $\sqrt{\frac{\sum_{i=1}^{n}(x_i-\bar{x})^2}{n-1}}$ is the standard deviation of x, and $\sqrt{\frac{\sum_{i=1}^{n}(y_i-\bar{y})^2}{n-1}}$ is the standard deviation of y. Denoting s_x and s_y are the standard deviations of x and y, respectively, we can write

$$r = \frac{cov(x,y)}{s_x s_y} \tag{9.28}$$

where r is known as the sample correlation coefficient which implies only the linear association or linear dependence between two variables. It has no meaning for describing nonlinear relation. This is ascribed to the fact that we have deduced the expression of the correlation coefficient by considering the linear regression only. The value of r can be positive or negative and it lies between the limits $-1 \leq r \leq 1$. Figure 9.5 depicts the scatter plots of various datasets with different correlation coefficients.

Example 9.2:

Time taken in minutes to replace a draft change pinion of a ring frame by a fitter in the morning and in the afternoon is given below.

Morning x	Afternoon y
8.2	8.7
9.6	9.6
7.0	6.9
9.4	8.5
10.9	11.3
7.1	7.6
9.0	9.2
6.6	6.3
8.4	8.4
10.5	12.3

Compute the correlation coefficient between x and y.

SOLUTION

From the data, we obtain the following table.

	x	y	x^2	y^2	xy
	8.2	8.7	67.24	75.69	71.34
	9.6	9.6	92.17	92.16	92.16
	7.0	6.9	49	47.61	48.30
	9.4	8.5	88.36	72.25	79.90
	10.9	11.3	118.81	127.69	123.17
	7.1	7.6	50.41	57.76	53.96
	9.0	9.2	81	84.64	82.80
	6.6	6.3	43.56	39.69	41.58
	8.4	8.4	70.56	70.56	70.56
	10.5	12.3	110.25	151.29	129.15
Total	86.7	88.8	771.35	819.34	792.92

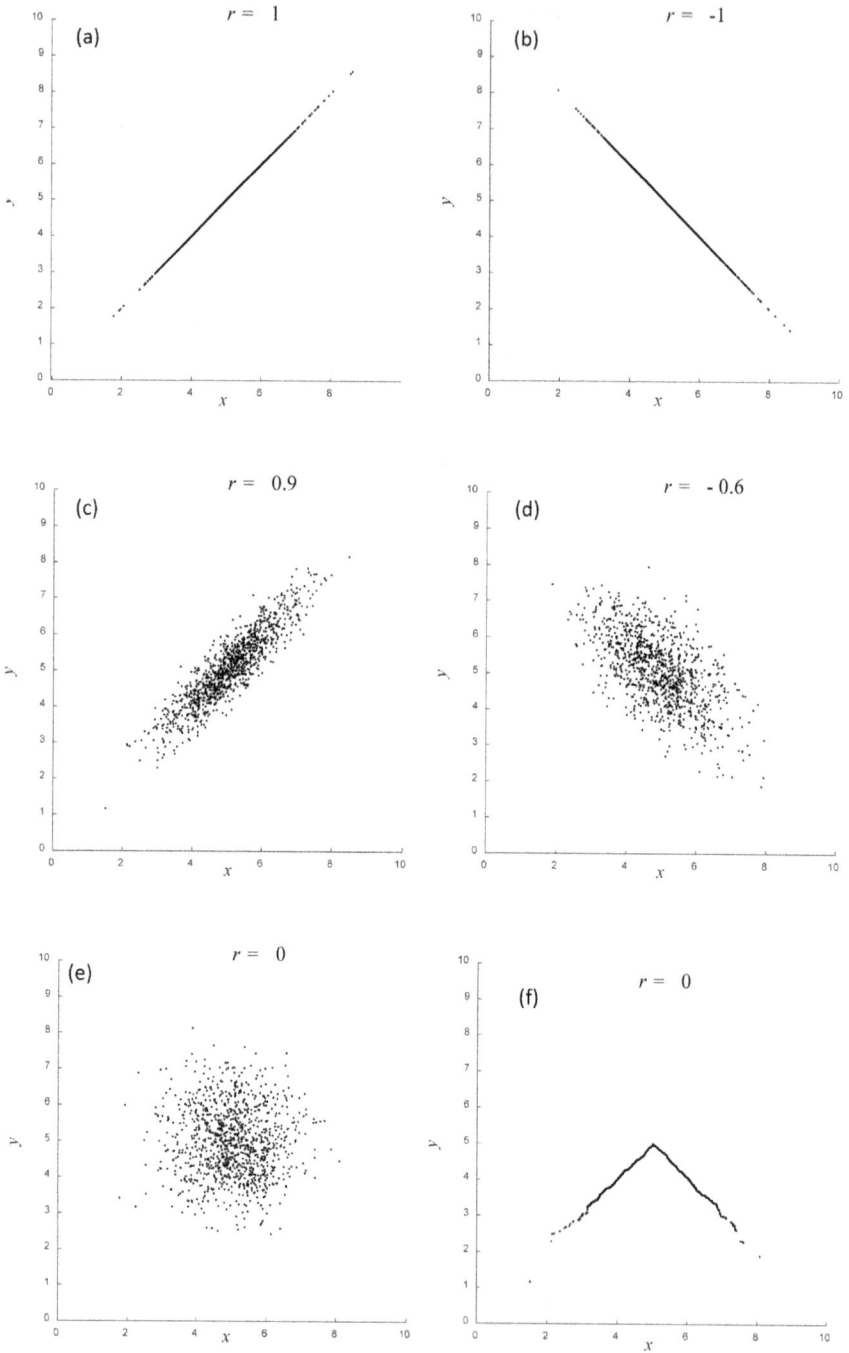

FIGURE 9.5 Scatter plots with different correlation coefficients.

The correlation coefficient is computed as follows:

$$n = 10$$

$$\bar{x} = \frac{86.7}{10} = 8.67$$

$$\bar{y} = \frac{88.8}{10} = 8.88$$

$$\sum (x - \bar{x})^2 = \sum x^2 - \frac{\left(\sum x\right)^2}{n} = 771.35 - \frac{86.7^2}{10} = 19.66$$

$$\sum (y - \bar{y})^2 = \sum y^2 - \frac{\left(\sum y\right)^2}{n} = 819.34 - \frac{88.8^2}{10} = 30.8$$

$$\sum (x - \bar{x})(y - \bar{y}) = \sum xy - \frac{\sum x \sum y}{n} = 792.92 - \frac{86.7 \times 88.8}{10} = 23.02$$

Thus using Equation (9.26) we get

$$r = \frac{\sum (x - \bar{x})(y - \bar{y})}{\sqrt{\sum (x - \bar{x})^2} \sqrt{\sum (y - \bar{y})^2}} = \frac{23.02}{\sqrt{19.66}\sqrt{30.8}} = 0.94$$

9.4 RANK CORRELATION

In many cases variables may not be directly measurable. For example, the efficiency of workers in a spinning mill cannot be measured numerically. However, a number of workers may be arranged in order of efficiency by their supervisors. Such type of ordered arrangement is called the ranking and the number indicating the place of an individual in the ranking is called its rank. The rank correlation coefficient measures the degree of association between two rankings.

Suppose that n individuals are ranked according to two properties x and y and the rankings have been made in the orders $x_1, x_2, \cdots, x_n$ and $y_1, y_2, \cdots, y_n$, respectively. Let us consider that there are no ties in the rankings, so x_i and y_i for $i = 1, 2, \cdots, n$ indicate the permutations of the integers from 1 to n. Making use of the fact that the sum and sum of the squares of the first n positive integers are $\frac{n(n+1)}{2}$ and $\frac{n(n+1)(2n+1)}{6}$, respectively, we find that

$$\sum_{i=1}^{n} x_i = \sum_{i=1}^{n} y_i = \frac{n(n+1)}{2}$$

and

$$\sum_{i=1}^{n} x_i^2 = \sum_{i=1}^{n} y_i^2 = \frac{n(n+1)(2n+1)}{6}$$

The mean $\bar{x}$ can be expressed as

$$\bar{x} = \frac{\sum_{i=1}^{n} x_i}{n} = \frac{1}{n} \times \frac{n(n+1)}{2} = \frac{n+1}{2}$$

Similarly,

$$\bar{y} = \frac{n+1}{2}$$

The variance s_x^2 is expressed as

$$s_x^2 = \frac{\sum_{i=1}^{n}(x_i - \bar{x})^2}{n-1}$$

$$= \frac{1}{n-1}\left(\sum_{n}^{i=1} x_i^2 - n\bar{x}^2\right)$$

$$= \frac{1}{n-1}\left\{\frac{n(n+1)(2n+1)}{6} - n\left(\frac{n+1}{2}\right)^2\right\}$$

$$= \frac{n}{n-1}\left\{\frac{2n^2+3n+1}{6} - \frac{n^2+2n+1}{4}\right\}$$

$$= \frac{n}{n-1}\left(\frac{n^2-1}{12}\right)$$

Thus,

$$s_x = \sqrt{\frac{n(n+1)}{12}} \qquad\qquad (9.29)$$

Similarly,

$$s_y = \sqrt{\frac{n(n+1)}{12}} \qquad\qquad (9.30)$$

Let d_i be the difference between the ranks assigned to x_i and y_i, so $d_i = x_i - y_i$. Now we have

$$\frac{\sum_{i=1}^{n} d_i^2}{n-1} = \frac{\sum_{i=1}^{n}(x_i - y_i)^2}{n-1}$$

$$= \frac{1}{n-1}\sum_{i=1}^{n}\left\{\left(x_i - \frac{n+1}{2}\right) - \left(y_i - \frac{n+1}{2}\right)\right\}^2$$

$$= \left(\frac{1}{n-1}\right)\sum_{i=1}^{n}\left\{(x_i - \bar{x}) - (y_i - \bar{y})\right\}^2$$

$$= \frac{\sum_{i=1}^{n}(x_i - \bar{x})^2}{n-1} + \frac{\sum_{i=1}^{n}(y_i - \bar{y})^2}{n-1} - 2\frac{\sum_{i=1}^{n}(x_i - \bar{x})(y_i - \bar{y})}{n-1}$$

$$= s_x^2 + s_y^2 - 2cov(x,y)$$

$$= \frac{n(n+1)}{12} + \frac{n(n+1)}{12} - 2cov(x,y)$$

Therefore,

$$cov(x,y) = \frac{n(n+1)}{12} - \frac{\sum_{i=1}^{n} d_i^2}{2(n-1)} \qquad (9.31)$$

Substituting the expressions of s_x, s_y and $cov(x,y)$ in the formula for the correlation coefficient in Equation (9.28), we obtain the rank correlation coefficient r_s as follows:

$$r_s = \frac{\dfrac{n(n+1)}{12} - \dfrac{\sum_{i=1}^{n} d_i^2}{2(n-1)}}{\sqrt{\dfrac{n(n+1)}{12}}\sqrt{\dfrac{n(n+1)}{12}}}$$

$$= \frac{\dfrac{n(n+1)}{12} - \dfrac{\sum_{i=1}^{n} d_i^2}{2(n-1)}}{\dfrac{n(n+1)}{12}}$$

or,

$$r_s = 1 - \frac{6\sum_{i=1}^{n} d_i^2}{n(n^2-1)} \qquad (9.32)$$

The rank correlation coefficient r_s indicates the relationships between the two variables when their values have been ranked according to some characteristic. It is also known as Spearman's rank correlation coefficient. The formula of r_s in (9.32) is valid when there are no ties in rank; however, the same formula may also be used if there are ties and then for each of the tied observations we substitute the average of the ranks that they jointly occupy.

Example 9.3:

The ranks of 10 Tossa jute lots based on two experts are given in the table below.

Lot number	1	2	3	4	5	6	7	8	9	10
Expert 1	2	1	3	7	6	5	4	10	9	8
Expert 2	2	1	4	8	6	4	5	10	9	7

Determine the rank correlation coefficient.

SOLUTION

The differences d_i between the ranks given by two experts are 0, 0, −1, −1, 0, 1, −1, 0, 0, 1. Hence, $\sum_{i=1}^{10} d_i^2 = 0+0+1+1+0+1+1+0+0+1 = 5$

Also, $n(n^2 - 1) = 10 \times (10^2 - 1) = 990$

Thus, using Equation (9.32) we get

$$r_s = 1 - \frac{6\sum_{i=1}^{n} d_i^2}{n(n^2 - 1)} = 1 - \frac{6 \times 5}{990} = 0.97$$

So, there is a strong degree of association between two sets of ranks.

Example 9.4:

Two supervisors ranked 7 workers of a spinning mill in order of efficiency as follows.

Workers	1	2	3	4	5	6	7
Supervisor 1	5.5	2	5.5	3	1	7	4
Supervisor 2	7	1	5	3.5	2	6	3.5

Calculate rank correlation coefficient between the two rankings.

SOLUTION

The differences d_i between the ranks given by two experts are −1.5, 1, 0.5, −0.5, −1, 1, 0.5. Hence, $\sum_{i=1}^{7} d_i^2 = 2.25 + 1 + 0.25 + 0.25 + 1 + 1 + 0.25 = 6$

Also, $n(n^2 - 1) = 7 \times (7^2 - 1) = 336$

Thus, using Equation (9.32) we get

$$r_s = 1 - \frac{6\sum_{i=1}^{n} d_i^2}{n(n^2 - 1)} = 1 - \frac{6 \times 6}{336} = 0.893$$

The result indicates that there is a reasonably good degree of association between two sets of ranks.

9.5 QUADRATIC REGRESSION

For quadratic fit,

$$\hat{y}_i = \hat{a}x_i^2 + \hat{b}x_i + \hat{c} \tag{9.33}$$

The least square method requires that we find a minimum of

$$\xi = \sum_{i=1}^{n}(y_i - \hat{y}_i)^2$$

$$= \sum_{i=1}^{n}(y_i - \hat{a}x_i^2 - \hat{b}x_i - \hat{c})^2 \tag{9.34}$$

The partial derivatives of ξ with respect to $\hat{c}, \hat{b}, \hat{a}$ are

$$\frac{\partial \xi}{\partial \hat{c}} = -2 \sum_{i=1}^{n} \left(y_i - \hat{a} x_i^2 - \hat{b} x_i - \hat{c} \right) \tag{9.35}$$

$$\frac{\partial \xi}{\partial \hat{b}} = -2 \sum_{i=1}^{n} \left(y_i - \hat{a} x_i^2 - \hat{b} x_i - \hat{c} \right) x_i \tag{9.36}$$

$$\frac{\partial \xi}{\partial \hat{a}} = -2 \sum_{i=1}^{n} \left(y_i - \hat{a} x_i^2 - \hat{b} x_i - \hat{c} \right) x_i^2 = 0 \tag{9.37}$$

Setting the partial derivatives in Equation (9.35) to Equation (9.37) equal to zero, the resulting normal equations becomes

$$\sum_{i=1}^{n} y_i = \hat{a} \sum_{i=1}^{n} x_i^2 + \hat{b} \sum_{i=1}^{n} x_i + n\hat{c} \tag{9.38}$$

$$\sum_{i=1}^{n} x_i y_i = \hat{a} \sum_{i=1}^{n} x_i^3 + \hat{b} \sum_{i=1}^{n} x_i^2 + \hat{c} \sum_{i=1}^{n} x_i \tag{9.39}$$

$$\sum_{i=1}^{n} x_i^2 y_i = \hat{a} \sum_{i=1}^{n} x_i^4 + \hat{b} \sum_{i=1}^{n} x_i^3 + \hat{c} \sum_{i=1}^{n} x_i^2 \tag{9.40}$$

Example 9.5:

The following table shows the data of loads (N) applied at the end of a cantilever made of textile composite and the corresponding deflections (cm). Fit a quadratic equation with these data.

Load	Deflection
x	y
20	0.52
25	1.02
30	1.41
35	1.6
40	1.78
45	1.85
50	1.97
55	2.1

SOLUTION

From the data, we obtain the following table.

x	y	xy	x^2	x^2y	x^3	x^4
20	0.52	10.4	400	208	8000	160,000
25	1.02	25.5	625	637.5	15,625	390,625
30	1.41	42.3	900	1269	27,000	810,000
35	1.6	56	1225	1960	42,875	1,500,625
40	1.78	71.2	1600	2848	64,000	2,560,000
45	1.85	83.25	2025	3746.25	91,125	4,100,625
50	1.97	98.5	2500	4925	125,000	6,250,000
55	2.1	115.5	3025	6352.5	166,375	9,150,625
Total 300	12.25	502.65	12,300	21,946.25	540,000	24,922,500

Here, $n = 8$. Using Equation (9.38) to Equation (9.40) we get

$$12300\hat{a} + 300\hat{b} + 8\hat{c} = 12.25$$

$$540,000\hat{a} + 12,300\hat{b} + 300\hat{c} = 502.65$$

$$24,922,500\hat{a} + 540,000\hat{b} + 12,300\hat{c} = 21,946.25$$

The above system of equations in equivalent matrix form can be written as

$$\begin{bmatrix} 12,300 & 300 & 8 \\ 540,000 & 12,300 & 300 \\ 24,922,500 & 540,000 & 12,300 \end{bmatrix} \begin{bmatrix} \hat{a} \\ \hat{b} \\ \hat{c} \end{bmatrix} = \begin{bmatrix} 12.25 \\ 502.65 \\ 21,946.25 \end{bmatrix}$$

Thus,

$$\begin{bmatrix} \hat{a} \\ \hat{b} \\ \hat{c} \end{bmatrix} = \begin{bmatrix} 12,300 & 300 & 8 \\ 540,000 & 12,300 & 300 \\ 24,922,500 & 540,000 & 12,300 \end{bmatrix}^{-1} \begin{bmatrix} 12.25 \\ 502.65 \\ 21,946.25 \end{bmatrix} = \begin{bmatrix} -0.00127 \\ 0.13675 \\ -1.6385 \end{bmatrix}$$

So, the equation of regression curve is estimated as

$$\hat{y} = -0.00127x^2 + 0.13675x - 1.6385$$

This curve is drawn in Figure 9.6.

9.6 MULTIPLE LINEAR REGRESSION

For linear fit of k number of multiple variables,

$$\hat{y}_i = \hat{m}_1 x_{1i} + \hat{m}_2 x_{2i} + \cdots + \hat{m}_k x_{ki} + \hat{c} \tag{9.41}$$

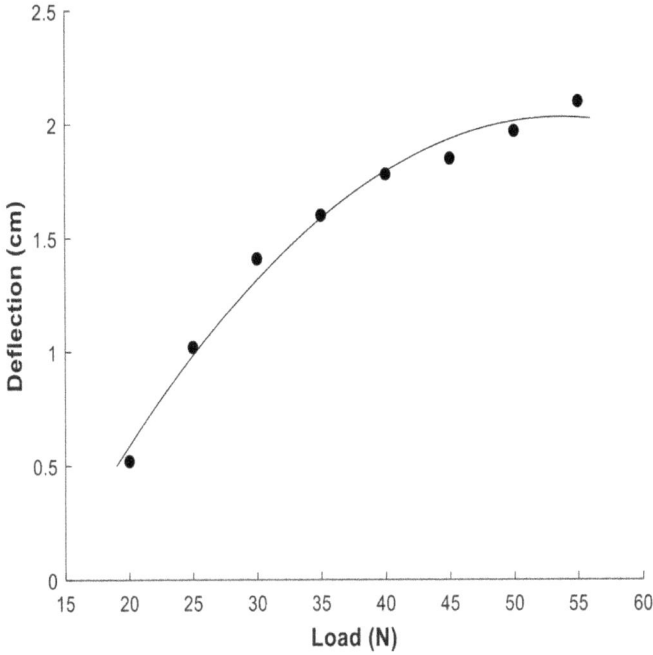

FIGURE 9.6 Fitting of quadratic equation with the data on load and deflection.

The least square method requires that we find a minimum of

$$\xi = \sum_{i=1}^{n} \left(y_i - \hat{y}_i \right)^2$$

$$= \sum_{i=1}^{n} \left(y_i - \hat{m}_1 x_{1i} - \hat{m}_2 x_{2i} - \cdots - \hat{m}_k x_{ki} - \hat{c} \right)^2$$

The partial derivatives of ξ with respect to $\hat{c}$, $\hat{m}_1$, $\hat{m}_2$, $\cdots$, $\hat{m}_k$ are

$$\frac{\partial \xi}{\partial \hat{c}} = -2 \sum_{i=1}^{n} \left(y_i - \hat{m}_1 x_{1i} - \hat{m}_2 x_{2i} - \cdots - \hat{m}_k x_{ki} - \hat{c} \right)$$

$$\frac{\partial \xi}{\partial \hat{m}_1} = -2 \sum_{i=1}^{n} \left(y_i - \hat{m}_1 x_{1i} - \hat{m}_2 x_{2i} - \cdots - \hat{m}_k x_{ki} - \hat{c} \right) x_{1i}$$

$$\frac{\partial \xi}{\partial \hat{m}_2} = -2 \sum_{i=1}^{n} \left(y_i - \hat{m}_1 x_{1i} - \hat{m}_2 x_{2i} - \cdots - \hat{m}_k x_{ki} - \hat{c} \right) x_{2i}$$

$$\vdots$$

$$\frac{\partial \xi}{\partial \hat{m}_k} = -2 \sum_{i=1}^{n} \left(y_i - \hat{m}_1 x_{1i} - \hat{m}_2 x_{2i} - \cdots - \hat{m}_k x_{ki} - \hat{c} \right) x_{ki}$$

Setting the above partial derivatives equal to zero, the resulting normal equations becomes

$$\sum_{i=1}^{n} y_i = \hat{m}_1 \sum_{i=1}^{n} x_{1i} + \hat{m}_2 \sum_{i=1}^{n} x_{2i} + \cdots \hat{m}_k \sum_{i=1}^{n} x_{ki} + n\hat{c}$$

$$\sum_{i=1}^{n} x_{1i} y_i = \hat{m}_1 \sum_{i=1}^{n} x_{1i}^2 + \hat{m}_2 \sum_{i=1}^{n} x_{1i} x_{2i} + \cdots \hat{m}_k \sum_{i=1}^{n} x_{1i} x_{ki} + \hat{c} \sum_{i=1}^{n} x_{1i}$$

$$\sum_{i=1}^{n} x_{2i} y_i = \hat{m}_1 \sum_{i=1}^{n} x_{1i} x_{2i} + \hat{m}_2 \sum_{i=1}^{n} x_{2i}^2 + \cdots \hat{m}_k \sum_{i=1}^{n} x_{2i} x_{ki} + \hat{c} \sum_{i=1}^{n} x_{2i}$$

$$\vdots$$

$$\sum_{i=1}^{n} x_{ki} y_i = \hat{m}_1 \sum_{i=1}^{n} x_{1i} x_{ki} + \hat{m}_2 \sum_{i=1}^{n} x_{2i} x_{ki} + \cdots \hat{m}_k \sum_{i=1}^{n} x_{ki}^2 + \hat{c} \sum_{i=1}^{n} x_{ki}$$

For two independent variables,

$$\hat{y}_i = \hat{m}_1 x_{1i} + \hat{m}_2 x_{2i} + \hat{c} \tag{9.42}$$

The resulting normal equations are

$$\sum_{i=1}^{n} y_i = \hat{m}_1 \sum_{i=1}^{n} x_{1i} + \hat{m}_2 \sum_{i=1}^{n} x_{2i} + n\hat{c} \tag{9.43}$$

$$\sum_{i=1}^{n} x_{1i} y_i = \hat{m}_1 \sum_{i=1}^{n} x_{1i}^2 + \hat{m}_2 \sum_{i=1}^{n} x_{1i} x_{2i} + \hat{c} \sum_{i=1}^{n} x_{1i} \tag{9.44}$$

$$\sum_{i=1}^{n} x_{2i} y_i = \hat{m}_1 \sum_{i=1}^{n} x_{1i} x_{2i} + \hat{m}_2 \sum_{i=1}^{n} x_{2i}^2 + \hat{c} \sum_{i=1}^{n} x_{2i} \tag{9.45}$$

Example 9.6:

The table below shows the yarn thick faults (x_1), yarn neps (x_2) and fabric faults (y). Find the linear least square regression equation of y on x_1 and x_2.

x_1	x_2	y
57	8	64
59	10	71
49	6	53
62	11	67
51	8	55
50	7	58
55	10	77
48	9	57
52	10	56
42	6	51
61	12	76
57	9	68

SOLUTION

From the data, we obtain the following table.

x_1	x_2	y_1	x_1^2	x_2^2	x_1x_2	x_1y	x_2y
57	8	64	3249	64	456	3648	512
59	10	71	3481	100	590	4189	710
49	6	53	2401	36	294	2597	318
62	11	67	3844	121	682	4154	737
51	8	55	2601	64	408	2805	440
50	7	58	2500	49	350	2900	406
55	10	77	3025	100	550	4235	770
48	9	57	2304	81	432	2736	513
52	10	56	2704	100	520	2912	560
42	6	51	1764	36	252	2142	306
61	12	76	3721	144	732	4636	912
57	9	68	3249	81	513	3876	612
Total 643	106	753	34,843	976	5779	40,830	6796

Here, $n = 12$. Using Equation (9.43) to Equation (9.45), we get

$$643\hat{m}_1 + 106\hat{m}_2 + 12\hat{c} = 753$$

$$34,843\hat{m}_1 + 5779\hat{m}_2 + 643\hat{c} = 40,830$$

$$5779\hat{m}_1 + 976\hat{m}_2 + 106\hat{c} = 6796$$

The above system of equations in equivalent matrix form can be written as

$$\begin{bmatrix} 643 & 106 & 12 \\ 34,843 & 5779 & 643 \\ 5779 & 976 & 106 \end{bmatrix} \begin{bmatrix} \hat{m}_1 \\ \hat{m}_2 \\ \hat{c} \end{bmatrix} = \begin{bmatrix} 753 \\ 40,830 \\ 6796 \end{bmatrix}$$

Thus,

$$\begin{bmatrix} \hat{m}_1 \\ \hat{m}_2 \\ \hat{c} \end{bmatrix} = \begin{bmatrix} 643 & 106 & 12 \\ 34,843 & 5779 & 643 \\ 5779 & 976 & 106 \end{bmatrix}^{-1} \begin{bmatrix} 753 \\ 40,830 \\ 6796 \end{bmatrix} = \begin{bmatrix} 0.8546 \\ 1.5063 \\ 3.6512 \end{bmatrix}$$

So, the regression equation is estimated as

$$\hat{y} = 0.8546x_1 + 1.5063x_2 + 3.6512$$

Figure 9.7 depicts the best fitted surface plot.

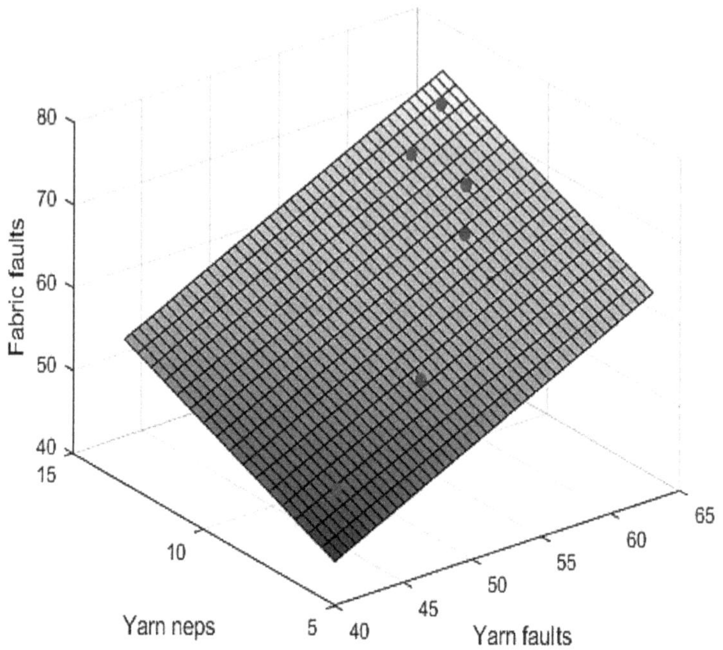

FIGURE 9.7   The surface plot of fabric faults with the data on yarn neps and yarn faults.

9.7   REGRESSION USING MATRIX APPROACH

9.7.1   LINEAR REGRESSION

For linear regression, the best fit line has the following equation

$$\hat{y}_i = \hat{m}x_i + \hat{c}$$

and

$$\hat{y}_i = y_i - e_i$$

For $i = 1, 2, \cdots n$, we thus write

$$\hat{m}x_1 + \hat{c} = y_1 - e_1$$
$$\hat{m}x_2 + \hat{c} = y_2 - e_2$$
$$\vdots$$
$$\hat{m}x_n + \hat{c} = y_n - e_n$$

Using the matrix notation we can write

$$\begin{bmatrix} x_1 & 1 \\ x_2 & 1 \\ \vdots & \vdots \\ x_n & 1 \end{bmatrix} \begin{bmatrix} \hat{m} \\ \hat{c} \end{bmatrix} = \begin{bmatrix} y_1 \\ y_2 \\ \vdots \\ y_n \end{bmatrix} - \begin{bmatrix} e_1 \\ e_2 \\ \vdots \\ e_n \end{bmatrix} \tag{9.46}$$

Suppose that X, B, Y and e denote the matrices as follows

$$X = \begin{bmatrix} x_1 & 1 \\ x_2 & 1 \\ \vdots & \vdots \\ x_n & 1 \end{bmatrix}, B = \begin{bmatrix} \hat{m} \\ \hat{c} \end{bmatrix}, Y = \begin{bmatrix} y_1 \\ y_2 \\ \vdots \\ y_n \end{bmatrix}, \text{ and } e = \begin{bmatrix} e_1 \\ e_2 \\ \vdots \\ e_n \end{bmatrix}$$

Thus Equation (9.46) becomes

$$XB = Y - e$$

or, $e = Y - XB$

Now the error sum of square can be expressed as

$$\sum_{i=1}^{n} e_i^2 = \begin{bmatrix} e_1 & e_2 \cdots e_n \end{bmatrix} \begin{bmatrix} e_1 \\ e_2 \\ \vdots \\ e_n \end{bmatrix}$$

$$= e'e$$

where e' is the transpose of e.

Thus, the error sum of square becomes

$$\xi = e'e$$
$$= (Y - XB)'(Y - XB) \qquad (9.47)$$
$$= (Y' - B'X')(Y - XB)$$
$$= Y'Y - B'X'Y - Y'XB + B'X'XB$$

The partial derivative of ξ with respect to B is

$$\frac{\partial \xi}{\partial B} = -X'Y - Y'X + 2X'XB$$

As $X'Y = Y'X$, we can write

$$\frac{\partial \xi}{\partial B} = -2X'Y + 2X'XB$$

Setting $\dfrac{\partial \xi}{\partial B} = 0$, we get

$$X'XB = X'Y$$

Thus,

$$B = (X'X)^{-1} X'Y \qquad (9.48)$$

where $(X'X)^{-1}$ is the inverse of $X'X$. Here we have assumed that $X'X$ is a non-singular matrix so that its inverse exists. $(X'X)^{-1} X'$ is called the pseudo-inverse of X.

Example 9.7:

The following data show the relationship between yarn count in the English system (Ne) and yarn twist per meter (TPM). Fit a straight-line equation that will enable us to predict the yarn TPM in terms of yarn count.

Yarn Count (Ne)	TPM
16	620
24	715
30	800
40	910
60	1100
80	1260

SOLUTION

From the data, we obtain the following matrices.

$$X = \begin{bmatrix} 16 & 1 \\ 24 & 1 \\ 30 & 1 \\ 40 & 1 \\ 60 & 1 \\ 80 & 1 \end{bmatrix}, Y = \begin{bmatrix} 620 \\ 715 \\ 800 \\ 910 \\ 1100 \\ 1260 \end{bmatrix}$$

Thus,

$$X'X = \begin{bmatrix} 13,332 & 250 \\ 250 & 6 \end{bmatrix} \text{ and } X'Y = \begin{bmatrix} 254,280 \\ 5405 \end{bmatrix}$$

Now, $(X'X)^{-1} = \begin{bmatrix} 0.00 & -0.01 \\ -0.01 & 0.76 \end{bmatrix}$

Therefore,

$$B = (X'X)^{-1} X'Y$$

$$= \begin{bmatrix} 0.00 & -0.01 \\ -0.01 & 0.76 \end{bmatrix} \begin{bmatrix} 254,280 \\ 5405 \end{bmatrix}$$

$$= \begin{bmatrix} 9.97 \\ 485.33 \end{bmatrix}$$

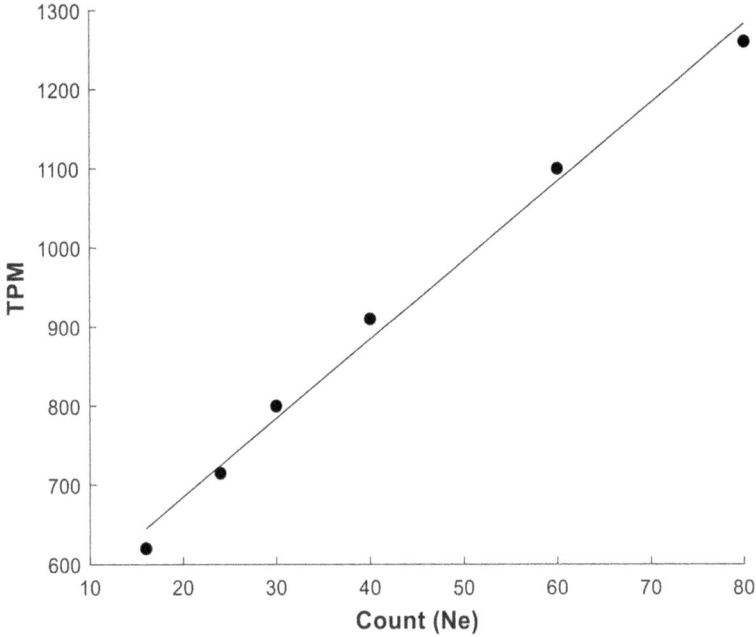

FIGURE 9.8 Fitted straight line with the data on yarn count and TPM.

So, the equation of regression line is estimated as

$$\hat{y} = 9.97x + 485.33$$

This line is drawn in Figure 9.8.

9.7.2 QUADRATIC REGRESSION

For the quadratic regression of single independent variable, the best fit curve has the following equation

$$\hat{y}_i = \hat{a}x_i^2 + \hat{b}x_i + \hat{c}$$

and

$$\hat{y}_i = y_i - e_i$$

Therefore, for $i = 1, 2, \cdots n$, we can write

$$\hat{a}x_1^2 + \hat{b}x_1 + \hat{c} = y_1 - e_1$$

$$\hat{a}x_2^2 + \hat{b}x_2 + \hat{c} = y_2 - e_2$$

$$\vdots$$

$$\hat{a}x_n^2 + \hat{b}x_n + \hat{c} = y_n - e_n$$

The matrix notation of quadratic regression thus becomes

$$
\begin{bmatrix} x_1^2 & x_1 & 1 \\ x_2^2 & x_2 & 1 \\ \vdots & \vdots & \vdots \\ x_n^2 & x_n & 1 \end{bmatrix} \begin{bmatrix} \hat{a} \\ \hat{b} \\ \hat{c} \end{bmatrix} = \begin{bmatrix} y_1 \\ y_2 \\ \vdots \\ y_n \end{bmatrix} - \begin{bmatrix} e_1 \\ e_2 \\ \vdots \\ e_n \end{bmatrix}
$$

Let us define the following matrices

$$
X = \begin{bmatrix} x_1^2 & x_1 & 1 \\ x_2^2 & x_2 & 1 \\ \vdots & \vdots & \vdots \\ x_n^2 & x_n & 1 \end{bmatrix}, B = \begin{bmatrix} \hat{a} \\ \hat{b} \\ \hat{c} \end{bmatrix}, Y = \begin{bmatrix} y_1 \\ y_2 \\ \vdots \\ y_n \end{bmatrix}, \text{ and } e = \begin{bmatrix} e_1 \\ e_2 \\ \vdots \\ e_n \end{bmatrix}.
$$

The least square estimates of the quadratic regression coefficients are given as

$$
B = (X'X)^{-1} X'Y
$$

Example 9.8:

The values of twist multiplier (TM) and respective yarn tenacity are given in the table below. Fit a quadratic equation with these data.

TM	Yarn Tenacity (cN/tex)
x	y
3.2	14.2
3.5	16.1
3.7	18.3
4	20.2
4.2	20.5
4.5	21.3
4.8	21.1
5	20.6
5.2	19.7
5.5	19.5

SOLUTION

From the data, we obtain the following matrices.

$$X = \begin{bmatrix} 10.24 & 3.2 & 1 \\ 12.25 & 3.5 & 1 \\ 13.69 & 3.7 & 1 \\ 16.00 & 4.0 & 1 \\ 17.64 & 4.2 & 1 \\ 20.25 & 4.5 & 1 \\ 23.04 & 4.8 & 1 \\ 25.00 & 5.0 & 1 \\ 27.04 & 5.2 & 1 \\ 30.25 & 5.5 & 1 \end{bmatrix}, Y = \begin{bmatrix} 14.2 \\ 16.1 \\ 18.3 \\ 20.2 \\ 20.5 \\ 21.3 \\ 21.1 \\ 20.6 \\ 19.7 \\ 19.5 \end{bmatrix}$$

Thus,

$$X'X = \begin{bmatrix} 4221.63 & 898.08 & 195.4 \\ 898.08 & 195.40 & 43.63 \\ 195.40 & 43.60 & 10.00 \end{bmatrix}, \text{ and } X'Y = \begin{bmatrix} 3833.01 \\ 846.22 \\ 191.50 \end{bmatrix}$$

Now $(X'X)^{-1} = \begin{bmatrix} 0.47 & -4.06 & 8.59 \\ -4.06 & 35.52 & -75.51 \\ 8.59 & -75.51 & 161.55 \end{bmatrix}$

Therefore,

$$B = (X'X)^{-1} X'Y$$

$$= \begin{bmatrix} 0.47 & -4.06 & 8.59 \\ -4.06 & 35.52 & -75.51 \\ 8.59 & -75.51 & 161.55 \end{bmatrix} \begin{bmatrix} 3833.01 \\ 846.22 \\ 191.50 \end{bmatrix}$$

$$= \begin{bmatrix} -3.28 \\ 30.65 \\ -50.43 \end{bmatrix}$$

So, the equation of regression curve is estimated as

$$\hat{y} = -3.28x^2 + 30.65x - 50.43$$

This regression curve is drawn in Figure 9.9.

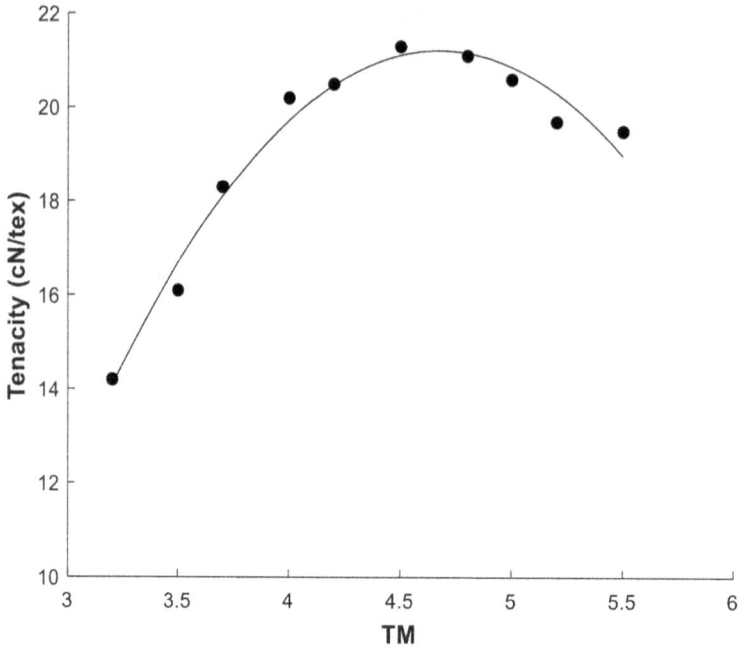

FIGURE 9.9 Fitting of quadratic equation with the data on TM and tenacity.

9.7.3 MULTIPLE LINEAR REGRESSION

For multiple linear regression of k number of independent variables, the best fitted equation has the following form

$$\hat{y}_i = \hat{m}_1 x_{1i} + \hat{m}_2 x_{2i} + \cdots + \hat{m}_k x_{ki} + \hat{c}$$

and

$$\hat{y}_i = y_i - e_i$$

Therefore, for $i = 1, 2, \cdots n$, we can write

$$\hat{m}_1 x_{11} + \hat{m}_2 x_{21} + \cdots + \hat{m}_k x_{k1} + \hat{c} = y_1 - e_1$$
$$\hat{m}_1 x_{12} + \hat{m}_2 x_{22} + \cdots + \hat{m}_k x_{k2} + \hat{c} = y_2 - e_2$$
$$\vdots$$
$$\hat{m}_1 x_{1n} + \hat{m}_2 x_{2n} + \cdots + \hat{m}_k x_{kn} + \hat{c} = y_n - e_n$$

The matrix notation of multiple linear regression thus becomes

$$\begin{bmatrix} x_{11} & x_{21} & \cdots & x_{k1} & 1 \\ x_{12} & x_{22} & \cdots & x_{k2} & 1 \\ \vdots & \vdots & \vdots & \vdots & \vdots \\ x_{1n} & x_{2n} & \cdots & x_{kn} & 1 \end{bmatrix} \begin{bmatrix} \hat{m}_1 \\ \hat{m}_2 \\ \vdots \\ \hat{m}_n \\ \hat{c} \end{bmatrix} = \begin{bmatrix} y_1 \\ y_2 \\ \vdots \\ y_n \end{bmatrix} - \begin{bmatrix} e_1 \\ e_2 \\ \vdots \\ e_n \end{bmatrix}$$

Let us define the following matrices

$$X = \begin{bmatrix} x_{11} & x_{21} & \cdots & x_{k1} & 1 \\ x_{12} & x_{22} & \cdots & x_{k2} & 1 \\ \vdots & \vdots & \vdots & \vdots & \vdots \\ x_{1n} & x_{2n} & \cdots & x_{kn} & 1 \end{bmatrix}, B = \begin{bmatrix} \hat{m}_1 \\ \hat{m}_2 \\ \vdots \\ \hat{m}_n \\ \hat{c} \end{bmatrix},$$

$$Y = \begin{bmatrix} y_1 \\ y_2 \\ \vdots \\ y_n \end{bmatrix}, \text{ and } e = \begin{bmatrix} e_1 \\ e_2 \\ \vdots \\ e_n \end{bmatrix}$$

The least square estimates of the multiple linear regression coefficients are given as

$$B = (X'X)^{-1} X'Y$$

Example 9.9:

An experiment was conducted in a ring frame to find out the effect of delivery speed and yarn count on yarn hairiness. The results of the experiment are shown in the table below. Find the linear least square regression equation of yarn hairiness as a function of delivery speed and yarn count.

Delivery Speed (m/min) x_1	Yarn Count (Ne) x_2	Hairiness Index y
20	16	4.9
20	24	4.7
20	30	4.5
23	16	5.3
23	24	5.2
23	30	5.0
26	16	6.0
26	24	5.7
26	30	5.5

SOLUTION

From the data, we obtain the following matrices.

$$X = \begin{bmatrix} 20 & 16 & 1 \\ 20 & 24 & 1 \\ 20 & 30 & 1 \\ 23 & 16 & 1 \\ 23 & 24 & 1 \\ 23 & 30 & 1 \\ 26 & 16 & 1 \\ 26 & 24 & 1 \\ 26 & 30 & 1 \end{bmatrix}, Y = \begin{bmatrix} 4.9 \\ 4.7 \\ 4.5 \\ 5.3 \\ 5.2 \\ 5 \\ 6 \\ 5.7 \\ 5.5 \end{bmatrix}$$

Thus,

$$X'X = \begin{bmatrix} 4815 & 4830 & 207 \\ 4830 & 5196 & 210 \\ 207 & 210 & 9 \end{bmatrix} \text{ and } X'Y = \begin{bmatrix} 1085.70 \\ 1083.60 \\ 46.80 \end{bmatrix}$$

Now, $(X'X)^{-1} = \begin{bmatrix} 0.02 & 0 & -0.43 \\ 0 & 0 & -0.08 \\ -0.43 & -0.08 & 11.75 \end{bmatrix}$

Therefore,

$$B = (X'X)^{-1} X'Y$$

$$= \begin{bmatrix} 0.02 & 0 & -0.43 \\ 0 & 0 & -0.08 \\ -0.43 & -0.08 & 11.75 \end{bmatrix} \begin{bmatrix} 1085.70 \\ 1083.60 \\ 46.80 \end{bmatrix}$$

$$= \begin{bmatrix} 0.17 \\ -0.03 \\ 1.90 \end{bmatrix}$$

So, the regression equation is estimated as

$$\hat{y} = 0.17x_1 - 0.03x_2 + 1.9$$

Figure 9.10 depicts the best fitted surface plot.

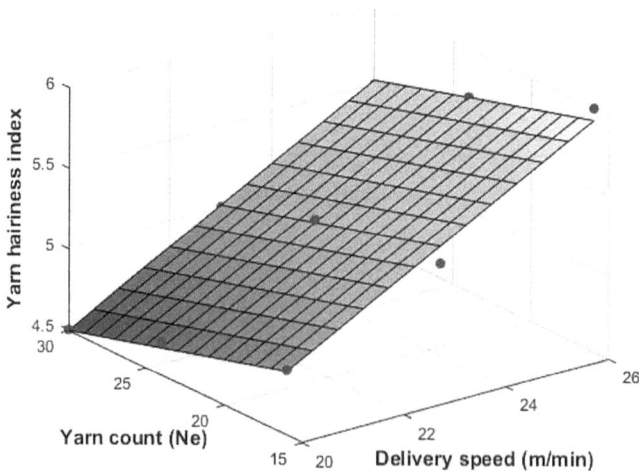

FIGURE 9.10 Surface plot of yarn hairiness index with the data on yarn count and delivery speed.

9.7.4 MULTIPLE QUADRATIC REGRESSION

For multiple quadratic regression, the best fitted equation with k number of independent variables has the following generalized form

$$\hat{y}_i = \hat{b}_0 + \sum_{j=1}^{k} \hat{b}_j x_{ji} + \sum_{j=1}^{k-1}\sum_{p=j+1}^{k} b_{jp} x_{ji} x_{pi} + \sum_{j=1}^{k} b_{jj} x_{ji}^2 \qquad (9.49)$$

Thus for three independent variables, x_1, x_2, and x_3, the generalized equation becomes

$$\hat{y}_i = \hat{b}_0 + \hat{b}_1 x_{1i} + \hat{b}_2 x_{2i} + \hat{b}_3 x_{3i} + \hat{b}_{12} x_{1i} x_{2i} + \hat{b}_{13} x_{1i} x_{3i} + \hat{b}_{23} x_{2i} x_{3i} + \hat{b}_{11} x_{1i}^2 + \hat{b}_{22} x_{2i}^2 + \hat{b}_{33} x_{3i}^2$$

and

$$\hat{y}_i = y_i - e_i$$

Therefore, for $i = 1, 2, \cdots n$, we can write

$$\hat{b}_0 + \hat{b}_1 x_{11} + \hat{b}_2 x_{21} + \hat{b}_3 x_{31} + \hat{b}_{12} x_{11} x_{21} + \hat{b}_{13} x_{11} x_{31} + \hat{b}_{23} x_{21} x_{31}$$
$$+ \hat{b}_{11} x_{11}^2 + \hat{b}_{22} x_{21}^2 + \hat{b}_{33} x_{31}^2 = y_1 - e_1$$

$$\hat{b}_0 + \hat{b}_1 x_{12} + \hat{b}_2 x_{22} + \hat{b}_3 x_{32} + \hat{b}_{12} x_{12} x_{22} + \hat{b}_{13} x_{12} x_{32} + \hat{b}_{23} x_{22} x_{32}$$
$$+ \hat{b}_{11} x_{12}^2 + \hat{b}_{22} x_{22}^2 + \hat{b}_{33} x_{32}^2 = y_2 - e_2$$
$$\vdots$$

$$\hat{b}_0 + \hat{b}_1 x_{1n} + \hat{b}_2 x_{2n} + \hat{b}_3 x_{3n} + \hat{b}_{12} x_{1n} x_{2n} + \hat{b}_{13} x_{1n} x_{3n} + \hat{b}_{23} x_{2n} x_{3n}$$
$$+ \hat{b}_{11} x_{1n}^2 + \hat{b}_{22} x_{2n}^2 + \hat{b}_{33} x_{3n}^2 = y_n - e_n$$

The matrix notation of multiple quadratic regression thus becomes

$$\begin{bmatrix} 1 & x_{11} & x_{21} & x_{31} & x_{11}x_{21} & x_{11}x_{31} & x_{21}x_{31} & x_{11}^2 & x_{21}^2 & x_{31}^2 \\ 1 & x_{12} & x_{22} & x_{32} & x_{12}x_{22} & x_{12}x_{32} & x_{22}x_{32} & x_{12}^2 & x_{22}^2 & x_{32}^2 \\ \vdots & \vdots & \vdots & \vdots & \vdots & \vdots & \vdots & \vdots & \vdots & \vdots \\ 1 & x_{1n} & x_{2n} & x_{3n} & x_{1n}x_{2n} & x_{1n}x_{3n} & x_{2n}x_{3n} & x_{1n}^2 & x_{2n}^2 & x_{3n}^2 \end{bmatrix} \begin{bmatrix} \hat{b}_0 \\ \hat{b}_1 \\ \hat{b}_2 \\ \hat{b}_3 \\ \hat{b}_{12} \\ \hat{b}_{13} \\ \hat{b}_{23} \\ \hat{b}_{11} \\ \hat{b}_{22} \\ \hat{b}_{33} \end{bmatrix} = \begin{bmatrix} y_1 \\ y_2 \\ \vdots \\ y_n \end{bmatrix} - \begin{bmatrix} e_1 \\ e_2 \\ \vdots \\ e_n \end{bmatrix}$$

Let us define the following matrices

$$
X = \begin{bmatrix}
1 & x_{11} & x_{21} & x_{31} & x_{11}x_{21} & x_{11}x_{31} & x_{21}x_{31} & x_{11}^2 & x_{21}^2 & x_{31}^2 \\
1 & x_{12} & x_{22} & x_{32} & x_{12}x_{22} & x_{12}x_{32} & x_{22}x_{32} & x_{12}^2 & x_{22}^2 & x_{32}^2 \\
\vdots & \vdots & \vdots & \vdots & \vdots & \vdots & \vdots & \vdots & \vdots & \vdots \\
1 & x_{1n} & x_{2n} & x_{3n} & x_{1n}x_{2n} & x_{1n}x_{3n} & x_{2n}x_{3n} & x_{1n}^2 & x_{2n}^2 & x_{3n}^2
\end{bmatrix}, \quad B = \begin{bmatrix}
\hat{b}_0 \\
\hat{b}_1 \\
\hat{b}_2 \\
\hat{b}_3 \\
\hat{b}_{12} \\
\hat{b}_{13} \\
\hat{b}_{23} \\
\hat{b}_{11} \\
\hat{b}_{22} \\
\hat{b}_{33}
\end{bmatrix},
$$

$$
Y = \begin{bmatrix} y_1 \\ y_2 \\ \vdots \\ y_n \end{bmatrix}, \text{ and } e = \begin{bmatrix} e_1 \\ e_2 \\ \vdots \\ e_n \end{bmatrix}.
$$

The least square estimates of the multiple linear regression coefficients are given as

$$
B = \left(X'X \right)^{-1} X'Y
$$

In the case of two independent variables, x_1 and x_2, the best fitted generalized equation becomes

$$
\hat{y}_i = \hat{b}_0 + \hat{b}_1 x_{1i} + \hat{b}_2 x_{2i} + \hat{b}_{12} x_{1i} x_{2i} + \hat{b}_{11} x_{1i}^2 + \hat{b}_{22} x_{2i}^2
$$

and the matrices X, B, Y and e have the following forms

$$
X = \begin{bmatrix}
1 & x_{11} & x_{21} & x_{11}x_{21} & x_{11}^2 & x_{21}^2 \\
1 & x_{12} & x_{22} & x_{12}x_{22} & x_{12}^2 & x_{22}^2 \\
\vdots & \vdots & \vdots & \vdots & \vdots & \vdots \\
1 & x_{1n} & x_{2n} & x_{1n}x_{2n} & x_{1n}^2 & x_{2n}^2
\end{bmatrix}, \quad B = \begin{bmatrix}
\hat{b}_0 \\
\hat{b}_1 \\
\hat{b}_2 \\
\hat{b}_{12} \\
\hat{b}_{11} \\
\hat{b}_{22}
\end{bmatrix},
$$

$$
Y = \begin{bmatrix} y_1 \\ y_2 \\ \vdots \\ y_n \end{bmatrix}, \text{ and } e = \begin{bmatrix} e_1 \\ e_2 \\ \vdots \\ e_n \end{bmatrix}.
$$

In the case of multiple quadratic regression, usually coded values are used for the independent variables. The conversion of the actual value to the coded value can be obtained by the following transformation

$$x_i = 2\left(\frac{A_i - A_{min}}{A_{max} - A_{min}}\right) - 1 \tag{9.50}$$

where x_i and A_i are the coded and original values of ith independent variable, respectively; A_{min} and A_{max} are the minimum and maximum of A_i, respectively. The coded value of the variable ranges from −1 to 1. The advantage of using the coded values of independent variables lies in the fact that all the variables have same domain of variation. When the independent variables are measured in different engineering units, one obtains different numerical results in comparison to coded unit analysis and often it may be difficult to interpret the interaction effects of independent variables on the dependent variable.

Example 9.10:

An experiment was carried out to find the effect of spindle rpm and yarn TM on yarn tenacity. Following table shows the results of the experiment. Find the quadratic least square regression equation of yarn tenacity as a function of spindle speed and yarn TM.

Spindle rpm	TM	Tenacity (cN/tex)
14,000	4	15.48
14,000	4.2	15.67
14,000	4.4	15.78
16,000	4	16.19
16,000	4.2	16.22
16,000	4.4	16.46
18,000	4	17.22
18,000	4.2	17.41
18,000	4.4	16.4

SOLUTION

From the data we obtain the following table where x_1 and x_2 are the coded values of spindle rpm and yarn TM, respectively, and y is the yarn tenacity.

x_1	x_2	y
−1	−1	15.48
−1	0	15.67
−1	1	15.78
0	−1	16.19
0	0	16.22
0	1	16.46
1	−1	17.22
1	0	17.41
1	1	16.4

Matrices **X** and **Y** are obtained as

$$X = \begin{bmatrix} 1 & -1 & -1 & 1 & 1 & 1 \\ 1 & -1 & 0 & 0 & 1 & 0 \\ 1 & -1 & 1 & -1 & 1 & 1 \\ 1 & 0 & -1 & 0 & 0 & 1 \\ 1 & 0 & 0 & 0 & 0 & 0 \\ 1 & 0 & 1 & 0 & 0 & 1 \\ 1 & 1 & -1 & -1 & 1 & 1 \\ 1 & 1 & 0 & 0 & 1 & 0 \\ 1 & 1 & 1 & 1 & 1 & 1 \end{bmatrix}, Y = \begin{bmatrix} 15.48 \\ 15.67 \\ 15.78 \\ 16.19 \\ 16.22 \\ 16.46 \\ 17.22 \\ 17.41 \\ 16.40 \end{bmatrix}$$

Thus,

$$X'X = \begin{bmatrix} 9 & 0 & 0 & 0 & 6 & 6 \\ 0 & 6 & 0 & 0 & 0 & 0 \\ 0 & 0 & 6 & 0 & 0 & 0 \\ 0 & 0 & 0 & 4 & 0 & 0 \\ 6 & 0 & 0 & 0 & 6 & 4 \\ 6 & 0 & 0 & 0 & 4 & 6 \end{bmatrix} \text{ and } X'Y = \begin{bmatrix} 146.83 \\ 4.10 \\ -0.25 \\ -1.12 \\ 97.96 \\ 97.53 \end{bmatrix}$$

Now, $(X'X)^{-1} = \begin{bmatrix} 0.56 & 0 & 0 & 0 & -0.33 & -0.33 \\ 0 & 0.17 & 0 & 0 & 0 & 0 \\ 0 & 0 & 0.17 & 0 & 0 & 0 \\ 0 & 0 & 0 & 0.25 & 0 & 0 \\ -0.33 & 0 & 0 & 0 & 0.5 & 0 \\ -0.33 & 0 & 0 & 0 & 0 & 0.5 \end{bmatrix}$

Therefore,

$$B = (X'X)^{-1} X'Y$$

$$= \begin{bmatrix} 0.56 & 0 & 0 & 0 & -0.33 & -0.33 \\ 0 & 0.17 & 0 & 0 & 0 & 0 \\ 0 & 0 & 0.17 & 0 & 0 & 0 \\ 0 & 0 & 0 & 0.25 & 0 & 0 \\ -0.33 & 0 & 0 & 0 & 0.5 & 0 \\ -0.33 & 0 & 0 & 0 & 0 & 0.5 \end{bmatrix} \begin{bmatrix} 146.83 \\ 4.10 \\ -0.25 \\ -1.12 \\ 97.96 \\ 97.53 \end{bmatrix}$$

$$= \begin{bmatrix} 16.41 \\ 0.68 \\ -0.04 \\ -0.28 \\ 0.04 \\ -0.18 \end{bmatrix}$$

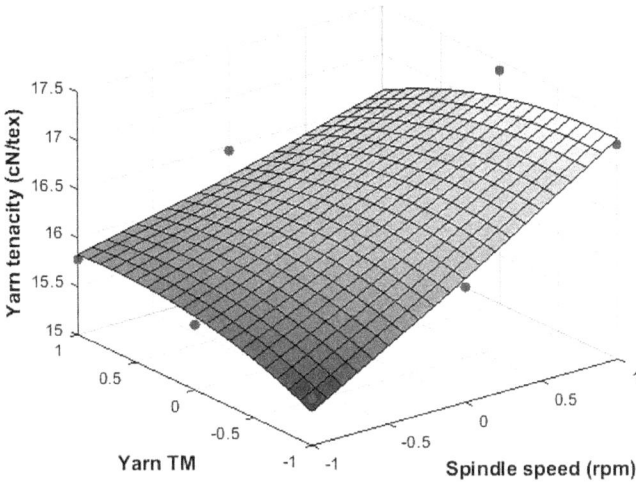

FIGURE 9.11   Surface plot of yarn tenacity with the coded values of spindle rpm and yarn TM.

So, the regression equation is estimated as

$$\hat{y} = 16.41 + 0.68x_1 - 0.04x_2 - 0.28x_1x_2 + 0.04x_1^2 - 0.18x_2^2$$

Figure 9.11 depicts the best fitted surface plot.

Example 9.11:

Following table shows the results of an experiment to study the influence of the proportion of polyester-cotton blend, yarn count and fabric sett (thread density) of a square fabric on its shear rigidity. Find the quadratic least square regression equation of fabric shear rigidity as a function of proportion of polyester, yarn count, and fabric sett.

Proportion of Polyester (%)	Yarn Count (Ne)	Fabric Sett (inch⁻¹)	Fabric Shear Rigidity (gf/cm.deg)
50	20	50	0.97
50	20	70	4.15
50	40	50	0.32
50	40	70	0.83
0	30	50	0.52
0	30	70	1.75
100	30	50	0.31
100	30	70	0.96
0	20	60	2.65
0	40	60	0.45
100	20	60	1.26
100	40	60	0.33
50	30	60	0.86
50	30	60	0.87
50	30	60	0.88

SOLUTION

From the data we obtain the following table where x_1, x_2 and x_3 are the coded values of proportion of polyester, yarn count and fabric sett, respectively, and y is the fabric shear rigidity.

x_1	x_2	x_3	y
0	−1	−1	0.97
0	−1	1	4.15
0	1	−1	0.32
0	1	1	0.83
−1	0	−1	0.52
−1	0	1	1.75
1	0	−1	0.31
1	0	1	0.96
−1	−1	0	2.65
−1	1	0	0.45
1	−1	0	1.26
1	1	0	0.33
0	0	0	0.86
0	0	0	0.87
0	0	0	0.88

Hence, the matrices X and Y are expressed as

$$X = \begin{bmatrix} 1 & 0 & -1 & -1 & 0 & 0 & 1 & 0 & 1 & 1 \\ 1 & 0 & -1 & 1 & 0 & 0 & -1 & 0 & 1 & 1 \\ 1 & 0 & 1 & -1 & 0 & 0 & -1 & 0 & 1 & 1 \\ 1 & 0 & 1 & 1 & 0 & 0 & 1 & 0 & 1 & 1 \\ 1 & -1 & 0 & -1 & 0 & 1 & 0 & 1 & 0 & 0 \\ 1 & -1 & 0 & 1 & 0 & -1 & 0 & 1 & 0 & 0 \\ 1 & 1 & 0 & -1 & 0 & -1 & 0 & 1 & 0 & 0 \\ 1 & 1 & 0 & 1 & 0 & 1 & 0 & 1 & 0 & 0 \\ 1 & -1 & -1 & 0 & 1 & 0 & 0 & 1 & 1 & 0 \\ 1 & -1 & 1 & 0 & -1 & 0 & 0 & 1 & 1 & 0 \\ 1 & 1 & -1 & 0 & -1 & 0 & 0 & 1 & 1 & 0 \\ 1 & 1 & 1 & 0 & 1 & 0 & 0 & 1 & 1 & 0 \\ 1 & 0 & 0 & 0 & 0 & 0 & 0 & 0 & 0 & 0 \\ 1 & 0 & 0 & 0 & 0 & 0 & 0 & 0 & 0 & 0 \\ 1 & 0 & 0 & 0 & 0 & 0 & 0 & 0 & 0 & 0 \end{bmatrix}, \ Y = \begin{bmatrix} 0.97 \\ 4.15 \\ 0.32 \\ 0.83 \\ 0.52 \\ 1.75 \\ 0.31 \\ 0.96 \\ 2.65 \\ 0.45 \\ 1.26 \\ 0.33 \\ 0.86 \\ 0.87 \\ 0.88 \end{bmatrix}$$

Thus,

$$X'X = \begin{bmatrix} 15 & 0 & 0 & 0 & 0 & 0 & 0 & 8 & 8 & 8 \\ 0 & 8 & 0 & 0 & 0 & 0 & 0 & 0 & 0 & 0 \\ 0 & 0 & 8 & 0 & 0 & 0 & 0 & 0 & 0 & 0 \\ 0 & 0 & 0 & 8 & 0 & 0 & 0 & 0 & 0 & 0 \\ 0 & 0 & 0 & 0 & 4 & 0 & 0 & 0 & 0 & 0 \\ 0 & 0 & 0 & 0 & 0 & 4 & 0 & 0 & 0 & 0 \\ 0 & 0 & 0 & 0 & 0 & 0 & 4 & 0 & 0 & 0 \\ 8 & 0 & 0 & 0 & 0 & 0 & 0 & 8 & 4 & 4 \\ 8 & 0 & 0 & 0 & 0 & 0 & 0 & 4 & 8 & 4 \\ 8 & 0 & 0 & 0 & 0 & 0 & 0 & 4 & 4 & 8 \end{bmatrix} \text{ and } X'Y = \begin{bmatrix} 17.11 \\ -2.51 \\ -7.10 \\ 5.57 \\ 1.27 \\ -0.58 \\ -2.67 \\ 8.23 \\ 10.96 \\ 9.81 \end{bmatrix}$$

Now

$$(X'X)^{-1} = \begin{bmatrix}
0.33 & 0 & 0 & 0 & 0 & 0 & 0 & -0.17 & -0.17 & -0.17 \\
0 & 0.13 & 0 & 0 & 0 & 0 & 0 & 0 & 0 & 0 \\
0 & 0 & 0.13 & 0 & 0 & 0 & 0 & 0 & 0 & 0 \\
0 & 0 & 0 & 0.13 & 0 & 0 & 0 & 0 & 0 & 0 \\
0 & 0 & 0 & 0 & 0.25 & 0 & 0 & 0 & 0 & 0 \\
0 & 0 & 0 & 0 & 0 & 0.25 & 0 & 0 & 0 & 0 \\
0 & 0 & 0 & 0 & 0 & 0 & 0.25 & 0 & 0 & 0 \\
-0.17 & 0 & 0 & 0 & 0 & 0 & 0 & 0.27 & 0.02 & 0.02 \\
-0.17 & 0 & 0 & 0 & 0 & 0 & 0 & 0.02 & 0.27 & 0.02 \\
-0.17 & 0 & 0 & 0 & 0 & 0 & 0 & 0.02 & 0.02 & 0.27
\end{bmatrix}$$

Therefore,

$$B = (X'X)^{-1} X'Y$$

$$= \begin{bmatrix}
0.33 & 0 & 0 & 0 & 0 & 0 & 0 & -0.17 & -0.17 & -0.17 \\
0 & 0.13 & 0 & 0 & 0 & 0 & 0 & 0 & 0 & 0 \\
0 & 0 & 0.13 & 0 & 0 & 0 & 0 & 0 & 0 & 0 \\
0 & 0 & 0 & 0.13 & 0 & 0 & 0 & 0 & 0 & 0 \\
0 & 0 & 0 & 0 & 0.25 & 0 & 0 & 0 & 0 & 0 \\
0 & 0 & 0 & 0 & 0 & 0.25 & 0 & 0 & 0 & 0 \\
0 & 0 & 0 & 0 & 0 & 0 & 0.25 & 0 & 0 & 0 \\
-0.17 & 0 & 0 & 0 & 0 & 0 & 0 & 0.27 & 0.02 & 0.02 \\
-0.17 & 0 & 0 & 0 & 0 & 0 & 0 & 0.02 & 0.27 & 0.02 \\
-0.17 & 0 & 0 & 0 & 0 & 0 & 0 & 0.02 & 0.02 & 0.27
\end{bmatrix} \begin{bmatrix}
17.11 \\
-2.51 \\
7.10 \\
5.57 \\
1.27 \\
-0.58 \\
-2.67 \\
8.23 \\
10.96 \\
9.81
\end{bmatrix}$$

$$= \begin{bmatrix}
0.87 \\
-0.31 \\
-0.89 \\
0.70 \\
0.32 \\
-0.15 \\
-0.67 \\
-0.19 \\
0.49 \\
0.21
\end{bmatrix}$$

So, the regression equation is estimated as

$$\hat{y} = 0.87 - 0.31x_1 - 0.89x_2 + 0.7x_3 + 0.32x_1x_2 - 0.15x_1x_3$$
$$- 0.67x_2x_3 - 0.19x_1^2 + 0.49x_2^2 + 0.21x_3^2$$

Figure 9.12(a) depicts the best fitted surface plot of fabric shear rigidity as a function of the proportion of polyester and yarn count, where the coded levels are used for both the independent variables. Similarly, the best fitted surface of fabric shear rigidity as a function of yarn count and fabric sett is shown in Figure 9.12(b).

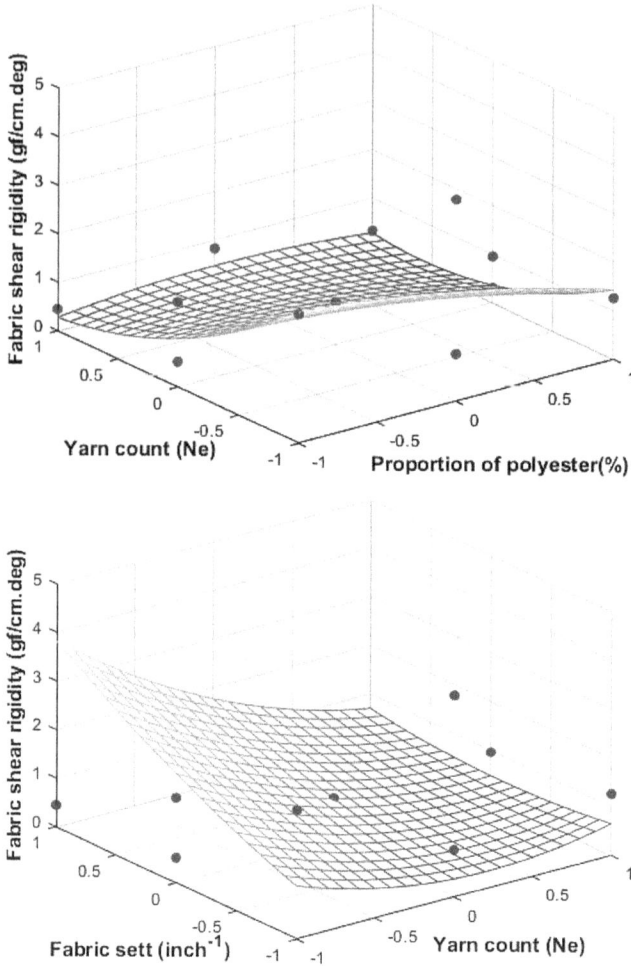

FIGURE 9.12 (a) Surface plot of fabric shear rigidity with the coded values of proportion of polyester and yarn count; (b) surface plot of fabric shear rigidity with the coded values of yarn count and fabric sett.

9.8 TEST OF SIGNIFICANCE OF REGRESSION COEFFICIENTS

In least square method, we assume that the errors (e_i) are independently normally distributed with means zero and common variance σ^2, for $i = 1, 2, \cdots, n$. From Equation (9.47), we get the error sum of square as

$$\xi = (Y - XB)'(Y - XB)$$

Thus, the estimate of σ^2 can be written as

$$S_e^2 = \frac{1}{n - k - 1}(Y - XB)'(Y - XB) \tag{9.51}$$

where S_e is termed as standard error of estimate, n is the number of observations, and k is the number of regression coefficients excluding the constant term. Now we have

$$B = (X'X)^{-1} X'Y$$

and

$$E(B) = \hat{B}$$

The observed value of $Y = E(Y) + e$

$$= E(XB + e) + e$$
$$= XE(B) + 0 + e$$
$$= X\hat{B} + e$$

Thus,

$$\left(B - \hat{B}\right)\left(B - \hat{B}\right)' = \left((X'X)^{-1} X'Y - \hat{B}\right)\left((X'X)^{-1} X'Y - \hat{B}\right)'$$
$$= \left((X'X)^{-1} X'\left(X\hat{B} + e\right) - \hat{B}\right)\left(Y'X(X'X)^{-1} - \hat{B}'\right)$$
$$= \left((X'X)^{-1} X'\left(X\hat{B} + e\right) - \hat{B}\right)\left(\left(X\hat{B} + e\right)' X(X'X)^{-1} - \hat{B}'\right)$$
$$= \left(\hat{B} + (X'X)^{-1} X'e - \hat{B}\right)\left(\hat{B}' + e'X(X'X)^{-1} - \hat{B}'\right)$$
$$= (X'X)^{-1} X'ee'X(X'X)^{-1}$$

It should be noted that

$$E\left((X'X)^{-1} X'\right) = (X'X)^{-1} X'$$

Therefore,

$$E\left(\left(B - \hat{B}\right)\left(B - \hat{B}\right)'\right) = E\left((X'X)^{-1} X'ee'X(X'X)^{-1}\right)$$
$$= (X'X)^{-1} X'E(ee')X(X'X)^{-1}$$

Now e is a multivariate normally distributed random variable with mean vector as null vector and $E(ee') = \sigma^2 I$ is the variance-covariance matrix of e. Hence, the above expression reduces to

$$E\left(\left(B - \hat{B}\right)\left(B - \hat{B}\right)'\right) = (X'X)^{-1} X'\left(\sigma^2 I\right)X(X'X)^{-1}$$
$$= \sigma^2 (X'X)^{-1}(X'X)(X'X)^{-1}$$
$$= \sigma^2 (X'X)^{-1}$$

Now an estimate of σ^2 is S_e^2. Therefore, the estimated variance-covariance matrix of the regression coefficients can be expressed as

$$var - cov(\mathbf{B}) = S_e^2 (\mathbf{X'X})^{-1} \qquad (9.52)$$

Thus, in order to obtain the estimated variance of B_j, for $j = 0, 1, \cdots, k$, the corresponding diagonal entry of $(\mathbf{X'X})^{-1}$ are to be multiplied by S_e^2. The standard error of B_j is the square root of its estimated variance. The t-statistic value of a regression coefficient is obtained by dividing each regression coefficient with its standard error. The probability of each regression coefficient is calculated from the corresponding t-statistic value at $(n - k - 1)$ degrees of freedom. For a regression coefficient, if the probability value is less than 0.05, then it is considered to be significant at 95% level of confidence.

Example 9.12:

Work out the test of significance of the regression coefficients of the Example 9.11.

SOLUTION

In Example 9.11, the values of n, k, X, B and Y are given as

$$n = 15, k = 9, \mathbf{X} = \begin{bmatrix}
1 & 0 & -1 & -1 & 0 & 0 & 1 & 0 & 1 & 1 \\
1 & 0 & -1 & 1 & 0 & 0 & -1 & 0 & 1 & 1 \\
1 & 0 & 1 & -1 & 0 & 0 & -1 & 0 & 1 & 1 \\
1 & 0 & 1 & 1 & 0 & 0 & 1 & 0 & 1 & 1 \\
1 & -1 & 0 & -1 & 0 & 1 & 0 & 1 & 0 & 0 \\
1 & -1 & 0 & 1 & 0 & -1 & 0 & 1 & 0 & 0 \\
1 & 1 & 0 & -1 & 0 & -1 & 0 & 1 & 0 & 0 \\
1 & 1 & 0 & 1 & 0 & 1 & 0 & 1 & 0 & 0 \\
1 & -1 & -1 & 0 & 1 & 0 & 0 & 1 & 1 & 0 \\
1 & -1 & 1 & 0 & -1 & 0 & 0 & 1 & 1 & 0 \\
1 & 1 & -1 & 0 & -1 & 0 & 0 & 1 & 1 & 0 \\
1 & 1 & 1 & 0 & 1 & 0 & 0 & 1 & 1 & 0 \\
1 & 0 & 0 & 0 & 0 & 0 & 0 & 0 & 0 & 0 \\
1 & 0 & 0 & 0 & 0 & 0 & 0 & 0 & 0 & 0 \\
1 & 0 & 0 & 0 & 0 & 0 & 0 & 0 & 0 & 0
\end{bmatrix},$$

$$\mathbf{B} = \begin{bmatrix}
0.87 \\
-0.31 \\
-0.89 \\
0.70 \\
0.32 \\
-0.15 \\
-0.67 \\
-0.19 \\
0.49 \\
0.21
\end{bmatrix}, \mathbf{Y} = \begin{bmatrix}
0.97 \\
4.15 \\
0.32 \\
0.83 \\
0.52 \\
1.75 \\
0.31 \\
0.96 \\
2.65 \\
0.45 \\
1.26 \\
0.33 \\
0.86 \\
0.87 \\
0.88
\end{bmatrix}$$

Now,

$$
\mathbf{XB} =
\begin{bmatrix}
1 & 0 & -1 & -1 & 0 & 0 & 1 & 0 & 1 & 1 \\
1 & 0 & -1 & 1 & 0 & 0 & -1 & 0 & 1 & 1 \\
1 & 0 & 1 & -1 & 0 & 0 & -1 & 0 & 1 & 1 \\
1 & 0 & 1 & 1 & 0 & 0 & 1 & 0 & 1 & 1 \\
1 & -1 & 0 & -1 & 0 & 1 & 0 & 1 & 0 & 0 \\
1 & -1 & 0 & 1 & 0 & -1 & 0 & 1 & 0 & 0 \\
1 & 1 & 0 & -1 & 0 & -1 & 0 & 1 & 0 & 0 \\
1 & 1 & 0 & 1 & 0 & 1 & 0 & 1 & 0 & 0 \\
1 & -1 & -1 & 0 & 1 & 0 & 0 & 1 & 1 & 0 \\
1 & -1 & 1 & 0 & -1 & 0 & 0 & 1 & 1 & 0 \\
1 & 1 & -1 & 0 & -1 & 0 & 0 & 1 & 1 & 0 \\
1 & 1 & 1 & 0 & 1 & 0 & 0 & 1 & 1 & 0 \\
1 & 0 & 0 & 0 & 0 & 0 & 0 & 0 & 0 & 0 \\
1 & 0 & 0 & 0 & 0 & 0 & 0 & 0 & 0 & 0 \\
1 & 0 & 0 & 0 & 0 & 0 & 0 & 0 & 0 & 0
\end{bmatrix}
\begin{bmatrix}
0.87 \\
-0.31 \\
-0.89 \\
0.70 \\
0.32 \\
-0.15 \\
-0.67 \\
-0.19 \\
0.49 \\
0.21
\end{bmatrix}
=
\begin{bmatrix}
1.09 \\
3.82 \\
0.65 \\
0.71 \\
0.36 \\
2.04 \\
0.02 \\
1.12 \\
2.69 \\
0.28 \\
1.43 \\
0.29 \\
0.87 \\
0.87 \\
0.87
\end{bmatrix}
$$

$$
\mathbf{Y} - \mathbf{XB} =
\begin{bmatrix}
0.97 \\
4.15 \\
0.32 \\
0.83 \\
0.52 \\
1.75 \\
0.31 \\
0.96 \\
2.65 \\
0.45 \\
1.26 \\
0.33 \\
0.86 \\
0.87 \\
0.88
\end{bmatrix}
-
\begin{bmatrix}
1.09 \\
3.82 \\
0.65 \\
0.71 \\
0.36 \\
2.04 \\
0.02 \\
1.12 \\
2.69 \\
0.28 \\
1.43 \\
0.29 \\
0.87 \\
0.87 \\
0.87
\end{bmatrix}
=
\begin{bmatrix}
-0.12 \\
0.33 \\
-0.33 \\
0.12 \\
0.16 \\
-0.29 \\
0.29 \\
-0.16 \\
-0.04 \\
0.17 \\
-0.17 \\
0.04 \\
-0.01 \\
0.00 \\
0.01
\end{bmatrix}
$$

$$
\xi = (\mathbf{Y} - \mathbf{XB})' (\mathbf{Y} - \mathbf{XB})
$$

$$
= \begin{bmatrix} -0.12 & 0.33 & -0.33 & 0.12 & 0.16 & -0.29 & 0.29 & -0.16 & -0.04 & 0.17 & -0.17 & 0.04 \end{bmatrix}
$$

$$
-0.01\ 0.00\ 0.01]
\begin{bmatrix}
-0.12 \\
0.33 \\
-0.33 \\
0.12 \\
0.16 \\
-0.29 \\
0.29 \\
-0.16 \\
-0.04 \\
0.17 \\
-0.17 \\
0.04 \\
-0.01 \\
0.00 \\
0.01
\end{bmatrix}
= 0.53
$$

Thus,

$$S_e^2 = \frac{1}{n-k-1}(Y-XB)'(Y-XB) = \frac{0.53}{15-9-1} = 0.11$$

$$(X'X)^{-1} = \begin{bmatrix} 0.33 & 0 & 0 & 0 & 0 & 0 & 0 & -0.17 & -0.17 & -0.17 \\ 0 & 0.13 & 0 & 0 & 0 & 0 & 0 & 0 & 0 & 0 \\ 0 & 0 & 0.13 & 0 & 0 & 0 & 0 & 0 & 0 & 0 \\ 0 & 0 & 0 & 0.13 & 0 & 0 & 0 & 0 & 0 & 0 \\ 0 & 0 & 0 & 0 & 0.25 & 0 & 0 & 0 & 0 & 0 \\ 0 & 0 & 0 & 0 & 0 & 0.25 & 0 & 0 & 0 & 0 \\ 0 & 0 & 0 & 0 & 0 & 0 & 0.25 & 0 & 0 & 0 \\ -0.17 & 0 & 0 & 0 & 0 & 0 & 0 & 0.27 & 0.02 & 0.02 \\ -0.17 & 0 & 0 & 0 & 0 & 0 & 0 & 0.02 & 0.27 & 0.02 \\ -0.17 & 0 & 0 & 0 & 0 & 0 & 0 & 0.02 & 0.02 & 0.27 \end{bmatrix}$$

The diagonal of the $(X'X)^{-1} = \begin{bmatrix} 0.33 \\ 0.13 \\ 0.13 \\ 0.13 \\ 0.25 \\ 0.25 \\ 0.25 \\ 0.27 \\ 0.27 \\ 0.27 \end{bmatrix}$

Thus,

$$var(B) = S_e^2 \times \text{diagonal of the } (X'X)^{-1}$$

$$= 0.11 \begin{bmatrix} 0.33 \\ 0.13 \\ 0.13 \\ 0.13 \\ 0.25 \\ 0.25 \\ 0.25 \\ 0.27 \\ 0.27 \\ 0.27 \end{bmatrix} = \begin{bmatrix} 0.04 \\ 0.01 \\ 0.01 \\ 0.01 \\ 0.03 \\ 0.03 \\ 0.03 \\ 0.03 \\ 0.03 \\ 0.03 \end{bmatrix}$$

The standard error of the regression coefficient is obtained by taking the square root of each element of the $var(B)$ and t-statistic values of the regression coefficients are obtained by dividing regression coefficients with corresponding standard errors. The regression coefficients and their corresponding standard errors, t-values and probabilities for all the terms are tabulated in the following table.

The terms with probabilities less than 0.05 are statistically significant at 95% confidence level. Thus, the results show that the terms x_1, x_2, x_3, x_2x_3 and x_2^2 are significant at 95% confidence level.

Term	Regression Coefficient	Standard Error of Coefficient	t-Value	Probability
Constant	0.87	0.20	4.35	0.01*
x_1	−0.31	0.10	−3.10	0.04*
x_2	−0.89	0.10	−8.90	0.00*
x_3	0.70	0.10	7.00	0.00*
x_1x_2	0.32	0.17	1.85	0.12
x_1x_3	−0.15	0.17	−0.87	0.43
x_2x_3	−0.67	0.17	−3.87	0.01*
x_1^2	−0.19	0.17	−1.10	0.32
x_2^2	0.49	0.17	2.83	0.04*
x_3^2	0.21	0.17	1.21	0.28

9.9 MATLAB® CODING

9.9.1 MATLAB® CODING OF EXAMPLE 9.2

```
clc
clear
close all
data=[ 8.2   8.7
       9.6   9.6
       7.0   6.9
       9.4   8.5
      10.9  11.3
       7.1   7.6
       9.0   9.2
       6.6   6.3
       8.4   8.4
      10.5  12.3];
x=data(:,1);
y=data(:,2);
r=corr(x,y)
```

9.9.2 MATLAB® CODING OF EXAMPLE 9.3

```
clc
close all
clear all
ranks=[2    1    3    7    6    5    4    10    9    8
       2    1    4    8    6    4    5    10    9    7];
x=ranks(1,:);
y=ranks(2,:);
```

```
n=length(x);
d=(x-y);
rs=1-(6*sum(d.^2)/(n*(n^2-1)))
```

9.9.3   MATLAB® CODING OF EXAMPLE 9.7

```
clc
clear
close all
data=[16      620
      24      715
      30      800
      40      910
      60      1100
      80      1260];
[row,col]=size(data);
x=data(:,1);
y=data(:,2);
scatter(x,y,'k','filled')
xlabel('Yarn count(Ne)')
ylabel('TPM')
hold on
X=[x ones(row,1)];
Y=y;
B=inv(X'*X)*X'*Y;
m_hat=B(1)
c_hat=B(2)
x1=min(x):max(x);
y1=m_hat*x1+c_hat;
plot(x1,y1)
set(gcf,'Color',[1,1,1])
```

9.9.4   MATLAB® CODING OF EXAMPLE 9.8

```
clc
clear
close all
data=[ 3.2   14.2
       3.5   16.1
       3.7   18.3
       4     20.2
       4.2   20.5
       4.5   21.3
       4.8   21.1
       5     20.6
       5.2   19.7
       5.5   19.5];
[row,col]=size(data);
x=data(:,1);
y=data(:,2);
```

```
scatter(x,y,'k','filled')
xlabel('TM')
ylabel('Tenacity (cN/tex)')
hold on
X=[x.^2 x ones(row,1)];
Y=y;
B=inv(X'*X)*X'*Y;
a_hat=B(1)
b_hat=B(2)
c_hat=B(3)
y_est=a_hat*x.^2+b_hat*x+c_hat;
x1=min(x):0.05:max(x);
y1=a_hat*x1.^2+b_hat*x1+c_hat;
plot(x1,y1,'k')
set(gcf,'Color',[1,1,1])
```

9.9.5   MATLAB® Coding of Example 9.9

```
clc
clear
close all
data=[20    16    4.9
      20    24    4.7
      20    30    4.5
      23    16    5.3
      23    24    5.2
      23    30    5.0
      26    16    6.0
      26    24    5.7
      26    30    5.5];
[row,col]=size(data);
x1=data(:,1);
x2=data(:,2);
y=data(:,3);
X=[x1 x2 ones(row,1)];
Y=y;
B=inv(X'*X)*X'*Y;
m1_hat=B(1)
m2_hat=B(2)
c_hat=B(3)
v1=min(x1):0.5:max(x1);
v2=min(x2):1:max(x2);
[X1,X2]=meshgrid(v1,v2);
y1=m1_hat*X1+m2_hat*X2+c_hat;
surf(X1,X2,y1)
hold on
scatter3(x1,x2,Y,'filled')
xlabel('Delivery speed (m/min)')
ylabel('Yarn count (Ne)')
zlabel('Yarn hairiness index')
set(gcf,'Color',[1,1,1])
```

9.9.6   MATLAB® Coding of Example 9.11

```
clc
clear
close all
data=[50      20   50   0.97
      50      20   70   4.15
      50      40   50   0.32
      50      40   70   0.83
      0       30   50   0.52
      0       30   70   1.75
      100     30   50   0.31
      100     30   70   0.96
      0       20   60   2.65
      0       40   60   0.45
      100     20   60   1.26
      100     40   60   0.33
      50      30   60   0.86
      50      30   60   0.87
      50      30   60   0.88];
for j=1:3
 x(:,j)=2*((data(:,j)-min(data(:,j)))/
(max(data(:,j))-min(data(:,j))))-1;
end
x1=x(:,1);
x2=x(:,2);
x3=x(:,3);
X=x2fx(x,'quadratic');
Y=data(:,4);
B=inv(X'*X)*X'*Y;
b0_hat=B(1)
b1_hat=B(2)
b2_hat=B(3)
b3_hat=B(4)
b12_hat=B(5)
b13_hat=B(6)
b23_hat=B(7)
b11_hat=B(8)
b22_hat=B(9)
b33_hat=B(10)
```

9.9.7   MATLAB® Coding of Example 9.12

```
clc
clear
close all
data=[50      20   50   0.97
      50      20   70   4.15
      50      40   50   0.32
      50      40   70   0.83
      0       30   50   0.52
```

```
       0      30   70   1.75
     100      30   50   0.31
     100      30   70   0.96
       0      20   60   2.65
       0      40   60   0.45
     100      20   60   1.26
     100      40   60   0.33
      50      30   60   0.86
      50      30   60   0.87
      50      30   60   0.88];
[row,col]=size(data);
n=row;
for j=1:3
 x(:,j)=2*((data(:,j)-min(data(:,j)))/
(max(data(:,j))-min(data(:,j))))-1;
end
x1=x(:,1);
x2=x(:,2);
x3=x(:,3);
X=x2fx(x,'quadratic');
y=data(:,4);
Y=y;
B=inv(X'*X)*X'*Y;
p=length(B);%Number of regression coefficients
dfe=n-p;%Degrees of freedom
k=p-1;
var_B=(Y-X*B)'*(Y-X*B)/(n-k-1);
se_B=sqrt(diag(var_B*inv(X'*X)));
t=B./se_B;% t-stats for regression coefficients
pval = 2*(tcdf(-abs(t), dfe)); % Probability value
Table=[B se_B t pval]
% regression coefficients which are significant at 95%
confidence level
j=0;
for i=1:length(pval)
    if pval(i)<=0.05
        j=j+1;
        row_pval(j)=i;
    end
end
Significant_B=row_pval'%Significant coefficients
```

Exercises

9.1 Bending lengths (cm) along warp and weft directions in cm for 10 fabric samples are given below.

| Bending Length Along Warp, x | 1.72 | 1.13 | 1.28 | 1.57 | 1.94 | 1.42 | 2.44 | 2.08 | 1.64 | 1.49 |
| Bending Length Along Weft, y | 1.65 | 1.10 | 1.41 | 1.53 | 1.88 | 1.39 | 2.18 | 2.21 | 1.57 | 1.61 |

Compute the correlation coefficient between x and y.

9.2 The table below shows the monthly measurements of biochemical oxygen demand (BOD) in mg/L at a textile wastewater treatment plant in both untreated and treated waste water.

Untreated Wastewater, x	330	285	260	401	238	380	240	290	310	420	365	296
Treated Wastewater, y	5.9	4.0	5.2	7.2	3.2	6.4	3.8	4.1	3.8	7.5	6.9	4.8

Compute the correlation coefficient between x and y.

9.3 Two experts ranked 8 cloths based on the sensorial comfort as given below.

Cloth Number	1	2	3	4	5	6	7	8
Expert 1	4	3	5	2	7	1	8	6
Expert 2	3	5	6	1	4	2	7	8

Determine the rank correlation coefficient between the rankings done by two experts.

9.4 In a fabric handle study by experts, an attempt was made to correlate fabric handle with bending rigidity for 9 fabric samples. The results are given below:

Fabric Number	1	2	3	4	5	6	7	8	9
Rigidity Rank	7	3	8	6	1	2	5	9	4
Handle Rank	8.5	3.5	6.5	6.5	1	2	3.5	8.5	5

Calculate rank correlation coefficient between the two rankings.

9.5 In the case of mélange yarn production, the percentage of dyed fibre in the mixing is called the shade percentage. The following data relates the tenacity of mélange yarns and shade percentage.

Shade Percentage, x	10	20	30	40	50	60	70	80
Tenacity (cN/tex), y	20.3	19.9	19.0	18.9	18.8	18.5	17.8	17.2

Fit a straight line with these data.

9.6 The following data relates the tenacity of polyester-cotton blended yarns and the polyester percentage in the blends.

Polyester percentage, x	0	20	40	60	80	100
Tenacity (cN/tex), y	18.0	18.4	20.2	24.6	27.5	27.8

Fit a straight line with these data.

9.7 The following data shows the result of a stress-strain curve for a polyester filament.

Strain (%), x	0.5	1	2	3	4	5	6	7	8	9	10	11	12
Stress (cN/denier), y	0.7	1.4	2.6	3.5	4.3	4.8	5.2	5.5	5.8	5.9	6.1	6.6	7

Fit a quadratic equation of stress as a function of strain.

9.8 The following data shows the relationship between size add on (%) and warp breakage rate (breaks/10,000 picks).

Size add on (%), x	2	4	6	8	10	12	14	16	18	20
Warp breakage rate, y	8	6	3	2	1	1	2	3	4	5

Fit a quadratic equation of warp breakage rate as a function of size add on.

9.9 The table below shows the effect of size add-on % (x_1) and size concentration % (x_2) on the amount of water to be evaporated per unit oven dry mass of yarn (y). Find the linear least square regression equation of y on x_1 and x_2.

x_1	8	10	12	14	8	10	12	14	8	10	12	14
x_2	10	10	10	10	12	12	12	12	14	14	14	14
y	0.70	0.90	1.10	1.25	0.60	0.75	0.90	1.05	0.50	0.60	0.75	0.85

9.10 The following table shows the effect of dyeing temperature (°C) and dyeing pressure (MPa) on the colour strength (K/S) of a dyed nylon fabric. Find the linear least square regression equation of colour strength as a function of dyeing temperature and dyeing pressure.

Temperature, x_1	80	100	120	80	100	120	80	100	120
Pressure, x_2	5	5	5	10	10	10	15	15	15
Colour strength, y	4	6	8	7	8	12	9	11	14

9.11 The minimum twist of cohesion (MTC) is inversely related to fibre cohesion in the yarn and it is defined as the minimum number of twists required to hold the fibres in the yarn structure. An experiment was carried out in a ring frame to find the effect of spindle rpm and traveller mass (ISO number) on MTC. The following table shows the results of the experiment. Find a quadratic least square regression equation of MTC as a function of spindle rpm and traveller mass.

Spindle rpm x_1	Traveller Mass x_2	MTC y
13,000	40	138
13,000	45	142
13,000	50	126
15,000	40	124
15,000	45	122
15,000	50	108
17,000	40	110
17,000	45	105
17,000	50	98

9.12 The following table shows the results of an experiment to study the influence of the proportion of polyester-cotton blend, yarn count and fabric sett (thread density) of a square fabric on its bending rigidity. Find the quadratic least square regression equation of fabric bending rigidity as a function of proportion of polyester, yarn count, and fabric sett.

Proportion of Polyester (%)	Yarn Count (Ne)	Fabric Sett (inch⁻¹)	Fabric Bending Rigidity (gf.cm²/cm)
x_1	x_2	x_3	y
50	20	50	0.049
50	20	70	0.119
50	40	50	0.019
50	40	70	0.032
0	30	50	0.035
0	30	70	0.070
100	30	50	0.018
100	30	70	0.033
0	20	60	0.103
0	40	60	0.028
100	20	60	0.047
100	40	60	0.018
50	30	60	0.038
50	30	60	0.037
50	30	60	0.036

9.13 An experiment was conducted to find out the effect of ring frame parameters such as spindle speed, top roller pressure and traveller mass on yarn imperfection. The following table shows the results of the experiment. Find the quadratic least square regression equation of yarn imperfection as a function of spindle speed, top roller pressure and traveller mass.

Spindle Speed (rpm)	Top Roller Pressure (Kgf/cm²)	Traveller Mass (ISO number)	Yarn Imperfection (per Km)
x_1	x_2	x_3	y
13,000	2.0	45	128
13,000	2.50	45	85
17,000	2.00	45	137
17,000	2.50	45	98
13,000	2.25	40	133
13,000	2.25	50	121
17,000	2.25	40	138
17,000	2.25	50	113
15,000	2.00	40	105
15,000	2.00	50	97
15,000	2.50	40	90
15,000	2.50	50	89
15,000	2.25	45	111
15,000	2.25	45	115
15,000	2.25	45	114

9.14 Using the results of Exercise 9.13, work out the test of significance of the regression coefficients.

10 Design of Experiments

10.1 INTRODUCTION

Investigators perform experiments, virtually in all fields to comprehend the effect of various input variables on the output. Such experiments are performed in a strategic way to understand, detect and identify the changes, and optimize the process. The input variables in such experimentations are chosen based on past experiences or literatures and they were deliberately varied up to certain extent to perceive the changes in the output. These input variables are known as 'parameters' or 'factors' or 'treatments', whereas the output variable is known as 'objective' or 'response'.

Let us consider an experiment in a spinning mill, which aims to optimize roving fineness to minimize yarn irregularity. Here, the roving fineness is the controllable factor and yarn irregularity is the response variable. Therefore, the mill would perform experiments with various roving fineness to make yarns of a specific count. It may be concluded that the roving fineness that produces yarn with minimum irregularity will be the best solution. However, one can raise the following questions during the aforementioned experimentation.

- To produce yarns of defined count from various fineness of rovings, change in draft in the spinning system is required. Does the roving fineness only affect the irregularity of the yarn? What are the impact of total draft and its interaction with roving fineness on yarn irregularity?
- Are there any other factors in the process or in the machinery that might affect the yarn irregularity?
- In which order does the experiment need to be conducted?
- To what extent do the uncontrollable and chance factors affect the outcome? Can an experiment be planned to minimize the effect of these uncontrollable and chance factors to understand the exact effect of factors of our interest?

For better understanding of manufacturing processes with controlled factors, uncontrolled factors (discrete and continuous) and response, a general model is illustrated in Figure 10.1.

The experimental data are used to develop empirical models which are mainly in linear or quadratic forms. Linear and quadratic models with two factors and three factors are shown below:

$$\text{Linear model with two factors: } Y = \beta_0 + \beta_1 x_1 + \beta_2 x_2 + \varepsilon \qquad (10.1)$$

$$\text{Linear model with three factors: } Y = \beta_0 + \beta_1 x_1 + \beta_2 x_2 + \beta_3 x_3 + \varepsilon \qquad (10.2)$$

DOI: 10.1201/9781003081234-10

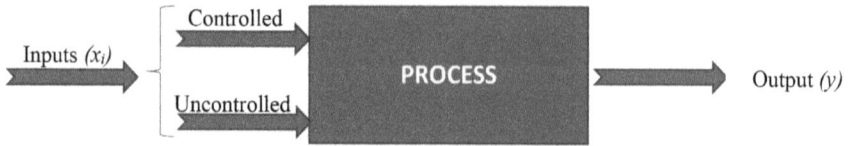

FIGURE 10.1 A general model of a manufacturing process.

Quadratic model with two factors: $Y = \beta_0 + \beta_1 x_1 + \beta_2 x_2 + \beta_3 x_1 x_2$
$$+ \beta_4 x_1^2 + \beta_5 x_2^2 + \varepsilon \qquad (10.3)$$

Quadratic model with three factors: $Y = \beta_0 + \beta_1 x_1 + \beta_2 x_2 + \beta_3 x_3 + \beta_4 x_1 x_2$
$$+ \beta_5 x_1 x_3 + \beta_6 x_2 x_3 + \beta_7 x_1^2 + \beta_8 x_2^2 + \beta_9 x_3^2 + \varepsilon \qquad (10.4)$$

where Y is the response variable and x_1, x_2 and x_3 are the input factors to the model. The terms $x_1 x_2$, $x_1 x_3$ and $x_2 x_3$ are the possible interaction effects of x_1 and x_2; x_1 and x_3; x_2 and x_3, respectively. The terms β_0, β_1, β_2, β_3, ... and β_9 are the regression coefficients, ε is the 'experimental error' or 'noise' which occurs due to the uncontrolled factors.

Therefore, looking into such complexity of manufacturing where various factors influences the response, a statistical design of experiment (DOE) is required to layout a detailed experimental plan with the following objectives:

- Choosing the best between alternatives: In manufacturing, sometimes alternatives or options as input factors are available and the best option needs to be chosen for better output. Comparative experiments are required to conclude the best among available input options. The data obtained from the experiments are compared and the best average outcome is preferred. For an example, sewing threads from three different suppliers, namely A, B and C, are available to stitch a garment. A supplier that provides sewing thread with maximum strength will be preferred. There might be two possibilities of designing this experiment. In the first type, comparison of A, B and C is performed under one common set of conditions, i.e., only the sewing threads from A, B and C will be changed to sew the garment, whereas other factors will remain constant. In the second type, comparison will be performed under various set of conditions i.e. by varying other factors (example – thread tension, stitch length etc.) purposefully and systematically along with the A, B and C.
- Selection of key factors affecting the response: In manufacturing, there lies multiple factors in which few are known to be significant and rest have little or no effect on the response. It is of utmost importance to identify these significant factors and reduce the number of factors into a small set so that consideration can be given to only controlling the significant factors. Therefore, planning a systematic set of experiments is required by

considering all the known factors initially so that significant factors may be identified ultimately by performing significant tests.

• Developing a response surface model to optimize a response: Once the significant factors are identified in a manufacturing process, various other objectives like hitting a target, maximizing or minimizing a response, reducing the variability, making a process robust etc., may be accomplished. These objectives are discussed below.

Hitting a target is a frequently encountered objective of an experiment, where various settings in a manufacturing is tried to achieve a specific or multiple responses simultaneously. For example, yarn fineness and thread density of a woven fabric is searched to achieve a target ultraviolet protection factor and stiffness of the said fabric.

Maximizing or minimizing a response means to search those input factors which maximize or minimize specific or multiple responses simultaneously. The same example mentioned before (hitting the target) can be cited to explain a maximize or minimize response problem. Yarn fineness and thread density of the same woven fabric can be determined to maximize ultraviolet protection factor (UPF) and minimize stiffness. It is to be noted that, the responses in this example namely, UPF and stiffness of a fabric are conflicting in nature and hence, a trade-off is required between these objectives. Such problems are known as multi-objective optimization and the reader may refer to the book *Advanced Optimization and Decision Making Techniques for Textile Manufacturing*, (CRC Press) authored by Ghosh, Mal and Majumdar (2019) for more details.

A manufacturing process may induce undue variations in the response due to variation of some key input factors during the process. Therefore, it is of utmost importance to identify the controlled factors which affects the response. Such factors should be adjusted and controlled during a process to reduce variability. For example, if A, B, C and D are the controlled factors in which the manufacturer is well aware only about the factors A and B and hence kept only these two factors (A and B) under control during the manufacturing. There occurred a variation in the factors C and D during the process which the manufacturer is not aware of, and that leads to objectionable variation in the response. If the manufacturer could have identified all the controlled factors, he could have controlled all of them to ensure reduce variability of the response.

Often an engineered product is designed with specific quality under controlled conditions in a lab. The product may fail outside in the field, while it is used by the consumer, due to some reasons or factors which are not considered during experimentation. For better understanding, let us take the example of designing a bullet proof vest developed in a lab to resist a bullet with a mass of 50 gm and maximum speed of 950 m/min. The testing was done under controlled conditions in a lab and it was found that the jacket is perfectly engineered to resist a bullet with specific mass and speed as mentioned. Unfortunately, in field the bullet proof vest failed which may be due to degradation of strength over time or due to different moisture level and temperature which were not considered during experimental planning. Therefore, a special experimental effort is required to design a product to make a process robust.

10.2 BASIC PRINCIPLES OF DOE

The data yielded from a set of experiments are subjected to experimental errors, due to the influence of some extraneous unknown factors or changes during experiments like operator fatigue, machine warm-up, variations in temperature, pressure, humidity etc., which are then analysed using various statistical tools to obtain valid and objective conclusions. Therefore, an experimental problem involves two basic aspects – design of the experiments and the statistical analysis of the yield data, which are closely related as the method of analysis dependent on the design employed. In order to increase the precision in the experiment and to reduce the biasness, three fundamental basic techniques or principles, namely replication, blocking and randomization (Montgomery 2009; Dean et al. 2014), are used during designing experiments. These techniques are discussed below.

10.2.1 REPLICATION

Replication is the repetition of an experiment by replicating the basic experimental unit. It is to be noted that 'replication' does not mean 'repeated measurements'. For an example, let us assume that a chemist wants to check the 'effect of pre-treatment' on 'whiteness of 100% cotton fabric'. Thus, the chemist planned to conduct two sets of experiments. In the first set, the fabric is bleached in alkali solution, whereas in the second set the fabric is bleached with enzyme. These pre-treatments on cotton fabrics may be termed as 'alkaline bleaching' and 'enzymatic bleaching'. Alkaline-bleached and enzymatic-bleached cotton fabrics are then checked for their whiteness (%) to draw a conclusion. If only one fabric in each set is bleached and the whiteness (%) is measured five times, then it is 'repeated measurement'. But, if five fabrics are bleached followed by evaluation of each of their whiteness (%), then we call that 'replication' where five replicates are obtained.

Replication is required, as it helps the experimenter to estimate the experimental error (ε) as shown in Equations (10.1) to (10.4). This experimental error is actually the unexplained random part of the variation of any experiment and determines if observed differences in the data are really statistically different. Without replication, it is not possible to estimate 'experimental error' or 'noise' in the estimates of effects of 'controlled factors' or 'treatments'.

Replication also helps in obtaining more reliable estimates by increasing the precision of an experiment. The precision of an estimate, obtained from experimentation results, increases with the increase in number of replications. If σ^2 is the population variance and n is the number of replications, the variance of the sample mean $(\bar{y})$ would be $\sigma_{\bar{y}}^2 = \frac{\sigma^2}{n}$. For instance, we may refer to the example discussed in the last paragraph in which a chemist is interested to see the effect of pre-treatment on whiteness (%). Two different sets of experiments are planned. In the first set of experiments, only one cotton fabric is taken each for alkaline and enzymatic bleaching $(n = 1)$ to check the effect of whiteness (%) and the results are $y_1 = 70$ and $y_2 = 60$. We can infer from these data that there is a reasonable difference in whiteness (%) due to the effect of pre-treatment. However, such inference is not satisfactory, as observed differences may be due to experimental error or noise. But if the replication (n) is

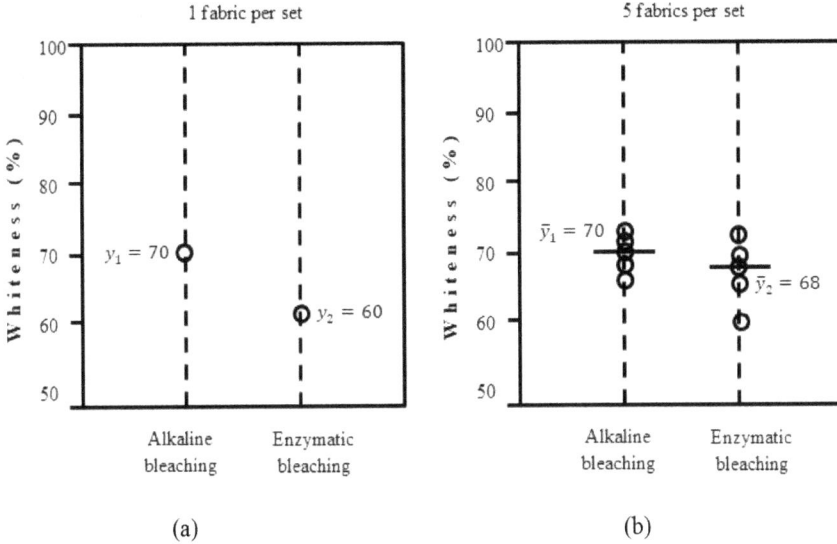

FIGURE 10.2 Whiteness % yielded due to pre-treatments (a) replication = 1; (b) replication = 5.

reasonably high, the experimental error or noise will be sufficiently small and we can effectively infer whether there is really a reasonable difference in whiteness (%) due to the effect of pre-treatment (alkaline and enzymatic bleaching). This is illustrated in Figure 10.2(a) and (b). Figure 10.2(a) shows the whiteness % yielded due to pre-treatments when the replication is 1. However, in Figure 10.2(b) it is observable that, when the replication is five $(n = 5)$ the whiteness % for the alkaline bleached and enzymatic bleached samples are $\bar{y}_1 = 70$ and $\bar{y}_2 = 68$ respectively, from which it is obvious that the difference is reasonably low.

10.2.2 RANDOMIZATION

DOE involves a set of experiments to be performed to yield some data which helps in concluding the effect of various factors, optimizing and making decisions based on the requirements. However, while conducting these experiments some extraneous but systematic unknown factors may affect the experiment. Therefore, in order to avoid any bias, the orders in which the individual trials of experimentation are conducted are randomly planned and conducted. Randomization distributes the errors independently and randomly and averages out the effects of the extraneous factors which may present.

Let us consider the example cited while discussing replication. We suppose that the two sets of fabrics (five fabrics per set) are collected from a single roll which has slightly different areal density along the length. Variation in areal density along the length of a fabric is quite obvious as yarns are not uniform in diameter. Different fabric weight is subjected to different whiteness during pre-treatments. If all the fabric samples for both the sets are collected from two specific areas, then systematic biasness is being introduced which will affect the test results. Random selection of fabrics from the roll for the pre-treatments will mitigate this problem. There are various ways of randomizing experiments, in which completely randomized design and randomized block design are most commonly used during experimentations. Both the completely randomized design and randomized block designs are discussed in section 10.3.

10.2.3 BLOCKING

Though the replication and randomization of experiments will reduce the experimental error or noise, the existence of uncontrolled sources of variations like time, place, material etc. may also influence the experimental response. Now, if the experiments are divided into different groups or blocks with respect to some uncontrolled sources of variations, such that the experiments are homogenous within the blocks, then the effects of the blocks and the treatments can be analysed separately. Therefore, we have

Response = constant + effect of treatments + effect of blocks + experimental error

For an example, experimentation is carried out in which 100% cotton woven fabric is treated with an anti-slip finish from four different suppliers to check their effect on tearing strength. The replication of each treatment is four, thus making the total number of experiments to sixteen. Anti-slip treatments are carried out for four days and the fabrics are evaluated for tearing strength. As each day varies from the other in respect to temperature and relative humidity that may affect the response, each day is considered as a block and four treatments are conducted each day. The experimental plan of such blocking along with replication is shown in Figure 10.3. A, B, C and D in the figure are the four different anti-slip treatments. It is obvious from Figure 10.3 that the fabrics are treated with all four anti-slip treatments every day. The treatments in each day, however, are not followed in a sequence, rather, they are randomized. The randomization and replication (here replication = 4) ensure minimization of experimental error.

Therefore, the response for the above planned experiment can be shown as:

$$\text{Tearing strength (due to anti slippage finish)}$$
$$= \text{constant} + \text{effect of antislip treatments}$$
$$+ \text{effect of blocks} + \text{experimental error}$$

Days	Treatments			
1	A	B	C	D
2	C	A	D	B
3	B	D	A	C
4	D	C	B	A

FIGURE 10.3 Randomized block experimental design (replications of each treatment = 4).

TABLE 10.1
Completely Randomized Design

Experimental Run	Treatment	Experimental Run	Treatment
1	D	9	D
2	A	10	C
3	C	11	B
4	C	12	D
5	B	13	A
6	D	14	B
7	B	15	C
8	A	16	A

10.3 COMPLETELY RANDOMIZED DESIGN AND BLOCK DESIGN

10.3.1 COMPLETELY RANDOMIZED DESIGN

Completely randomized design is used when the experimenter wants to choose the best method from several alternatives. Such design of experiment can be utilized only when one controlled factor is considered with different levels and involves no blocking factors. As the name suggests, the levels of the controlled factors are randomly assigned from random number tables or by some other physical mechanism. The random sequence of experimentation cited in the earlier example (where 100% cotton woven fabrics are treated with four various types of anti-slip treatments namely A, B, C and D to evaluate its effect on tearing strength) is shown in Table 10.1.

Therefore, the response for such experimental design is given below:

$$\text{Response} = \text{constant} + \text{effect of treatments} + \text{experimental error}$$

Example 10.1:

The results of four different anti-slip treatments on fabric tearing strength are shown in Figure 10.4. The treatments are coded as A, B, C and D. The tearing strengths of fabrics along with their corresponding codes are shown in individual blocks of the table. The tearing strength is measured in gf. It is obvious from

D	A	C	C
4600	4850	4800	4750
B	D	B	A
4650	4500	4600	4500
D	C	B	D
4550	4650	4550	4500
A	B	C	A
4800	4650	4700	4750

FIGURE 10.4 Completely randomized design for anti-slip treatment experiment.

Figure 10.4 that this experimental design is a completely randomized design. Test at the 0.05 level of significance whether there are significant differences among the anti-slip treatments.

SOLUTION

In this example,

$$\text{Response} = \text{constant} + \text{effect of treatments} + \text{experimental error}$$

or, $Y_{ij} = \mu + \alpha_i + \varepsilon_{ij}$

where α_i are the population treatment effects, i stands for four anti-slip treatments, j is the number of replications and ε_{ij} are the error terms. The random errors ε_{ij} are assumed to be independent, normally distributed random variables, each with zero mean and variance σ^2. (Reader may refer to Chapter 8 of this book for the statistical analysis).

From the given data we have: number of treatments $(k) = 4$, number of replications $(n) = 4$, sum of tearing strength for treatment A $(T_1) = 18,900$, sum of tearing strength for treatment B $(T_2) = 18450$, sum of tearing strength for treatment C $(T_3) = 18,900$, sum of tearing strength for treatment D $(T_4) = 18,150$, total sum of tearing strengths $(T) = 74,400$, $\sum_{i=1}^{k}\sum_{j=1}^{n} y_{ij}^2 = 346,160,000$.

Step 1: $H_0: \alpha_i = 0$, for $i = 1,2,3,4$
$\qquad\quad H_1: \alpha_i \neq 0$, for at least one value of i
$\qquad\quad \alpha = 0.05$.

Step 2: The critical value is $f_{0.05,(4-1),4\times(4-1)} = f_{0.05,3,\ 12} = 3.5$ (Table A5).

Step 3: SST_r and SSE are calculated as

$$SST = \sum_{i=1}^{k}\sum_{j=1}^{n} y_{ij}^2 - \frac{T^2}{k\times n} = 346,160,000 - \frac{74,400^2}{4\times 4} = 200,000$$

$$SST_r = \frac{1}{n}\sum_{i=1}^{k} T_i^2 - \frac{T^2}{kn} = \frac{1}{4}\left(18,900^2 + 18,450^2 + 18,900^2 + 18,150^2\right)$$

$$-\frac{74,400^2}{4\times 4} = 101,250$$

By subtraction we obtain, $SSE = SST - SST_r = 200,000 - 101,250 = 98,750$
Therefore,

$$MST_r = \frac{SST_r}{(k-1)} = \frac{101,250}{4-1} = 33,750$$

and

$$MSE = \frac{SSE}{k(n-1)} = \frac{98,750}{4\times(4-1)} = 8,229.17$$

Hence, the calculated value of $f = \frac{MST_r}{MSE} = \frac{33,750}{8,229.17} = 4.1$.
Thus we get the following analysis of variance table (Table 10.2).

TABLE 10.2
The Analysis of Variance Table for Single Factor Completely Randomized Design Model

Source of Variation	Sum of Square	Degrees of Freedom	Mean Square	f
Treatments	101,250	3	33,750	4.1
Error	98,750	12	8,229.17	
Total	200,000	15		

Step 4: As $4.1 > 3.5$, the null hypothesis is rejected. Therefore, we do have statistical evidence that there are significant differences among the anti-slip treatments.

10.3.2 BLOCK DESIGN

It is obvious that, apart from the primary factors of our interest, there are other nuisance factors which are unavoidable and affecting the response. For example, while conducting an experiment, nuisance factors may be the chemist or the operator, the time or day of conducting the experiment, the atmospheric condition of the plant (like relative humidity and temperature), the equipment etc. Therefore, the experimenter should study and decide the nuisance factors which are to be considered and controlled during the experimentation to minimize the experimental error. 'Blocking' is a technique to eliminate or reduce the impact of nuisance factors to the experimental error. In block designs, homogeneous blocks are formed in which the nuisance factors are kept constant and the primary factors are varied. Moreover, randomization and replication are considered along with the blocking. Where blocks are used to eliminate or reduce the effect of most important nuisance factors, the randomization and replication reduce the effect of remaining nuisance factors. The response of such experimental design can be expressed as:

Response = constant + effect of treatments + effect of blocks + experimental error

10.3.2.1 Randomized Block Design

In experimentation, randomized block design is applicable when one factor is of primary interest with only one nuisance factor. The benefit of blocking is discussed before. Randomization of experimentations in blocks ensures further elimination or minimization of experimental error. A simple block design (without randomization) and randomized block design are illustrated in Figure 10.5(a) and (b), where the treatments are coded as A, B, C and D whereas the blocks are coded as I, II, III and IV. In case of block design without randomization, only blocking and replication will eliminate or minimize the experimental error whereas in case of randomized block design all the three principles of experimental design, i.e. blocking, replication and randomization will eliminate or minimize the experimental error to further extent.

Blocks

I	II	III	IV
A	A	A	A
B	B	B	B
C	C	C	C
D	D	D	D

(a)

Blocks

I	II	III	IV
D	B	C	A
B	D	A	C
C	A	D	B
A	C	B	D

(b)

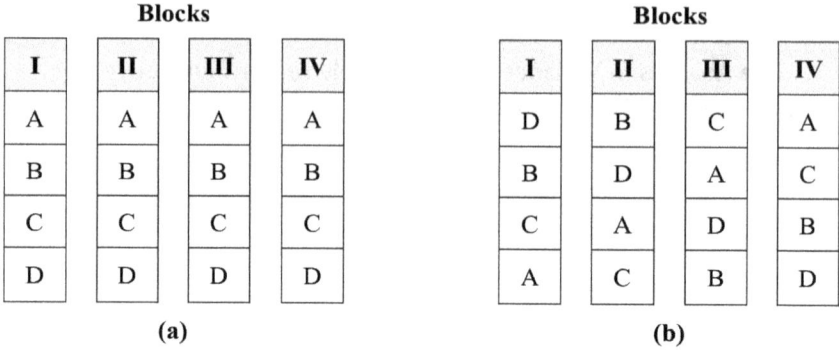

FIGURE 10.5 Block design (a) without randomization, (b) with randomization.

Example 10.2:

A similar experiment discussed in Example 10.1 is exemplified to understand the effect of randomized block design. Four anti-slip treatments are conducted on poplin fabrics for four days to observe the effect on tearing strength of poplin fabrics. The randomized block design illustrated in Figure 10.5(b) is considered here. The anti-slip treatments are coded as A, B, C and D. The tearing strength of the fabrics is measured in gf and the values are shown in Figure 10.6. It is obvious from Figure 10.6 that this experimental design is an example of randomized block design. Test at the 0.05 level of significance (i) whether there are significant differences in the tearing strength due to anti-slip treatments, and (ii) whether there are significant differences in the tearing strength due to anti-slip treatments applied on different days.

SOLUTION

In this example,

Response = constant + effect of treatment + effect of block + experimental error

or, $Y_{ij} = \mu + \alpha_i + \beta_j + \varepsilon_{ij}$

Blocks

Day 1	Day 2	Day 3	Day 4
D	B	C	A
4,600	4,600	4,650	4,750
B	D	A	C
4,650	4,500	4,800	4,700
C	A	D	B
4,800	4,500	4,550	4,650
A	C	B	D
4,850	4,750	4,550	4,500

FIGURE 10.6 Randomized block design for anti-slip treatment experiment.

where α_i are the population treatment effects for $i = 4$ anti-slip treatments; β_j are the population block effects for $j = 4$ different days; and ε_{ij} are the error terms. The random errors ε_{ij} are independent, normally distributed random variables, each with zero mean and variance σ^2.

From the given data we have: number of treatments $(a) = 4$, number of day $(b) = 4$, sum of tearing strength for treatment A $(T_{1.}) = 18,900$; sum of tearing strength for treatment B $(T_{2.}) = 18,450$; sum of tearing strength for treatment C $(T_{3.}) = 18,900$; sum of tearing strength for treatment D $(T_{4.}) = 18,150$; sum of tearing strength in day 1 $(T_{.1}) = 18,900$; sum of tearing strength for treatments in day 2 $(T_{.2}) = 18,350$; sum of tearing strength for treatments in day 3 $(T_{.3}) = 18,550$; sum of tearing strength for treatments in day 4 $(T_{.4}) = 18,600$; total sum of tearing strengths $(T) = 74,400$; total sum of squares of tearing strengths $= \sum_{i=1}^{a}\sum_{j=1}^{b} y_{ij}^2 = 346,160,000$.

Step 1: H_0: $\alpha_i = 0$, for $i = 1,2,3,4$
H_0': $\beta_j = 0$, for $j = 1,2,3,4$
H_1: $\alpha_i \neq 0$, for at least one value of i
H_1': $\beta_j \neq 0$, for at least one value of j
$\alpha = 0.05$.

Step 2: For treatments, the critical value is $f_{0.05,(4-1),(4-1)\times(4-1)} = f_{0.05,3,9} = 3.9$ (Table A5). For blocks, the critical value is also equal to 3.9, because $a = b$.

Step 3: SST_r, SSB_{days} and SSE are calculated as

$$SST_r = \frac{1}{b}\sum_{i=1}^{a} T_{i.}^2 - \frac{T^2}{ab}$$

$$= \frac{1}{4}\left(18,900^2 + 18,450^2 + 18,900^2 + 18,150^2\right) - \frac{74,400^2}{4\times4} = 101,250$$

$$SSB_{days} = \frac{1}{a}\sum_{i=1}^{b} T_{.j}^2 - \frac{T^2}{ab}$$

$$= \frac{1}{4}\left(18,900^2 + 18,350^2 + 18,550^2 + 18,600^2\right) - \frac{74,400^2}{4\times4} = 38,750$$

$$SST = \sum_{i=1}^{a}\sum_{j=1}^{b} y_{ij}^2 - \frac{T^2}{ab} = 346,160,000 - \frac{74,400^2}{4\times4} = 200,000$$

By subtraction we obtain, $SSE = SST - SST_r - SSB_{days} = 200,000 - 101,250 - 38,750 = 60,000$
Therefore,

$$MST_r = \frac{SST_r}{(a-1)} = \frac{101,250}{4-1} = 33,750$$

$$MSB_{days} = \frac{SSB_{days}}{(b-1)} = \frac{38,750}{4-1} = 12,916.67$$

TABLE 10.3

The Analysis of Variance Table for Single Factor Randomized Block Design for Anti-Slip Treatment Experiment

Source of Variation	Sum of Square	Degrees of Freedom	Mean Square	f
Treatments	101,250	3	33,750	5.06
Blocks	38,750	3	12,916.67	1.94
Error	60,000	9	6666.67	
Total	200,000	15		

and

$$MSE = \frac{SSE}{(a-1)(b-1)} = \frac{60,000}{(4-1)\times(4-1)} = 6666.67$$

Hence, $f_{T_r} = \frac{MST_r}{MSE} = \frac{33,750}{6666.67} = 5.06$

and

$$f_B = \frac{MSB_{days}}{MSE} = \frac{12,916.67}{6666.67} = 1.94$$

Thus, we get the following analysis of variance table (Table 10.3).

Step 4: For treatments, as 5.06 > 3.9, the null hypothesis is rejected. Therefore, there is statistical evidence that there are significant differences in tearing strength due to anti-slip treatments. For blocks, as 1.94 < 3.9, the null hypothesis cannot be rejected. Therefore, there are no significant differences in tearing strength due to anti-slip treatments applied on different days.

10.3.2.2 Latin Square Design

This type of experimental design is applicable when one factor is of primary interest with two different nuisance factors which cannot be combined into a single factor. However, Latin square design has certain limitations. In this model, the number of levels of individual block should be equal to the number of levels of the treatment. Also, this model does not consider the effect of interaction between the blocks or between the treatment and block. A 3×3 and 4×4 Latin square experimental block designs are illustrated in Figure 10.7(a) and (b), where the treatments are coded as A, B, C and D whereas the blocks are presented along the rows and columns.

Example 10.3:

The experiment in Example 10.2 is further modified to Latin square design by introducing one more block, i.e. the time of conducting experiments in each day. Poplin fabrics are treated with four different types' of anti-slip treatments for four

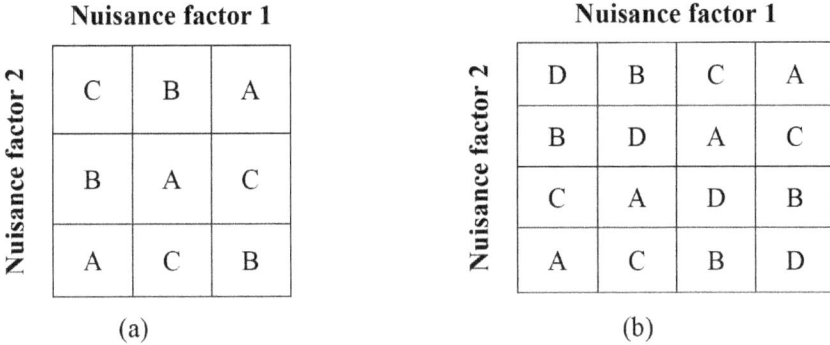

FIGURE 10.7 Latin square block designs (a) 3 × 3; (b) 4 × 4.

days to observe their effect on fabric tearing strength. Every day four treatments were given each at 9 am, 11 am, 1 pm and 3 pm. The Latin square design along with the results of tearing strength of poplin fabrics is shown in Figure 10.8. The anti-slip treatments are coded as A, B, C and D. The tearing strength of the fabrics is measured in gf. Test at the 0.05 level of significance whether there are significant differences in the tearing strength due to anti-slip treatments and the effect of the blocks.

SOLUTION

In this example,

$$\text{Response} = \text{constant} + \text{effect of treatment} + \text{effect of block 1}$$
$$+ \text{effect of block 2} + \text{experimental error}$$

or, $Y_{ijk} = \mu + \alpha_i + \beta_j + \gamma_k + \varepsilon_{ijk}$

where α_i are the population treatment effects for $i = 4$ anti-slip treatments; β_j are the population block 1 effects for $j = 4$ different days; γ_k are the population block 2 effects for $k = 4$ different times in a day and ε_{ijk} are the error terms. The random

Blocks (days)

Blocks (Time of day)		Day 1	Day 2	Day 3	Day 4
	9.00 am	D 4,600	B 4,600	C 4,650	A 4,750
	11.00 am	B 4,650	D 4,500	A 4,800	C 4,700
	1.00 pm	C 4,800	A 4,500	D 4,550	B 4,650
	3.00 pm	A 4,850	C 4,750	B 4,550	D 4,500

FIGURE 10.8 Latin square block design for anti-slip treatment experiment.

316   Textile Engineering

TABLE 10.4
Experimental Data of Anti-Slip Treatment

Time in a Day	Days				$T_{..k}$
	1	2	3	4	
9 am	D = 4,600	B = 4,600	C = 4,650	A = 4,750	18,600
11 am	B = 4,650	D = 4,500	A = 4,800	C = 4,700	18,650
1 pm	C = 4,800	A = 4,500	D = 4,550	B = 4,650	18,500
3 pm	A = 4,850	C = 4,750	B = 4,550	D = 4,500	18,650
$T_{.j.}$	18,900	18,350	18,550	18,600	T = 74,400

---

errors ε_{ijk} are independent, normally distributed random variables, each with zero mean and variance σ^2.

The experimental data for single factor Latin square block design for anti-slip treatment experiment is shown in Table 10.4.

From the given data we have: number of treatments = number of days = number of times in a day = a = 4, $T_{1..} = 18,900$, $T_{2..} = 18,450$, $T_{3..} = 18,900$, $T_{4..} = 18,150$; $T_{.1.} = 18,900$, $T_{.2.} = 18,350$, $T_{.3.} = 18,550$, $T_{.4.} = 18,600$; $T_{..1} = 18,600$, $T_{..2} = 18,650$, $T_{..3} = 18,500$, $T_{..4} = 18,650$; $T = 74,400$; $\sum_{i=1}^{a}\sum_{j=1}^{a}\sum_{k=1}^{a} y_{ijk}^2 = 346,160,000$.

Step 1: $H_0: \alpha_i = 0$, for $i = 1,2,3,4$
$H_0: \beta_j = 0$, for $j = 1,2,3,4$
$H_0'': \gamma_k = 0$, for $k = 1,2,3,4$
$H_1: \alpha_i \neq 0$, for at least one value of i
$H_1': \beta_j \neq 0$, for at least one value of j
$H_1'': \gamma_k \neq 0$, for at least one value of k
$\alpha = 0.05$.

Step 2: For treatments, the critical value is $f_{\alpha,(a-1),(a-2)(a-1)} = f_{0.05,(4-1),(4-2)\times(4-1)} = f_{0.05,3,6} = 4.8$ (Table A5).

Step 3: SST_r, SSB_{day}, $SSB_{time\ in\ a\ day}$ and SSE are calculated as

$$SST_r = \frac{1}{a}\sum_{i=1}^{a} T_{i..}^2 - \frac{T^2}{a\times a}$$

$$= \frac{1}{4}\left(18,900^2 + 18,450^2 + 18,900^2 + 18,150^2\right) - \frac{74,400^2}{4\times 4} = 101,250$$

$$SSB_{days} = \frac{1}{a}\sum_{j=1}^{a} T_{.j.}^2 - \frac{T^2}{a\times a}$$

$$= \frac{1}{4}\left(18,900^2 + 18,350^2 + 18,550^2 + 18,600^2\right) - \frac{74,400^2}{4\times 4} = 38,750$$

$$SSB_{\text{time in a day}} = \frac{1}{a} \sum_{k=1}^{a} T_{..k}^2 - \frac{T^2}{a \times a}$$

$$= \frac{1}{4}\left(18,600^2 + 18,650^2 + 18,500^2 + 18,650^2\right) - \frac{74,400^2}{4 \times 4} = 3750$$

$$SST = \sum_{i=1}^{a}\sum_{j=1}^{a}\sum_{k=1}^{a} y_{ijk}^2 - \frac{T^2}{a \times a} = 346,160,000 - \frac{74,400^2}{4 \times 4} = 200,000$$

By subtraction we obtain, $SSE = SST - SST_r - SSB_{\text{days}} - SSB_{\text{time of day}} = 200,000 - 101,250 - 38,750 - 3750 = 56,250$
Therefore,

$$MST_r = \frac{SST_r}{(a-1)} = \frac{101,250}{4-1} = 33,750$$

$$MSB_{\text{days}} = \frac{SSB_{\text{days}}}{(a-1)} = \frac{38,750}{4-1} = 12,916.67$$

$$MSB_{\text{time in a day}} = \frac{SSB_{\text{time in a day}}}{(a-1)} = \frac{3750}{4-1} = 1250$$

and

$$MSE = \frac{SSE}{(a-2)(a-1)} = \frac{56,250}{(4-2)\times(4-1)} = 9375$$

Hence, $f_{T_r} = \dfrac{MST_r}{MSE} = \dfrac{33,750}{9375} = 3.6$

$$f_{\text{days}} = \frac{MSB_{\text{days}}}{MSE} = \frac{12,916.67}{9375} = 1.38$$

$$f_{\text{time of day}} = \frac{MSB_{\text{time in a day}}}{MSE} = \frac{1250}{9375} = 0.13$$

Thus, we get the following analysis of variance table (Table 10.5).

TABLE 10.5
The Analysis of Variance Table for Single Factor Latin Square Block Design for Anti-Slip Treatment Experiment

Source of Variation	Sum of Square	Degrees of Freedom	Mean Square	f
Treatments	101,250	3	33,750	3.6
Days	38,750	3	12,916.67	1.38
Time of day	3750	3	1250	0.13
Error	56,250	6	9375	
Total	200,000	15		

Step 4:   For treatments, as 3.6 < 4.8, the null hypothesis cannot be rejected. Therefore, there are no significant differences in the tearing strength due to anti-slip treatments. For the blocks, namely days and time of day, f_{days} (1.38) and $f_{time\ of\ day}$ (0.13) < 4.8, the null hypothesis cannot be rejected. Therefore, there are no significant differences in tearing strength due to anti-slip treatments applied on different days and different times of a day.

10.3.3.3   Graeco-Latin Square Design

Graeco-Latin square experimental design is used when an experiment involves one factor of primary interest and three different nuisance factors. Alike Latin square, this experimental design is also valid for equal number of levels for the blocks and the treatments. Also, this model does not consider the interaction between the blocks or between treatments and blocks. A 3 × 3 and 4 × 4 Graeco-Latin square experimental designs are shown in Figure 10.9 (a) and (b), where the symbols A, B, C and D are the treatments. In Figure 10.9 two blocking factors are shown along the rows and columns, whereas the third blocking factor is indicated by the suffix symbols *p, q, r* and *s*.

Example 10.4:

The experiment in Example 10.3 is further modified to Graeco-Latin square design by introducing the third block, i.e. the equipment. Poplin fabrics are treated with four different types' anti-slip treatments for four days to observe their effect on fabric tearing strength. Every day four treatments were given each at 9 am, 11 am, 1 pm and 3 pm. Each day, four pieces of equipment are used for the treatments. The Graeco-Latin square design along with the results of the tearing strength of poplin fabrics is shown in Figure 10.10. The anti-slip treatments are coded as A, B, C and D. The equipment used for each treatment in a day are coded as *p, q, r* and *s*, which are shown in the suffix of each treatment. The tearing strength of the fabrics is measured in gf. Test at the 0.05 level of significance whether there are significant differences in the tearing strength due to anti-slip treatments and the effect of the blocks.

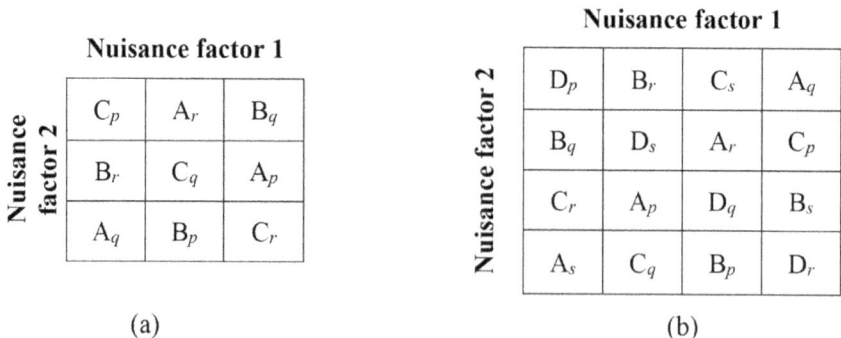

FIGURE 10.9   Graeco-Latin square experimental design (a) 3 × 3; (b) 4 × 4.

Blocks (days)

		Day 1	Day 2	Day 3	Day 4
Blocks (Time of day)	9.00 am	D_p 4,600	B_r 4,600	C_s 4,650	A_q 4,750
	11.00 am	B_q 4,650	D_s 4,500	A_r 4,800	C_p 4,700
	1.00 pm	C_r 4,800	A_p 4,500	D_q 4,550	B_s 4,650
	3.00 pm	A_s 4,850	C_q 4,750	B_p 4,550	D_r 4,500

FIGURE 10.10   Latin square block design for anti-slip treatment experiment.

SOLUTION

In this example,

$$\text{Response} = \text{constant} + \text{effect of treatment} + \text{effect of block 1}$$
$$+ \text{effect of block 2} + \text{effect of block 3} + \text{experimental error}$$

Or, $Y_{ijkl} = \mu + \alpha_i + \beta_j + \gamma_k + \delta_l + \varepsilon_{ijkl}$
where α_i are the population treatment effects for $j = 4$ anti-slip treatments; β_j are the population block 1 effects for $j = 4$ different days; γ_k are the population block 2 effects for $k = 4$ different times in a day, δ_l are the population block 3 effects for $l = 4$ different equipemnet used per day for the treatments and ε_{ijkl} are the error terms. The random errors ε_{ijkl} are independent, normally distributed random variables, each with zero mean and variance σ^2.

The experimental data of single factor Graeco-Latin square block design for anti-slip treatment experiment is shown in Table 10.6.

From the given data we have: number of treatments = number of days = number of times in a day = $a = 4$, $T_{1...} = 18,900$ $T_{2...} = 18,450$, $T_{3...} = 18,900$, $T_{4...} = 18,150$; $T_{.1..} = 18,900$, $T_{.2..} = 18,350$, $T_{.3..} = 18,550$, $T_{.4..} = 18,600$; $T_{..1.} = 18,600$, $T_{..2.} = 18,650$, $T_{..3.} = 18,500$, $T_{..4.} = 18,650$; $T_{...1} = 18,350$, $T_{...2} = 18,700$, $T_{...3} = 18,700$, $T_{...4} = 18,650$; $T = 74,400$; $\sum_{i=1}^{a} \sum_{j=1}^{a} \sum_{k=1}^{a} \sum_{l=1}^{a} y_{ijkl}^2 = 346,160,000$.

Step 1: H_0: $\alpha_i = 0$, for $i = 1, 2, 3, 4$
H_0': $\beta_j = 0$, for $j = 1, 2, 3, 4$
H_0'': $\gamma_k = 0$, for $k = 1, 2, 3, 4$
H_0''': $\delta_l = 0$, for $l = 1, 2, 3, 4$
H_1: $\alpha_i \neq 0$, for at least one value of i
H_1': $\beta_j \neq 0$, for at least one value of j
H_1'': $\gamma_k \neq 0$, for at least one value of k
H_1''': $\delta_l \neq 0$, for at least one value of l
$\alpha = 0.05$.
Step 2: For treatments, the critical value is $f_{\alpha,(a-1),(a-3)(a-1)} = f_{0.05,(4-1),(4-3)\times(4-1)} = f_{0.05,3,3} = 9.3$ (Table A5).

TABLE 10.6

Experimental Data of Single Factor Graeco-Latin Square Block for Anti-Slip Treatment

	Days				
Time of Day	**1**	**2**	**3**	**4**	$T_{..k.}$
9 am	$D_p = 4,600$	$B_r = 4,600$	$C_s = 4,650$	$A_q = 4,750$	18,600
11 am	$B_q = 4,650$	$D_s = 4,500$	$A_r = 4,800$	$C_p = 4,700$	18,650
1 pm	$C_r = 4,800$	$A_p = 4,500$	$D_q = 4,550$	$B_s = 4,650$	18,500
3 pm	$A_s = 4,850$	$C_q = 4,750$	$B_p = 4,550$	$D_r = 4,500$	18,650
$T_{.j..}$	18,900	18,350	18,550	18,600	$T = 74,400$

	Equipment			
	p	**q**	**r**	**s**
	$D_p = 4,600$	$B_q = 4,650$	$C_r = 4,800$	$A_s = 4,850$
	$A_p = 4,500$	$C_q = 4,750$	$B_r = 4,600$	$D_s = 4,500$
	$B_p = 4,550$	$D_q = 4,550$	$A_r = 4,800$	$C_s = 4,650$
	$C_p = 4,700$	$A_q = 4,750$	$D_r = 4,500$	$B_s = 4,650$
$T_{...l}$	18,350	18,700	18,700	18,650

Step 3: SST_r, SSB_{day}, $SSB_{time\ of\ day}$, $SSB_{equipment}$ and SSE are calculated as

$$SST_r = \frac{1}{a}\sum_{i=1}^{a} T_{i...}^2 - \frac{T^2}{a \times a}$$

$$= \frac{1}{4}\left(18,900^2 + 18,450^2 + 18,900^2 + 18,150^2\right) - \frac{74,400^2}{4 \times 4} = 101,250$$

$$SSB_{days} = \frac{1}{a}\sum_{j=1}^{a} T_{.j..}^2 - \frac{T^2}{a \times a}$$

$$= \frac{1}{4}\left(18,900^2 + 18,350^2 + 18,550^2 + 18,600^2\right) - \frac{74,400^2}{4 \times 4} = 38,750$$

$$SSB_{time\ in\ a\ day} = \frac{1}{a}\sum_{k=1}^{a} T_{..k.}^2 - \frac{T^2}{a \times a}$$

$$= \frac{1}{4}\left(18,600^2 + 18,650^2 + 18,500^2 + 18,650^2\right) - \frac{74,400^2}{4 \times 4} = 3750$$

$$SSB_{equipment} = \frac{1}{a}\sum_{k=1}^{a} T_{...l}^2 - \frac{T^2}{a \times a}$$

$$= \frac{1}{4}\left(18,350^2 + 18,700^2 + 18,700^2 + 18,650^2\right) - \frac{74,400^2}{4 \times 4} = 21,250$$

$$SST = \sum_{i=1}^{a}\sum_{j=1}^{a}\sum_{k=1}^{a}\sum_{l=1}^{a} y_{ijkl}^2 - \frac{T^2}{a \times a} = 346,160,000 - \frac{74,400^2}{4 \times 4} = 200,000$$

By subtraction we obtain, $SSE = SST - SST_r - SSB_{days} - SSB_{time\ of\ day} = 200,000 - 101,250 - 38,750 - 3750 - 21,250 = 35,000$

Therefore,

$$MST_r = \frac{SST_r}{(a-1)} = \frac{101,250}{4-1} = 33,750$$

$$MSB_{days} = \frac{SSB_{days}}{a-1} = \frac{38,750}{4-1} = 12,916.67$$

$$MSB_{time\ of\ day} = \frac{SSB_{time\ of\ day}}{a-1} = \frac{3750}{4-1} = 1250$$

$$MSB_{equipment} = \frac{SSB_{equipment}}{a-1} = \frac{21,250}{4-1} = 7083.33$$

and

$$MSE = \frac{SSE}{(a-3)(a-1)} = \frac{35,000}{(4-3)\times(4-1)} = 11,666.67$$

Hence, $f_{T_r} = \dfrac{MST_r}{MSE} = \dfrac{33,750}{11,666.67} = 2.89$

$$f_{days} = \frac{MSB_{days}}{MSE} = \frac{12,916.67}{11,666.67} = 1.11$$

$$f_{time\ of\ day} = \frac{MSB_{time\ of\ day}}{MSE} = \frac{1250}{11,666.67} = 0.11$$

$$f_{equipment} = \frac{MSB_{equipment}}{MSE} = \frac{7083.33}{11,666.67} = 0.61$$

Thus, we get the following analysis of variance table (Table 10.7).

TABLE 10.7

The Analysis of Variance Table for Single Factor Graeco-Latin Square Block Design for Anti-Slip Treatment Experiment

Source of Variation	Sum of Square	Degrees of Freedom	Mean Square	f
Treatments	101,250	3	33,750	2.89
Days	38,750	3	12,916.67	1.11
Time of day	3750	3	1250	0.11
Equipment	21,250	3	7083.33	0.61
Error	35,000	3	11,666.67	
Total	200,000	15		

Nuisance factor 1

Nuisance factor 1

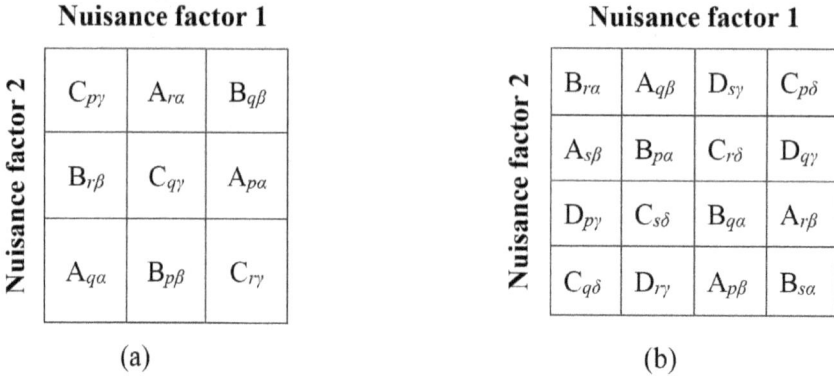

Nuisance factor 2	$C_{p\gamma}$	$A_{r\alpha}$	$B_{q\beta}$
	$B_{r\beta}$	$C_{q\gamma}$	$A_{p\alpha}$
	$A_{q\alpha}$	$B_{p\beta}$	$C_{r\gamma}$

(a)

Nuisance factor 2	$B_{r\alpha}$	$A_{q\beta}$	$D_{s\gamma}$	$C_{p\delta}$
	$A_{s\beta}$	$B_{p\alpha}$	$C_{r\delta}$	$D_{q\gamma}$
	$D_{p\gamma}$	$C_{s\delta}$	$B_{q\alpha}$	$A_{r\beta}$
	$C_{q\delta}$	$D_{r\gamma}$	$A_{p\beta}$	$B_{s\alpha}$

(b)

FIGURE 10.11 Hyper-Graeco-Latin square experimental design (a) 3×3; (b) 4×4.

Step 4: For treatments, as $2.89 < 9.3$, the null hypothesis cannot be rejected. Therefore, there are no significant differences in the tearing strength due to anti-slip treatments. For the blocks, namely days and time of day, f_{days} (1.11), $f_{time\ of\ day}$ (0.11) and $f_{equipment}$ (0.61) < 9.3, the null hypothesis cannot be rejected. Therefore, there are no significant differences in tearing strength due to anti-slip treatments applied on different days, different times of a day and using different equipment.

It is to be noted that in Examples 10.2, 10.3 and 10.4, the critical values have increased due to changes in the degrees of freedom. Therefore, though the effect of anti-slip treatment in Example 10.2 was found significant, in subsequent Examples 10.3 and 10.4 it was insignificant.

10.3.3.4 Hyper-Graeco-Latin Square Design

When an experimental design demands inclusion of four nuisance factors along with a single factor of primary interest, Hyper-Graeco-Latin square design is applied. This model has the similar limitations of Latin square and Graeco-Latin square. A 3×3 and 4×4 Hyper-Graeco-Latin square experimental design are shown in Figure 10.11(a) and (b), respectively, where the symbols A, B, C and D are the treatments. In Figure 10.11 two blocking factors are shown along the rows and columns, whereas the third and fourth blocking factors are indicated by the suffix symbols p, q, r, s and $\alpha, \beta, \gamma, \delta$, respectively.

10.4 FACTORIAL DESIGN

So far, we were concerned with only one factor as primary interest along with one or more nuisance or blocking factors. But many experiments involve two or more factors that affect the response variable. In such cases, the most efficient experimental design is 'factorial design', which is often called 'complete factorial design'. In each complete trial of a factorial design, all possible combinations of factors and levels are considered. This means that, the effects of both the primary factors and

TABLE 10.8
IPI of 20^s Ne Combed Cotton Yarn

		Card Cylinder Speed (rpm)	
		−	+
Card Production	−	20	10
Rate (kg/h)	+	32	22

their interactions on response variables are examined. So if an experiment involves two primary factors A and B, with levels *a* and *b* respectively, the complete trial will contain all permutations of *ab* treatments. When a change in level of either of the factors A or B affects the response, it is called 'main effect'. But if the response variable is affected by various permutations of levels of A and B, then it is known as 'interaction effect'. A factorial design with two factors each at two levels is known as 2^2 factorial design; similarly, factorial design with three factors each at two levels is known as 2^3 factorial design. Therefore, a 2^n factorial design means a design with *n* factors, each with two levels.

For better understanding of 'main effect' and 'interaction effect', let us consider two simple 2^2 factorial experiments. In the first experiment, a spinner (say, spinner A) wants to study the effect of card cylinder speed and card production rate on the yarn imperfections (IPI) which are tabulated in Table 10.8 and illustrated in Figures 10.12(a) and (b). He has considered the IPI for 20^s Ne combed cotton yarn for this experiment and both the factors are taken at two levels (card cylinder speed = 400 rpm and 500 rpm; card production rate = 50 kg/h and 60 kg/h). All possible combinations of these two factors are considered in the experiment. In such two level factorial designs the lower and higher levels are usually expressed by the signs '−' and '+' respectively. Therefore, '−' and '+' signs for card cylinder speed indicate 400 rpm and 500 rpm respectively; whereas, that for card production rate indicate

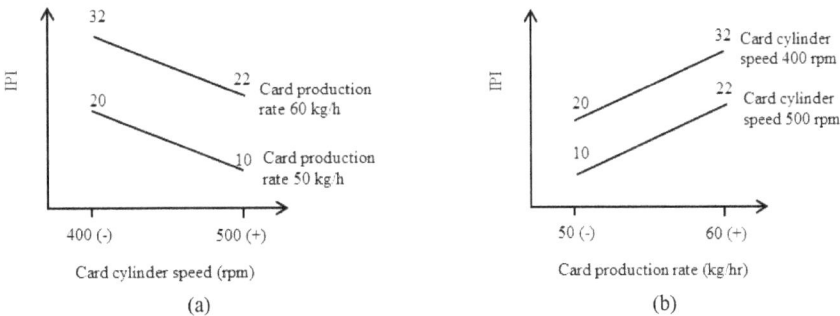

FIGURE 10.12 Effect of (a) card cylinder speed on the IPI of 20^s Ne cotton yarn; (b) card production rate on the IPI of 20^s Ne cotton yarn.

TABLE 10.9

IPI of 30ˢ Ne Combed Cotton Yarn

		Ring Frame Spindle Speed (rpm)	
		−	+
Roving	−	40	56
Fineness(Ne)	+	48	44

50 kg/h and 60 kg/h respectively. In this example of 20ˢ Ne combed cotton yarn, the 'main effects' and their 'interactions' on yarn IPI can be computed as follows:

The effect of card cylinder speed is the difference between the average IPI at higher level (+) and average IPI at lower level (−). Therefore, the effect of card cylinder speed $= \frac{10+22}{2} - \frac{20+32}{2} = -10$.

Similarly, the effect of card production rate $= \frac{32+22}{2} - \frac{20+10}{2} = 12$.

The interaction effect of card cylinder speed and card production rate is the average of the difference between the effects of card cylinder speed at two levels of card production rate. The effect of card cylinder speed at higher card production rate (+) = $22 - 32 = -10$. Again, the effect of card cylinder speed at lower card production rate (−) $= 10 - 20 = -10$. Therefore, the average of differences between the effects of the card cylinder at higher and lower levels of card production rate is $= \frac{(-10)-(-10)}{2} = 0$. Hence, the interaction effect of card cylinder speed and card production rate on the IPI of 20ˢ Ne yarn is nil which is also obvious in Figure 10.12, where the lines are parallel and not interacting with each other.

In a second experiment, a spinner B in the same spinning mill wants to study the effect of 'roving fineness' and 'ring frame spindle speed' on yarn imperfections (IPI) of 30ˢ Ne cotton yarn. The IPI of 30ˢ Ne cotton yarns are tabulated in Table 10.9 and illustrated in Figure 10.13(a) and (b). In this experiment both the factors are also taken at two levels (roving fineness = 0.9 Ne and 1.2 Ne; ring frame spindle speed = 15,000 rpm and 18,000 rpm) with all possible combinations. In this example

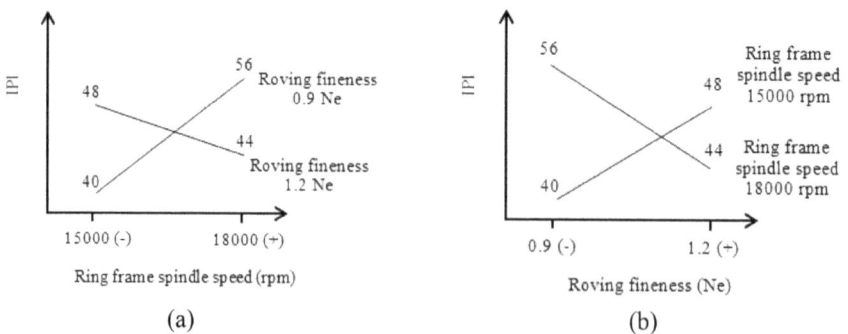

FIGURE 10.13 Effect of (a) ring frame spindle speed on IPI of 30ˢ Ne cotton yarn; (b) roving hank on IPI of 30ˢ Ne cotton yarn.

of 30^s Ne combed cotton yarn, the 'main effects' and their 'interactions' on yarn IPI can be computed as follows:

$$\text{The effect of ring frame spindle speed} = \frac{56+44}{2} - \frac{48+40}{2} = 6.$$

$$\text{The effect of roving fineness} = \frac{48+44}{2} - \frac{56+40}{2} = -2$$

The interaction effect of ring frame spindle speed and roving fineness is the average of difference between the effects of ring frame spindle at two levels of roving fineness. The effect of ring frame spindle speed at higher roving fineness $(+) = 44 - 48 = -4$. Again, the effect of ring frame spindle speed at lower roving fineness $(-) = 56 - 40 = 16$. Therefore, the average of differences between effects of ring frame spindle speed at higher and lower levels of roving fineness is $= \frac{(-4)-(16)}{2} = -10$. Hence, there exists an interaction effect of ring frame spindle speed and roving fineness on the IPI of 30^s Ne cotton yarn. Therefore, the response for this experiment (conducted by spinner B) can be shown as:

Response = constant + effect of ring frame spindle speed + effect of
roving fineness + interaction effect of ring frame spindle speed
and roving fineness + experimental error

Example 10.5:

A 2^2 experiment was conducted by a dyer to find the effect of salt concentration and dyeing temperature on direct dye fixation in cotton fabric. Two levels of salt concentrations viz. 30 gpl and 60 gpl were used. Dyeing temperatures considered in this experiment were 60°C and 90°C. The results of dye fixation (%) are given in Table 10.10. The lower and higher values of salt concentrations (gpl) and dyeing temperatures are denoted by '−' and '+' signs respectively. Test at the 0.05 level of significance whether there are significant differences in the dye fixation (%) by the cotton fabric due to salt concentrations, dyeing temperature and their interactions.

TABLE 10.10
Dye Fixation (%) in Cotton Fabric

		Salt Concentration (gpl)	
		−	+
		50	62
	−	48	60
Dyeing		47	63
Temperature		48	57
(°C)	+	54	60
		50	56

SOLUTION

This is an example of a two-factors and two-levels experiment with three replications. In this example, we have

Response (dye fixation) = constant + effect of salt concentration + effect of dyeing temperature + interaction effect of salt concentration and dyeing temperature + experimental error

or, $Y_{ij} = \mu + \alpha_i + \beta_j + (\alpha\beta)_{ij} + \varepsilon_{ij}$

where α_i are the effects of salt concentration for $i = 2$ levels; β_j are the effects of dyeing temperature for $j = 2$ levels; $(\alpha\beta)_{ij}$ is the interaction between α_i and β_j; and ε_{ij} are the error terms. The random errors ε_{ij} are independent, normally distributed random variables, each with zero mean and variance σ^2.

From the given data we have: Levels of salt concentrations = $a = 2$, levels of dyeing temperatures = $b = 2$, number of replications = $n = 3$, $\sum_{i=1}^{a} \sum_{j=1}^{b} \sum_{k=1}^{n} y_{ijk}^2 = 36{,}131$. The required sums and sum of squares can be calculated by forming the following two-way table giving the sums T_{ij} (Table 10.11).

Thus, we obtain, $T_{1..} = 297$, $T_{2..} = 358$, $T_{.1.} = 330$, $T_{.2.} = 325$, $T = 655$, and $\sum_{i=1}^{a} \sum_{j=1}^{b} T_{ij.}^2 = 108{,}283$.

Step 1: H_0: $\alpha_i = 0$, for $i = 1, 2$
H_0': $\beta_j = 0$, for $j = 1, 2$
H_0'': $(\alpha\beta)_{ij} = 0$, for $i = 1, 2$ and $j = 1, 2$
H_1: $\alpha_i \neq 0$, for at least one i
H_1': $\beta_j \neq 0$, for at least one j
H_1'': $(\alpha\beta)_{ij} \neq 0$, for at least one (i, j)
$\alpha = 0.05$.

Step 2:   For salt concentration (gpl), the critical value is $f_{\alpha,(a-1),ab(n-1)} = f_{0.05,1,8} = 5.3$.
For dyeing temperature (°C), the critical value is $f_{\alpha,(b-1),ab(n-1)} = f_{0.05,1,8} = 5.3$.
For interaction of salt concentration (gpl) and dyeing temperature (°C), the critical value is $f_{\alpha,(a-1)(b-1),ab(n-1)} = f_{0.05,1,8} = 5.3$. (Table A5).

TABLE 10.11
Sum of the Dye Fixation Values of Three Replications

		Salt Concentration (gpl)		
		–	+	Total
Dyeing	–	145	185	330
Temperature (°C)	+	152	173	325
Total		297	358	655

Step 3: $SS_{\text{Salt concentration}}$, $SS_{\text{Dyeing temperature}}$, $SS_{\text{Salt concentration} \times \text{Dyeing temperature}}$ and SSE are calculated as

$$SS_{\text{Salt concentration}} = \frac{1}{bn}\sum_{i=1}^{a}T_{i..}^2 - \frac{T^2}{abn} = \frac{1}{2\times 3}\left(297^2 + 358^2\right) - \frac{655^2}{2\times 2\times 3}$$

$$= 36{,}062.2 - 35{,}752.1 = 310.1$$

$$SS_{\text{Dyeing temperature}} = \frac{1}{an}\sum_{j=1}^{b}T_{.j.}^2 - \frac{T^2}{abn} = \frac{1}{2\times 3}\left(330^2 + 325^2\right) - \frac{655^2}{2\times 2\times 3}$$

$$= 35{,}754.2 - 35{,}752.1 = 2.1$$

$$SS_{\text{Salt concentration} \times \text{Dyeing temperature}} = \frac{1}{n}\sum_{i=1}^{a}\sum_{j=1}^{b}T_{ij.}^2 - \frac{1}{bn}\sum_{i=1}^{a}T_{i..}^2 - \frac{1}{an}\sum_{j=1}^{b}T_{.j.}^2 + \frac{T^2}{abn}$$

$$= \frac{108{,}283}{3} - 36{,}062.2 - 35{,}754.2 + 35{,}752.1 = 30$$

$$SST = \sum_{i=1}^{a}\sum_{j=1}^{b}\sum_{k=1}^{n}y_{ijk}^2 - \frac{T^2}{abn} = 36{,}131 - 35{,}752.1 = 378.9$$

By subtraction we obtain,

$$SSE = SST - SS_{\text{Salt concentration}} - SS_{\text{Dyeing temperature}}$$
$$- SS_{\text{Salt concentration} \times \text{Dyeing temperature}}$$
$$= 378.9 - 310.1 - 2.1 - 30 = 36.7$$

Therefore,

$$MS_{\text{Salt concentration}} = \frac{SS_{\text{Salt concentration}}}{(a-1)} = \frac{310.1}{2-1} = 310.1$$

$$MS_{\text{Dyeing temperature}} = \frac{SS_{\text{Dyeing temperature}}}{(b-1)} = \frac{2.1}{2-1} = 2.1$$

$$MS_{\text{Salt concentration} \times \text{Dyeing temperature}} = \frac{SS_{\text{Salt concentration} \times \text{Dyeing temperature}}}{(a-1)(b-1)}$$

$$= \frac{30}{(2-1)\times(2-1)} = 30$$

and

$$MSE = \frac{SSE}{ab(n-1)} = \frac{36.7}{2 \times 2 \times (3-1)} = 4.59$$

Hence,

$$f_{\text{Salt concentration}} = \frac{MS_{\text{Salt concentration}}}{MSE} = \frac{310.1}{4.59} = 67.56$$

$$f_{\text{Dyeing temperature}} = \frac{MS_{\text{Dyeing temperature}}}{MSE} = \frac{2.1}{4.59} = 0.46$$

and

$$f_{\text{Salt concentration} \times \text{Dyeing temperature}} = \frac{MS_{\text{Salt concentration} \times \text{Dyeing temperature}}}{MSE}$$

$$= \frac{30}{4.59} = 6.54$$

Thus, we get the following analysis of variance table (Table 10.12).

Step 4: For salt concentration (gpl), as $67.56 > 5.3$, the null hypothesis is rejected. Therefore, we do have statistical evidence that there are significant differences in the dye fixation (%) due to the salt concentration (gpl). For dyeing temperature, as $0.46 < 5.3$, the null hypothesis cannot be rejected. Therefore, there is no sufficient statistical evidence to conclude that the dyeing temperature (°C) has influence on dye fixation (%) in cotton fabric. For interaction, salt concentration (gpl) × dyeing temperature (°C), as $6.54 > 5.3$, the null hypothesis is rejected. Therefore, there is statistical evidence to conclude that the interaction between salt concentration (gpl) and dyeing temperature (°C) has significant influence on dye fixation (%) in cotton fabric.

TABLE 10.12

The Analysis of Variance Table for 2^2 Factorial Designs for Dye Fixation Experiment

Source of Variation	Sum of Square	Degrees of Freedom	Mean Square	f
Salt concentration (gpl)	310.1	1	310.1	67.56
Dyeing temperature (°C)	2.1	1	2.1	0.46
Interaction (Salt concentration × Dyeing temperature)	30	1	30	6.54
Error	36.7	8	4.59	
Total	378.9	11		

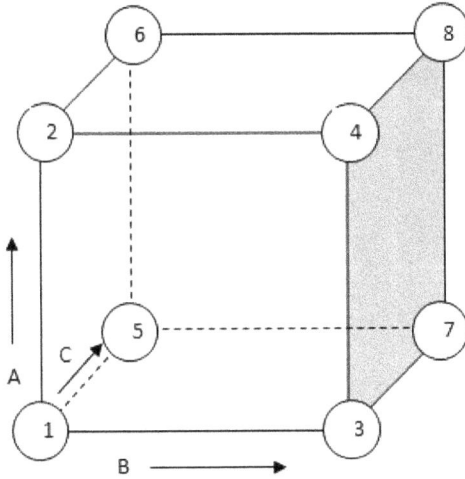

FIGURE 10.14   A 2^3 full factorial design with factors A, B and C.

A 2^2 full factorial design which is exemplified in Example 10.6 has two factors, each with two levels; whereas 2^3 full factorial designs have three factors, each with two levels. A 2^3 full factorial design is shown in Figure 10.14, where the experiment runs are displayed at the corners of the cube. The 'runs' of a three-factor two-level full factorial design in a 'standard order' is given in Table 10.13.

Example 10.6:

A 2^3 experiment is conducted by the quality control (QC) department in a knitting unit to understand the effect of three factors on the ultra-protection factor (UPF) of single jersey knitted fabrics. Three factors, namely loop length (A), yarn linear density (B) and knitting speed (C) are considered, each at two levels, viz., low (−)

TABLE 10.13
Experimental Run of A 2^3 Full Factorial Design

Experimental Run	A	B	C
1	−1	−1	−1
2	+1	−1	−1
3	−1	+1	−1
4	+1	+1	−1
5	−1	−1	+1
6	+1	−1	+1
7	−1	+1	+1
8	+1	+1	+1

TABLE 10.14

UPF of Single Jersey Knitted Fabrics

		A−		A+	
		B−	B+	B−	B+
C−		22	28	14	25
		21	26	17	26
		20	30	17	24
		63	84	48	75
C+		22	26	17	24
		21	26	15	23
		23	29	16	25
		66	81	48	72

and high (+). The results obtained are given in Table 10.14. Test at the 0.5% level of significance whether there are significant differences in UPF by the loop length, yarn linear density, knitting speed and their interactions.

SOLUTION

In this example,

Response (UPF) = constant + effect of A + effect of B + effect of

C + interaction effect of A and B + interaction effect of A and

C + interaction effect of B and C + interaction effect of A, B and

C + experimental error

or, $Y_{ijk} = \mu + \alpha_i + \beta_j + \gamma_k + (\alpha\beta)_{ij} + (\alpha\gamma)_{ik} + (\beta\gamma)_{ik} + (\alpha\beta\gamma)_{ijk} + \varepsilon_{ijk}$
where α_i are the effects of A for $i = 2$ levels; β_j are the effects of B for $j = 2$ levels; γ_k are the effects of C for $j = 2$ levels; $(\alpha\beta)_{ij}$ is the interaction between α_i and β_j; $(\alpha\gamma)_{ik}$ is the interaction between α_i and γ_k; $(\beta\gamma)_{ik}$ is the interaction between β_j and γ_k; $(\alpha\beta\gamma)_{ijk}$ is the interaction between α_i, β_j and γ_k; and ε_{ijk} are the error terms. The random errors ε_{ijk} are independent, normally distributed random variables, each with zero mean and variance σ^2.

From the given data we have: levels of loop length (A) = $a = 2$; levels of yarn linear density (B) = $b = 2$; levels of knitting speed (C) = $c = 2$; number of replications = $n = 3$, $\sum_{i=1}^{a}\sum_{j=1}^{b}\sum_{k=1}^{c}\sum_{l=1}^{n} y_{ijkl}^2 = 12,483$. The required sums and sum of squares can be calculated by forming the following table giving the sums T_{ijk} (Table 10.15).

Thus we obtain, $T_{1..} = 294, T_{2..} = 243, T_{.1.} = 225, T_{.2.} = 312, T_{..1} = 270, T_{..2} = 267$, $T = 537$, and $\sum_{i=1}^{a}\sum_{j=1}^{b}\sum_{k=1}^{c} T_{ijk}^2 = 37,359$.

Step 1: $H_0: \alpha_i = 0$, for $i = 1, 2$
                $H_0': \beta_j = 0$, for $j = 1, 2$

TABLE 10.15

Sum of the UPF Values of Three Replications

	B−		B+		Total
	C−	C+	C−	C+	
A−	63	66	84	81	294
A+	48	48	75	72	243

					537

	A−		A+		Total
	C−	C+	C−	C+	
B−	63	66	48	48	225
B+	84	81	75	72	312

					537

	A−		A+		Total
	B−	B+	B−	B+	
C−	63	84	48	75	270
C+	66	81	48	72	267

					537

$H_0'' : \gamma_k = 0$, for $k = 1, 2$

$H_0''' : (\alpha\beta)_{ij} = 0$, for $i = 1, 2$ and $j = 1, 2$

$H_0'''' : (\alpha\gamma)_{ik} = 0$, for $i = 1, 2$ and $k = 1, 2$

$H_0''''' : (\beta\gamma)_{jk} = 0$, for $j = 1, 2$ and $k = 1, 2$

$H_0'''''' : (\alpha\beta\gamma)_{ijk} = 0$, for $i = 1, 2$; $j = 1, 2$ and $k = 1, 2$

$H_1 : \alpha_i \neq 0$, for at least one i

$H_1' : \beta_j \neq 0$, for at least one j

$H_1'' : \gamma_k \neq 0$, for at least one k

$H_1''' : (\alpha\beta)_{ij} \neq 0$, for at least one (i, j)

$H_1'''' : (\alpha\gamma)_{ik} \neq 0$, for at least one (i, k)

$H_1''''' : (\beta\gamma)_{jk} \neq 0$, for at least one (j, k)

$H_1'''''' : (\alpha\beta\gamma)_{ijk} \neq 0$, for at least one (i, j, k)

$\alpha = 0.05$.

Step 2: For A, the critical value is $f_{\alpha,(a-1),abc(n-1)} = f_{0.05,1,16} = 4.48$; for B, the critical value is $f_{\alpha,(b-1),abc(n-1)} = f_{0.05,1,16} = 4.48$; for C, the critical value is $f_{\alpha,(c-1),abc(n-1)} = f_{0.05,1,16} = 4.48$; for interaction A × B, the critical value is $f_{\alpha,(a-1)(b-1),abc(n-1)} = f_{0.05,1,16} = 4.48$; for interaction A × C, the critical value is $f_{\alpha,(a-1)(c-1),abc(n-1)} = f_{0.05,1,16} = 4.48$; for interaction B × C, the critical value is $f_{\alpha,(b-1)(c-1),abc(n-1)} = f_{0.05,1,16} = 4.48$; for interaction A × B × C, the critical value is $f_{\alpha,(a-1)(b-1)(c-1),abc(n-1)} = f_{0.05,1,16} = 4.48$. (Table A5).

Step 3: SS_A, SS_B, SS_C, $SS_{A \times B}$, $SS_{A \times C}$, $SS_{B \times C}$, $SS_{A \times B \times C}$ and SSE are calculated as

$$SS_A = \frac{1}{bcn}\sum_{i=1}^{a}T_{i...}^2 - \frac{T^2}{abcn} = \frac{1}{2 \times 2 \times 3}\left(294^2 + 243^2\right) - \frac{537^2}{2 \times 2 \times 2 \times 3}$$

$$= 12{,}123.75 - 12{,}015.38 = 108.37$$

$$SS_B = \frac{1}{acn}\sum_{b}^{j=1}T_{.j..}^2 - \frac{T^2}{abcn} = \frac{1}{2 \times 2 \times 3}\left(225^2 + 312^2\right) - \frac{537^2}{2 \times 2 \times 2 \times 3}$$

$$= 12{,}330.75 - 12{,}015.38 = 315.37$$

$$SS_C = \frac{1}{abn}\sum_{c}^{k=1}T_{..k.}^2 - \frac{T^2}{abcn} = \frac{1}{2 \times 2 \times 3}\left(270^2 + 267^2\right) - \frac{537^2}{2 \times 2 \times 2 \times 3}$$

$$= 12{,}015.75 - 12{,}015.38 = 0.38$$

$$SS_{A \times B} = \frac{1}{cn}\sum_{i=1}^{a}\sum_{j=1}^{b}T_{ij..}^2 - \frac{T^2}{abcn} - SS_A - SS_B$$

$$= \frac{74{,}691}{6} - 12{,}015.38 - 108.37 - 315.37 = 9.38$$

$$SS_{A \times C} = \frac{1}{bn}\sum_{i=1}^{a}\sum_{k=1}^{c}T_{i.k.}^2 - \frac{T^2}{abcn} - SS_A - SS_C$$

$$= \frac{72{,}747}{6} - 12{,}015.38 - 108.37 - 0.37 = 0.38$$

$$SS_{B \times C} = \frac{1}{an}\sum_{j=1}^{b}\sum_{k=1}^{c}T_{.jk.}^2 - \frac{T^2}{abcn} - SS_B - SS_C$$

$$= \frac{74{,}007}{6} - 12{,}015.38 - 315.37 - 0.37 = 3.38$$

$$SS_{A \times B \times C} = \frac{1}{n}\sum_{i=1}^{a}\sum_{j=1}^{b}\sum_{k=1}^{c}T_{ijk.}^2 - \frac{T^2}{abcn} - SS_A - SS_B - SS_C - SS_{A \times B} - SS_{A \times C} - SS_{B \times C}$$

$$= \frac{37{,}359}{3} - 12{,}015.38 - 108.37 - 315.37 - 0.37 - 9.38 - 0.38 - 3.38 = 0.38$$

$$SST = \sum_{i=1}^{a}\sum_{j=1}^{b}\sum_{k=1}^{c}\sum_{l}^{n}y_{ijkl}^2 - \frac{T^2}{abcn} = 12{,}483 - 12{,}015.38 = 467.62$$

By subtraction we obtain,

$$SSE = SST - SS_A - SS_B - SS_C - SS_{A \times B} - SS_{A \times C} - SS_{B \times C} - SS_{A \times B \times C}$$

$$= 467.62 - 108.37 - 315.37 - 0.38 - 9.38 - 0.38 - 3.38 - 0.38 = 30$$

Therefore,

$$MS_A = \frac{SS_A}{(a-1)} = \frac{108.37}{2-1} = 108.37$$

$$MS_B = \frac{SS_B}{(b-1)} = \frac{315.37}{2-1} = 315.37$$

$$MS_C = \frac{SS_C}{(c-1)} = \frac{0.38}{2-1} = 0.38$$

$$MS_{A \times B} = \frac{SS_{A \times B}}{(a-1)(b-1)} = \frac{9.38}{(2-1) \times (2-1)} = 9.38$$

$$MS_{A \times C} = \frac{SS_{A \times C}}{(a-1)(c-1)} = \frac{0.38}{(2-1) \times (2-1)} = 0.38$$

$$MS_{B \times C} = \frac{SS_{B \times C}}{(b-1)(c-1)} = \frac{3.38}{(2-1) \times (2-1)} = 3.38$$

$$MS_{A \times B \times C} = \frac{SS_{A \times B \times C}}{(a-1)(b-1)(c-1)} = \frac{0.38}{(2-1) \times (2-1)(2-1)} = 0.38$$

and

$$MSE = \frac{SSE}{abc(n-1)} = \frac{30}{2 \times 2 \times 2 \times (3-1)} = 1.88$$

Hence,

$$f_A = \frac{MS_A}{MSE} = \frac{108.37}{1.88} = 57.64$$

$$f_B = \frac{MS_B}{MSE} = \frac{315.37}{1.88} = 167.75$$

$$f_C = \frac{MS_C}{MSE} = \frac{0.38}{1.88} = 0.2$$

$$f_{A \times B} = \frac{MS_{A \times B}}{MSE} = \frac{9.38}{1.88} = 4.99$$

TABLE 10.16

Analysis of Variance Table for Example 10.6

Source of Variation	Sum of Squares	Degree of Freedom	Mean Square	f
A	108.37	1	108.37	57.64
B	315.37	1	315.37	167.75
C	0.37	1	0.38	0.2
AB	9.38	1	9.38	4.99
AC	0.38	1	0.38	0.2
BC	3.38	1	3.38	1.8
ABC	0.37	1	0.38	0.2
Error	30	16	1.88	
Total	467.62	23		

$$f_{A\times C} = \frac{MS_{A\times C}}{MSE} = \frac{0.38}{1.88} = 0.2$$

$$f_{B\times C} = \frac{MS_{B\times C}}{MSE} = \frac{3.38}{1.88} = 1.8$$

and

$$f_{A\times B\times C} = \frac{MS_{A\times B\times C}}{MSE} = \frac{0.38}{1.88} = 0.2$$

Thus, we get the following analysis of variance table (Table 10.16).

Step 4:   For A (loop length), as 57.64 > 4.48, the null hypothesis is rejected; therefore, we do have statistical evidence that there are significant differences in the UPF due to loop length. For B (yarn linear density), 167.75 > 4.48, the null hypothesis is rejected; therefore, we do have statistical evidence that there are significant differences in the UPF due to yarn linear density. For C (knitting speed), as 0.2 < 4.48, the null hypothesis cannot be rejected; therefore, there is no sufficient statistical evidence to conclude that knitting speed has influence on UPF of single jersey cotton fabric. For interaction, A × B (loop length × yarn linear density), as 4.99 > 4.48, the null hypothesis is rejected; therefore, there is statistical evidence to conclude that the interaction between loop length and yarn linear density has significant influence on the UPF. For the interactions A × C, B × C and A × B × C, the f values are less than 4.48 (refer to Table 10.16) the null hypothesis cannot be rejected; therefore, there is no sufficient statistical evidence to conclude that the interactions A × C, B × C and A × B × C have influence on the UPF of single jersey cotton fabric.

The computations of sum of squares shown above (for 2^n factorial design) can also be done in a systematic and quick method using Yates' algorithm. This algorithm was named after the English statistician Frank Yates. Yates' algorithm with examples is discussed next.

10.4.1 Yates' Algorithm

A 2^n factorial design experiment can also be analysed using Yates' algorithm. Before performing this analysis, the data obtained through 2^n factorial design experimentation should be arranged in 'Yates' order' which denotes the treatment combinations and these columns at the beginning of the table is known as 'treatment column'. In Yates' order, there should be total n columns and 2^n rows. The columns should be filled with higher (+) and lower (−) levels in a specific order. The first column in the treatment column should start with lower level (−) followed by higher level '+' and all the 2^n rows will be filled with alternative '−' and '+' levels. The second column should start with two levels of '−' followed by two levels of '+' alternatively till 2^n rows. The other columns will be filled accordingly and the last column i.e. n^{th} column will be filled with 2^{n-1} levels of '−' followed by 2^{n-1} levels of '+'. Yates' order of 2^3 and 2^4 designs are shown in Table 10.17. Yates' procedure for 2^3 factorial design is shown in Table 10.18.

The Yates' algorithm and ANOVA for Example 10.6 is shown in Tables 10.19 and 10.20, respectively. It is to be noted that the higher order interactions (here $A \times B \times C$) are often difficult to interpret and are combined to give 'residual' or 'error'. The 'error mean square' is calculated by dividing 'error' with the degrees of freedom. f values are then calculated by diving the mean squares by 'error mean square'.

TABLE 10.17
Yates' Order for 2^3 and 2^4 Factorial Designs

2^3 Factorial Design			2^4 Factorial Design			
−	−	−	−	−	−	−
+	−	−	+	−	−	−
−	+	−	−	+	−	−
+	+	−	+	+	−	−
−	−	+	−	−	+	−
+	−	+	+	−	+	−
−	+	+	−	+	+	−
+	+	+	+	+	+	−
			−	−	−	+
			+	−	−	+
			−	+	−	+
			+	+	−	+
			−	−	+	+
			+	−	+	+
			−	+	+	+
			+	+	+	+

TABLE 10.18

Yates' Procedure for 2^3 Factorial Design

Row No.	A	B	C	y (Total yield from all 'r' replicates)	1	2	3	Factor Sum of Squares	Factor
1	–	–	–	y_1	$y_1 + y_2$ + $(y_3 + y_4)$	$(y_1 + y_2)$ + $(y_5 + y_6)$	$\{(y_1 + y_2) + (y_3 + y_4)\}$ + $\{(y_5 + y_6) + (y_7 + y_8)\}$	–	Mean
2	+	–	–	y_2	$y_3 + y_4$ + $(y_7 + y_8)$	$(y_5 + y_6)$ + $(y_6 - y_5) + (y_8 - y_7)$	$\{(y_2 - y_1) + (y_4 - y_3)\}$ + $\{(y_6 - y_5) + (y_8 - y_7)\}$	$\left[\dfrac{\{(y_2 - y_1) + (y_4 - y_3)\} + \{(y_6 - y_5) + (y_8 - y_7)\}}{r \times 8}\right]^2$	A
3	–	+	–	y_3	$y_5 + y_6$ + $(y_4 - y_3)$	$(y_2 - y_1)$ + $\{(y_7 + y_8) - (y_5 + y_6)\}$	$\{(y_3 + y_4) - (y_1 + y_2)\}$ + $\{(y_7 + y_8) - (y_5 + y_6)\}$	$\left[\dfrac{\{(y_3 + y_4) - (y_1 + y_2)\} + \{(y_7 + y_8) - (y_5 + y_6)\}}{r \times 8}\right]^2$	B
4	+	+	–	y_4	$y_7 + y_8$ + $(y_8 - y_7)$	$(y_6 - y_5)$ + $\{(y_8 - y_7) - (y_6 - y_5)\}$	$\{(y_4 - y_3) - (y_2 - y_1)\}$ + $\{(y_8 - y_7) - (y_6 - y_5)\}$	$\left[\dfrac{\{(y_4 - y_3) - (y_2 - y_1)\} + \{(y_8 - y_7) - (y_6 - y_5)\}}{r \times 8}\right]^2$	AB
5	–	–	+	y_5	$y_2 - y_1$ – $(y_1 + y_2)$	$(y_3 + y_4)$ – $\{(y_1 + y_2) + (y_3 + y_4)\}$	$\{(y_5 + y_6) + (y_7 + y_8)\}$ – $\{(y_1 + y_2) + (y_3 + y_4)\}$	$\left[\dfrac{\{(y_5 + y_6) + (y_7 + y_8)\} + \{(y_1 + y_2) + (y_3 + y_4)\}}{r \times 8}\right]^2$	C
6	+	–	+	y_6	$y_4 - y_3$ – $(y_5 + y_6)$	$(y_7 + y_8)$ – $\{(y_2 - y_1) + (y_4 - y_3)\}$	$\{(y_6 - y_5) + (y_8 - y_7)\}$ – $\{(y_2 - y_1) + (y_4 - y_3)\}$	$\left[\dfrac{\{(y_6 - y_5) + (y_8 - y_7)\} + \{(y_2 - y_1) + (y_4 - y_3)\}}{r \times 8}\right]^2$	AC
7	–	+	+	y_7	$y_6 - y_5$ – $(y_2 - y_1)$	$(y_4 - y_3)$ – $\{(y_3 + y_4) - (y_1 + y_2)\}$	$\{(y_7 + y_8) - (y_5 + y_6)\}$ – $\{(y_3 + y_4) - (y_1 + y_2)\}$	$\left[\dfrac{\{(y_7 + y_8) - (y_5 + y_6)\} - \{(y_3 + y_4) - (y_1 + y_2)\}}{r \times 8}\right]^2$	BC
8	+	+	+	y_8	$y_8 - y_7$ – $(y_6 - y_5)$	$(y_8 - y_7)$ – $\{(y_4 - y_3) - (y_2 - y_1)\}$	$\{(y_8 - y_7) - (y_6 - y_5)\}$ – $\{(y_4 - y_3) - (y_2 - y_1)\}$	$\left[\dfrac{\{(y_8 - y_7) + (y_6 - y_5)\} + \{(y_4 - y_3) + (y_2 - y_1)\}}{r \times 8}\right]^2$	ABC

TABLE 10.19
Yates' Algorithm for Example 10.6

Row No.	A	B	C	y	1	2	3	Factor Sum of Squares	Factor
1	–	–	–	63	$63+48=111$	$111+159=270$	$270+267=537$	–	Mean
2	+	–	–	48	$84+75=159$	$114+153=267$	$(-24)+(-27)=-51$	$\dfrac{(-51)^2}{3\times 8}=108.38$	A
3	–	+	–	84	$66+48=114$	$(-15)+(-9)=-24$	$48+39=87$	$\dfrac{(87)^2}{3\times 8}=315.38$	B
4	+	+	–	75	$81+72=153$	$(-18)+(-9)=-27$	$6+9=15$	$\dfrac{(15)^2}{3\times 8}=9.38$	AB
5	–	–	+	66	$48-63=-15$	$159-111=48$	$267-270=-3$	$\dfrac{(-3)^2}{3\times 8}=0.38$	C
6	+	–	+	48	$75-84=-9$	$153-114=39$	$-27-(-24)=-3$	$\dfrac{(3)^2}{3\times 8}=0.38$	AC
7	–	+	+	81	$48-66=-18$	$(-9)-(-15)=6$	$39-48=-9$	$\dfrac{(-9)^2}{3\times 8}=3.38$	BC
8	+	+	+	72	$72-81=-9$	$(-9)-(-18)=9$	$9-6=3$	$\dfrac{(3)^2}{3\times 8}=0.38$	ABC

TABLE 10.20
ANOVA Table for Example 10.6

Source	Sum of Squares	Degree of Freedom	Mean Squares	f
A	108.38	1	108.38	285.21
B	315.38	1	315.38	829.95
C	0.38	1	0.38	1
AB	9.38	1	9.38	24.68
AC	0.38	1	0.38	1
BC	3.38	1	3.38	8.90
ABC	0.38	1	0.38	
Total	437.63	7		

TABLE 10.21

Seam Efficiency of Woven Fabrics

		A−		A+	
		B−	B+	B−	B+
C−	D−	78	80	85	88
	D+	76	78	80	82
C+	D−	77	82	82	84
	D+	75	79	78	81

An example on application of Yates' algorithm for a 2^4 design is cited below.

Example 10.7:

A 2^4 experiment was carried out to investigate the effect of four variables on seam efficiency of woven 1/3 twill cotton fabrics. The four variables are as follows:

A: whether the seam is of single yarn (−) or 3 ply yarn (+) of same fineness
B: whether the seam type is 201 (−) or 301 (+)
C: whether the yarn tension is low (−) or high (+)
D: whether the stitch length is low (−) or high (+)

The results are shown in Table 10.21.
Analyse these data and test at the 0.5% level of significance whether there are significant differences in A, B, C and D and their interactions on seam efficiency of woven fabric.

SOLUTION

The Yates' algorithm and ANOVA for this example is shown in Tables 10.22 and 10.23, respectively.

It is to be noted that the experiment discussed in Example 10.7 is not replicated and hence there is no unbiased estimate of the experimental error. In such cases, it is assumed that the interaction effect of three or more factors have no practical significance and hence are combined to give 'residual' or 'error', which has a value of 3.8 in this example. The 'error mean square' is calculated by dividing 'error' with the degrees of freedom (5 in this case) which is 0.76 (Table 10.24). All other mean squares are divided by 'error mean square' to provide f values. The f values are then compared with critical values of f (from Table A5) to test the significance of the main and two factor interaction effects. From Table A5 we get, $f_{0.05,1,5} = 6.6$.

Now, it is obvious from Table 10.23 that the f values of the factors A, B, C, D, A × C and A × D are more than their corresponding critical values ($f_{0.05,1,5} = 6.6$) and hence, the null hypothesis is rejected. Therefore, we do have statistical evidence that there are significant differences in the seam efficiency due to A, B, C, D, A × C and A × D. Now, for A × B, B × C, B × D and C × D as their f values are less than their corresponding critical values, the null hypothesis cannot be rejected; therefore, there is no sufficient statistical evidence to conclude that A × B, B × C, B × D and C × D have influence on seam efficiency on 1/3 twill woven cotton fabric.

TABLE 10.22
Yates' Algorithm for Example 10.7

Row No.	A	B	C	D	y	1	2	3	4	Factor Sum of Squares	Factor
1	−	−	−	−	78	163	331	656	1285	−	Mean
2	+	−	−	−	85	168	325	629	35	76.56	A
3	−	+	−	−	80	159	316	22	23	33.06	B
4	+	+	−	−	88	166	313	13	−3	0.56	AB
5	−	−	+	−	77	156	15	12	−9	5.06	C
6	+	−	+	−	82	160	7	11	−11	7.56	AC
7	−	+	+	−	82	153	8	−2	5	1.56	BC
8	+	+	+	−	84	160	5	−1	−5	1.56	ABC
9	−	−	−	+	76	7	5	−6	−27	45.56	D
10	+	−	−	+	80	8	7	−3	−9	5.06	AD
11	−	+	−	+	78	5	4	−8	−1	0.06	BD
12	+	+	−	+	82	2	7	−3	1	0.06	ABD
13	−	−	+	+	75	4	1	2	3	0.56	CD
14	+	−	+	+	78	4	−3	3	5	1.56	ACD
15	−	+	+	+	79	3	0	−4	1	0.06	BCD
16	+	+	+	+	81	2	−1	−1	3	0.56	ABCD

TABLE 10.23
ANOVA Table for Example 10.7

Source	Sum of Squares	Degree of Freedom	Mean Squares	f
A	76.56	1	76.56	100.73
B	33.06	1	33.06	43.5
C	5.06	1	5.06	6.66
D	45.56	1	45.56	59.9
AB	0.56	1	0.56	0.74
AC	7.56	1	7.56	9.94
AD	5.06	1	5.06	6.66
BC	1.56	1	1.56	2.05
BD	0.06	1	0.06	0.08
CD	0.56	1	0.56	0.74
ABC	1.56 ⎤	1 ⎤		
ABD	0.06 ⎥	1 ⎥		
ACD	1.56 ⎬ = 3.8	1 ⎬ = 5	0.76	
BCD	0.06 ⎥	1 ⎥		
ABCD	0.56 ⎦	1 ⎦		
Total	179.4	15		

TABLE 10.24

Number of Runs of Full Factorial Design with Various Factors, Each at Two Levels

Number of Factors	Number of Runs
2	4
3	8
4	16
5	32
6	64
7	128
8	256

10.4.2 FRACTIONAL FACTORIAL DESIGN

Till now, we have discussed 2^k full factorial designs which have k factors, each with two levels. Such design has 2^k runs. The number of runs of a full factorial design with various factors, each at two levels is given in Table 10.24. It is quite obvious from Table 10.24 that for a higher number of factors, a full factorial design requires a large number of runs, which outgrows the resources of most of the experimenters to conduct all the runs. For example, a mere seven factors and two levels full factorial design (2^7) requires 128 runs in which 7 of the 127 degrees of freedom corresponds to the main effect, 21 degrees of freedom corresponds to two factor interactions and the remaining 99 relates to three and higher interactions of factors. Generally, it is assumed that effect of higher order interactions (three and higher interaction of factors) to the response is negligible and hence, may not be considered while designing the experiment. Therefore, the information on the effect of main factors and low order interactions to the response may be obtained by running only a fraction of complete runs of the full factorial design of the experiment. Such design in which only the fraction of the complete runs of a full factorial design is conducted is known as 'fractional factorial design'.

Let us refer to Example 10.6 in which the Quality Control (QC) department of a knitting unit conducted a 2^3 full factorial design to investigate the effect of loop length (A), yarn linear density (B) and knitting speed (C) on fabric UPF. The average UPF is considered here and the design is shown in both tabular and graphical form in Table 10.25 and Figure 10.15, respectively. Now, let us consider that the same QC department has the similar objective to find the effect of A, B and C on fabric UPF, but cannot afford eight treatment combinations (2^3 runs); it can only afford four treatment combinations (2^{3-1}) i.e. a fraction of 2^3 runs. Since this design contains only $2^{3-1} = 4$ treatment combinations (1/2 fraction of 2^3 design), it is known as '2^{3-1} design of experimentation'. Let us see how a full factorial design can be substituted by a fractional factorial design without losing information at a significant level.

TABLE 10.25

A 2³ Full Factorial Design and UPF as the Response

Experimental Run	Loop Length (A)	Yarn Linear Density (B)	Knitting Speed (C)	UPF (y)
1	−	−	−	21
2	+	−	−	16
3	−	+	−	28
4	+	+	−	25
5	−	−	+	22
6	+	−	+	16
7	−	+	+	27
8	+	+	+	24

From Table 10.25 and Figures 10.15 and 10.16, the effects of main factors (A, B and C), first-order interaction effects (A × B, A × C and B × C), and second-order interaction effect (A × B × C) can be computed as follows:

$$\text{The effect of factor A} = \frac{1}{4}(16 + 25 + 16 + 24) - \frac{1}{4}(21 + 28 + 22 + 27) = -4.25$$

$$\text{The effect of factor B} = \frac{1}{4}(28 + 25 + 27 + 24) - \frac{1}{4}(21 + 16 + 22 + 16) = 7.25$$

$$\text{The effect of factor C} = \frac{1}{4}(22 + 16 + 27 + 24) - \frac{1}{4}(21 + 16 + 28 + 25) = -0.25$$

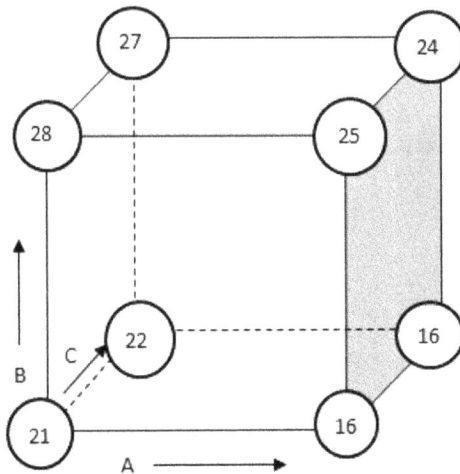

FIGURE 10.15   A 2³ full factorial design.

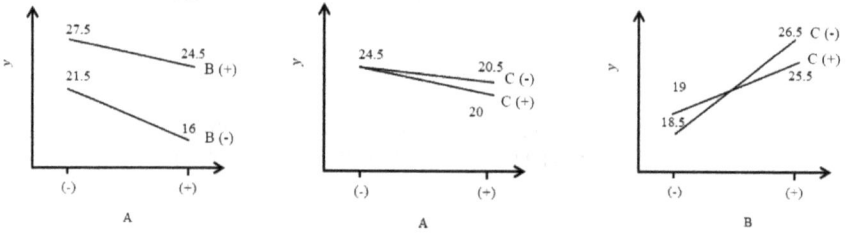

FIGURE 10.16 Effect of A and B, A and C, B and C on y (A: loop length, B: yarn linear density, C: knitting speed, y: UPF).

The interaction effect of $A \times B$

$$= \frac{1}{2}[\text{average effect of A for B(+)} - \text{average effect of A for (B−)}]$$

$$= \frac{1}{2}(24.5 - 27.5) - \frac{1}{2}(16 - 21.5) = 1.25$$

The interaction effect of $A \times C$

$$= \frac{1}{2}[\text{average effect of A for C(+)} - \text{average effect of A for C(−)}]$$

$$= \frac{1}{2}(20 - 24.5) - \frac{1}{2}(20.5 - 24.5) = -0.25$$

The interaction effect of $B \times C$

$$= \frac{1}{2}[\text{average effect of B for C(+)} - \text{average effect of B for C(−)}]$$

$$= \frac{1}{2}(25.5 - 19) - \frac{1}{2}(26.5 - 18.5) = -0.75.$$

The second order interaction effect i.e. $A \times B \times C$

$$= \frac{1}{2}[A \times B \text{ interaction for C (+)} - A \times B \text{ interaction for C(−)}]$$

$$= \frac{1}{2}[\frac{1}{2}\{\text{effect of A for B(+)} - \text{effect of A for B(−)}\} \text{for C(+)} - \frac{1}{2}$$
$$\{\text{effect of A for B(+)} - \text{effect of A for B(−)}\} \text{for C(−)}]$$

$$= \frac{1}{2}[\frac{1}{2}\{(24 - 27) - (16 - 22)\} - \frac{1}{2}\{(25 - 28) - (16 - 21)\}]$$

$$= 0.25$$

TABLE 10.26

Full Factorial Design Matrix Along with Responses and Contrast Values

	A	B	C	A × B	A × C	B × C	A × B × C	y
	−1	−1	−1	+1	+1	+1	−1	21
	+1	−1	−1	−1	−1	+1	+1	16
	−1	+1	−1	−1	+1	−1	+1	28
	+1	+1	−1	+1	−1	−1	−1	25
	−1	−1	+1	+1	−1	−1	+1	22
	+1	−1	+1	−1	+1	−1	−1	16
	−1	+1	+1	−1	−1	+1	−1	27
	+1	+1	+1	+1	+1	+1	+1	24
Contrast values	−4.25	7.25	−0.25	1.25	−0.25	−0.75	0.25	

The contrast values or the effect of the prime factors and their interactions are tabulated in Table 10.26.

It is to be noted that the higher order interactions are usually insignificant in comparison to the prime factors and lower order interaction effects. Thus, fractional factorial design ignores the possible importance of higher order interactions and considers only a fraction of the experimental runs. To transform this example of full factorial design into fractional factorial design, four runs in Table 10.26 with positive A × B × C interactions are chosen (highlighted in grey colour). The contrast values of the prime factors of this fractional factorial design of four runs are calculated as follows:

$$\text{Effect of factor A} = \frac{1}{2}(16+24)-\frac{1}{2}(28+22) = -5$$

$$\text{Effect of factor B} = \frac{1}{2}(28+24)-\frac{1}{2}(16+22) = 7$$

$$\text{Effect for the factor C} = \frac{1}{2}(22+24)-\frac{1}{2}(16+28) = 1$$

The contrast table is tabulated in Table 10.27. Figure 10.17 illustrates the fractional factorial design of the said example.

From Tables 10.26 and 10.27, it may be noted that column 1 is the product of B and C; column 2 is the product of A and B; column 3 is the product of A and B. Hence, column 1, column 2 and column 3 measure the interactions B × C, A × C and A × B, respectively. Also, the contrast values of column 1, column 2 and column 3 in Table 10.27 (fractional factorial design) are the sum of contrasts of A and B × C, B and A × C, C and A × B, respectively of Table 10.26 (full factorial design). This is known as 'confounding' or 'aliasing', in which the main effects for the factors A,

TABLE 10.27

Fractional Factorial Design Matrix Along with Responses and Contrast Values

	Column 1	Column 2	Column 3	y
	+1	−1	−1	16
	−1	+1	−1	28
	−1	−1	+1	22
	+1	+1	+1	24
Contrast values	−5	7	1	

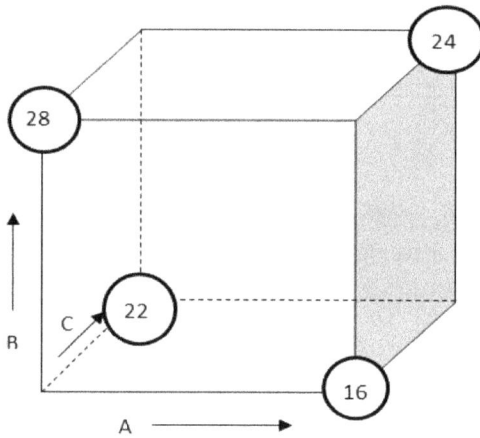

FIGURE 10.17   A 2^3 fractional factorial design.

B and C are confounded with the interaction effects of B × C, A × C and A × B, respectively. Therefore, a fractional factorial design is a solution of designing an experiment with less number of runs in comparison to full factorial design without losing much information.

10.5   RESPONSE SURFACE DESIGN

Response surface methodology (RSM) was first developed by Box and Wilson (1951) for optimizing the manufacturing process. RSM is used to design an experiment, help the experimenter to estimate the interaction effects of the prime factors and generate the response surface which can be used to optimize the response. A response surface is the geometric representation of a response variable (Y) plotted as a function of one or more prime factors. The general form of a response surface model with n factors is represented by a quadratic equation as follows:

$$Y = \beta_0 + \sum_{i=1}^{n} \beta_i x_i + \sum_{i=1}^{n-1}\sum_{j=i+1}^{n} \beta_{ij} x_i x_j + \sum_{i=1}^{n} \beta_{ii} x_i^2 + \varepsilon \qquad (10.5)$$

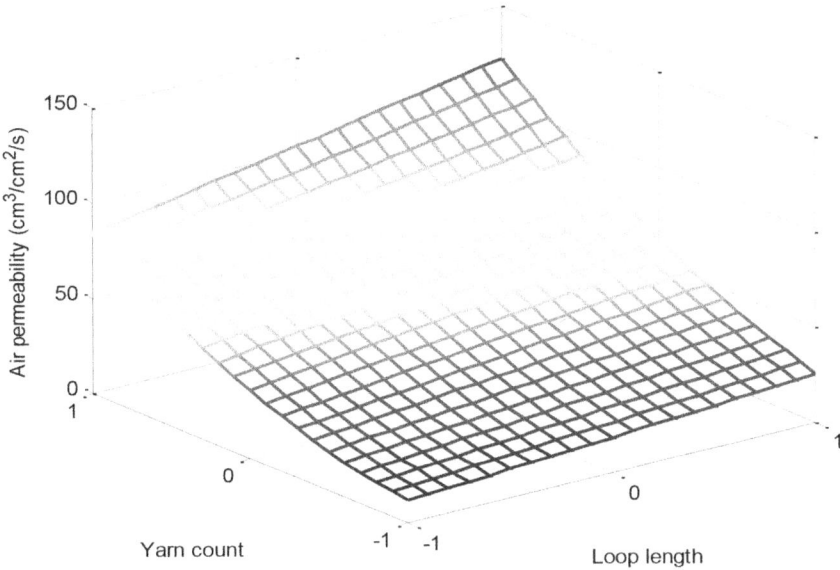

FIGURE 10.18 Response surface for two factors. (From Ghosh, Mal, Majumdar and Banerjee, 2016.)

where β_0, β_i, β_{ij} and β_{ii} are the regression coefficients of interception, linear terms, interaction terms and quadratic terms, respectively and ε is the experimental error.

An example of response surface is shown in Figure 10.18 where the response is air permeability and the prime factors are yarn count and loop length. The factors are represented by the coded levels in Figure 10.18. The response surface plot deciphers the relationship between the response variable and prime factors and hence leads to an optimal process setting. 'Central composite design' and 'Box-Behnken design' are the two most commonly used response surface designs.

10.5.1 Central Composite Design

Box and Wilson described 'Box-Wilson central composite design' in 1951, commonly known as 'central composite design' and is one of the most popular response surface designs. Each central composite design has a 'centre point', which is augmented with a group of axial points, known as 'star points'. The star points correspond to the extreme value of each factor and are coded as $\pm\alpha$. Therefore, the distance between the 'centre point' and 'star points' in the design space is α ($|\alpha| > 1$). The value of α depends on number of factors considered in the experiment and is calculated by the following equation:

$$\alpha = \left[2^{\text{number of factors}} \right]^{1/4} \tag{10.6}$$

Values of α, corresponding to various factors are given in Table 10.28. Central composite design for two and three factors are shown in Figure 10.19(a) and (b),

---

TABLE 10.28

Values of α for Various Factors

Number of Factors	α
2	$2^{2/4} = 1.414$
3	$2^{3/4} = 1.682$
4	$2^{4/4} = 2.000$

---

respectively. The structure of a three factors central composite design is shown in Table 10.29.

There are three types of centre composite designs, namely 'circumscribed' (CCC), 'inscribed' (CCI) and 'face centred' (CCF). CCC designs are the original form of central composite design and are explained above. It is obvious from the above discussion that CCC has five levels, with the edge points at the design limits and the star points at some distance from the centre points, depending upon the number of factors (refer to Table 10.29). For two factor central composite design, the distance of the star points from the centre is ±1.414, whereas for a three factor central composite design the distance is ±1.682. It is to be noted that in CCC, the star points extend the range outside the low and high setting of the factors, thus providing a high quality prediction over the entire design space. However, the star points should remain at feasible levels.

In certain situations, the limits specified for factors (i.e. design limits) are true limits and beyond that, level experiments are not feasible/possible. In such cases, the star points are set at these design limits and the edge points are inside the range. Such central composite designs are known as CCI. In some ways, a CCI design is a scaled down CCC design which also has five levels for each factor. Unlike CCC, the prediction space for CCI design is limited.

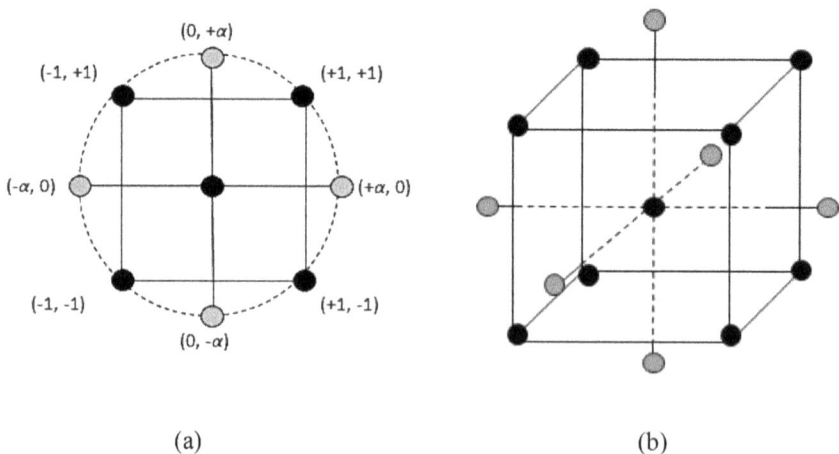

(a)   (b)

FIGURE 10.19 Central composite design: (a) 2 factors; (b) 3 factors.

TABLE 10.29

Two and Three Factors Central Composite Designs

	2 Factors			3 Factors		
Runs	A	B	Runs	A	B	C
1	−1	−1	1	−1	−1	−1
2	+1	−1	2	+1	−1	−1
3	−1	+1	3	−1	+1	−1
4	+1	+1	4	+1	+1	−1
5	−1.414	0	5	−1	−1	+1
6	+1.414	0	6	+1	−1	+1
7	0	−1.414	7	−1	+1	+1
8	0	+1.414	8	+1	+1	+1
9	0	0	9	−1.682	0	0
10	0	0	10	+1.682	0	0
11	0	0	11	0	−1.682	0
12	0	0	12	0	+1.682	0
			13	0	0	−1.682
			14	0	0	+1.682
			15	0	0	0
			16	0	0	0
			17	0	0	0
			18	0	0	0
			19	0	0	0
			20	0	0	0
	Total runs = 12				Total runs = 20	

In the case of CCF, the star points are at the centre of each face of the factorial space, such that $\alpha = \pm 1$. The axial points are at the centre of each side of factorial space. Unlike CCC and CCI design, CCF uses three levels. CCF designs provide a high quality prediction over the entire design space. Figure 10.20 illustrates CCC, CCI and CCD designs for two factors. It is noticeable from Figure 10.20 that CCC explores the largest design space; whereas, the CCI explores the smallest design space. It is also to be noted that both CCC and CCI are rotatable design, whereas CCF is not.

A design is known to be rotatable if it produces information that predicts the response with the same precision at all points equidistant from the design centre point. Rotatability is a desired property for a response surface design. As the objective of a response surface design is optimization and the location of optimum value in design space is unknown prior to running the experiment, a design with equal precision of estimation in all direction is desirable. A central composite design is made rotatable by the choice of α.

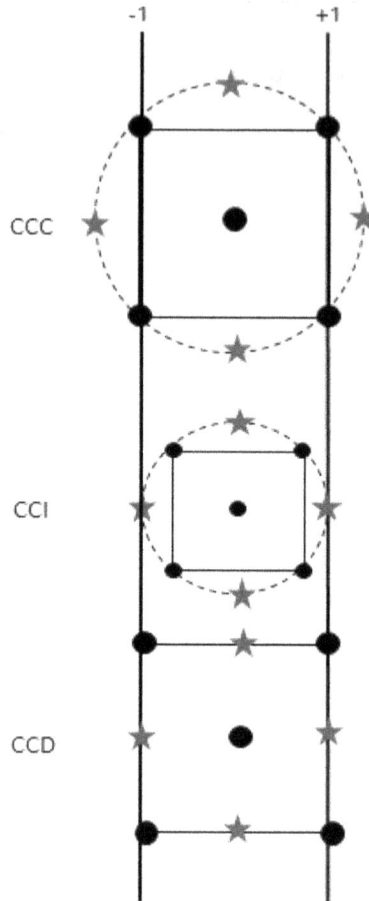

FIGURE 10.20   CCC, CCI and CCD designs for 2 factors.

10.5.2   BOX-BEHNKEN DESIGN

Central composite design is applicable when each factor has five levels, viz., $0, \pm1, \pm\alpha$. There are experiments where fewer levels are of interest (at least 3) of the experimenter. In such cases 3^n full factorial design or 3^{n-p} fractional factorial design is considered. The number of runs for such design may be large for even four factors (Number of run for 3^4 full factorial experimental design = 81 and that for 3^{4-1} fractional factorial design = 27). So for such cases, a different type of experimental design known as Box-Behnken design is preferable. The Box-Behnken (Box and Behnken 1960) design is an independent quadratic design, and the treatment combinations are at the centre and midpoints of the edges of the process space (Figure 10.21). Unlike central composite design, Box-Behnken design requires three levels for each factor and is rotatable. A Box-Behnken design for three factors and three levels is shown in Figure 10.21.

It is obvious from Figures 10.19 and 10.21 that the central composite design comprises the treatment combinations at the extreme corners of the cube; whereas, in case

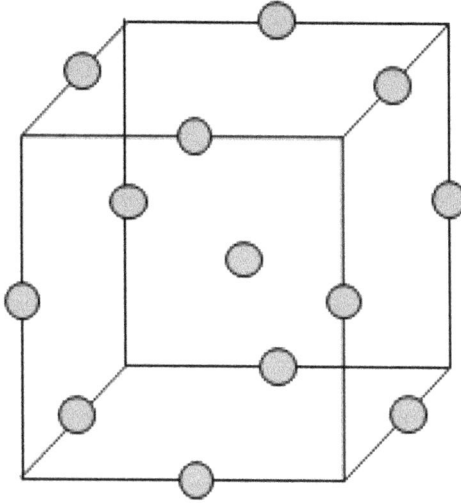

FIGURE 10.21 A Box-Behnken design for 3 factors.

of Box-Behnken design, all the design points are at the middle edge of the cube. The factor level combinations which are represented by the corners of a cube in case of central composite design are often difficult to test. This limitation of central composite design is avoided in the Box-Behnken design. Also, the number of runs in the Box-Behnken design is comparatively less than the central composite design. Structures of Box Behnken designs are given in Table 10.30 and the number of runs for a given number of factors for central composite and Box-Behnken designs is shown in Table 10.31.

10.5.2.1 An Application of the Box-Behnken Design

Alam et al. (2019) used the Box-Behnken design of experiment to study the influence of fibre, yarn and fabric parameters on the bending and shearing behaviour of plain woven fabrics. Ring spun yarns of three different counts, each of three different blends, were considered in the study. Yarn fineness of 20^s Ne, 30^s Ne and 40^s Ne were considered, whereas fibre composition of 100% cotton, 100% polyester and 50:50 polyester-cotton blends were considered in this experiment. Woven fabrics of three sets of thread counts (equal in warp and weft) were made in a rapier sample loom. A total of fifteen samples were made as per the Box-Behnken experimental design for three factors and three levels. Table 10.32 shows the actual values of the controlled factors corresponding to their coded values. The controlled factors x_1, x_2 and x_3 correspond to proportion of polyester, yarn count and fabric set in the woven fabric. The fabrics were conditioned for 48 hours and then evaluated for bending rigidity and shear rigidity. To develop response surface models, quadratic regression equation models were used to relate the controlled factors with the response variables. The regression coefficients of the response surface equations for bending and shear rigidity were determined and are tabulated in Table 10.33. Significance test of regression coefficients in the fitted regression equation models was conducted and only those regression coefficients which are significant at 95% limits were considered.

TABLE 10.30

Three Factors and Four Factors Box-Behnken Designs

	3 Factors				4 Factors			
Runs	A	B	C	Runs	A	B	C	D
1	−1	−1	0	1	−1	−1	0	0
2	+1	−1	0	2	−1	+1	0	0
3	−1	+1	0	3	+1	−1	0	0
4	+1	+1	0	4	+1	+1	0	0
5	0	0	0	5	0	0	−1	−1
6	−1	0	−1	6	0	0	−1	+1
7	+1	0	−1	7	0	0	+1	−1
8	−1	0	+1	8	0	0	+1	+1
9	+1	0	+1	9	0	0	0	0
10	0	0	0	10	−1	0	−1	0
11	0	−1	−1	11	−1	0	+1	0
12	0	+1	−1	12	+1	0	−1	0
13	0	−1	+1	13	+1	0	+1	0
14	0	+1	+1	14	0	−1	0	−1
15	0	0	0	15	0	−1	0	+1
				16	0	+1	0	−1
				17	0	+1	0	+1
				18	0	0	0	0
				19	−1	0	0	−1
				20	−1	0	0	+1
				21	+1	0	0	−1
				22	+1	0	0	+1
				23	0	−1	−1	0
				24	0	−1	+1	0
				25	0	+1	−1	0
				26	0	+1	+1	0
				27	0	0	0	0

TABLE 10.31

**Number of Runs for Central Composite Designs
and Box-Behnken Designs**

Number of Factors	Central Composite Design	Box-Behnken Design
2	12 (including 4 centre points)	—
3	20 (including 6 centre points)	15 (including 3 centre points)
4	30 (including 6 centre points)	27 (including 3 centre points)

TABLE 10.32

Actual Values Corresponding to Coded Levels of Yarn and Fabric Parameters for Woven Fabrics

Controlled Factors	Coded Levels		
	−1	0	+1
Proportion of polyester (x_1), %	0	50	100
Yarn count (x_2), Ne	20	30	40
Fabric sett (x_3), inch^{-1}	50	60	70

Source: Alam, Md. S et al. 2019. *Indian Journal of Fibre and Textile Research*, 44, 9–15.

It was found that all the controlled factors, namely proportion of polyester, yarn count and fabric sett have significant influence on both bending and shear rigidities of woven fabrics. The interaction $x_2 \times x_3$ has significant influence on both the responses, whereas the interaction $x_1 \times x_3$ has significant influence on shear rigidity. Among the quadratic terms, x_2^2 was found to be statistically significant for both the responses. The response surface equations for bending and shear rigidities are shown in Table 10.34. Figures 10.22 and 10.23 depict the response surfaces for bending and shear rigidities.

TABLE 10.33

Regression Coefficients for Different Independent Parameters Using Coded Values

Term	For Bending Rigidity		For Shear Rigidity	
	Coefficient	p-Value	Coefficient	p-Value
Constant	0.038	0.0004*	0.88	0.004*
x_1	−0.015	0.0008*	−0.31	0.0415*
x_2	−0.028	<0.0001*	−0.89	0.0006*
x_3	0.017	0.0005*	0.70	0.0018*
$x_1 \times x_2$	0.011	0.154	0.32	0.4141
$x_2 \times x_3$	−0.014	0.0049*	−0.67	0.0094*
$x_1 \times x_3$	−0.005	0.012*	−0.15	0.1087
x_1^2	0.0019	0.571	−0.19	0.3133
x_2^2	0.014	0.007*	0.49	0.0336*
x_3^2	0.0039	0.267	0.20	0.2806

Source: Alam, Md. S et al. 2019. *Indian Journal of Fibre and Textile Research*, 44, 9–15.

* *Statistically significant at 95% confidence level*

TABLE 10.34

Response Surface Equations for Bending and Shear Rigidities

Parameters	Response Surface Equation
Bending rigidity (gf.cm²/cm)	$0.038 - 0.015x_1 - 0.028x_2 + 0.017x_3 - 0.011x_1x_2 - 0.014x_2x_3 + 0.014x_2^2$
Shear rigidity (gf/cm.deg)	$0.88 - 0.31x_1 - 0.89x_2 + 0.7x_3 - 0.67x_2x_3 + 0.49x_2^2$

Source: Alam, Md. S et al. 2019. *Indian Journal of Fibre and Textile Research*, 44, 9–15.

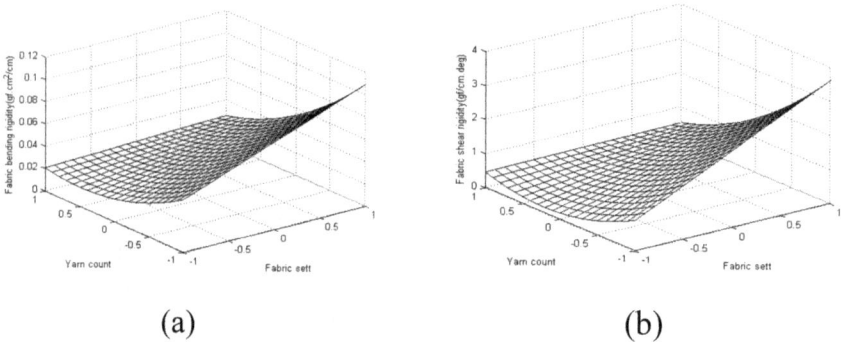

(a)   (b)

FIGURE 10.22 Effect of yarn count and fabric sett in (a) bending rigidity; (b) shear rigidity. (From Alam, Md. S et al. 2019. *Indian Journal of Fibre and Textile Research*, 44, 9–15.)

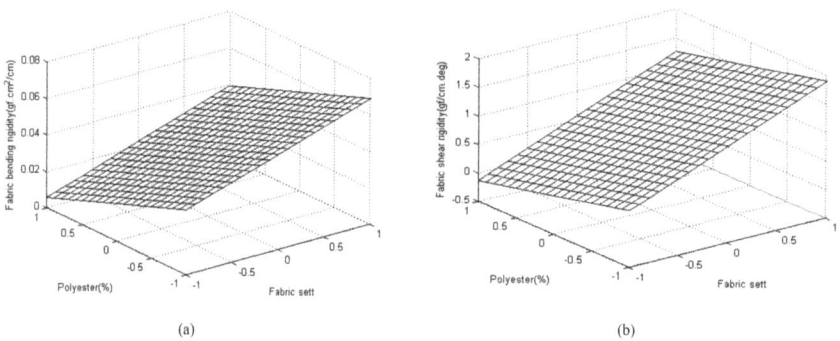

(a)   (b)

FIGURE 10.23 Effect of polyester proportion and fabric sett in (a) bending rigidity; (b) shear rigidity. (From Alam, Md. S et al. 2019. *Indian Journal of Fibre and Textile Research*, 44, 9–15.)

10.5.3 BLOCKING OF RESPONSE SURFACE DESIGNS

In order to eliminate or reduce the nuisance factors, response surface design may be blocked. When an experiment comprises a large number of runs, conducting experiments in a steady state is not possible. For example, experiments are conducted on

$$
\begin{bmatrix}
-1 & -1 & -1 & 0 \\
-1 & -1 & +1 & 0 \\
-1 & +1 & -1 & 0 \\
-1 & +1 & +1 & 0 \\
+1 & -1 & -1 & 0 \\
+1 & -1 & +1 & 0 \\
+1 & +1 & -1 & 0 \\
+1 & +1 & +1 & 0 \\
0 & 0 & 0 & 0
\end{bmatrix}
\begin{bmatrix}
-1 & -1 & 0 & -1 \\
-1 & -1 & 0 & +1 \\
-1 & +1 & 0 & -1 \\
-1 & +1 & 0 & +1 \\
+1 & -1 & 0 & -1 \\
+1 & -1 & 0 & +1 \\
+1 & +1 & 0 & -1 \\
+1 & +1 & 0 & +1 \\
0 & 0 & 0 & 0
\end{bmatrix}
\begin{bmatrix}
-1 & 0 & -1 & -1 \\
-1 & 0 & -1 & +1 \\
-1 & 0 & +1 & -1 \\
-1 & 0 & +1 & +1 \\
+1 & 0 & -1 & -1 \\
+1 & 0 & -1 & +1 \\
+1 & 0 & +1 & -1 \\
+1 & 0 & +1 & +1 \\
0 & 0 & 0 & 0
\end{bmatrix}
\begin{bmatrix}
0 & -1 & -1 & -1 \\
0 & -1 & -1 & +1 \\
0 & -1 & +1 & -1 \\
0 & -1 & +1 & +1 \\
0 & +1 & -1 & -1 \\
0 & +1 & -1 & +1 \\
0 & +1 & +1 & -1 \\
0 & +1 & +1 & +1 \\
0 & 0 & 0 & 0
\end{bmatrix}
$$

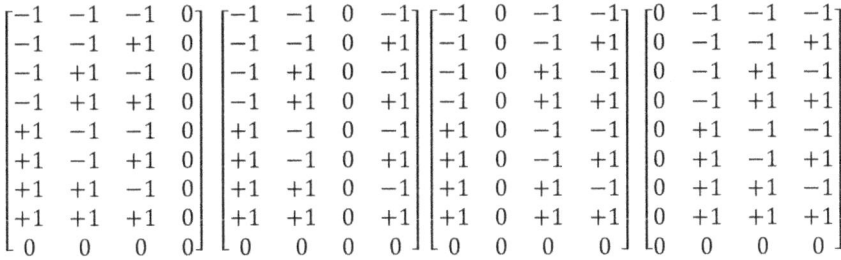

FIGURE 10.24 Orthogonal block of Box-Behnken design for 4 factors and 3 levels. (From Ghosh et al. 2016. *Journal of the Institute of Engineers (India): Series E*, 97(2), 89–98.)

different days by different operators and at different times of a day etc. These factors may affect the response, and in order to estimate the block effect (apart from the prime factors), experiments are conducted in different blocks. Response surface designs are blocked orthogonally such that block effects do not affect the ability to estimate parameters independently, and the number of blocks depends on the number of prime factors and the design fraction.

Both central composite design and Box-Behnken designs can be blocked orthogonally. A 3^4 orthogonally blocked Box-Behnken design is illustrated in Figure 10.24.

10.5.3.1 An Application of Blocked Response Surface Design

Ghosh et al. (2016) applied orthogonal block Box-Behnken design to investigate the effect of various knitting parameters and yarn count on air permeability of 1×1 rib knitted fabrics. 100% cotton yarns of three different counts (5^s, 7.5^s and 10^s Ne) were used to prepare 1×1 rib knitted fabrics in a computerized flat knitting machine according to a 3^4 (four factors and three levels) orthogonal block design proposed by Box and Behnken as shown in Figure 10.24. Four controlled factors, namely loop length, carriage speed, yarn input tension and yarn count, were considered. Table 10.35 shows the actual values of the controlled factors corresponding to their

TABLE 10.35

Actual Values Corresponding to Coded Levels for 1×1 Rib Knitted Fabrics

Controlled Factors	Coded Level		
	−1	0	+1
Loop length (x_1), mm	5.09	5.39	5.69
Carriage speed (x_2), m/s	0.25	0.45	0.65
Yarn input tension (x_3), gf	6	8	10
Yarn count (x_4), Ne	5	7.5	10

Source: Ghosh et al. 2016. *Journal of the Institute of Engineers (India): Series E*, 97(2), 89–98.

TABLE 10.36
ANOVA between Blocks

Sources of Variation	Sum of Squares	Degrees of Freedom	Mean of Squares	f
Between blocks	660.5	3	220.17	0.15
Error	48,520.5	32	1516.27	
Total	49,181	35	–	

Source: Ghosh et al. 2016. *Journal of the Institute of Engineers (India): Series E,* 97(2), 89–98.

coded levels. The controlled factors x_1, x_2, x_3 and x_4 correspond to loop length, carriage speed, yarn input tension and yarn count, respectively. All the samples were washed in a washing machine for complete relaxation, dried and then evaluated for air permeability.

The analysis of variance between blocks was conducted to investigate the effect of blocks. The ANOVA results are given in Table 10.36. It was concluded that there were no significant differences among the blocks.

TABLE 10.37
Regression Coefficients Using Coded Values of Input Variables

Term	Coefficient	p-value
Constant	48.93	1.56×10^{-19}*
x_1	13.05	9.21×10^{-16}*
x_2	−0.34	0.5836
x_3	−0.83	0.1855
x_4	42.33	2.57×10^{-26}*
$x_1 x_2$	−0.85	0.2686
$x_1 x_3$	−0.38	0.6195
$x_1 x_4$	6.26	3.71×10^{-8}*
$x_2 x_3$	−1.29	0.0980
$x_2 x_4$	−0.12	0.8788
$x_3 x_4$	−0.36	0.6317
x_1^2	0.55	0.6042
x_2^2	0.31	0.7700
x_3^2	0.55	0.6075
x_4^2	12.28	1.26×10^{-10}*

Source: Majumdar et al. 2016. *The Journal of the Textile Institute,* 108(1), 110–116.

Note:
* *Statistically significant at 95% confidence level*

TABLE 10.38

Response Surface Equations for Air Permeability of 1 × 1 Rib-Knitted Fabrics

Fabric Properties	Response Surface Equation
Air permeability (cm³/cm²/s)	$48.93 + 13.05x_1 + 42.33x_4 + 6.26x_1x_4 + 12.28x_4^2$

Source: Ghosh et al. 2016. *Journal of the Institute of Engineers (India): Series E,* 97(2), 89–98.

Quadratic regression equation models were used to relate the controlled factors with the response variables. The regression coefficients of the response surface equation of air permeability were determined and are given in Table 10.37. Significance test of regression coefficients in the fitted regression equation models was conducted and only those regression coefficients which are significant at 95% limits were considered. It was found that x_1 (loop length), x_4 (yarn count), interaction $x_1 \times x_4$ and x_4^2 have significant influence on air permeability. The response surface equation for air permeability is given in Table 10.38. Figure 10.25 illustrates the response surface for air permeability as a function of loop length and yarn count.

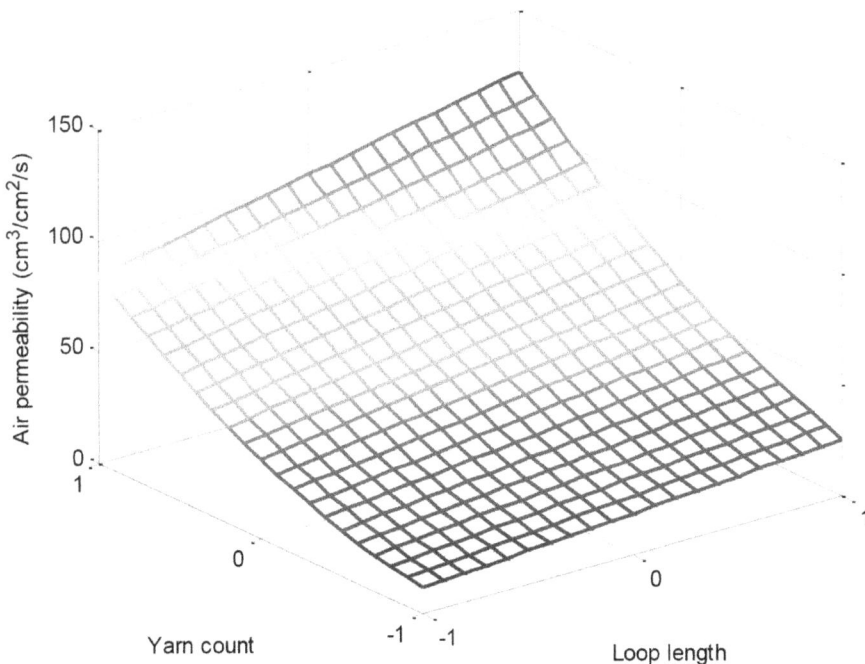

FIGURE 10.25 Response surface of air permeability as a function of loop length and yarn count (yarn count and loop length are in coded levels). (From Ghosh et al. 2016. *Journal of the Institute of Engineers (India): Series E*, 97(2), 89–98.)

10.6 TAGUCHI'S DESIGN

During experimentation, it is desired to design the experiment in such a way that it should reduce or eliminate undesirable 'nuisance factors' or 'experimental errors'. Blocking techniques are often used to reduce such undesirable 'nuisance factors' or 'experimental errors'. However, blocking of experimental designs may induce unwanted block effects on the response surface. Therefore, to reduce the effect of blocks on factor estimation of response surface models, experimental designs are 'orthogonally blocked'. Example of an orthogonal block Box-Behnken design was cited earlier. Again, there exist certain unavoidable nuisance or noise factors that may affect the response. In an industrial setup, these noise factors could be huge in number, which may be standard atmospheric conditions during experimentation (temperature and humidity), machine parameters (machine number and machine vibration), operators etc. Therefore, it is desirable to incorporate such unavoidable noise factors along with the prime factors to investigate their effect on the response. Genichi Taguchi (Taguchi et al. 2005) developed a statistical approach to improve the quality of manufactured products by minimizing the effect of causes of variation. This approach, also known as Taguchi method, uses a special set of arrays known as orthogonal arrays and signal to noise ratio (S/N). In an orthogonal array, a special set of arrays of factors and level combinations are used to conduct experimental runs that provide complete information of all the factors that affect the response. Table 10.39 shows a typical L9 orthogonal array of 4 factors, each with three levels. The value '9' specifies nine prototypes that need to be tested. In Table 10.39 x_1, x_2, x_3 and x_4 are the controllable factors, each with three levels; y_1 and y_2 are the noise factors, each with two levels. Tables 10.40 and 10.41 show orthogonal arrays for factors ranging from two to four, each with two levels and three levels, respectively.

The outputs of these experimental runs are converted into signal to noise ratio (S/N) ratio which signifies the ratio of sensitivity to variability. A higher S/N ratio is desirable as it indicates better quality. Therefore, maximization of S/N

TABLE 10.39
L9 (3^4) Orthogonal Array

Experimental Run	Controlled Factors				Noise Factors	
	x_1	x_2	x_3	x_4	y_1	y_2
1	−1	−1	−1	−1	±1	±1
2	−1	0	0	0	±1	±1
3	−1	+1	+1	+1	±1	±1
4	0	−1	0	+1	±1	±1
5	0	0	+1	−1	±1	±1
6	0	+1	−1	0	±1	±1
7	+1	−1	+1	0	±1	±1
8	+1	0	−1	+1	±1	±1
9	+1	+1	0	−1	±1	±1

TABLE 10.40

Taguchi Orthogonal Arrays for Factors Ranging from Two to Four Each at Two Levels

2 Factors and 2 Levels

Runs	Controlled Factors		Noise Factor
	x_1	x_2	y
1	−1	−1	±1
2	−1	+1	±1
3	+1	−1	±1
4	+1	+1	±1

3 Factors and 2 Levels

Runs	Controlled Factors			Noise Factor
	x_1	x_2	x_3	y
1	−1	−1	−1	±1
2	−1	+1	+1	±1
3	+1	−1	+1	±1
4	+1	+1	−1	±1

4 Factors and 2 Levels

Runs	Controlled Factors				Noise Factor
	x_1	x_2	x_3	x_4	y
1	−1	−1	−1	−1	±1
2	−1	−1	−1	+1	±1
3	−1	+1	+1	−1	±1
4	−1	+1	+1	+1	±1
5	+1	−1	+1	−1	±1
6	+1	−1	+1	+1	±1
7	+1	+1	−1	−1	±1
8	+1	+1	−1	+1	±1

ratio eventually minimizes the effect of uncontrolled noise factors on the response. Depending on the nature of responses viz., 'nominal the best', 'larger the better' or 'smaller the better' the S/N ratios are calculated using the following equations (Ross, 1996; Roy, 2001; Taguchi et al., 2005).

For 'nominal the best'

$$\frac{S}{N} = 10 \ \log\left[\frac{\bar{y}}{S_y^2}\right] \tag{10.7}$$

For 'larger the better'

$$\frac{S}{N} = -10 \ \log\left[\frac{1}{n}\sum_{i=1}^{n}\frac{1}{y_i^2}\right] \tag{10.8}$$

TABLE 10.41

Taguchi Orthogonal Arrays for Factors Ranging from 2 to 4 Each at Three Levels

2 Factors and 3 Levels

Runs	Controlled Factors		Noise Factor
	x_1	x_2	y
1	−1	−1	±1
2	−1	0	±1
3	−1	+1	±1
4	0	−1	±1
5	0	0	±1
6	0	+1	±1
7	+1	−1	±1
8	+1	0	±1
9	+1	+1	±1

3 Factors and 3 Levels

Runs	Controlled Factors			Noise Factor
	x_1	x_2	x_3	y
1	−1	−1	−1	±1
2	−1	0	0	±1
3	−1	+1	+1	±1
4	0	−1	0	±1
5	0	0	+1	±1
6	0	+1	−1	±1
7	+1	−1	+1	±1
8	+1	0	−1	±1
9	+1	+1	0	±1

4 Factors and 3 Levels

Runs	Controlled Factors				Noise Factor
	x_1	x_2	x_3	x_4	y
1	−1	−1	−1	−1	±1
2	−1	0	0	0	±1
3	−1	+1	+1	+1	±1
4	0	−1	0	+1	±1
5	0	0	+1	−1	±1
6	0	+1	−1	0	±1
7	+1	−1	+1	0	±1
8	+1	0	−1	+1	±1
9	+1	+1	0	−1	±1

For 'smaller the better'

$$\frac{S}{N} = -10\log\left[\frac{1}{n}\sum_{i=1}^{n}y_i^2\right] \tag{10.9}$$

where, $\bar{y}$ is mean of observed data, S_y^2 is the variance of y, n is the number of experiments in the orthogonal array and y_i is the ith value measured. From the S/N ratio, optimum process is calculated. For each significant factor, the level corresponding to the highest S/N ratio is the optimum level. ANOVA of S/N ratio may also be done to compute the percentage contribution of each of the controlled factors on the response variable.

10.6.1 An Application of the Taguchi Design

Ghosh et al. (2017) used the Taguchi experimental design to study the effect of knitting process variables and yarn count on air permeability of single jersey-knitted fabrics in the presence of two unavoidable noise factors, namely fabric production from two different cones and two sides of a knitting machine (right side and left side). One hundred percent gas mercerized, combed, ring spun cotton yarns of different fineness were used to manufacture single jersey fabrics on a 12-gauge computerized flat knitting machine. Four controlled factors, (loop length, carriage speed, yarn input tension and yarn count) each at three levels and two noise factors were considered in the study. Single jersey samples were prepared according to the L9 orthogonal array as shown in Table 10.42. Two noise factors, each at two different levels generate four repetitions for each of the nine experimental runs. Hence, the

TABLE 10.42
L9 (3⁴) Orthogonal Array

	Controlled Factors				Noise Factors	
Experimental Run	x_1 (loop length, mm)	x_2 (carriage speed, m/s)	x_3 (yarn input tension, Ne)	x_4 (yarn count, Ne)	Cone Number	Machine Side
1	−1	−1	−1	−1	±1	±1
2	−1	0	0	0	±1	±1
3	−1	+1	+1	+1	±1	±1
4	0	−1	0	+1	±1	±1
5	0	0	+1	−1	±1	±1
6	0	+1	−1	0	±1	±1
7	+1	−1	+1	0	±1	±1
8	+1	0	−1	+1	±1	±1
9	+1	+1	0	−1	±1	±1

Source: Ghosh et al. 2017. *Autex Research Journal*, 17(2), 152–163.

TABLE 10.43

Actual and Coded Levels of Noise Factors

	Coded Levels		
Controlled Factors	**−1**	**0**	**+1**
Loop length (x_1), mm	6.6	7.0	7.4
Carriage speed (x_2), m/s	0.25	0.6	0.95
Yarn input tension (x_3), cN	6	8	10
Yarn count (x_4), Ne	5	7.5	10

Source: Ghosh et al. 2017. *Autex Research Journal*, 17(2), 152–163.

total number of single jersey samples in the study was $9 \times 4 = 36$. The actual values of controlled and noise factors corresponding to their coded levels are given in Tables 10.43 and 10.44, respectively.

All 36 single jersey fabrics were washed and conditioned for complete relaxation and then evaluated for air permeability. The experimental results were transformed into *S/N* ratio. In this study, the *S/N* ratio was calculated using Equation (10.8) i.e. 'larger the better', and is shown in Table 10.45. It is obvious from the Table 10.45 that the maximum *S/N* ratio of air permeability is 48.83. Therefore, the optimum combinations of the controlled factors that lead to the maximum *S/N* ratio of air permeability are 7.4 mm, 0.6 m/s, 6 cN and 10 Ne for loop length, carriage speed, yarn input tension and yarn count, respectively. The response for *S/N* ratio of air permeability was also calculated and is given in Table 10.46.

It is obvious from Table 10.46 that yarn count has maximum influence on air permeability, followed by loop length, yarn input tension and carriage speed.

ANOVA analysis was conducted on *S/N* ratios of air permeability and is tabulated in Table 10.47. It is evident from the table that yarn count is the most dominating factor influencing the air permeability with a contribution of 93.8%. The second dominating factor influencing the air permeability is loop length which contributes only 4.5%. Yarn input tension and carriage speed have little effect on the air permeability.

TABLE 10.44

Actual and Coded Levels of Noise Factors

	Coded Levels	
Noise Factors	**−1**	**+1**
Cone number	1	2
Machine's Side	Left side	Right side

Source: Ghosh et al. 2017. *Autex Research Journal*, 17(2), 152–163.

TABLE 10.45
Average Values of Air Permeability and S/N Ratios

Exp. No.	x_1	x_2	x_3	x_4	Average Air Permeability $(cm^3/cm^2/s)$	S/N Ratio
1	−1	−1	−1	−1	43.18	32.62
2	−1	0	0	0	111.59	40.93
3	−1	+1	+1	+1	232.65	47.33
4	0	−1	0	+1	274.52	48.74
5	0	0	+1	−1	51.74	34.23
6	0	+1	−1	0	102.34	40.19
7	+1	−1	+1	0	168.1	44.50
8	+1	0	−1	+1	277.17	48.83
9	+1	+1	0	−1	66.03	36.35

Source: Ghosh et al. 2017. *Autex Research Journal*, 17(2), 152–163.

TABLE 10.46
Response for S/N Ratios of Air Permeability

Factors	Average S/N Ratio Level −1	Level 0	Level +1	Range	Rank
x_1	40.29	41.05	43.23	2.94	2
x_2	41.95	41.33	41.29	0.66	4
x_3	40.55	42.01	42.02	1.47	3
x_4	34.40	41.87	48.30	13.90	1

Source: Ghosh et al. 2017. *Autex Research Journal*, 17(2), 152–163.

TABLE 10.47
Summary of ANOVA Conducted on S/N Ratios of Air Permeability

Factors	Sum of Squares	Degrees of Freedom	Mean Square	Percentage Contribution
x_1	13.91	2	6.96	4.5
x_2	0.84	2	0.42	0.3
x_3	4.31	2	2.16	1.4
x_4	290.47	2	145.24	93.8
Total	309.53	8		

Source: Ghosh et al. 2017. *Autex Research Journal*, 17(2), 152–163.

10.7   MATLAB® CODING

10.7.1   MATLAB® Coding of Example 10.1

```
clc
clear
close all
A=[4850 4500 4800 4750];
B=[4650 4600 4550 4650];
C=[4800 4750 4650 4700];
D=[4600 4500 4550 4500];
y=[A' B' C' D'];
[p,tbl]=anova1(y)
```

10.7.2   MATLAB® Coding of Example 10.2

```
clc
clear
close all
A=[4850 4500 4800 4750];
B=[4650 4600 4550 4650];
C=[4800 4750 4650 4700];
D=[4600 4500 4550 4500];
y=[A' B' C' D'];
[p,tbl]=anova2(y)
```

10.7.3   MATLAB® Coding of Example 10.3

```
clc
clear
close all
y=[4600 4650 4800 4850 4600 4500 4500 4750 4650 4800 4550 4550
4750 4700 4650 4500];
g1 = ['4';'2';'3';'1';'2';'4';'1';'3';'3';'1';'4';'2';'1';'3';
'2';'4'];
g2 = ['1';'1';'1';'1';'2';'2';'2';'2';'3';'3';'3';'3';'4';'4';
'4';'4'];
g3 = ['1';'2';'3';'4';'1';'2';'3';'4';'1';'2';'3';'4';'1';'2';
'3';'4'];
[p,tbl] = anovan(y,{g1 g2 g3})
```

10.7.4   MATLAB® Coding of Example 10.4

```
clc
clear
close all
y=[4600 4650 4800 4850 4600 4500 4500 4750 4650 4800 4550 4550
4750 4700 4650 4500];
g1 = ['4';'2';'3';'1';'2';'4';'1';'3';'3';'1';'4';'2';'1';'3';
'2';'4'];
```

```
g2 = ['1';'1';'1';'1';'2';'2';'2';'2';'3';'3';'3';'3';'4';'4';
'4';'4'];
g3 = ['1';'2';'3';'4';'1';'2';'3';'4';'1';'2';'3';'4';'1';'2';
'3';'4'];
g4=['1';'2';'3';'4';'3';'4';'1';'2';'4';'3';'2';'1';'2';'1';
'4';'3'];
[p,tbl] = anovan(y,{g1 g2 g3 g4})
```

10.7.5   MATLAB® CODING OF EXAMPLE 10.5

```
clc
clear
close all
y=[50   62
    48   60
    47   63
    48   57
    54   60
    50   56];
[p,tbl]=anova2(y,3)
```

10.7.6   MATLAB® CODING OF EXAMPLE 10.6

```
clc
clear
close all
y=[22 21 20 28 26 30 14 17 17 25 26 24 22 21 23 26 26 29 17 15
16 24 23 25];
g1 = {'-';'-';'-';'-';'-';'-';'+';'+';'+';'+';'+';'+';'-';'-';
'-';'-';'-';'-';'+';'+';'+';'+';'+';'+'};
g2 = {'-';'-';'-';'+';'+';'+';'-';'-';'-';'+';'+';'+';'-';'-';
'-';'+';'+';'+';'-';'-';'-';'+';'+';'+'};
g3 = {'-';'-';'-';'-';'-';'-';'-';'-';'-';'-';'-';'-';'+';'+';
'+';'+';'+';'+';'+';'+';'+';'+';'+';'+'};
[p,tbl]= anovan(y,{g1 g2 g3},'model','full','varnames',{'g1',
'g2','g3'})
```

10.7.7   MATLAB® CODING OF EXAMPLE 10.7

```
clc
clear
close all
y=[78 85 80 88 77 82 82 84 76 80 78 82 75 78 79 81];
g1 = {'-';'+';'-';'+';'-';'+';'-';'+';'-';'+';'-';'+';'-';'+';
'-';'+'};
g2 = {'-';'-';'+';'+';'-';'-';'+';'+';'-';'-';'+';'+';'-';'-';
'+';'+'};
g3 = {'-';'-';'-';'-';'+';'+';'+';'+';'-';'-';'-';'-';'+';'+';
'+';'+'};
```

```
g4 = {'-';'-';'-';'-';'-';'-';'-';'-';'-';'+';'+';'+';'+';'+';'+';
'+';'+'};
[p,tbl]= anovan(y,{g1 g2 g3 g4},'model','interaction',
'varnames',{'g1','g2','g3','g4'})
```

Exercises

10.1 The results of 4 different spindle speeds ($X_1 = 15,000$ rpm; $X_2 = 18,000$ rpm; $X_3 = 20,000$ rpm; $X_4 = 22,000$ rpm) on yarn breakage rate are shown in the table below. The average value of yarn breakage rate (number of breaks/100 spindle/hour) were calculated on four different days. Test at 0.05 level of significance whether there are significant differences in yarn breakage rate due to spindle speed and day of observation.

Day 1	Day 2	Day 3	Day 4
$\dfrac{X_1}{2.8}$	$\dfrac{X_2}{2.8}$	$\dfrac{X_3}{3}$	$\dfrac{X_4}{5}$
$\dfrac{X_3}{3}$	$\dfrac{X_4}{5.5}$	$\dfrac{X_1}{2.7}$	$\dfrac{X_2}{3}$
$\dfrac{X_2}{3.2}$	$\dfrac{X_1}{2.6}$	$\dfrac{X_4}{5.5}$	$\dfrac{X_3}{3.2}$
$\dfrac{X_4}{6}$	$\dfrac{X_3}{3.4}$	$\dfrac{X_2}{2.8}$	$\dfrac{X_1}{2.5}$

10.2 An experimenter wanted to investigate the effect of bleaching agents supplied by four different suppliers (X_1, X_2, X_3 and X_4) on 100% cotton drill fabrics. The reflectance degree of the cotton drill fabrics after bleaching with these bleaching agents are given in the table below. The experimenter has considered two different blocks, 'day of treatment' and 'equipment' (A, B, C and D). Test at 0.05 level of significance whether there are significant differences in fabric whiteness due to supplier, day of bleaching and equipment.

Blocks (Days)

Blocks (Equipment)	Day 1	Day 2	Day 3	Day 4
A	$\dfrac{X_4}{89}$	$\dfrac{X_2}{88}$	$\dfrac{X_3}{89}$	$\dfrac{X_1}{85}$
B	$\dfrac{X_2}{88}$	$\dfrac{X_4}{90}$	$\dfrac{X_1}{84}$	$\dfrac{X_3}{88}$
C	$\dfrac{X_3}{86}$	$\dfrac{X_1}{81}$	$\dfrac{X_4}{84}$	$\dfrac{X_2}{87}$
D	$\dfrac{X_1}{83}$	$\dfrac{X_3}{89}$	$\dfrac{X_2}{90}$	$\dfrac{X_4}{88}$

10.3 A 2^3 experiment was carried out to investigate the effect of three variables on thermal conductivity of single jersey-knitted fabrics. The three variables are as follows:

A: Yarn fineness: 10 Ne (−) or 12 Ne (+)
B: Loop length: 6.7 mm (−) or 7 mm (+)
C: Knitting speed: 0.5 m/s (−) or 0.8 m/s (+)

The results are shown below.

	A−		A+	
	B−	B+	B−	B+
C−	55	44	40	38
C+	57	42	43	36

Analyse these data and test at the 0.5% level of significance whether there are significant differences in A, B and C and their interactions on yarn imperfection.

10.4 A 2^4 experiment was carried out to investigate the effect of four variables on yarn imperfection of 30^s Ne cotton yarn. The four variables are as follows:

A: Roving fineness: 0.8 Ne (−) or 1.1 Ne (+)
B: Break draft: 1.14 (−) or 1.21 (+)
C: Average spindle speed: 15,000 rpm (−) or 21,000 rpm (+)
D: Traveller: 3/0 (−) or 4/0 (+)

The results are shown below.

		A−		A+	
		B−	B+	B−	B+
C−	D−	56	58	63	65
	D+	54	55	54	59
C+	D−	55	62	61	60
	D+	51	56	55	58

Analyse these data and test at the 0.5% level of significance whether there are significant differences in A, B, C and D and their interactions on yarn imperfection using Yates' algorithm.

10.5 (i) Determine the value of α for a rotatable central composite design for 5 factors. Also find the levels.

(ii) Determine the value of α for a rotatable central composite design for 6 factors. Also find the levels.

10.6 Construct a Box-Behnken design for five factors and three levels. Find out the number of runs required.

11 Statistical Quality Control

11.1 INTRODUCTION

Quality control is a method of inspecting the quality of manufactured articles by means of testing. In most situations, 100% inspection, that is inspecting each and every article, is practically impossible and therefore it is most suitable to inspect a random sample of a few articles. Sampling inspection can be carried out at many stages in a manufacturing process. There may be inspection of raw materials, process inspection at various points in the manufacturing operation, inspection of semi-finished products and finally the inspection of finished products. When the sampling inspection is done for the purpose of acceptance or rejection of a batch of raw materials or a batch of products, the statistical technique employed in dealing such situation is termed as acceptance sampling. When the sampling inspection is done at the conversion stage for detecting whether the manufacturing process is under control or out of control, the statistical technique employed in dealing such situation is called the control chart. The control charts are widely used for monitoring the manufacturing process. In order to achieve the targeted quality of products, the manufacturing process must be kept under control. If the control chart exhibits the evidence that the process is out of control due to some assignable causes, then it is the job of the quality control engineer to determine the causes and take remedial action.

11.2 ACCEPTANCE SAMPLING

A typical application of acceptance sampling is discussed in the following few lines. A manufacturer receives a shipment of product from a vendor. A sample is taken from the lot for inspecting some quality characteristics. On the basis of the information in this sample, a decision is made whether the lot will be accepted or rejected. Sometimes the articles are inspected simply on the basis of either defective or non-defective and the sampling scheme dealing such type of inspection is called the acceptance sampling for attributes. When the quality characteristics of the articles are measured on a numerical scale, then the adopted sampling scheme is called the acceptance sampling for variables.

11.2.1 ACCEPTANCE SAMPLING FOR ATTRIBUTES

In attribute sampling plan, a sample of size n is taken at random from a batch and if the number of defectives found in the sample is greater than a given acceptance number c, the batch is rejected. A rejected batch may be merely returned to the producer, or the batch may be accepted with a lower price deal as negotiated by the producer and consumer, or the batch would be 100% inspected and defectives

DOI: 10.1201/9781003081234-11

are replaced by non-defectives. Let p be the proportion of defective articles in the batch. Thus, if a single article is randomly chosen from a batch, the probability that it will be defective is p. If the batch size is large relative to the sample size n, this probability is the same for each article in the sample. Hence, using binomial distribution, the probability of finding exactly k number of defectives in a sample of size n is

$$P(r=k) = \binom{n}{k} p^k (1-p)^{n-k} \tag{11.1}$$

where r denotes the number of defectives in the sample. The probability of accepting a batch, that is, the probability of finding c or fewer defectives is

$$P_a(p) = P(r \le c)$$

$$= P(r=0) + P(r=1) + \cdots + P(r=c)$$

$$= \binom{n}{0} p^0 (1-p)^{n-0} + \binom{n}{1} p^1 (1-p)^{n-1} + \cdots + \binom{n}{c} p^c (1-p)^{n-c}$$

$$= \sum_{k=0}^{c} \binom{n}{k} p^k (1-p)^{n-k} \tag{11.2}$$

It is obvious that the probability that a batch will be accepted by a given sampling plan (n and c) will depend upon p. As the value of p is unknown, we calculate the probability that a batch will be accepted for several different values of p. Thus, a curve can be drawn that provides the probability of accepting a batch as a function of batch proportion defective p. This curve is called the operating characteristic curve or OC curve which defines the characteristics of the sampling plan. A typical OC curve is depicted in Figure 11.1. It shows that when $p=0$, $P_a(p)=1$ and when $p=1$, $P_a(p)=0$. It is quite evident from the OC curve that for small values of p, the probability of acceptance is high. On the other hand, for higher values of p, the probability of acceptance is low. Figure 11.2 shows an ideal OC curve. For an ideal OC curve it is certain that a batch will be accepted until a level of proportion defective which is considered to be 'bad' is obtained.

As there is always some statistical error associated with sampling, the OC curve of a sampling plan can never be like the ideal curve of Figure 11.2. Nevertheless, a sampling plan can be appraised by selecting two values of p on which the probabilities of accepting a batch are calculated. First a value of $p = p_1$ is selected so that a batch containing a proportion of defectives less than or equal to p_1 is preferred to be accepted. This value of p is called the acceptable quality level (AQL). Ideally, a producer tries to produce batches of quality better than p_1. Then, another value of $p = p_2 (> p_1)$ is selected so that a batch containing a proportion of defectives

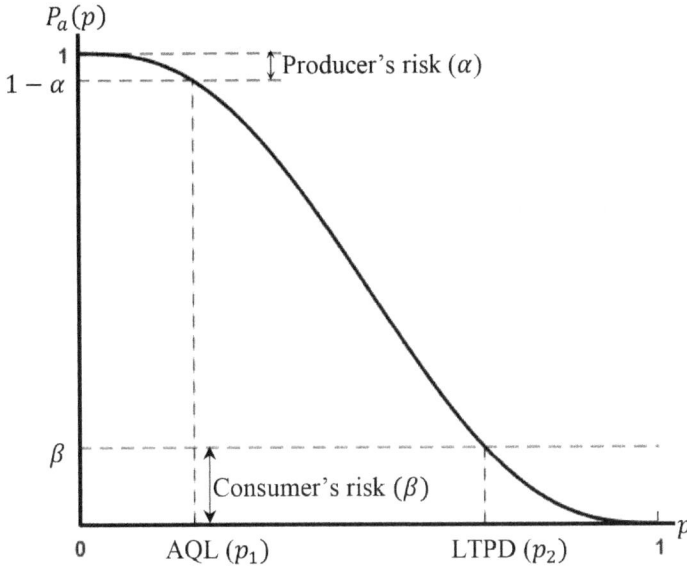

FIGURE 11.1 A typical OC curve.

greater than p_2 is preferred to be rejected. This value of p is called the lot tolerance proportion defective (LTPD). We apprise a sampling plan by calculating the probability that a 'good' batch (with $p \leq p_1$) will be rejected and the probability that a 'bad' batch (with $p \geq p_2$) will be accepted. The probability that a 'good' batch will be rejected is termed as producer's risk. The probability that a 'bad' batch will be

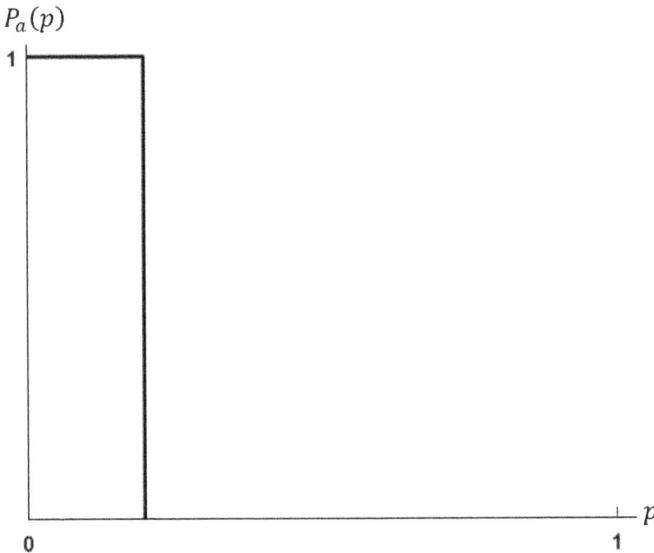

FIGURE 11.2 An ideal OC curve.

accepted is termed as consumer's risk. The producer's and consumer's risks are the consequences of the sampling variability. Figure 11.1 shows that when $p = p_1$, $P_a(p_1) = 1 - \alpha$. Therefore, when $p = p_1$, the probability that a batch will be rejected = $1 - P_a(p_1) = \alpha$. Hence, α is the producer's risk of the sampling plan. It is also evident from Figure 11.1 that when $p = p_2$, $P_a(p_2) = \beta$. Thus β is the consumer's risk of the sampling plan. In order to design an acceptance sampling plan, it is required to know the values of n and c. For given values of p_1, p_2, α and β, the values of n and c can be obtained by solving the following two equations as obtained from the binomial distribution

$$P_a(p_1) = 1 - \alpha = \sum_{k=0}^{c} \binom{n}{k} p_1^k (1 - p_1)^{n-k} \tag{11.3}$$

and

$$P_a(p_2) = \beta = \sum_{k=0}^{c} \binom{n}{k} p_2^k (1 - p_2)^{n-k} \tag{11.4}$$

The two simultaneous Equations (11.3) and (11.4) are nonlinear and exact solution is difficult to obtain since n and c must be the integers. A convenient method to get the approximate solution is based on the chi-square distribution. If a random variable X has a chi-square distribution with $2(c+1)$ degrees of freedom, its probability density function is given by

$$f(x) = \frac{x^c e^{-\frac{x}{2}}}{2^{c+1} \Gamma(c+1)} \tag{11.5}$$

Now $\Gamma(c+1) = c!$
   Thus, Equation (11.5) becomes

$$f(x) = \frac{x^c e^{-\frac{x}{2}}}{2^{c+1} c!}$$

For chi-square distribution, we have

$$P(X \geq \lambda) = \frac{1}{2^{c+1} c!} \int_{\lambda}^{\infty} u^c e^{-\frac{u}{2}} du \tag{11.6}$$

where u is the variable over which the integral is taken. Taking the transformation $\frac{u}{2} = z$, we find $du = 2dz$. Therefore,

$$P(X \geq \lambda) = \frac{1}{c!} \int_{\frac{\lambda}{2}}^{\infty} z^c e^{-z} dz$$

$$= \frac{1}{c!} \left[\left(\frac{\lambda}{2}\right)^c e^{-\frac{\lambda}{2}} + c \int_{\frac{\lambda}{2}}^{\infty} z^{c-1} e^{-z} dz \right]$$

$$= e^{-\frac{\lambda}{2}} \left[\frac{\left(\frac{\lambda}{2}\right)^c}{c!} + \frac{\left(\frac{\lambda}{2}\right)^{c-1}}{(c-1)!} + \cdots + \frac{\lambda}{2} + 1 \right]$$

$$= \sum_{k=0}^{c} \frac{e^{-\frac{\lambda}{2}} \left(\frac{\lambda}{2}\right)^k}{k!}$$

$$= P(r \leq c) \tag{11.7}$$

So, it is evident from (11.7) that if X has a chi-square distribution with $2(c+1)$ degrees of freedom then $P(X \geq \lambda) = P(r \leq c)$ where r follows Poisson distribution with the parameter $\frac{\lambda}{2}$. Now when n is large and p is small, a binomial distribution approaches to a Poisson distribution having mean $= np$.

At $p = p_1$,

$$P_a(p_1) = P(r \leq c) = P(X \geq 2np_1) = 1 - \alpha \left[\text{Using } (11.7)\right]$$

where r follows binomial distribution with parameters n and p_1 and X follows chi-square distribution with $2(c+1)$ degrees of freedom.

Now the definition of $\chi^2_{\alpha,v}$ and the last equation yield,

$$P\left(X \geq \chi^2_{1-\alpha,2(c+1)}\right) = 1 - \alpha = P(X \geq 2np_1)$$

Thus,

$$\chi^2_{1-\alpha,2(c+1)} = 2np_1 \tag{11.8}$$

Again at $p = p_2$,

$$P_a(p_2) = P(r \leq c) = \beta \text{ and proceeding similarly as above we have}$$

$$\chi^2_{\beta,2(c+1)} = 2np_2 \tag{11.9}$$

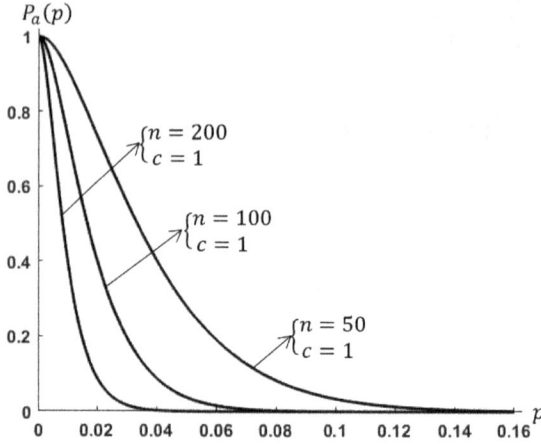

FIGURE 11.3 Effect of sample size on OC curve.

By dividing (11.9) by (11.8), we have

$$\frac{\chi^2_{\beta,2(c+1)}}{\chi^2_{1-\alpha,2(c+1)}} = \frac{p_2}{p_1} \qquad (11.10)$$

First the value of c is determined from Equation (11.10) and then n is obtained from Equation (11.8) and Equation (11.9). Once the values of n and c for a sampling plan are obtained, an OC curve can be drawn using Equation (11.2). Figure 11.3 depicts the effect of sample size (n) on the OC curve. As n becomes larger, the slope of the OC curve turns out to be steeper. The effect of acceptance number (c) on OC curve is shown in Figure 11.4. Changing the values of c does not change the slope of the OC curve considerably. As c is reduced the OC curve is shifted towards the left.

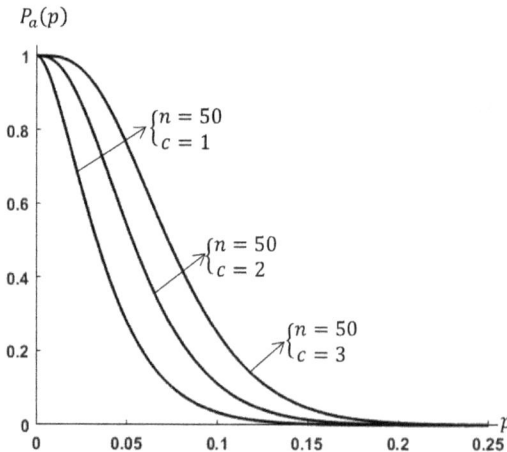

FIGURE 11.4 Effect of acceptance number on OC curve.

Example 11.1:

In a sampling plan, sample size = 100, acceptance number = 1, and lot proportion defective = 0.01. Determine the probability of accepting the lot.

SOLUTION

Using Equation (11.2) we get the probability of accepting the lot

$$P_a(p) = \sum_{k=0}^{c} \binom{n}{k} p^k (1-p)^{n-k}$$

Substituting $p = 0.01$, $n = 100$, and $c = 1$, we have

$$P_a(0.01) = \sum_{k=0}^{1} \binom{100}{k} (0.01)^k (1-0.01)^{100-k}$$

$$= \binom{100}{0} (0.01)^0 (0.99)^{100} + \binom{100}{1} (0.01)^1 (0.99)^{99}$$

$$= 0.366 + 0.369 = 0.735$$

Example 11.2:

Suppose the acceptance number in Example 11.1 is changed from 1 to 2. What will be the probability of accepting the lot?

SOLUTION

In Example 11.1, by changing the value of c from 1 to 2, we have

$$P_a(0.01) = \sum_{k=0}^{2} \binom{100}{k} (0.01)^k (1-0.01)^{100-k}$$

$$= \binom{100}{0} (0.01)^0 (0.99)^{100} + \binom{100}{1} (0.01)^1 (0.99)^{99} + \binom{100}{2} (0.01)^2 (0.99)^{98}$$

$$= 0.366 + 0.369 + 0.185 = 0.92$$

Example 11.3:

A needle maker gets an order from a knitwear manufacturer to deliver a batch of needles. The agreement made between the needle maker and knitwear manufacturer on the values of AQL (p_1), LTPD (p_2), producer's risk (α) and consumer's risk (β) are as follows:

$$p_1 = 0.02, \; p_2 = 0.06, \; \alpha = 0.05, \; \beta = 0.1$$

Determine the sampling plan.

SOLUTION

Substituting the values of p_1, p_2, α, β in (11.10), we have

$$\frac{\chi^2_{0.1,2(c+1)}}{\chi^2_{0.95,2(c+1)}} = \frac{0.06}{0.02} = 3$$

From χ^2 table (Table A3), we find $\chi^2_{0.1,16} = 23.54$ and $\chi^2_{0.95,16} = 7.96$, hence

$$\frac{\chi^2_{0.1,16}}{\chi^2_{0.95,16}} = \frac{23.54}{7.96} = 2.96 \cong 3$$

Thus, $2(c+1) = 16$
　or, $c = 7$
　From Equation (11.8), we find

$$\chi^2_{1-\alpha,2(c+1)} = 2np_1$$

or, $2np_1 = 2n \times 0.02 = 7.96$
　or, $n = \frac{7.96}{2 \times 0.02} = 199$
　Again, from Equation (11.9), we find

$$\chi^2_{\beta,2(c+1)} = 2np_2$$

or, $2np_2 = 2n \times 0.06 = 23.54$
　or, $n = \frac{23.54}{2 \times 0.06} = 196.2$
　Thus, $n = \frac{199+196.2}{2} = 197.6 \cong 198$.

The required sampling plan is to take a sample of size 198 and to reject the batch if there are more than 7 defective needles. Figure 11.5 shows the OC curve of this sampling plan.

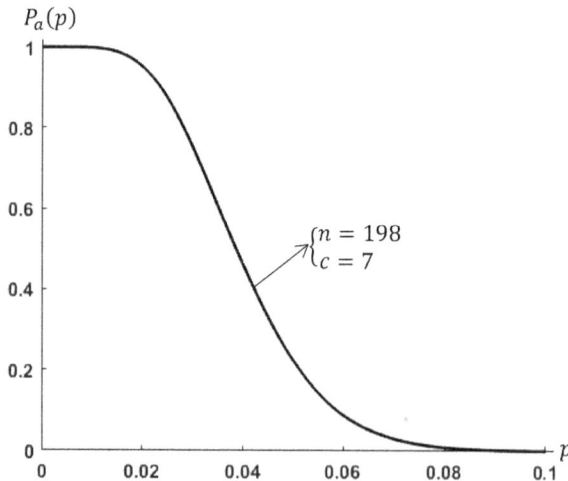

FIGURE 11.5　OC curve with sampling plan $n = 198$ and $c = 7$.

Example 11.4:

In Example 11.3 if producer's risk is changed from 0.05 to 0.1, then what will be the sampling plan?

SOLUTION

Substituting $p_1 = 0.02$, $p_2 = 0.06$, $\alpha = 0.1$, $\beta = 0.1$ in Equation (11.10), we have

$$\frac{\chi^2_{0.1,2(c+1)}}{\chi^2_{0.9,2(c+1)}} = \frac{0.06}{0.02} = 3$$

From χ^2 table (Table A3), we find $\chi^2_{0.1,12} = 18.55$ and $\chi^2_{0.9,12} = 6.3$, hence

$$\frac{\chi^2_{0.1,12}}{\chi^2_{0.9,12}} = \frac{18.55}{6.3} = 2.944 \cong 3$$

Thus, $2(c+1) = 12$
  or, $c = 5$
  From Equation (11.8), we find

$$\chi^2_{1-\alpha,2(c+1)} = 2np_1$$

or, $2np_1 = 2n \times 0.02 = 6.3$
  or, $n = \frac{6.3}{2 \times 0.02} = 157.5$
  Again from Equation (11.9), we find

$$\chi^2_{\beta,2(c+1)} = 2np_2$$

or, $2np_2 = 2n \times 0.06 = 18.55$
  or, $n = \frac{18.55}{2 \times 0.06} = 154.6$
  Thus, $n = \frac{157.5 + 154.6}{2} = 156.05 \cong 156$
  Therefore, the required sampling plan is to take a sample of size 156 and to reject the batch if there are more than 5 defective needles. Figure 11.6 shows the OC curve of this sampling plan.

Example 11.5:

A spindle manufacturer delivers spindles to a ring frame manufacturer. They have made an agreement on the values of AQL (p_1), LTPD (p_2), producer's risk (α) and consumer's risk (β) as follows:

$$p_1 = 0.01, \ p_2 = 0.07, \ \alpha = 0.04, \ \beta = 0.1$$

  a. Determine the sampling plan.
  b. Calculate the probability of acceptance of a batch that contains 2% defective spindles.

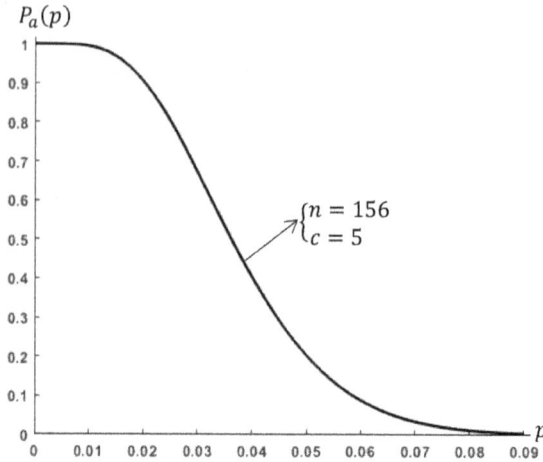

FIGURE 11.6 OC curve with sampling plan $n = 156$ and $c = 5$.

SOLUTION

a. Substituting the values of p_1, p_2, α, β in Equation (11.10), we have

$$\frac{\chi^2_{0.1,2(c+1)}}{\chi^2_{0.96,2(c+1)}} = \frac{0.07}{0.01} = 7$$

From χ^2 table (Table A3), we find $\chi^2_{0.1,6} = 10.65$ and $\chi^2_{0.96,6} = 1.49$, hence

$$\frac{\chi^2_{0.1,6}}{\chi^2_{0.96,6}} = \frac{10.65}{1.49} = 7.147 \cong 7$$

Thus, $2(c + 1) = 6$
   or, $c = 2$
   From Equation (11.8), we find

$$\chi^2_{1-\alpha,2(c+1)} = 2np_1$$

or, $2np_1 = 2n \times 0.01 = 1.49$
   or, $n = \frac{1.49}{2 \times 0.01} = 74.5$
   Again from Equation (11.9), we find

$$\chi^2_{\beta,2(c+1)} = 2np_2$$

or, $2np_2 = 2n \times 0.07 = 10.65$
   or, $n = \frac{10.65}{2 \times 0.07} = 76.1$
   Thus, $n = \frac{74.5+76.1}{2} = 75.3 \cong 75$
   The required sampling plan is to take a sample of size 75 and to reject the batch if there are more than 2 defective needles. Figure 11.7 shows the OC curve of this sampling plan.

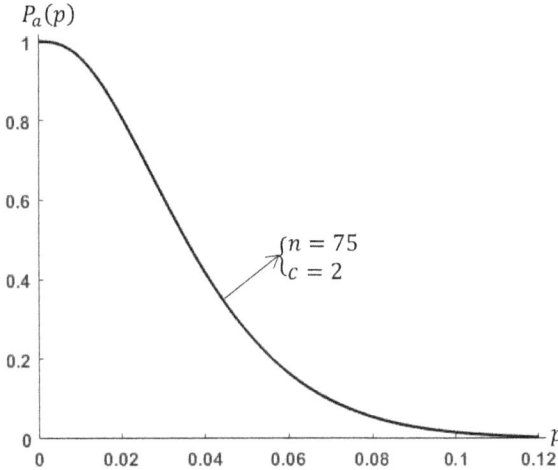

FIGURE 11.7 OC curve with sampling plan $n = 75$ and $c = 2$.

b. Substituting $p = 0.02$, $n = 75$, and $c = 2$ in (11.2), we get

$$P_a(0.02) = \sum_{k=0}^{2} \binom{75}{k}(0.02)^k (1-0.02)^{75-k}$$

$$= \binom{75}{0}(0.02)^0 (0.98)^{75} + \binom{75}{1}(0.02)^1 (0.98)^{74}$$

$$+ \binom{75}{2}(0.02)^2 (0.98)^{73}$$

$$= 0.22 + 0.336 + 0.254 = 0.81$$

11.2.1.1 Average Outgoing Quality (AOQ) and Average Total Inspection (ATI)

In acceptance sampling scheme, corrective action is usually required when batches are rejected. In general, if a batch is rejected, 100% inspection is done and the defective items are replaced from a stock of good items. This type of acceptance sampling scheme is called the rectifying inspection. The average outgoing quality (AOQ) is the quality in the batch resulting from the application of rectifying inspection. AOQ is the average value of batch quality that would be achieved over a long sequence of batches with fraction defective p. Assuming the batch size is N for a rectifying inspection, there will be n items in the batch that contain no defectives because all discovered defectives in the sample are replaced with good ones. If this batch is rejected, the remaining $N - n$ items also contain no defectives. However, if the batch is accepted, the remaining $N - n$ items contain $(N - n)p$ defectives. Therefore,

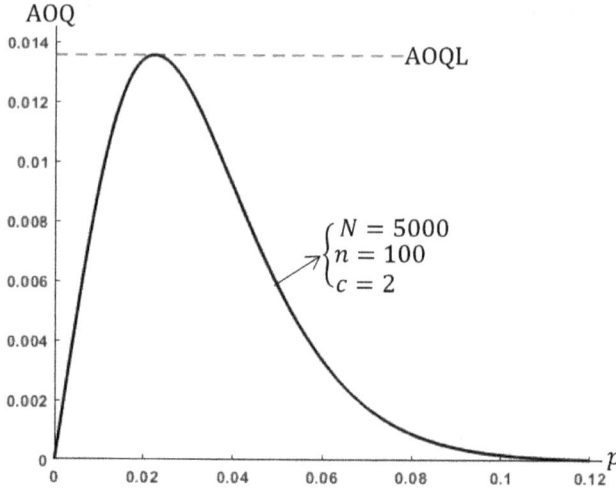

FIGURE 11.8   AOQ curve with $N = 5000$, $n = 100$ and $c = 2$.

the batches in the outgoing stages of inspection would have expected number of
defectives equal to $P_a(p)(N-n)p$, which can be expressed as an average proportion
defective or the average outgoing quality as follows

$$\text{AOQ} = \frac{P_a(p)(N-n)p}{N} \tag{11.11}$$

AOQ depends upon p for a given sampling plan. The curve that plots AOQ against
p is called an AOQ curve. Figure 11.8 depicts the AOQ curve for the sample plan
$N = 5000$, $n = 100$ and $c = 2$. From Figure 11.8 It is quite clear that as p increases
AOQ initially increases, reaches to maximum and then decreases. The maximum
value of AOQ is called average outgoing quality limit (AOQL). In Figure 11.8, AOQL
is observed to be approximately 0.0135. It means that on an average the outgoing
batches will never have a worse quality level than 1.35% defective, no matter how
bad the proportion defective is in the incoming batches.

The average total inspection (ATI) required by a sampling scheme is another
important measure relative to the rectifying inspection. If the batches contain no
defectives ($p = 0$), they are not rejected and the total number of inspections per batch
equals to the sample size n. On the other extreme, if the batches contain all defectives ($p = 1$), each batch will be subjected to 100% inspection and the total number of
inspections per batch equals to batch size N. For $0 < p < 1$, the expected total inspection per batch will be between n and N. Thus, average total inspection per batch can
be expressed as

$$\text{ATI} = n + \{1 - P_a(p)\}(N-n) \tag{11.12}$$

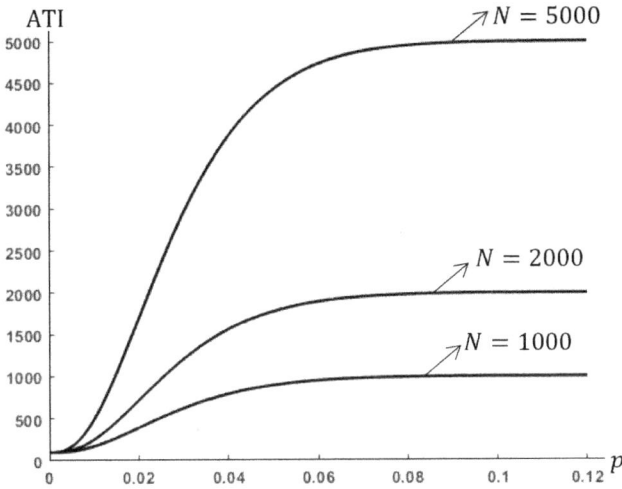

FIGURE 11.9 ATI curves for sampling plan $n = 100$ and $c = 2$ with batch sizes with $N = 1000$, 2000 and 5000.

The curve that plots ATI against p is called an ATI curve. Figure 11.9 depicts the ATI curves for the sample plan $n = 100$ and $c = 2$ with batch sizes with $N = 1000$, 2000 and 5000.

Example 11.6:

In a rectifying inspection plan, lot size = 10000, sample size = 80 and acceptance number = 1. If the lot proportion defective = 0.01, what will be the AOQ and ATI?

SOLUTION

Substituting $p = 0.01$, $n = 80$, and $c = 1$ in Equation (11.2), we have

$$P_a(0.01) = \sum_{k=0}^{1} \binom{80}{k} (0.01)^k (1-0.01)^{80-k}$$

$$= \binom{80}{0} (0.01)^0 (0.99)^{80} + \binom{80}{1} (0.01)^1 (0.99)^{79}$$

$$= 0.809$$

Since $N = 10,000$, from Equations (11.11) and (11.12), we have

$$AOQ = \frac{0.809(10,000 - 80)0.01}{10,000} = 0.008$$

$$ATI = 80 + (1 - 0.809)(10000 - 80) \cong 1973$$

11.2.2 Acceptance Sampling for Variables

11.2.2.1 Assurance About a Minimum Value

Consider a variable sampling plan where a producer is supplying batches of articles to a consumer with an assurance of a lower specification limit. Let X denote a continuous variable being measured which has a normal distribution with unknown mean μ but the standard deviation σ is known based on the past experience. We would like to define a sampling plan with sample size n and a decision rule which is subjected to the conditions that AQL = p_1, LTPD = p_2, producer's risk = α, consumer's risk = β, and a lower specification limit of $x = L$. Let μ_1 and μ_2 be the means corresponding to the batches of AQL and LTPD, respectively as shown in Figure 11.10, from which we can write

$$\mu_1 = L + z_{p_1}\sigma \tag{11.13}$$

and

$$\mu_2 = L + z_{p_2}\sigma \tag{11.14}$$

where z_{p_1} and z_{p_2} are the standard normal variates corresponding to the probabilities p_1 and p_2, respectively. By subtracting Equation (11.14) from Equation (11.13), we find

$$\mu_1 - \mu_2 = \sigma\left(z_{p_1} - z_{p_2}\right) \tag{11.15}$$

Now consider that a large number of random samples, each of size n, are prepared from the batch. As per central limit theorem, the sample mean $\bar{X}$ will be normally distributed with standard deviation $\sigma/\sqrt{n}$. We define a critical value $\bar{x}_L$ such that if $\bar{x} \geq \bar{x}_L$ the batch will be accepted, otherwise it is rejected. Therefore, if a batch of AQL is submitted, the probability of accepting the batch is

$$P\left(\bar{X} \geq \bar{x}_L\right) = 1 - \alpha$$

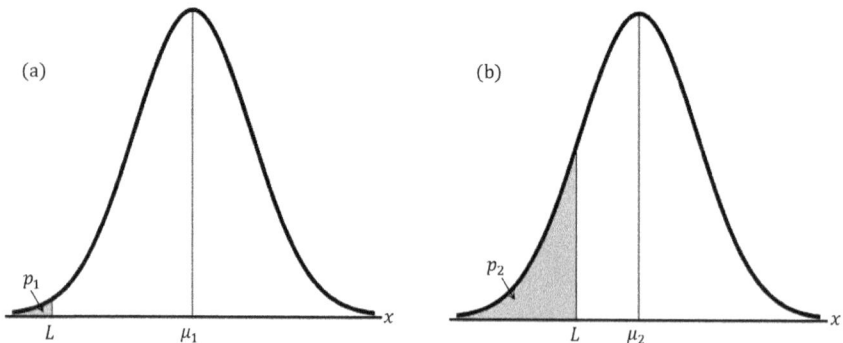

FIGURE 11.10 (a) AQL batch; (b) LTPD batch.

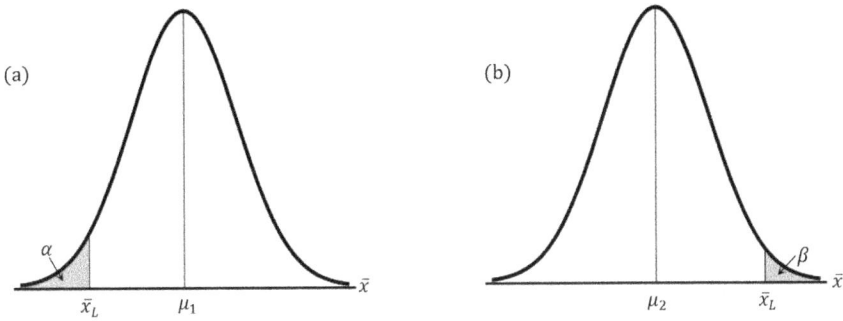

FIGURE 11.11   (a) Producer's risk condition; (b) consumer's risk condition.

Figure 11.11(a) shows the producer's risk condition, from which we can write

$$\mu_1 = \bar{x}_L + z_\alpha \sigma / \sqrt{n} \tag{11.16}$$

where z_α is the standard normal variate corresponding to the probability α. Similarly, if a batch of LTPD is submitted, the probability of accepting the batch is

$$P\left(\bar{X} \geq \bar{x}_L\right) = \beta$$

Figure 11.11(b) shows the consumer's risk condition, from which we can write

$$\mu_2 = \bar{x}_L - z_\beta \sigma / \sqrt{n} \tag{11.17}$$

where z_β is the standard normal variate corresponding to the probability β. By subtracting Equation (11.17) from Equation (11.16), we find

$$\mu_1 - \mu_2 = \frac{\sigma}{\sqrt{n}} \left(z_\alpha + z_\beta \right) \tag{11.18}$$

Using Equation (11.15), we get

$$n = \left(\frac{z_\alpha + z_\beta}{z_{p_1} - z_{p_2}} \right)^2 \tag{11.19}$$

Using Equations (11.16) and (11.17), we obtain

$$\frac{\mu_1 - \bar{x}_L}{\bar{x}_L - \mu_2} = \frac{z_\alpha}{z_\beta}$$

By replacing μ_1 and μ_2 from Equations (11.13) and (11.14), the above equation leads to

$$\bar{x}_L = L + \sigma \left(\frac{z_{p_1} z_\beta + z_{p_2} z_\alpha}{z_\alpha + z_\beta} \right) \tag{11.20}$$

Example 11.7:

A yarn manufacturer supplies batches of yarn to a weaving mill. For a satisfactory loom operation, the yarn has to meet a specification that its tenacity should not be less than 16 cN/tex. Past experience shows that the standard deviation of yarn tenacity is 1 cN/tex. The values of AQL (p_1), LTPD (p_2), producer's risk (α) and consumer's risk (β) are as follows:

$$p_1 = 0.01, \ p_2 = 0.04, \ \alpha = 0.05, \ \beta = 0.02.$$

Device the sampling scheme.

SOLUTION

The values of standard normal variates as found from Table A2 are as follows:

$$z_{p_1} = 2.326, \ z_{p_2} = 1.751, \ z_\alpha = 1.645 \text{ and } z_\beta = 2.054$$

From Equation (11.19), we obtain

$$n = \left(\frac{z_\alpha + z_\beta}{z_{p_1} - z_{p_2}} \right)^2 = \left(\frac{1.645 + 2.054}{2.326 - 1.751} \right)^2 = 41.38 \cong 42$$

Furthermore, from Equation (11.20), we obtain

$$\bar{x}_L = L + \sigma \left(\frac{z_{p_1} z_\beta + z_{p_2} z_\alpha}{z_\alpha + z_\beta} \right) = 16 + 1 \times \left(\frac{2.326 \times 2.054 + 1.751 \times 1.645}{1.645 + 2.054} \right) = 18.07$$

Hence, 42 strength tests are required to perform and if their mean value $\geq$ 18.07 cN/tex, the batch of yarn is accepted, otherwise it is rejected.

11.2.2.2 Assurance About a Maximum Value

A similar approach to the assurance about a minimum value as discussed in the foregoing section is also applicable for assurance about a maximum value. The main difference is that instead of the lower specification limit, we will use the upper specification limit $x = U$. For assurance about a maximum value, the AQL and LTPD batches are shown in Figure 11.12, from which we can write

$$\mu_1 = U - z_{p_1} \sigma \tag{11.21}$$

and

$$\mu_2 = U - z_{p_2} \sigma \tag{11.22}$$

By subtracting Equation (11.22) from Equation (11.21), we find the equation shown below:

$$\mu_1 - \mu_2 = \sigma \left(z_{p_2} - z_{p_1} \right) \tag{11.23}$$

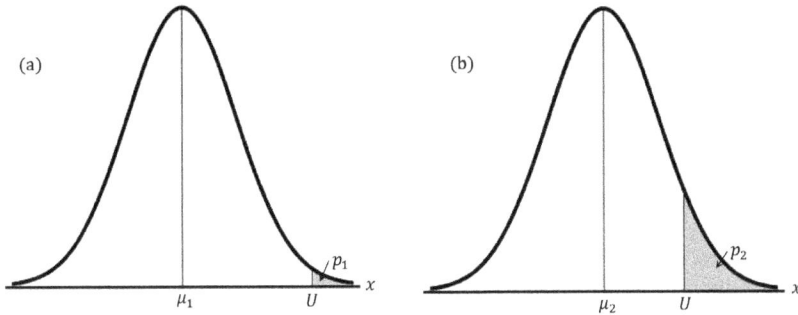

FIGURE 11.12   (a) AQL batch; (b) LTPD batch.

Here we define a critical value $\bar{x}_U$ such that if $\bar{x} \le \bar{x}_U$ the batch will be accepted, otherwise it is rejected. Therefore, if a batch of AQL is submitted, the probability of accepting the batch is

$$P\left(\bar{X} \le \bar{x}_U\right) = 1 - \alpha$$

Figure 11.13 (a) shows the producer's risk condition, from which we can write

$$\mu_1 = \bar{x}_U - z_\alpha \sigma / \sqrt{n} \qquad (11.24)$$

Similarly, if a batch of LTPD is submitted, the probability of accepting the batch is

$$P\left(\bar{X} \le \bar{x}_U\right) = \beta$$

Figure 11.13(b) shows the consumer's risk condition, from which we can write

$$\mu_2 = \bar{x}_U + z_\beta \sigma / \sqrt{n} \qquad (11.25)$$

By subtracting Equation (11.25) from Equation (11.24), we find

$$\mu_1 - \mu_2 = -\frac{\sigma}{\sqrt{n}}\left(z_\alpha + z_\beta\right) \qquad (11.26)$$

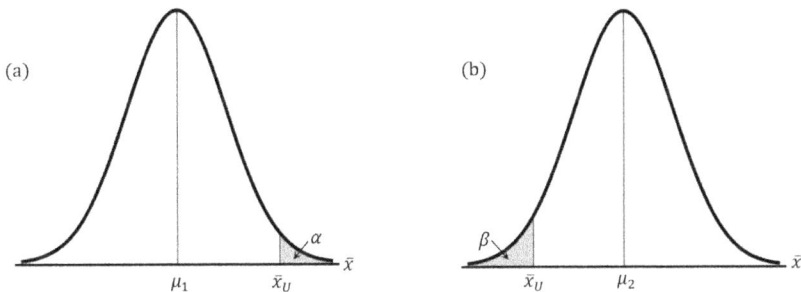

FIGURE 11.13   (a) Producer's risk condition; (b) consumer's risk condition.

Thus, we obtain

$$n = \left(\frac{z_\alpha + z_\beta}{z_{p_1} - z_{p_2}} \right)^2 \tag{11.27}$$

Using Equations (11.24) and (11.25), we find

$$\frac{\bar{x}_U - \mu_1}{\mu_2 - \bar{x}_U} = \frac{z_\alpha}{z_\beta}$$

Ultimately the above equation leads to

$$\bar{x}_U = U - \sigma \left(\frac{z_{p_1} z_\beta + z_{p_2} z_\alpha}{z_\alpha + z_\beta} \right) \tag{11.28}$$

Example 11.8:

A fabric manufacturer supplies batches of curtain cloth made by cotton to a vendor. For a satisfactory aesthetic appearance of the curtain, the cloth has to meet a specification that its drape coefficient should not be greater than 45%. Past experience shows that the standard deviation of drape coefficient is 2.4%. The values of AQL (p_1), LTPD (p_2), producer's risk (α) and consumer's risk (β) are as follows:

$$p_1 = 0.01, \ p_2 = 0.05, \ \alpha = 0.1, \ \beta = 0.05.$$

Device the sampling scheme.

SOLUTION

The values of standard normal variates as found from Table A2 are as follows:

$$z_{p_1} = 2.326, \ z_{p_2} = 1.645, \ z_\alpha = 1.282 \text{ and } z_\beta = 1.645$$

From Equation (11.27), we obtain

$$n = \left(\frac{z_\alpha + z_\beta}{z_{p_1} - z_{p_2}} \right)^2 = \left(\frac{1.282 + 1.645}{2.326 - 1.645} \right)^2 = 18.47 \cong 19$$

Moreover, from Equation (11.28), we obtain

$$\bar{x}_U = U - \sigma \left(\frac{z_{p_1} z_\beta + z_{p_2} z_\alpha}{z_\alpha + z_\beta} \right) = 45 - 2.4 \times \left(\frac{2.326 \times 1.645 + 1.645 \times 1.282}{1.282 + 1.645} \right) = 40.13$$

Hence, 19 tests of drape coefficient are required to perform and if their mean value $\leq 40.13\%$, the batch of curtain is accepted, otherwise it is rejected.

11.2.2.3 Assurance About the Mean Value

Consider a variable sampling plan where a producer is supplying batches of articles whose mean critical measurement is nominally equal to μ_0 with a standard deviation of σ. The producer and consumer has agreed that a batch will be acceptable if

$$\mu_0 - T < \mu < \mu_0 + T$$

where μ is the actual mean and T is the allowable tolerance for the mean value. Here we define the producer's risk α as the probability of rejecting a perfect batch for which $\mu = \mu_0$. The consumer's risk β is defined as the probability of accepting a just-imperfect batch for which $\mu = \mu_0 \pm T$.

Now consider that a large number of random samples, each of size n, are prepared from the batch and the sample mean $\bar{X}$ is normally distributed with mean μ and standard deviation $\sigma/\sqrt{n}$. For producer's risk condition, the sampling distribution of $\bar{X}$ from a delivery has a mean μ which is exactly equal to μ_0 as depicted in Figure 11.14. Such a perfect batch will be rejected if the observed value of sample mean falls outside the range $\mu_0 \pm t$, where t is the tolerance for the sample mean. Under producer's risk condition, this probability is equal to α. Thus, from Figure 11.14 we have

$$z_{\alpha/2} = \frac{\mu_0 + t - \mu_0}{\sigma/\sqrt{n}} = \frac{t\sqrt{n}}{\sigma} \tag{11.29}$$

where $z_{\alpha/2}$ is the standard normal variate corresponding to the probability $\alpha/2$.

For consumer's risk condition, the sampling distribution of $\bar{X}$ from the deliveries have means which are just equal to the allowed tolerances $\mu_0 \pm T$ as depicted in Figure 11.15. Such a poor batch will be accepted if the observed value of sample

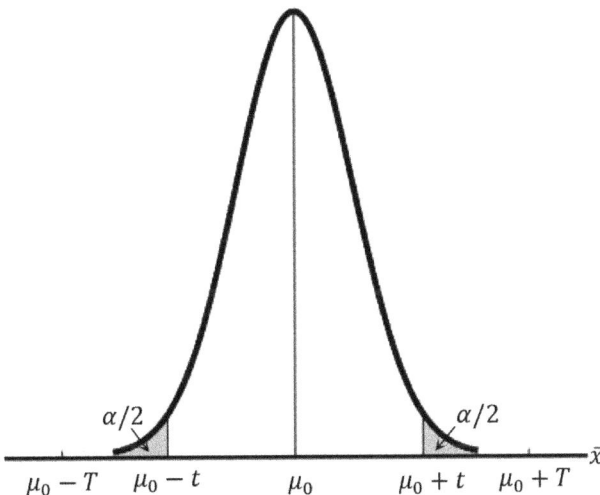

FIGURE 11.14   Producer's risk condition.

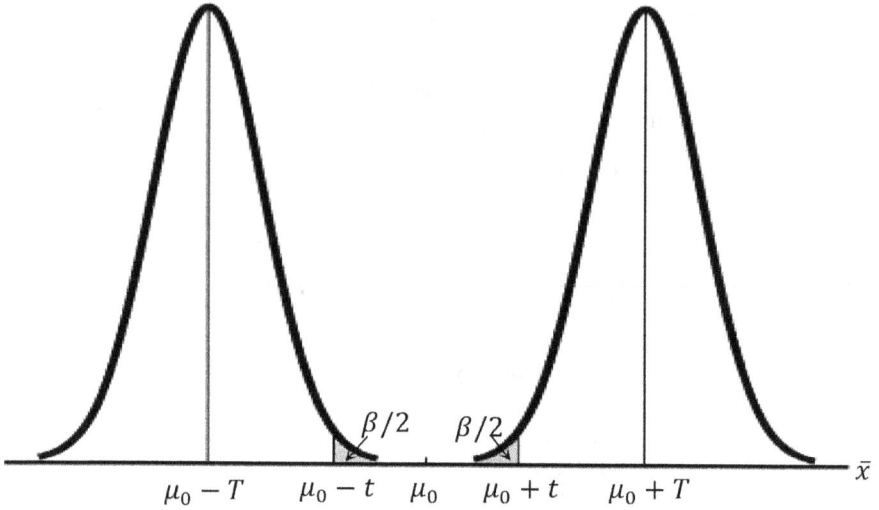

FIGURE 11.15   Consumer's risk condition.

mean falls inside the range $\mu_0 \pm t$. Under consumer's risk condition, this probability is equal to β. Thus, from Figure 11.15 we have

$$z_{\beta/2} = \frac{(\mu_0 - t) - (\mu_0 - T)}{\sigma/\sqrt{n}} = \frac{(T - t)\sqrt{n}}{\sigma} \tag{11.30}$$

where $z_{\beta/2}$ is the standard normal variate corresponding to the probability $\beta/2$.
  By solving Equations (11.29) and (11.30) simultaneously, we find

$$n = \sigma^2 \frac{\left(z_{\alpha/2} + z_{\beta/2}\right)^2}{T^2} \tag{11.31}$$

and

$$t = \frac{T z_{\alpha/2}}{z_{\alpha/2} + z_{\beta/2}} \tag{11.32}$$

Example 11.9:

A yarn manufacturer supplies yarn of nominal linear density equal to 20 tex to a fabric manufacturer. The delivery of the yarn is acceptable if the mean linear density lies within a range of 20 ± 1 tex. Device a sampling scheme if the producer's risk and consumer's risk are 0.075 and 0.05, respectively. Assume the standard deviation of yarn linear density is 0.8 tex.

SOLUTION

Given data: $\alpha = 0.075$, $\beta = 0.05$, $\mu_0 = 20$, $T = 1$, $\sigma = 0.8$.
The values of standard normal variates as found from Table A2 are as follows:

$$z_{\alpha/2} = 1.78 \text{ and } z_{\beta/2} = 1.96$$

From Equation (11.31), we obtain

$$n = \sigma^2 \frac{(z_{\alpha/2} + z_{\beta/2})^2}{T^2} = 0.8^2 \times \left(\frac{1.78 + 1.96}{1}\right)^2 = 8.95 \cong 9$$

Furthermore, from Equation (11.32), we obtain

$$t = \frac{T z_{\alpha/2}}{z_{\alpha/2} + z_{\beta/2}} = \frac{1 \times 1.78}{1.78 + 1.96} = 0.476$$

Hence, 9 tests of yarn linear density are required to perform and if their mean value lies in the range 20 ± 0.476 tex, the batch of yarn is accepted, otherwise it is rejected.

11.3 CONTROL CHART

In manufacturing, articles are produced in industry with an aim to make them uniform. However, with all precision of modern engineering and automations, every manufacturing process is susceptible to certain variations, which may be due to chance or assignable causes. The variations due to chance causes are inherent and are generated due to innumerous independent factors. This class of causes are impossible to trace and do not affect the quality at a significant level. The assignable causes, contrarily, are those which consequence large and significant variations due to the faulty machines, process, improper material handling, etc. Though it is inevitable to eliminate the chance causes, it is possible to eliminate the assignable causes, and if successfully eliminated, the process is then said to be 'under statistical control'. To ensure a process to be in a state of statistical control, the quality control department in the manufacturing organization every so often checks the process, detects the significant deviations in quality level, inspects the causes of defects (if any) and removes the causes. 'Control chart' is a graphical representation of information on the quality of samples, drawn from a lot that represents the population on a regular basis and thus enables us to know the presence of any assignable causes, which may be eliminated by rectifying the process.

11.3.1 CONCEPT

Let us assume a manufacturing process in which n samples are collected by their quality department to measure or calculate the quality parameters, viz., mean, standard deviation, range, number of defective pieces, etc. Whether the said manufacturing

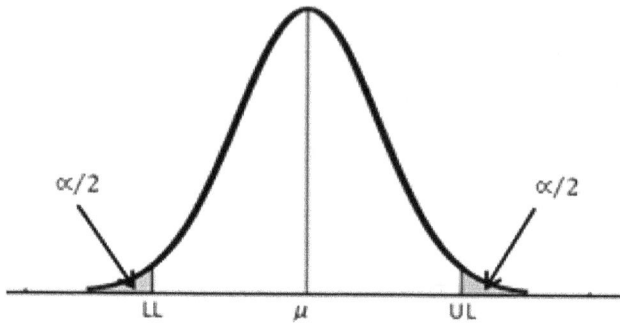

FIGURE 11.16 Distribution of X.

process is in control or out of control may be assessed by the test of null hypothesis against the alternative hypothesis which is given below.

H_0: the process is in control
H_1: the process is out of control

The principle of the above test of significance is already explained in Chapter 7 and the same principle of significance test may be deployed to check whether the process is in control or out of control. Suppose a sample of size n is selected from a process and a quality characteristic X is measured. Figure 11.16 shows the distribution of X. If the null hypothesis is not rejected, then most of the observed values of X will cluster around the mean μ of the distribution and the process is said to be in control. Now, if any one of the observed values falls in one of the definite tails (Figure 11.16), then the null hypothesis will be rejected which leads to the conclusion that the process is out of control. Therefore, the nature of distribution of tails is of utmost importance. Figure 11.16 depicts that each corresponding tail has an area equal to $\alpha/2$, assuming the test is done at a significance level of α. Therefore, it is obvious that there are two control limits, lower limit (LL) and upper limit (UL). If the observed values are in between LL and UL, then the process is assumed to be in control, else the process is out of control.

Now, the above figure may be turned by 90° and the control limits along with the mean may be extrapolated to get a control chart, which is shown in Figure 11.17. The values of regular testing or inspection are plotted in this chart. As long as these values lie within the UL and LL, the process is assumed to be in control; otherwise, the process is out of control.

11.3.2 CENTRAL LINE, ACTION AND WARNING LIMITS

A control chart contains a central line and two control limits (upper limit and lower limit). The central line represents the mean μ of the quality characteristic X. Generally, each control limit is set up at 0.001 probability level, which means the values of $\alpha/2$ representing the area of each tail in Figure 11.16 is 0.001. Therefore, the possibility of a point falling beyond the upper limit is 1 out of 1000. Similarly, the possibility of a

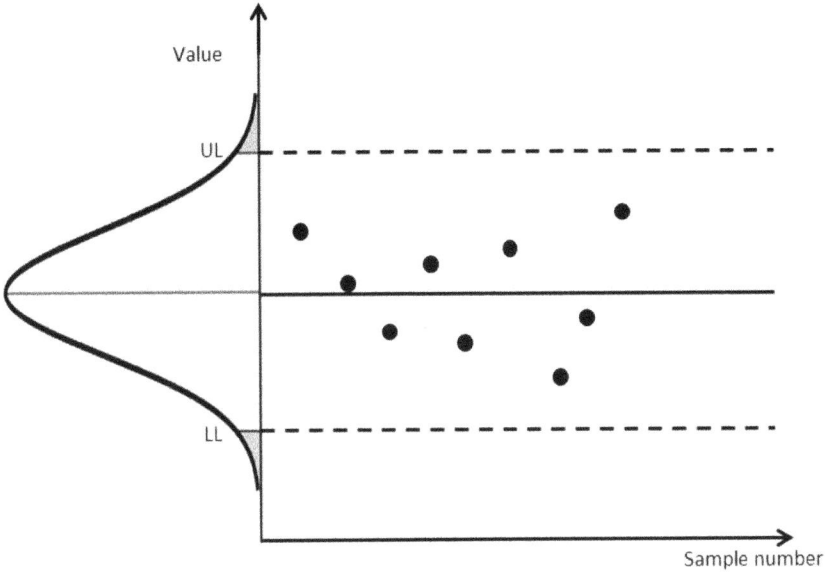

FIGURE 11.17 Control chart showing the process is in control.

point falling beyond the lower limit is also 1 out of 1000. Thus, α equals 0.002, representing the probability that a point falling beyond upper and lower limits is 0.002. Since the falling of two points out of 1000 beyond the control limits is a small risk, it is rational to say that in case a point falls beyond these limits, the deviation is due to some assignable cause. If X is assumed to be normally distributed with mean μ and standard deviation σ, the position of the control limits can be easily found. Now from Table A2, $z_{0.001} = 3.09$, therefore, the upper and lower control limits would be set at $\mu \pm 3.09\sigma$. These limits are often known as action limits. Similarly, two more limits, known as warning limits are defined at 0.025 probability level. From Table A2, $z_{0.025} = 1.96$, thus, the warning limits would be set at $\mu \pm 1.96\sigma$. Thus, we have

$$\text{Action limits} = \mu \pm 3.09\sigma \qquad (11.33)$$

$$\text{Warning limits} = \mu \pm 1.96\sigma \qquad (11.34)$$

A format of a control chart with central lines and upper and lower limits is illustrated in Figure 11.18.

11.3.3 INTERPRETING CONTROL CHARTS

In general, if any value falls outside either UL or LL, the process is concluded to be out of control. There are, however, other indications in which, although all values lie in between the UL and LL, the process may be assumed as lack of control. In occasions when two consecutive points fall between the same action and warning limits or sequence of points lying between mean and anyone of the warning limits or

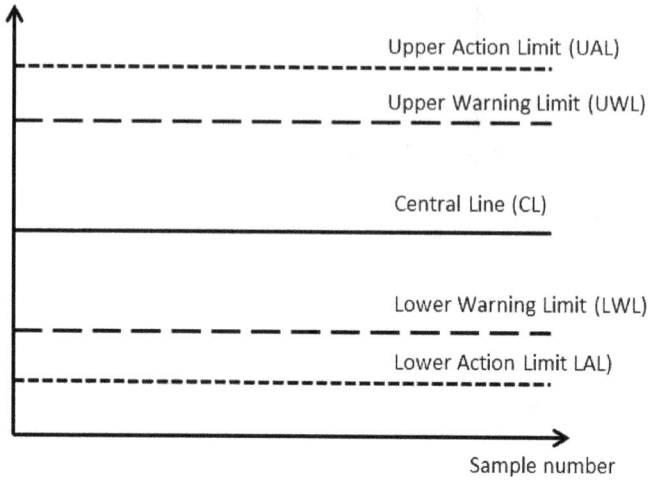

FIGURE 11.18   Central line, warning and action limits.

points, showing an upward/downward trend or points showing a periodic nature or any non-random pattern, indicates that the process is subjected to some assignable causes and hence should be investigated to eliminate it. Some indications of a lack of control of a process are shown in Figure 11.19(a)–(e).

Example 11.10:

If the samples are independent, calculate (a) the probability that two successive points will fall between the warning and action limits of same side; (b) the probability that two successive points will fall between any of the warning and action limits; (c) nine consecutive points will fall within a particular side of action limit.

SOLUTION

a. Figure 11.20(a) depicts the falling of two successive points between the warning and action limits of same side. The probability that a point will fall between the warning and action limits of same side is given by

$$P(X > \mu + 1.96\sigma) - P(X > \mu + 3.09\sigma)$$

$$= P\left(Z > \frac{\mu + 1.96\sigma - \mu}{\sigma}\right) - PP\left(Z > \frac{\mu + 3.09\sigma - \mu}{\sigma}\right)$$

$$= P(Z > 1.96) - P(Z > 3.09)$$

$$= 0.025 - 0.001 = 0.024.$$

Thus, the probability that two successive points will fall between the warning and action limits of same side $= (0.024)^2 = 0.00058$.

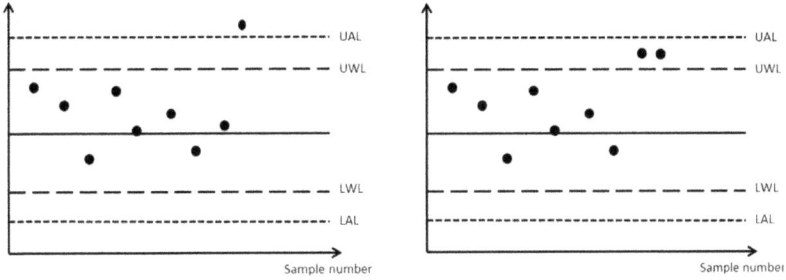

(a) One point lying beyond action limit

(b) Two consecutive points lying between same action and warning limits

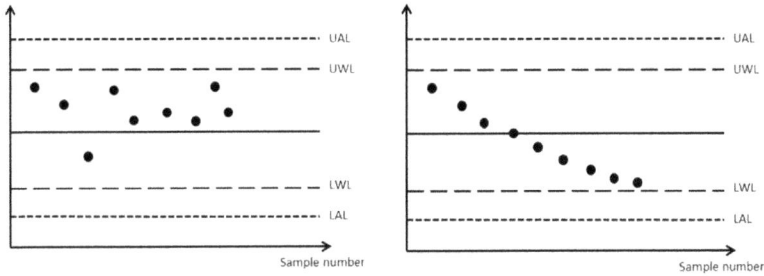

(c) Sequence of points lying between central line and upper warning limits

(d) points showing a downward trend

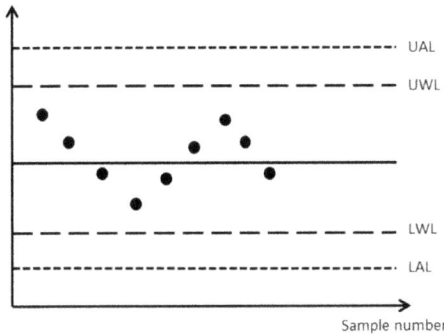

(e) Points showing a periodic trend

FIGURE 11.19   (a)–(e) Indications showing the lack of control of a process.

b. Figure 11.20(b) shows the falling of two successive points between any of the warning and action limits. The probability that a point will fall between the warning and action limits of same side = 0.024. Thus, the probability that a point will fall between any of the warning and action limits = $2 \times 0.024 = 0.048$. Hence, the probability that two successive points will fall between any of the warning and action limits = $(0.048)^2 = 0.0023$.

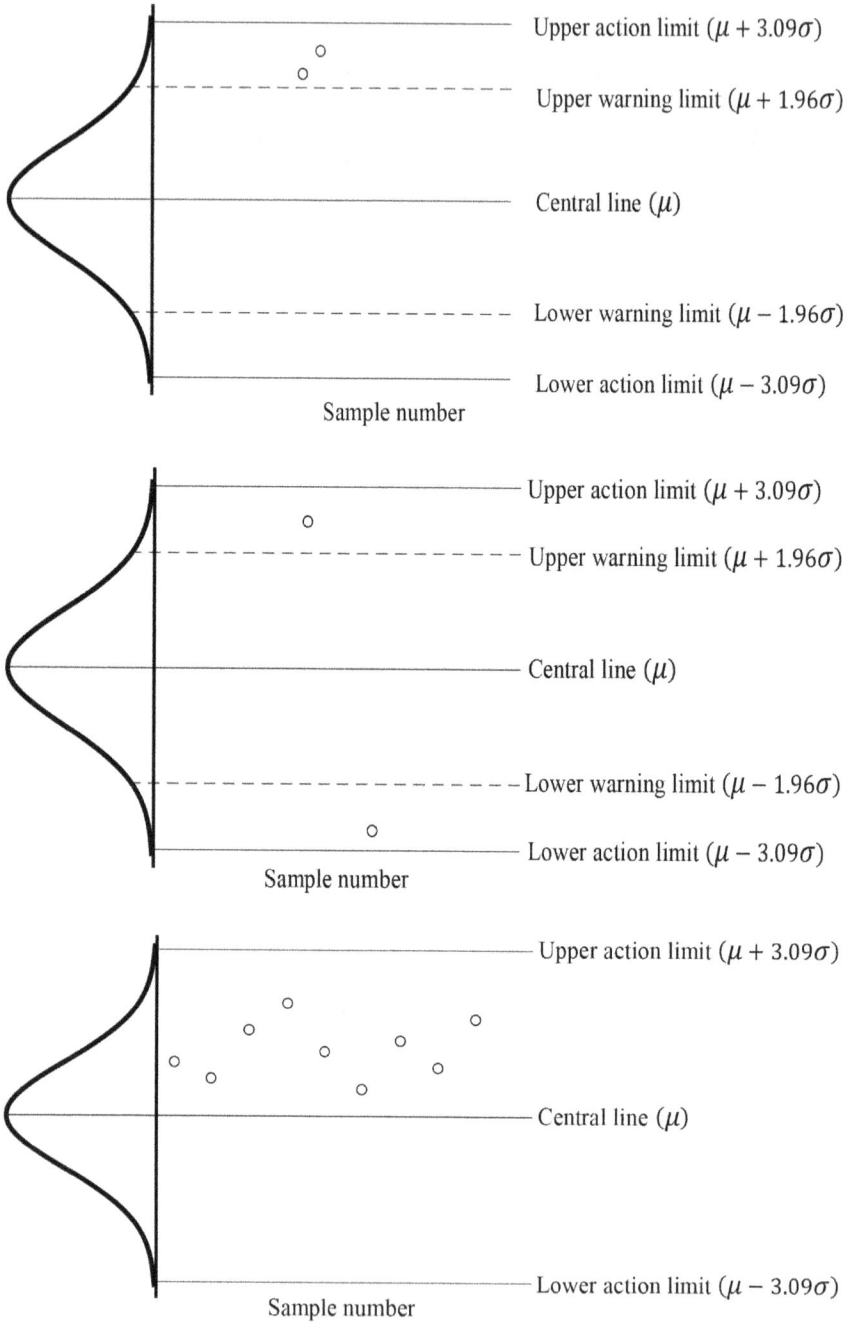

FIGURE 11.20 (a) Control chart showing two successive points fall between the warning and action limits of same side; (b) control chart showing two successive points fall between any of the warning and action limits; (c) control chart showing nine consecutive points fall within a particular side of action limit.

c. Figure 11.20(c) represents the falling of nine consecutive points within a particular side of action limit. The probability that a point will fall within a particular side of action limit is given by

$$P(X > \mu) - P(X > \mu + 3.09\sigma)$$

$$= P\left(Z > \frac{\mu - \mu}{\sigma}\right) - P\left(Z > \frac{\mu + 3.09\sigma - \mu}{\sigma}\right)$$

$$= P(Z > 0) - P(Z > 3.09)$$

$$= 0.5 - 0.001 = 0.499.$$

Thus, the probability that nine consecutive points will fall within a particular side of action limit $= (0.499)^9 = 0.00192$.

11.3.4 TYPES OF CONTROL CHARTS

Control charts are classified based on variables (actual measurements like dimension, mass, strength, etc.) and attributes ('yes' or 'no'). Control charts for variables involve the measurement of certain parameters during the manufacturing process and the manufactured product is accepted or rejected if the measured values lie within or beyond the tolerance limits. The control chart for attributes is applicable when the quality cannot be measured directly and are classified either as 'good or bad', defective or non-defective', etc. The most common control charts for variables and attributes are shown in Figure 11.21.

11.3.4.1 Control Chart for Average and Range

Control charts for average and range are applicable when the variables are continuous. Yarn linear density, fabric areal density, fabric tensile strength, etc. are some examples where these properties are measured in a continuous scale. For such properties, both the mean and range of measured values can be calculated. Mean measures the level of central tendency of the process, whereas the range indicates the variation. Control charts for sample mean and range are among the most important and useful statistical quality control techniques.

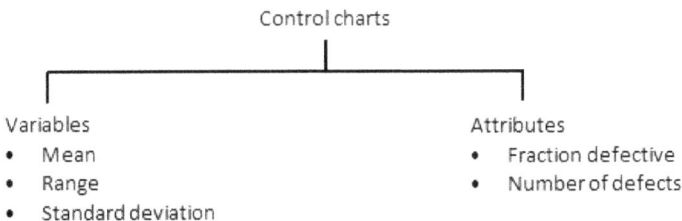

Control charts

Variables
• Mean
• Range
• Standard deviation

Attributes
• Fraction defective
• Number of defects

FIGURE 11.21 Control charts for variables and attributes.

Let us assume a process in which a quality characteristic X is measured to investigate whether the process is under control. Suppose that $X_1, X_2, \cdots, X_n$ constitute a random sample of size n drawn from a normally distributed population, having a known mean and standard deviation μ and σ, respectively, then if the process is in control, the sample mean $\bar{X}$ will follow a normal distribution with mean $\mu_{\bar{x}} = \mu$ and standard deviation $\sigma_{\bar{x}} = \sigma/\sqrt{n}$. Therefore, if μ and σ are known, the warning and action limits for sample mean can be written as

$$\text{Action limits} = \mu \pm 3.09\sigma/\sqrt{n} \qquad (11.35)$$

$$\text{Warning limits} = \mu \pm 1.96\sigma/\sqrt{n} \qquad (11.36)$$

Here we assume that the quality characteristic follows a normal distribution. Even if the quality characteristic does not follow a normal distribution, the above results in Equations (11.35) and (11.36) are still valid by virtue of the central limit theorem.

If μ and σ are unknown, they are usually estimated from a preliminary experiment based on m samples each containing n observations. If $\bar{x}_1, \bar{x}_2, \cdots, \bar{x}_m$ are the means of each sample, then the best estimator of μ is given by

$$\bar{\bar{x}} = \frac{\bar{x}_1 + \bar{x}_2 + \cdots + \bar{x}_m}{m}.$$

We generally estimate σ from the ranges of m samples due to the fact that the range of a sample from a normal distribution and standard deviation are related. If $x_1, x_2, \cdots, x_n$ are the values of a sample of size n, then the range of the sample is the difference between the largest and smallest values. If $\bar{R}$ is the mean range of m samples, then an estimate of σ is given by

$$\hat{\sigma} = \bar{R}/d_2$$

where d_2 depends on the sample size n. For a small sample size ($n = 3$ to 6), $\bar{R}/d_2$ gives a satisfactory estimation of σ.

If μ and σ are unknown, we use $\bar{\bar{x}}$ as an estimator of μ and $\bar{R}/d_2$ as the estimator of σ. Therefore, the action and warning limits in Equations (11.35) and (11.36) may be written as

$$\text{Action limits} = \bar{\bar{x}} \pm 3.09\,\bar{R}/d_2\sqrt{n}$$
$$= \bar{\bar{x}} \pm A_2\bar{R} \qquad (11.37)$$

$$\text{Warning limits} = \bar{\bar{x}} \pm 1.96\,\bar{R}/d_2\sqrt{n}$$
$$= \bar{\bar{x}} \pm A_2'\bar{R} \qquad (11.38)$$

where $A_2 = 3.09/d_2\sqrt{n}$, $A_2' = 1.96/d_2\sqrt{n}$.

The values of d_2, A_2 and A_2' for sample size n are given in Table 11.1.

As the sample range is related to the standard deviation, the variability of a process may be controlled by constructing a control chart for range (R) with the central

TABLE 11.1
Values of d_2, A_2 and A_2'

n	d_2	A_2	A_2'
2	1.128	1.94	1.23
3	1.693	1.05	0.67
4	2.059	0.75	0.48
5	2.326	0.59	0.38
6	2.534	0.50	0.32
7	2.704	0.43	0.27
8	2.847	0.38	0.24
9	2.970	0.35	0.22
10	3.078	0.32	0.20

line $\bar{R}$. We denote the standard deviation of R by σ_R. In order to determine control limits of R, we need an estimate of σ_R. If a quality characteristic is normally distributed, an estimation of σ_R is given by

$$\hat{\sigma}_R = d_3\hat{\sigma} = d_3 \frac{\bar{R}}{d_2}$$

where d_3 depends on the sample size n.

Therefore, the action and warning limits of the R chart are given by

$$\text{Upper action limit} = \bar{R} + 3.09 \frac{d_3}{d_2} \bar{R}$$

$$= \left(1 + 3.09 \frac{d_3}{d_2}\right) \bar{R} \qquad (11.39)$$

$$= D_4 \bar{R}$$

$$\text{Lower action limit} = \bar{R} - 3.09 \frac{d_3}{d_2} \bar{R}$$

$$= \left(1 - 3.09 \frac{d_3}{d_2}\right) \bar{R} \qquad (11.40)$$

$$= D_3 \bar{R}$$

$$\text{Upper warning limit} = \bar{R} + 1.96\frac{d_3}{d_2}\bar{R}$$

$$= \left(1 + 1.96\frac{d_3}{d_2}\right)\bar{R} \qquad (11.41)$$

$$= D_4'\bar{R}$$

$$\text{Lower warning limit} = \bar{R} - 1.96\frac{d_3}{d_2}\bar{R}$$

$$= \left(1 - 1.96\frac{d_3}{d_2}\right)\bar{R} \qquad (11.42)$$

$$= D_3'\bar{R}$$

The values of d_2, d_3, D_3', D_4', D_3 and D_4 for sample size n are given in Table 11.2.

Example 11.11:

4 cops were randomly taken on each day over a period of 10 days from a ring frame. The following yarn counts were obtained.

Day	1	2	3	4	5	6	7	8	9	10
Count	39.85	41.46	41.50	39.84	39.50	39.42	39.77	40.48	40.92	38.93
(Ne)	40.49	39.41	40.48	39.54	39.91	41.08	39.84	40.32	41.35	40.50
	39.41	40.32	40.21	39.66	40.17	40.97	39.92	40.17	39.72	40.62
	40.05	40.53	39.29	39.69	40.09	41.32	39.64	40.25	39.68	39.17

Use these data to construct control charts for averages and ranges of samples of size 4.

TABLE 11.2
Values of D_3', D_4', D_3 and D_4

n	d_2	d_3	D_3'	D_4'	D_3	D_4
2	1.128	0.853	0	2.482	0	3.337
3	1.693	0.888	0	2.028	0	2.621
4	2.059	0.880	0.162	1.838	0	2.321
5	2.326	0.864	0.272	1.728	0	2.148
6	2.534	0.848	0.344	1.656	0	2.034
7	2.704	0.833	0.396	1.604	0.048	1.952
8	2.847	0.820	0.435	1.565	0.110	1.890
9	2.970	0.808	0.467	1.533	0.159	1.841
10	3.078	0.797	0.492	1.508	0.200	1.800

SOLUTION

The sample average and sample range of yarn count for each day is calculated as follows.

Day	1	2	3	4	5	6	7	8	9	10
$\bar{x}$	39.95	40.43	40.37	39.68	39.92	40.70	39.79	40.31	40.42	39.81
R	1.08	2.05	2.21	0.30	0.67	1.90	0.28	0.31	1.67	1.69

From the above table, we have

$$\bar{\bar{x}} = \frac{\sum \bar{x}}{10} = 40.138$$

$$\bar{R} = \frac{\sum R}{10} = \frac{12.16}{10} = 1.216$$

Substituting $A_2' = 0.48$ and $A_2 = 0.75$ for $n = 4$, we get

Warning limits for averages $= \bar{\bar{x}} \pm A_2' \bar{R} = 40.138 \pm 0.48 \times 1.216 = (39.554, 40.722)$

Action limits for averages $= \bar{\bar{x}} \pm A_2 \bar{R} = 40.138 \pm 0.75 \times 1.216 = (39.226, 41.05)$

Substituting $D_3' = 0.162$, $D_4' = 1.838$, $D_3 = 0$ and $D_4 = 2.321$ for $n = 4$, we get

Warning limits for ranges $= D_3' \bar{R}, D_4' \bar{R} = (0.162 \times 1.216, 1.838 \times 1.216) = (0.20, 2.24)$

Action limits for ranges $= D_3 \bar{R}, D_4 \bar{R} = (0 \times 1.216, 2.321 \times 1.216) = (0, 2.82)$

Figure 11.22(a) and (b) depicts the control charts for averages and ranges, respectively. The result shows that the process is in control.

11.3.4.2 Control Chart for Defectives and Fraction Defectives

Many parameters of our interest in a production process cannot be measured in a continuous scale. These parameters are known as attributes. A common attribute in manufacturing process is 'yes' or 'no' to a question 'if an item is defective or not?'. Such items are classified as 'defective' or 'non-defective'. In such cases, the quality of the process is expressed by the number of defectives. Let a random variable X be the number of defectives in a sample of size n. If p is the proportion of defectives produced under the condition that the process is in control, then X has a binomial distribution with mean np and the standard deviation $\sqrt{np(1-p)}$. If n is large,

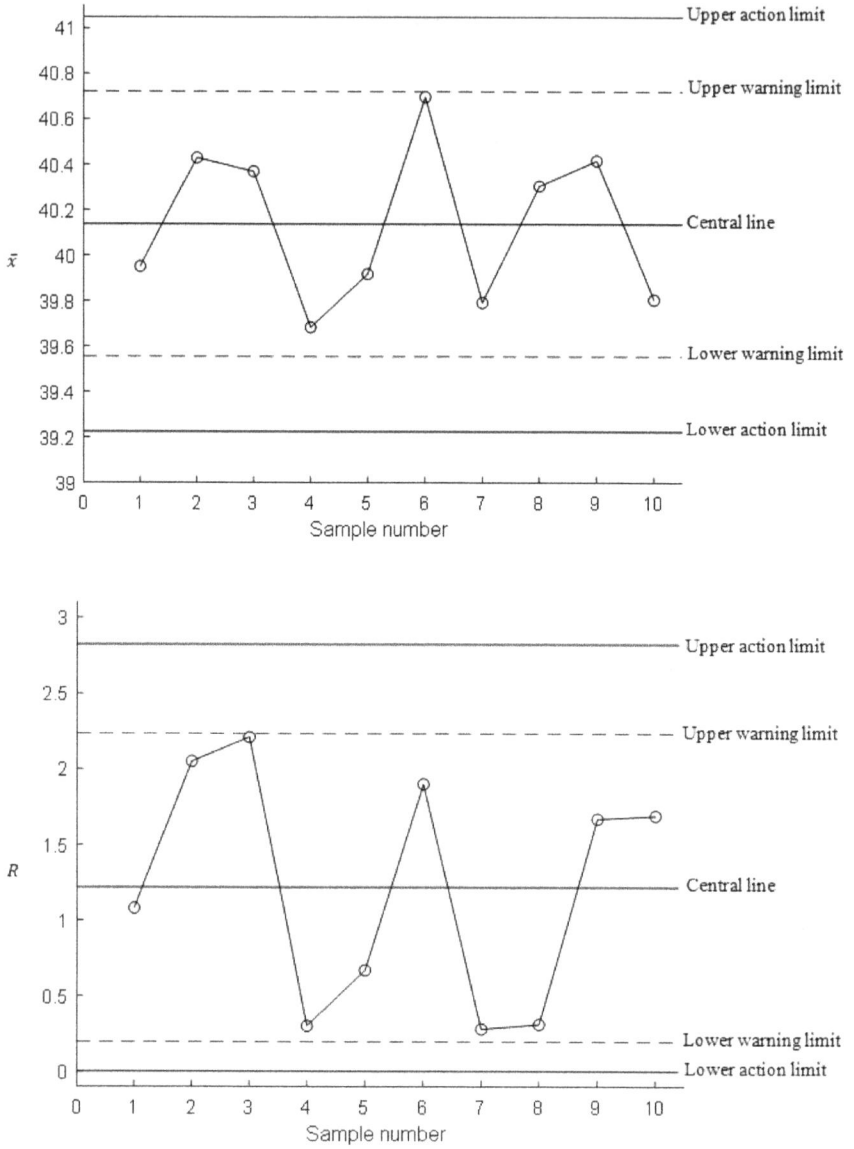

FIGURE 11.22   (a) Control chart for average; (b) control chart for range.

then the binomial distribution approaches to the normal distribution. Therefore, the action and warning limits for defectives are computed as

$$\text{Action limit} = np \pm 3.09\sqrt{np(1-p)} \qquad (11.43)$$

$$\text{Warning limit} = np \pm 1.96\sqrt{np(1-p)} \qquad (11.44)$$

If p is not known, then it is estimated from the preliminary experiment based on m samples each of size n. If $x_1, x_2, \cdots, x_m$ are the values of defectives for m samples, then an estimate of p is given by

$$\bar{p} = \frac{\sum_{i=1}^{m} x_i}{mn} = \frac{\sum_{i=1}^{m} \hat{p}_i}{m}$$

where $\hat{p}_i$ is the observed proportion of defectives for ith sample. Therefore, the action and warning limits for defectives in Equations (11.43) and (11.44) may be written as

$$\text{Action limit} = n\bar{p} \pm 3.09\sqrt{n\bar{p}(1-\bar{p})} \qquad (11.45)$$

$$\text{Warning limit} = n\bar{p} \pm 1.96\sqrt{n\bar{p}(1-\bar{p})} \qquad (11.46)$$

It is often recommended to use the control chart based on fraction or proportion defectives. The estimates of mean and standard deviation of fraction defectives are $\bar{p}$ and $\sqrt{\frac{\bar{p}(1-\bar{p})}{n}}$, respectively. Therefore, the action and warning limits for fraction defectives are computed as

$$\text{Action limit} = \bar{p} \pm 3.09\sqrt{\frac{\bar{p}(1-\bar{p})}{n}} \qquad (11.47)$$

$$\text{Warning limit} = \bar{p} \pm 1.96\sqrt{\frac{n\bar{p}(1-\bar{p})}{n}} \qquad (11.48)$$

Example 11.12:

The numbers of defective cones in samples of size 400 were counted on each day from a winding section over a period of 15 days. The results were as follows.

Day	1	2	3	4	5	6	7	8	9	10	11	12	13	14	15
Defective Cones	18	22	19	24	18	17	16	19	17	21	23	18	17	20	22

Construct a control chart for defective cones.

SOLUTION

Total number of defective cones = 291

$$\text{Sample size } n = 400$$

$$\text{Total number of cones inspected} = 400 \times 15 = 6000$$

$$\text{Mean proportion of defective cones} = \bar{p} = \frac{291}{6000} = 0.0485$$

$$\text{The mean number of defective cones} = n\bar{p} = 400 \times 0.0485 = 19.4$$

$$\text{Standard deviation of defective cones} = \sqrt{n\bar{p}(1-\bar{p})}$$

$$= \sqrt{400 \times 0.0485 \times (1-0.0485)} = 4.3$$

$$\text{The warning limits of the defective cones} = n\bar{p} \pm 1.96\sqrt{n\bar{p}(1-\bar{p})}$$

$$= 19.4 \pm 1.96 \times 4.3 = (10.97,\ 27.83)$$

$$\text{The action limits of the defective cones} = n\bar{p} \pm 3.09\sqrt{n\bar{p}(1-\bar{p})}$$

$$= 19.4 \pm 3.09 \times 4.3 = (6.11,\ 32.69)$$

Figure 11.23 depicts the control chart for defective cones. It is evident from the result that the process is in control.

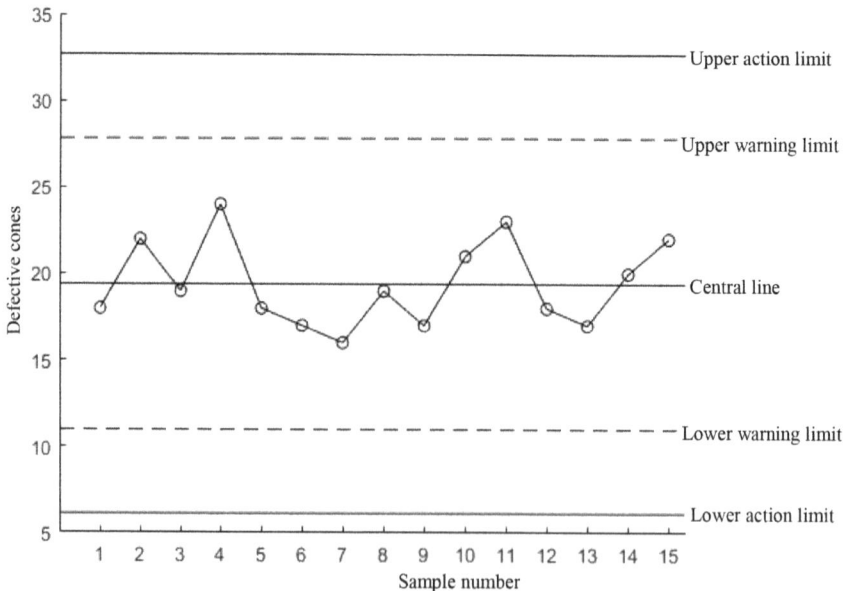

FIGURE 11.23 Control chart for defective cones.

Example 11.13:

The numbers of heavy laps in samples of size 80 were measured from a blow room scutcher on each day over a period of 12 days. The results were as follows.

Day	1	2	3	4	5	6	7	8	9	10	11	12
Heavy Laps	3	2	4	6	5	4	2	3	7	4	3	6

Construct a control chart for proportion of heavy laps.

SOLUTION

Total number of heavy laps = 49

$$\text{Sample size } n = 80$$

$$\text{Total number of laps inspected} = 80 \times 12 = 960$$

$$\text{Mean proportion of heavy laps} = \bar{p} = \frac{49}{960} = 0.051$$

$$\text{Standard deviation of proportion of heavy laps} = \sqrt{\frac{\bar{p}(1-\bar{p})}{n}}$$

$$= \sqrt{\frac{0.051 \times (1-0.051)}{80}} = 0.0246$$

$$\text{The warning limits of the proportion of heavy laps} = \bar{p} \pm 1.96\sqrt{\frac{\bar{p}(1-\bar{p})}{n}}$$

$$= 0.051 \pm 1.96 \times 0.0246$$

$$= (0.0028, 0.1)$$

$$\text{The action limits of the proportion of heavy laps} = \bar{p} \pm 3.09\sqrt{\frac{\bar{p}(1-\bar{p})}{n}}$$

$$= 0.051 \pm 3.09 \times 0.0246$$

$$= (-0.025, 0.127)$$

As the lower action limit cannot be negative, action limits are 0 and 0.127.
Figure 11.24 depicts the control chart for proportion of heavy laps, from which it can be concluded that the process is in control.

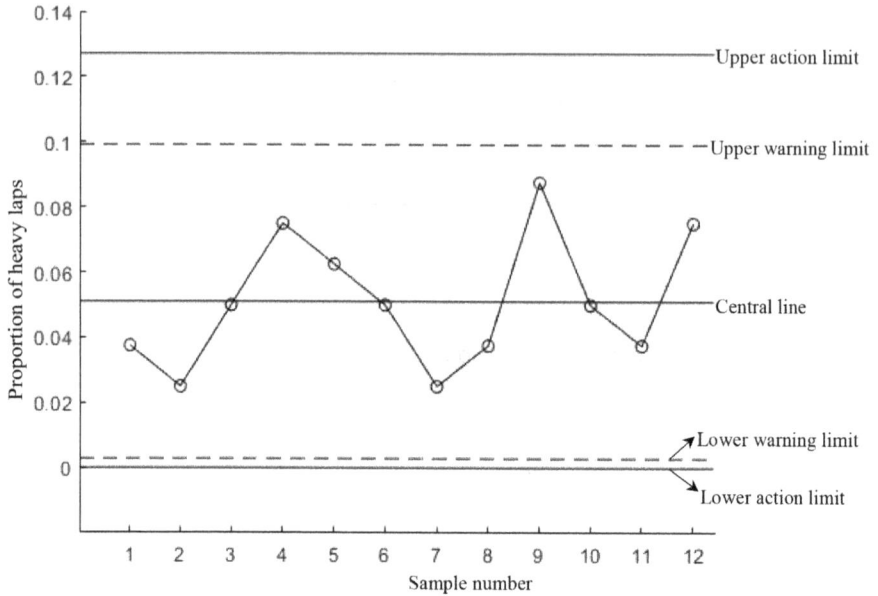

FIGURE 11.24   Control chart for proportion of heavy laps.

11.3.4.3 Control Chart for Defects

Sometimes, in a manufacturing process, the numbers of defects per unit of product, for example, yarn imperfections per km, number of thick lines per yard of fabric, etc., are quite useful as the objects of control. Such defects per unit of product that occur at random follow a Poisson distribution. Let a random variable X be the number of defects per unit. If λ is the mean number of defects per unit under the condition that the process is in control, then X has a Poisson distribution with mean λ and the standard deviation $\sqrt{\lambda}$. If λ is large, then the Poisson distribution can be approximated to normal distribution. Therefore, the action and warning limits for a control chart for defects are given by

$$\text{Action limit} = \lambda \pm 3.09\sqrt{\lambda} \qquad (11.49)$$

$$\text{Warning limit} = \lambda \pm 1.96\sqrt{\lambda} \qquad (11.50)$$

If λ is not known, then it is estimated as the observed average number of defects per unit from a preliminary experiment based on m samples. If $\bar{\lambda}$ is an estimate of λ, then the action and warning limits for defects in Equations (11.49) and (11.50) may be written as

$$\text{Action limit} = \bar{\lambda} \pm 3.09\sqrt{\bar{\lambda}} \qquad (11.51)$$

$$\text{Warning limit} = \bar{\lambda} \pm 1.96\sqrt{\bar{\lambda}} \qquad (11.52)$$

Example 11.14:

The number of neps per Km length of yarn was measured. The results of 20 tests were

55, 42, 36, 45, 41, 43, 43, 43, 29, 39, 35, 41, 32, 40, 46, 28, 54, 45, 47 and 30.

Construct a control chart for number of neps/Km length of yarn.

SOLUTION

The number of neps/Km length of yarn can be assumed to follow Poisson distribution.

$$\text{The mean number of neps/Km length of yarn} = \bar{\lambda} = 40.7$$

$$\text{Standard deviation of neps/Km length of yarn} = \sqrt{\bar{\lambda}} = \sqrt{40.7} = 6.38$$

$$\text{The warning limits of neps/Km length of yarn} = \bar{\lambda} \pm 1.96\sqrt{\bar{\lambda}}$$

$$= 40.7 \pm 1.96 \times 6.38 = \left(28.2, 53.2\right)$$

$$\text{The action limits of neps/Km length of yarn} = \bar{\lambda} \pm 3.09\sqrt{\bar{\lambda}}$$

$$= 40.7 \pm 3.09 \times 6.38 = \left(20.99, 60.41\right)$$

Figure 11.25 shows the control chart for neps/Km length of yarn. From the control chart plot, it is evident that the process is in control.

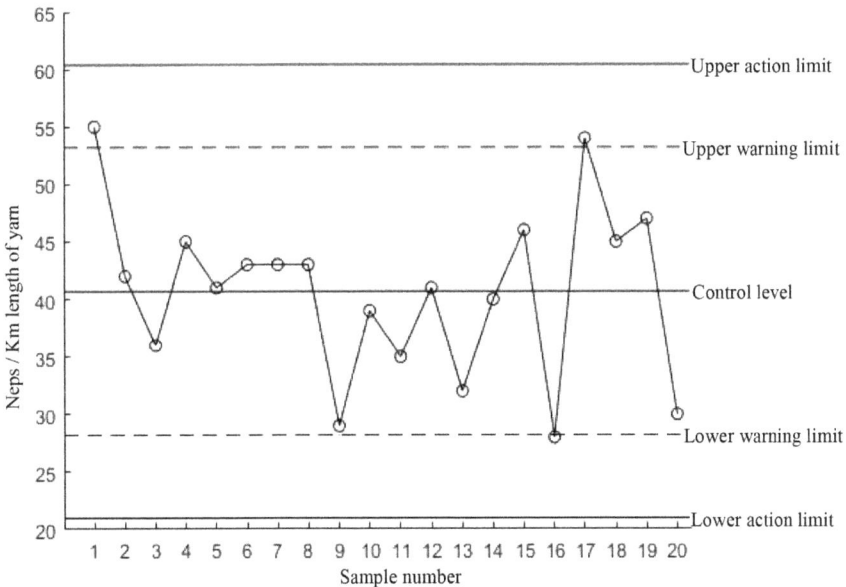

FIGURE 11.25 Control chart for neps/Km length of yarn.

Example 11.15:

In fabric grading technique, it is a common practice to assign point values to each defect. The point values are varied with the intensity of the defect and are added to give a total point value for each piece being graded. A fabric manufacturer has established from market research that the maximum number of points/piece acceptable for first quality fabric is 30, beyond which the fabric is treated as second quality. What is the average number of points/piece of first quality fabric in order that fabric manufacturer can be assured that practically all the fabric will be the first quality?

SOLUTION

The upper action limit of points/piece = 30.

Let $\bar{\lambda}$ be the average number of points/piece.

Assuming the number of points/piece to follow Poisson distribution, the upper action limit is $\bar{\lambda} + 3.09\sqrt{\bar{\lambda}}$.

Hence, $\bar{\lambda} + 3.09\sqrt{\bar{\lambda}} = 30$

or, $\left(\bar{\lambda} - 30\right)^2 = 3.09^2\bar{\lambda}$

or, $\bar{\lambda}^2 - 60\bar{\lambda} + 900 = 9.55\bar{\lambda}$

or, $\bar{\lambda}^2 - 69.55\bar{\lambda} + 900 = 0$

or, $\bar{\lambda} = \frac{69.55 \pm \sqrt{69.55^2 - 4 \times 900}}{2}$

or, $\bar{\lambda} = 17.19$ or 52.36

Here the feasible solution of $\bar{\lambda}$ is 17.19 since the upper action limit is 30. So, the average points/piece that should warrant approximately all first quality fabric is 17.19.

11.4   MATLAB® CODING

11.4.1   MATLAB® CODING OF EXAMPLE 11.1

```
clc
clear
close all
c=1;
n=100;
p=0.01;
Pa=binocdf(c,n,p)
```

11.4.2   MATLAB® CODING OF EXAMPLE 11.3

```
clc
clear
close all
format short g
p1=0.02;
p2=0.06;
alpha=0.05;
beta=0.1;
r=p2/p1;
```

```
for i=1:100
    x1 = chi2inv(alpha,i);
    x2 = chi2inv(1-beta,i);
    R(i)=x2/x1;
    y(i)=abs(R(i)-r);
end
[d,j]=min(y);
c=(j/2)-1
n1=chi2inv(alpha,j)/(2*p1)
n2=chi2inv(1-beta,j)/(2*p2)
n=round((n1+n2)/2)
```

11.4.3 MATLAB® CODING OF EXAMPLE 11.6

```
clc
clear
close all
c=1;
n=80;
N=10000;
p=0.01;
pra=binocdf(c,n,p)
AOQ=p*pra*(N-n)/N
ATI=n+(1-pra)*(N-n)
```

11.4.4 MATLAB® CODING OF EXAMPLE 11.7

```
clc
clear
close all
p1=0.01;
p2=0.04;
alpha=0.05;
beta=0.02;
L=16;
sigma=1;
z_p1=norminv(1-p1,0,1)
z_p2=norminv(1-p2,0,1)
z_alpha=norminv(1-alpha,0,1)
z_beta=norminv(1-beta,0,1)
n=((z_alpha+z_beta)/(z_p1-z_p2))^2
xbar_L=L+sigma*(z_p1*z_beta+z_p2*z_alpha)/(z_alpha+z_beta)
```

11.4.5 MATLAB® CODING OF EXAMPLE 11.8

```
clc
clear
close all
p1=0.01;
p2=0.05;
```

```
alpha=0.1;
beta=0.05;
U=45;
sigma=2.4;
z_p1=norminv(1-p1,0,1);
z_p2=norminv(1-p2,0,1);
z_alpha=norminv(1-alpha,0,1);
z_beta=norminv(1-beta,0,1);
n=((z_alpha+z_beta)/(z_p1-z_p2))^2
xbar_U=U-sigma*(z_p1*z_beta+z_p2*z_alpha)/(z_alpha+z_beta)
```

11.4.6   MATLAB® Coding of Example 11.9

```
clc
clear
close all
alpha=0.075;
beta=0.05;
T=1;
sigma=0.8;
z_alphaby2=norminv(1-alpha/2,0,1)
z_betaby2=norminv(1-beta/2,0,1)
n=sigma^2*(z_alphaby2+z_betaby2)^2/T^2
t=T*z_alphaby2/(z_alphaby2+z_betaby2)
```

11.4.7   MATLAB® Coding of Example 11.11

```
clc
clear
close all
x=[39.85  41.46  41.50  39.84  39.50  39.42  39.77  40.48
40.92  38.93
   40.49  39.41  40.48  39.54  39.91  41.08  39.84  40.32
41.35  40.50
   39.41  40.32  40.21  39.66  40.17  40.97  39.92  40.17
39.72  40.62
   40.05  40.53  39.29  39.69  40.09  41.32  39.64  40.25
39.68  39.17];
st=controlchart(x','charttype',{'xbar' 'r'})
```

11.4.8   MATLAB® Coding of Example 11.12

```
clc
clear
close all
x=[18    22  19  24  18  17  16  19  17  21  23  18  17  20
22];
st=controlchart(x,'unit',400,'charttype','np')
```

11.4.9   MATLAB® CODING OF EXAMPLE 11.13

```
clc
clear
close all
format short g
x= [3    2    4    6    5    4    2    3    7    4    3    6];
st=controlchart(x,'unit',80,'charttype','p')
```

11.4.10   MATLAB® CODING OF EXAMPLE 11.14

```
clc
clear
close all
format short g
x= [55 42 36 45 41 43 43 43 29 39 35 41 32 40 46 28 54 45 47
30];
st=controlchart(x,'charttype','c')
```

Exercises

11.1 In a sampling plan, sample size = 250, acceptance number = 5, and lot proportion defective = 0.02. Determine the probability of accepting the lot. If
the acceptance number is changed from 5 to 3, what will be the probability
of accepting the lot?

11.2 A knitwear producer supplies batches of garments to a chain store. The
agreement made between the knitwear producer and chain store on the val-
ues of AQL (p_1), LTPD (p_2), producer's risk (α) and consumer's risk (β) are
as follows:

$p_1 = 0.015$, $p_2 = 0.045$, $\alpha = 0.01$, $\beta = 0.15$.

Determine the sampling plan.

11.3 A ring traveller manufacturer supplies batches of rings to a ring frame manufacturer. They made an agreement on the values of AQL (p_1), LTPD (p_2),
producer's risk (α) and consumer's risk (β) as follows:

$p_1 = 0.025$, $p_2 = 0.1$, $\alpha = 0.01$, $\beta = 0.1$.

Determine the sampling plan.

11.4 In a rectifying inspection plan, lot size = 12,000, sample size = 150 and
acceptance number = 2. If the lot proportion defective = 0.02, what will be
the average outgoing quality (AOQ) and average total inspection (ATI)?

11.5 A fabric manufacturer supplies batches of fabrics to a garment manufacturer. The fabric should meet a specification that its warp strength should not be less than 60 N/cm. Past experience shows that the standard deviation of the fabric warp strength is 2.5 N/cm. The values of AQL (p_1), LTPD (p_2), producer's risk (α) and consumer's risk (β) are as follows:

$p_1 = 0.005$, $p_2 = 0.05$, $\alpha = 0.05$, $\beta = 0.05$.

Device the sampling scheme for accepting the batch.

11.6 A cotton fibre supplier supplies batches of cotton bales to a yarn manufacturer. The specification of short fibre content should not be more than 10%. Past experience shows that the standard deviation of short fibre content is 1.2%. The values of AQL (p_1), LTPD (p_2), producer's risk (α) and consumer's risk (β) are as follows:

$p_1 = 0.012$, $p_2 = 0.06$, $\alpha = 0.075$, $\beta = 0.1$.

Device the sampling scheme for accepting the batch.

11.7 A yarn manufacturer supplies yarn of nominal linear density equal to 36 tex to a knitwear manufacturer. The delivery of the yarn is acceptable if the mean linear density lies within a range of 36 ± 1.2 tex. Device a sampling scheme if the producer's risk and consumer's risk are 0.1 and 0.05, respectively. Assume the standard deviation of yarn linear density is 1 tex.

11.8 A fabric manufacturer supplies batches of fabrics to a garment manufacturer. The delivery of the fabric is acceptable if the mean GSM lies within a range of 110 ± 2.5 g/m². Device a sampling scheme if the producer's risk and consumer's risk are 0.06 and 0.07, respectively. Assume the standard deviation of fabric GSM is 3 g/m².

11.9 Five cops were randomly taken on each day over a period of 12 days from a ring frame for measuring the breaking length. The results of yarn breaking lengths were as follows.

Day	1	2	3	4	5	6	7	8	9	10	11	12
Breaking	19.8	17.6	21.0	19.5	20.1	19.4	19.7	20.8	17.9	18.3	18.2	20.6
Length	18.4	18.9	20.7	21.0	18.9	20.3	20.8	19.7	18.2	20.5	19.8	20.3
(Km)	20.4	18.3	20.2	19.9	21.1	20.6	19.3	21.1	17.5	20.8	18.9	19.8
	20.1	17.8	20.9	20.4	19.7	21.2	20.6	19.2	18.4	19.7	20.3	20.9
	20.4	18.0	20.5	20.8	20.7	19.8	20.2	20.4	18.3	21.0	20.7	20.4

Use these data to construct control charts for averages and ranges of samples of size 5 and verify whether the process is under control or not.

11.10 Four samples of nonwoven fabrics were collected from a nonwoven machine daily over a period of 10 days for the measurement of GSM with the following results.

Day	1	2	3	4	5	6	7	8	9	10
GSM	50.24	49.51	50.02	49.66	50.42	50.29	49.96	49.66	49.87	49.89
(g/m²)	49.68	50.27	49.93	50.09	50.23	50.01	49.88	49.82	49.53	50.17
	49.87	49.92	50.94	49.87	49.98	50.46	50.24	49.73	50.17	49.64
	49.49	49.81	50.22	50.23	50.37	50.50	50.18	50.04	49.68	50.38

Use these data to construct control charts for averages and ranges of samples of size 4 and verify whether the process is under control or not.

11.11 The numbers of defective garment buttons were observed in the 10 different samples of size 100 randomly selected from a production line with the following results.

Sample Number	1	2	3	4	5	6	7	8	9	10
Defective Cones	4	2	1	3	0	4	1	0	2	2

Construct a control chart for the defective buttons. Draw conclusion about the state of the process.

11.12 The numbers of defective silk cocoons were observed in 12 different samples of size 150 with the following results.

Sample Number	1	2	3	4	5	6	7	8	9	10	11	12
Defective Cocoons	6	11	8	13	10	8	7	6	8	12	14	9

Construct a control chart for the proportion of defective cocoons. Write a conclusion about the state of the process.

11.13 The following data represents the number of warp breaks per 100,000 picks observed on a weaving machine for 10 consecutive days.

Day	1	2	3	4	5	6	7	8	9	10	11	12
Warp Breaks	3	2	5	1	8	4	6	2	1	7	3	5

Construct a control chart for the number of warp breaks per 100,000 picks. Draw a conclusion about the state of the process.

11.14 The number of thin places (−50%) per Km length of yarn was measured. The results of 15 tests were: 5, 2, 1, 4, 0, 6, 3, 2, 2, 1, 5, 1, 0, 4, 3.

Construct a control chart for number of thin places/Km length of yarn. Write a conclusion about the state of the process.

12 Stochastic Modelling

12.1 INTRODUCTION

In the real world, every system is associated with noise and, hence, it is only natural to model it using the help of statistical techniques. In many situations, in particular when the system is time dependent, a sequence of random variables which are functions of time may be necessary to model it. This sequence of random variables is termed as the stochastic process. The stochastic modelling techniques find applications in many areas including the textile manufacturing process. For example, one can apply the Markovian process to model the mechanism of a carding machine in the textile industry. Dynamical processes in the real world are in general prone to external disturbances and noises and, hence, a stochastic model is more appropriate than its deterministic counterpart. We start this chapter with some basic concepts of Markov chains.

12.2 MARKOV CHAINS

In many situations, when we need to predict the future state of a system, the information about the present situation is more relevant than that about the past. For example, in a railway ticket counter, the probability that at any given time point the queue will be empty depends on the number of customers present in the queue in that moment. To be precise, in such cases the present state of the system carries relevant information which is used to predict the future state and the information about the past is not necessary at all. This property is called Markov property. In the following, we give the formal definition of the Markov chain (Prabhu, 1966).

Definition 12.1: A sequence of random variables $X_1, X_2, \cdots, X_n, \cdots$ is called a Markov chain if, for a given finite set of random variables $X_{n_1}, X_{n_2}, \cdots, X_{n_r}$ with $n_1 < n_2 < \cdots < n_r < n$, we have

$$P\left[X_n \mid X_{n_1}, X_{n_2}, \cdots, X_{n_r}\right] = P\left[X_n \mid X_{n_r}\right]$$

Let,

$$P_{ij}(m) = P\{X_{m+1} = j \mid X_m = i\}$$

Thus, $P_{ij}(m)$ is the probability that the stochastic process will be in the state j at $(m+1)$th step starting from the state i at mth step. If $P_{ij}(m)$ is independent of m for all i, j, we say that the chain is homogeneous. For a homogeneous Markov chain, we define

$$P_{ij}^{(n)} = P\{X_{m+n} = j \mid X_m = i\}; \; n = 1, 2, \cdots \tag{12.1}$$

From this definition one can write

$$P_{ij}^{(0)} = \delta_{ij}$$

where δ_{ij} is Kronecker's delta. The probabilities in Equation (12.1) can be expressed as

$$P_{ij}^{(n)} = \sum_{i_1, i_2, \cdots, i_{n-1}} P(X_{m+1} = i_1 \mid X_m = i) P(X_{m+2} = i_2 \mid X_{m+1} = i_1) \cdots$$

$$P(X_{m+n} = j \mid X_{m+n-1} = i_{n-1}) \tag{12.2}$$

$$= \sum_{i_1, i_2, \cdots, i_{n-1}} P_{ii_1} P_{i_1 i_2} \cdots P_{i_{n-1} j}$$

The matrix $P = (P_{ij})$ is called the transition matrix. From the relation in Equation (12.2) we see that $P^n = (P_{ij}^{(n)})$. Again, from the relation $P^{m+n} = P^m P^n$, we obtain

$$P_{ij}^{(m+n)} = \sum_k P_{ik}^{(m)} P_{kj}^{(n)} \text{ for } m, n \geq 0.$$

For any two states i and j, the first passage time to the state j starting from the state i can be defined by the random variable

$$T_{ij} = min\{n \mid X_n = j \mid X_0 = i\}$$

Let,

$$f_{ij}^{(n)} = P(T_{ij} = n)$$

$$= P(X_n = j, X_m \neq j \ (m = 1, 2, 3, \cdots, n-1) \mid X_0 = i)$$

$$= \sum_{i_1, i_2, \cdots, i_{n-1} \neq j} P_{ii_1} P_{i_1 i_2} \cdots P_{i_{n-1} j}$$

We define

$$f_{ij} = P(T_{ij} < \infty) = \sum_{n=1}^{\infty} f_{ij}^{(n)}$$

From the definition it is clear that, f_{ij} is the probability that the process will reach the state j starting from the state i in a finite time. Thus, $1 - f_{ij}$ is the probability that the process will never reach the state j starting from the state i. In particular, the random variable T_{ii} represents the return time to the state i and f_{ii} is the probability that the return to the state i takes place in a finite number of steps. Thus, a state i is classified as persistent or transient according as

$$f_{ii} = 1 \text{ or } < 1.$$

On the other hand, a state i is called absorbing if

$$P_{ij} = 1 \text{ if } i = j$$
$$= 0 \text{ if } i \neq j$$

In the domain of textile technology, the working process of a carding machine can be modelled by a Markov chain (Monfort, 1962, Singh and Swani, 1973). In this example, we will discuss how the working of a carding machine can be represented by the Markov chain. The useful notations in this context are as follows:

v_{ij} = number of times the system is in state j starting from the state i.
P_{ij} = transition probability from the state i to the state j.
P = transition matrix having entries P_{ij}.
I = identity matrix.
Q = square sub-matrix of P having the transition probabilities between the transient states as its entries.
N = fundamental matrix obtained from the matrix P.
n_{ij} = entries of N which are the means of v_{ij}.
E_i = states corresponding to the locations of the state.
R_i = average number of collections at a carding point i.

A state in a Markov chain is called absorbing if the system cannot leave the state once it reaches the said state. A Markov chain is said to be an absorbing Markov chain if it contains at least one absorbing state, and the absorbing state can be reached from any other state, not necessarily in one stage. In case of an absorbing Markov chain, the states of the chain can be rearranged to put it in the canonical form where all the absorbing states precede all the transient states. Suppose an absorbing Markov chain has r absorbing states and s transient states. In that case, the matrix P can be divided into four parts as shown below.

$$P = \begin{bmatrix} I & O \\ R & Q \end{bmatrix} \tag{12.3}$$

The matrix I is an $r \times r$ identity matrix, O is $r \times s$ zero matrix, R is a nonzero $s \times r$ submatrix, and Q is an $s \times s$ square nonzero submatrix which is the most important submatrix for this example, because the fundamental matrix N is obtained from it using the following relation:

$$N = (I - Q)^{-1} \tag{12.4}$$

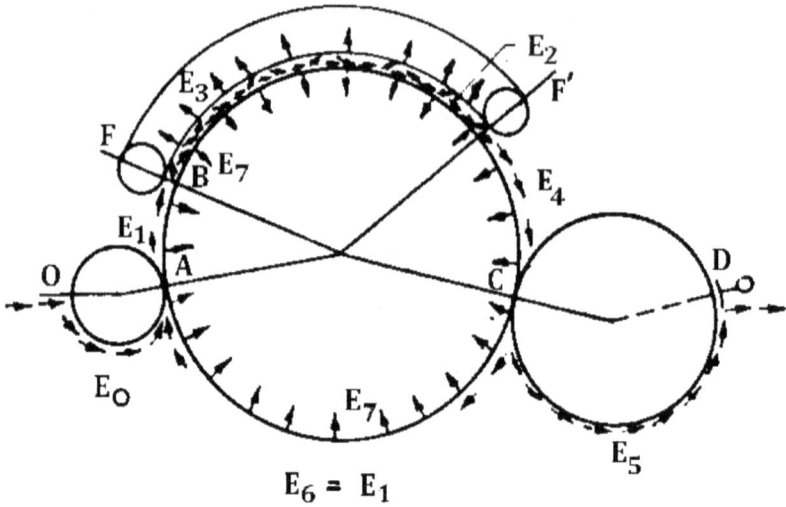

FIGURE 12.1 Schematic diagram of a conventional revolving flat card. (From Singh et al. *The Journal of the Textile Institute*, 1973).

The vector-row of N has the components which is the average number of times the system visits the non-absorbing states, starting from the state corresponding to the row vector under consideration. The sum of the components of each row vector of N represents the mean number of stages before being absorbed. Figure 12.1 represents different transitions of fibres from one state to another for a conventional revolving flat card. The card is divided into seven states through which the fibres may transit.

The different states of the card are as follows:

E_0 = arc OA on the taker-in
E_1 = arc AB on the cylinder
E_2 = arc FF' on the flats' active region
E_3 = arc FF' on the flats' inactive region
E_4 = arc BC on the cylinder
E_5 = arc CD on the doffer
E_6 = arc CA on the cylinder
E_7 = inactive zone in the cylinder wires around the cylinder.

In this case, the state E_6 may be taken as the state E_1 because all the fibres in state E_6 go to state E_1. The transition probabilities between different states are furnished in Figure 12.2. The fibres transit from state E_0 to E_1 with a probability of 1. From the state E_1, the fibres may transit to any of the states E_2, E_3, E_4, or E_7 with respective probabilities P_{12}, P_{13}, P_{14} and P_{17}, such that

$$P_{12} + P_{13} + P_{14} + P_{17} = 1 \qquad (12.5)$$

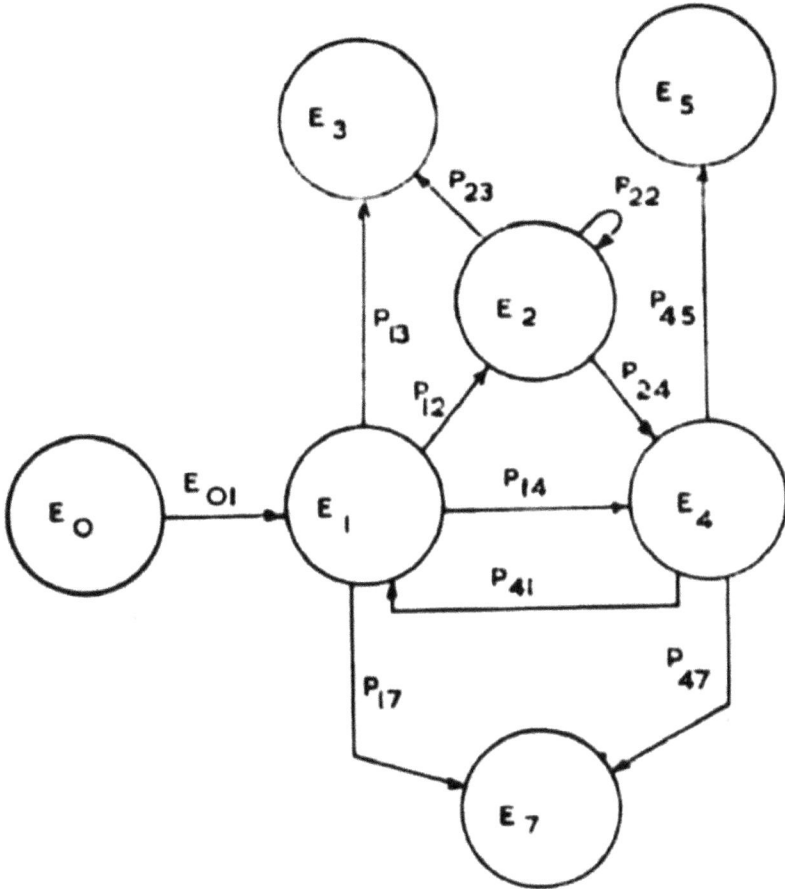

FIGURE 12.2 Transition probabilities between various states of a card. (From Singh et al. *The Journal of the Textile Institute*, 1973).

Similarly, from the state E_2, the fibres may transit either to the state E_4 with a probability P_{24}, or to the state E_3 with a probability P_{23}, or temporarily they may stay in the state E_2 with a probability P_{22}, before finally transiting to the state E_4, such that

$$P_{22} + P_{23} + P_{24} = 1 \tag{12.6}$$

Again, from the state E_4, the fibres may transit to any of the states E_5, E_1, or E_7 with respective probabilities P_{45}, P_{41}, and P_{47}, such that

$$P_{41} + P_{45} + P_{47} = 1 \tag{12.7}$$

In the process of carding, E_1, E_2 and E_4 are the transient states and E_3, E_5 and E_7 are the absorbing states. There are two possible outcomes while a fibre is transiting from one transient state to the others, viz., it may either transit to one of the adjoining states or it may stay in the same state. The probability P_{ij} that a fibre is transiting from the state i to the state j is governed by the conditions prevailing in the preceding state i alone. The transition matrix P is given as

$$
\begin{array}{c c}
 & \begin{array}{cccccc} 1 & 2 & 3 & 4 & 5 & 7 \end{array} \\
\begin{array}{c} 1 \\ 2 \\ 3 \\ 4 \\ 5 \\ 7 \end{array} &
\left[\begin{array}{cccccc}
0 & P_{12} & P_{13} & P_{14} & 0 & P_{17} \\
0 & P_{22} & P_{23} & P_{24} & 0 & 0 \\
0 & 0 & 1 & 0 & 0 & 0 \\
P_{41} & 0 & 0 & 0 & P_{45} & P_{47} \\
0 & 0 & 0 & 0 & 1 & 0 \\
0 & 0 & 0 & 0 & 0 & 1
\end{array} \right]
\end{array}
$$

The canonical form of the above transition matrix can be written as

$$
\begin{array}{c c}
 & \begin{array}{cccccc} 7 & 5 & 3 & 1 & 2 & 4 \end{array} \\
\begin{array}{c} 7 \\ 5 \\ 3 \\ 1 \\ 2 \\ 4 \end{array} &
\left[\begin{array}{cccccc}
1 & 0 & 0 & 0 & 0 & 0 \\
0 & 1 & 0 & 0 & 0 & 0 \\
0 & 0 & 1 & 0 & 0 & 0 \\
P_{17} & 0 & P_{13} & 0 & P_{12} & P_{14} \\
0 & 0 & P_{23} & 0 & P_{22} & P_{24} \\
P_{47} & P_{45} & 0 & P_{41} & 0 & 0
\end{array} \right]
\end{array}
$$

We assume that the probabilities P_{17}, P_{13}, P_{23} and P_{47} are small and we neglect it. Thus, from Equations (12.5) to (12.7), we get

$$P_{14} = 1 - P_{12}$$

$$P_{22} = 1 - P_{24}$$

and

$$P_{41} = 1 - P_{45}$$

Thus, in Equation (12.3) we have

$$I = \begin{bmatrix} 1 & 0 & 0 \\ 0 & 1 & 0 \\ 0 & 0 & 1 \end{bmatrix}$$

$$O = \begin{bmatrix} 0 & 0 & 0 \\ 0 & 0 & 0 \\ 0 & 0 & 0 \end{bmatrix}$$

$$R = \begin{bmatrix} 0 & 0 & 0 \\ 0 & 0 & 0 \\ 0 & P_{45} & 0 \end{bmatrix}$$

and

$$Q = \begin{bmatrix} 0 & P_{12} & 1-P_{12} \\ 0 & 1-P_{24} & P_{24} \\ 1-P_{45} & 0 & 0 \end{bmatrix}$$

Therefore, from Equation (12.4) we get

$$N = (I-Q)^{-1} = \begin{bmatrix} \dfrac{1}{P_{45}} & \dfrac{P_{12}}{P_{45}P_{24}} & \dfrac{1}{P_{45}} \\[2ex] \dfrac{1-P_{45}}{P_{45}} & \dfrac{P_{12}+P_{45}-P_{12}P_{45}}{P_{45}P_{24}} & \dfrac{1}{P_{45}} \\[2ex] \dfrac{1-P_{45}}{P_{45}} & \dfrac{P_{12}(1-P_{45})}{P_{45}P_{24}} & \dfrac{1}{P_{45}} \end{bmatrix} \qquad (12.8)$$

The mean number of times the fibre passes through each transient state is given by the components of the first-row vector of N. Therefore, the mean number of times a fibre passes through the state E_1 before it is absorbed is expressed as

$$v_{11} = \frac{1}{P_{45}}$$

Hence, on average, a fibre recycles $1/P_{45}$ revolutions in the cylinder before it is transferred to the doffer. The transition probability P_{45} is nothing but the transfer coefficient between cylinder and doffer. Therefore, P_{45} can be expressed as a ratio of the fibre load transferred to the doffer during one revolution of cylinder to the fibre load on the operational layer on the cylinder. In a steady state condition, only a fraction of the fibre load from the operational layer of the cylinder is transferred to the doffer and the same amount is replenished from the licker-in. In a conventional card, the transfer coefficient is around 0.05; hence' on average a fibre recycles $1/0.05 = 20$ revolutions with the cylinder before it is transferred to the doffer. Whereas, in a modern card, the transfer coefficient is around 0.25; hence for a modern card a fibre approximately recycles $1/0.25 = 4$ revolutions with the cylinder before it is transferred to the doffer. Also, the mean number of times a fibre passes through the state E_2 before it is absorbed is expressed as

$$v_{12} = \frac{P_{12}}{P_{45}P_{24}}$$

Thus, the mean time spent by a fibre for the carding action in the cylinder-flat region before it is absorbed is given by $\frac{P_{12}}{P_{45}P_{24}}$ cylinder revolutions.

Similarly,

$$v_{14} = \frac{1}{P_{45}}$$

In terms of cylinder revolutions, the mean carding time T_C is the sum total of average recycle time in the cylinder and average carding time in the cylinder-flat region. Hence,

$$T_C = \frac{1}{P_{45}} + \frac{P_{12}}{P_{45}P_{24}}.$$

For example, if we assume that $P_{45} = 0.25$, $P_{12} = 0.2$, and $P_{24} = 0.8$, we have $T_C = \frac{1}{0.25} + \frac{0.2}{0.25 \times 0.8} = 5$ revolutions of cylinder.

If cylinder rotates at 500 rpm, T_C becomes 0.6 sec.

12.3   STOCHASTIC MODELLING

The random dynamical systems in nature are frequently modelled by the Ito stochastic differential equations. Among different modelling procedures, here we discuss one which is developed by the discrete stochastic modelling based on the changes in the system components over a small interval of time (Allen et al. 2008). This modelling technique finds its application in general science, engineering and in many others. In this section we focus our attention to the application of this modelling techniques in the field of textile engineering, particularly in the cotton fibre breakage problem which is common in the manufacturing of yarn. The stochastic model approach is capable of providing clear understanding of fibre breakage phenomena.

12.3.1   FIRST METHOD OF CONSTRUCTING THE STOCHASTIC DIFFERENTIAL EQUATION

The set of all possible values a random variable can take on is called the state space. Consider a dynamic process in which d number of system variables say

$X_1, X_2, \cdots, X_d$ are involved and we denote $X = [X_1, X_2, \cdots, X_d]^T$. Suppose there are m number of possible changes denoted by the random change vector say, $e_1, e_2, \cdots, e_m$ which can occur in the next small-time interval Δt. Furthermore, we assume that the probability of the jth change is $p_j\Delta t = p_j(t, X(t))\Delta t$, for $j = 1, 2, \cdots, m$ and the jth change alters the ith component of X by the amount α_{ji}. If $\Delta X(t) = (\Delta X_1(t), \Delta X_2(t), \cdots, \Delta X_d(t))$ represents the changes during the time interval Δt, then one can show that

$$E(\Delta X(t)) = \sum_{j=1}^{m} e_j p_j(t, X(t))\Delta t = f(t, X(t))\Delta t \ (\text{say})$$

Let the vector $e = (e_1, e_2, \cdots, e_m)$ where each e_j is a vector random variable representing the changes during the small-time interval Δt and the distribution of this vector random variable is given by

$$e_j = (\alpha_{j1}, \alpha_{j2}, \cdots, \alpha_{jd})^T \text{ with probability } p_j\Delta t$$
$$= (0,0,\cdots,0)^T \text{ with probability } 1 - p_j\Delta t$$

Note that the mean of the change in the ith component of the system variable vector in small time interval Δt is $\alpha_{ji}p_j\Delta t$ and the corresponding variance is $\alpha_{ji}^2 p_j\Delta t$. We neglected the terms containing $(\Delta t)^n; n \geq 2$

In the first procedure, $X(t + \Delta t)$ given $X(t)$ can be approximated as

$$X(t + \Delta t) = X(t) + \sum_{j=1}^{m} e_j \tag{12.9}$$

$$X(0) = X_0$$

Each component of the vector random variable $X(t)$ follows

$$X_i(t + \Delta t) = X_i(t) + \sum_{j=1}^{m} \alpha_{ji} \text{ for } i = 1,2,\cdots, d.$$

It can be shown that (Allen et. al. 2008), the stochastic process generated using Equation (12.9) follows the stochastic differential equation given by

$$dX(t) = f(t, X(t))dt + B(t, X(t))dW(t) \tag{12.10}$$

$$X(0) = X_0$$

In this case the function f is a vector valued function having components $f_1, f_2, \cdots, f_d$ and

$$f_i(t) = \sum_{j=1}^{m} \alpha_{ji} p_j(t, X(t)) \qquad (12.11)$$

and

$$B = V^{\frac{1}{2}}$$

where V is the variance covariance matrix and it is obtained as

$$V = E\left(\Delta X (\Delta X)^T\right) = \sum_{j=1}^{m} p_j e_j (e_j)^T$$

The vector valued random variable $W(t)$ consists of d independent Weiner processes (see Remark 12.1). Here, the vector f is called the drift coefficient and the matrix B is called the diffusion matrix which is the square root of the covariance matrix V. One can see that the drift vector is the expected change during time Δt, divided by Δt.

Remark 12.1: The Weiner process $W(t)$ is a stochastic process that is a set of random variables having the following properties:

i. The difference $W(t) - W(s)$ for $0 < s < t$ is independent on the random variable $W(\rho)$ for any $\rho \le s$.
ii. The random variable $W(t + t_1) - W(t_1)$ is independent of t_1 and it follows the same distribution as $W(t)$.
iii. The random variable $W(t)$ follows the normal distribution with mean 0 and variance t.

12.3.2 SECOND METHOD OF CONSTRUCTION OF STOCHASTIC DIFFERENTIAL EQUATION

In the second method, the m random changes are approximated by m independent normal random variables $\gamma_j \sim N(0, 1)(j = 1, 2, 3, \cdots, m)$. This normal approximation can be justified by the normal approximation of Poisson distribution. Thus, the equivalent form of the Equation (12.9) is (Allen, 2008):

$$X(t + \Delta t) = X(t) + f(t, X(t))\Delta t + \sum_{j=1}^{m} \alpha_{ji} p_j^{\frac{1}{2}} (\Delta t)^{\frac{1}{2}} \gamma_j \qquad (12.12)$$

where each component of f i.e. f_i is defined in Equation (12.11). Equation (12.12) is actually the Euler-Maruyama approximation and it converges to the following stochastic differential equation as $\Delta t \to 0$.

$$dX(t) = f(t, X(t))dt + G(t, X(t))dW(t) \qquad (12.13)$$

$$X(0) = X_0$$

where $W(t)$ is a vector random variable having m independent Weiner processes as its components. The matrix G is a $d \times m$ matrix having entries

$$g_{ij} = \alpha_{ji} p_j^{\frac{1}{2}} \text{ for } i = 1,2,\cdots,d \text{ and } j = 1,2,\cdots,m.$$

It can be shown that

$$V = GG^T$$

and thus, the entries of the matrix V can be written as

$$v_{ij} = \sum_{l=1}^{m} g_{il} g_{jl} = \sum_{l=1}^{m} p_l \alpha_{li} \alpha_{lj}$$

The second method is more convenient to use as the entries of the matrix G can be obtained easily with the help of the expression $g_{ij} = \alpha_{ji} p_j^{\frac{1}{2}}$, although the sample path solutions of both the systems are same. On the contrary, in case of the first method, it is not easy to get the square root of the matrix V to obtain matrix B. One possible way of obtaining the square root, is to diagonalize the matrix V. Assume that $\lambda_1, \lambda_2, \cdots, \lambda_d$ are d different eigen values of V and $v_1, v_2, \cdots, v_d$ are the corresponding d eigen vectors, then

$$P^{-1}VP = diag(\lambda_i)$$

where $diag(\lambda_i)$ is the diagonal matrix with diagonal entries λ_i $(i = 1,2,...,d)$ and the matrix P is obtained by taking its ith column to be v_i. Thus,

$$V = Pdiag(\lambda_i)P^{-1}$$

and hence

$$B = V^{\frac{1}{2}} = Pdiag\left(\lambda_i^{\frac{1}{2}}\right)P^{-1}$$

This method is convenient to use as long as the dimension of the matrix is not too high, otherwise it is difficult to get all the eigen vectors. It should be noted that the vector valued function $f(t, X(t))$ generates the differential equation

$$\frac{dX}{dt} = f(t, X(t))$$

in the deterministic set up.

12.3.3   STOCHASTIC MODELLING OF COTTON FIBRE BREAKAGE

In cotton processing, fibre breakage occurs in ginning, blow room and carding. Higher breakage of the fibres in cotton processing generally leads to lower quality yarn. The development of a stochastic model for fibre-length distributions provides a better insight about the fibre breakage phenomenon and the origin of different fibre-length distributions. By comparing calculations of the stochastic model with fibre-length data, fibre breakage parameters can be estimated and the distribution characteristics can be investigated. In this section, the stochastic modelling of cotton fibre breakage during the spinning operation has been discussed.

Suppose there are N number of fibres classified in to m classes such that the kth class consists of all fibres of length L_k, $k = 1,2,3,\cdots,m$. We assume that $L_k = kh$, where h is the smallest possible length of fibres for which no further breakage can take place. Let $N_k(t)$ for $k = 1,2,3,\cdots,m$ be the number of fibres having length L_k at time t. In the next Δt time, a fibre of length L_k ($k = 2,3,\cdots,m$) can break only once into the fibres of length L_j, $j = 1,\ 2,\ 3,\cdots,k-1$ or it remains unaltered. The probability that a fibre of length L_k will break in time Δt is $b_k \Delta t$ which is proportional to the fibre length. The probability $b_k \Delta t$ represents the fraction of fibres of length L_k broken in time Δt. Let $S_{k,l}$ be the fraction of fibres of length L_l formed due to the breakage of fibre of length L_k. This fraction is basically the conditional probability of having a fibre of length L_l given the breakage of a fibre of length L_k. We denote $p_{k,l} \Delta t$ as the probability of formation of a fibre of length L_l from the breakage of a fibre of length L_k i.e., $p_{k,l} \Delta t = N_k(t) b_k \Delta t S_{k,l}$. From these definitions, we observe that $\sum_{l=1}^{k-1} S_{k,l}$, is the probability that the fibre having lengths between L_1 to L_{k-1} will be formed from the fibre of length L_k provided that the break of it will take place. Further, the formation of a fibre of length L_l due to the breakage of a fibre of length L_k also results in the formation of a fibre of length L_{k-l}. Analogously, formation of a fibre of length L_{k-l} due to the breakage of a fibre of length L_k results in the formation of a fibre of length L_l. Therefore, $S_{k,l} = S_{k,k-l}$.

The total number of possible changes is an important parameter particularly in the second modelling process. We can obtain this number as follows.

Suppose, $\Delta N^{k,l}$ = change in the number of fibres due to the breakage of a fibre of length L_k into fibres of length L_l and L_{k-l} in time Δt. The ith element of $\Delta N^{k,l}$ is given by

$$\left(\Delta N^{k,l}\right)_i = -1, \text{ if } i = k$$
$$= 1, \text{ if } i = l \text{ or } i = k - l$$
$$= 0, \text{ otherwise.}$$

It should be noted that $\Delta N^{k,l} = \Delta N^{k,k-l}$. Now $k \in \{1, 2, \cdots, m\}$ and for any given k, $l \in \{1, 2, \cdots, k-1\}$. Thus, the total number of possible cases is $m(m-1)$. But since $\Delta N^{k,l} = \Delta N^{k,k-l}$, the total number of distinct cases is $\frac{m(m-1)}{2}$. For example, suppose there are 7 groups of fibres. Let a fibre in group 5 breaks into two fibres, one in the group 3 and other in the group 2. So in this case, the change which occurs is $\Delta N^{5,3} = [0,1,1,0,-1,0,0]^T = \Delta N^{5,2}$ with probability $p_{5,3} = N_5(t) S_{5,3} b_5 \Delta t$.

The expectation of $\Delta N(t)$ is obtained by taking the summation of the products of the changes in time Δt and the corresponding probability. Thus,

$$E(\Delta N(t)) = \sum_{k=1}^{m} \sum_{l=1}^{k-1} \Delta N^{k,l} p_{k,l}$$

More explicitly, we can write the expectation of the ith component of $\Delta N(t)$ as

$$E(\Delta N(t))_i = \sum_{k=i+1}^{m} p_{k,i} - \sum_{k=1}^{i-1} p_{i,k}$$

Furthermore, the covariance matrix is given by

$$E\left((\Delta N(t))(\Delta N(t))^T\right) = \sum_{k=1}^{m} \sum_{l=1}^{k-1} C^{k,l} p_{k,l}(t) \Delta t$$

where $C^{k,l}$ is the matrix obtained as

$$C^{k,l} = \left(\Delta N^{k,l}\right)\left(\Delta N^{k,l}\right)^T.$$

For example, if there are 7 groups of fibres, $\Delta N^{5,3} = \begin{bmatrix} 0 \\ 1 \\ 1 \\ 0 \\ -1 \\ 0 \\ 0 \end{bmatrix}$

Therefore,

$$C^{5,3} = \begin{bmatrix} 0 \\ 1 \\ 1 \\ 0 \\ -1 \\ 0 \\ 0 \end{bmatrix} [0,1,1,0,-1,0,0] = \begin{bmatrix} 0 & 0 & 0 & 0 & 0 & 0 & 0 \\ 0 & 1 & 1 & 0 & -1 & 0 & 0 \\ 0 & 1 & 1 & 0 & -1 & 0 & 0 \\ 0 & 0 & 0 & 0 & 0 & 0 & 0 \\ 0 & -1 & -1 & 0 & 1 & 0 & 0 \\ 0 & 0 & 0 & 0 & 0 & 0 & 0 \\ 0 & 0 & 0 & 0 & 0 & 0 & 0 \end{bmatrix}$$

Thus, the expected change in time Δt and the covariance matrix are given by

$$E(\Delta N(t)) = f(t, N(t)) \Delta t$$

and

$$E\left((\Delta N(t))(\Delta N(t))^T\right) = V(t, N(t)) \Delta t = \sum_{k=1}^{m}\sum_{l=1}^{k-1} C^{k,l} p_{k,l}(t) \Delta t \qquad (12.14)$$

So, in the first modelling procedure, the stochastic differential equation is given by

$$dN(t) = f(t, N(t)) dt + (V(t, N(t)))^{\frac{1}{2}} dW(t), \text{ as } \Delta t \to 0 \qquad (12.15)$$

In this example, we assume that the breakage occurs randomly and the probability of the breakage of a fibre is proportional to its length. Under this assumption,

$$b_k \Delta t = \mu\left(\frac{L_k}{L_{max}}\right)\Delta t = \mu\left(\frac{kh}{L_{max}}\right)\Delta t \qquad (12.16)$$

where μ is a constant determining the rate of fibre breakage. We further assume the probability that the breakage of one fibre of length L_k will yield a fibre of length L_l is the same for all $l = 1, 2, \cdots, k$. Hence,

$$S_{k,l} = \frac{1}{k} \qquad (12.17)$$

where $S_{k,l}$ is the probability that a fibre of length l is formed due to the breakage of a fibre of length L_k. For better understanding, we set $L_{max} = mh$, where m is the number of groups. Therefore,

$$p_{k,l}(t)\Delta t = N_k(t) S_{k,l} b_k \Delta t = N_k(t)\frac{\mu h}{L_{max}}\Delta t.$$

For $L_{max} = mh$, the above expression for $p_{k,l}(t)\Delta t$ reduces to

$$p_{k,l}(t)\Delta t = N_k(t)\frac{\mu}{m}\Delta t \qquad (12.18)$$

The frequency in the lth class will be increased by 1 in time Δt due to the breakage of a fibre belonging to the kth class for $k > l$. It can be shown that for $k = 2n$ or $2n+1$, a fibre in the kth class can break in n different ways and the number of fibres in the kth class will decrease by 1. For example, a fibre in class 6 after breakage can be added to classes (1,5), (2,4) or (3,3) showing that the total number of ways the fibre can break is 3 and the number of fibres in class 6 will decrease by 1. Similarly, a fibre in class 7 after breakage can be added to classes (1,6), (2,5) or (3,4) showing that total number of ways the fibre can break is also 3 and the number of fibres in class 7 will decrease by 1.

So, we have

$$E\left(\Delta N(t)\right)_l = \sum_{k=l+1}^{m} p_{k,l}\Delta t - \sum_{k=1}^{n} p_{l,l-k}\Delta t = \sum_{k=l+1}^{m} p_{k,l}(t)\Delta t - \sum_{k=1}^{n} N_l(t)S_{l,l-k}b_l\Delta t$$

$$= \Delta t \sum_{k=l+1}^{m} N_k(t)\frac{\mu}{m} - N_l(t) \sum_{k=1}^{n} S_{l,l-k}b_l\Delta t$$

[Using Equation 12.18]

when l is of the form $2n$ or $2n+1$ for some n.

Now from Equations (12.16) and (12.17) we have $S_{l,l-k}b_l = \dfrac{\mu}{m}$
Therefore,

$$E\left(\Delta N(t)\right)_l = \frac{\mu}{m}\Delta t \sum_{k=l+1}^{m} N_k(t) - N_l(t)n\frac{\mu}{m}\Delta t$$

or, $E\left(\Delta N(t)\right)_l = -N_l(t)\frac{n\mu}{m}\Delta t + \frac{\mu}{m}\Delta t \sum_{k=l+1}^{m} N_k(t); \; (l=1,2,\cdots,m)$

Therefore, we can use matrix notation for the expression of $E\left(\Delta N(t)\right)$ as follows.

$$E\left(\Delta N(t)\right) = AN(t)\Delta t$$

where $A = \begin{pmatrix} 0 & 1 & 1 & 1 & \cdots & 1 \\ 0 & -1 & 1 & 1 & \cdots & 1 \\ 0 & 0 & -1 & 1 & \cdots & 1 \\ 0 & 0 & 0 & -2 & \cdots & 1 \\ \cdots & \cdots & \cdots & \cdots & \cdots & \cdots \\ 0 & 0 & 0 & 0 & \cdots & -n \end{pmatrix} \dfrac{\mu}{m}$

for $m = 2n$ or $2n+1$

Thus, for the first modelling process the stochastic differential Equation (12.15) can be written as

$$dN(t) = AN(t)dt + \left(V\left(t, N(t)\right)\right)^{\frac{1}{2}} dW(t)$$

where $V\left(t, N(t)\right)$ can be obtained using Equation (12.14).

In the second modelling process, the all-possible independent changes are explicitly incorporated in the modelling process. Suppose, $\Delta N^{k,l}$ denotes the change in frequency of fibres due to the breakage of a fibre in the kth group and this breakage contributes one fibre in each group l and $k-l$. This approach produces the following stochastic model.

$$dN(t) = f\big(t,\, N(t)\big)dt + \sum_{k=1}^{m}\sum_{l=1}^{k-1}(\Delta N)^{k,l}\big(p_{k,l}(t)\big)^{\frac{1}{2}} dW_{k,l}(t) \qquad (12.19)$$

where $W_{k,l}(t)$ for $k=1,2,\cdots,m$ and $l=1,2,\cdots,k-1$ are $\frac{m(m-1)}{2}$ independent Weiner processes.

To demonstrate the application, we consider 11 classes of fibre length as given in the following frequency distribution table.

Length Class (mm)	Frequency
4	128
8	3184
12	4224
16	15920
20	23096
24	27200
28	17128
32	8060
36	3384
40	912
44	288

For the given data the number of class $m=11$ and $h=4\text{ mm}$. We assume that $dt = 0.01$ and $\mu = 1$. Figure 12.3 shows the initial distribution of fibre length. We

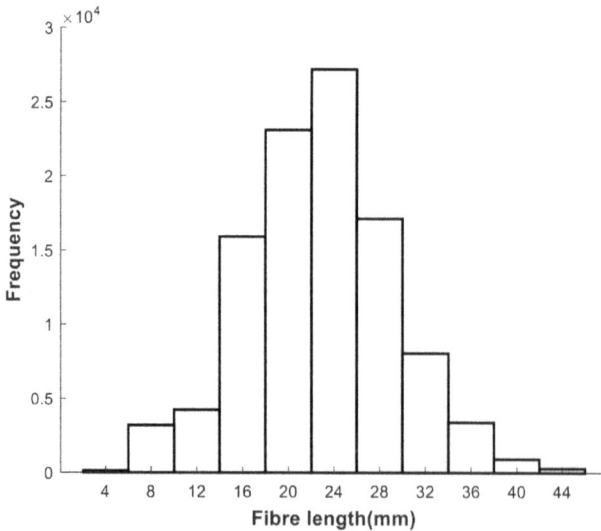

FIGURE 12.3   Initial distribution of fibre length.

TABLE 12.1
Result of Average Fibre Length Before and After Breakage

	Average Fibre Length (mm)
Before breakage	22.74
After breakage using first method	18.99
After breakage using second method	18.86

estimate the distribution of fibre length after breakage at time 1 based on 300 sample paths following the two methods discussed in Sections 12.3.1 and 12.3.2. First, we generate 300 sample paths following Equation (12.15) corresponding to first method and then perform the same using Equation (12.19) corresponding to second method with the help of MATLAB® coding. The average fibre length using both the methods is also computed and the results are shown in Table 12.1. Figure 12.4 depicts the distribution of fibre length after breakage using the first method. The distribution of fibre length after breakage following the second method is pretty similar to that of the first method and hence, we omit the figure corresponding to second method. The reader can verify it using the MATLAB® code provided in this chapter.

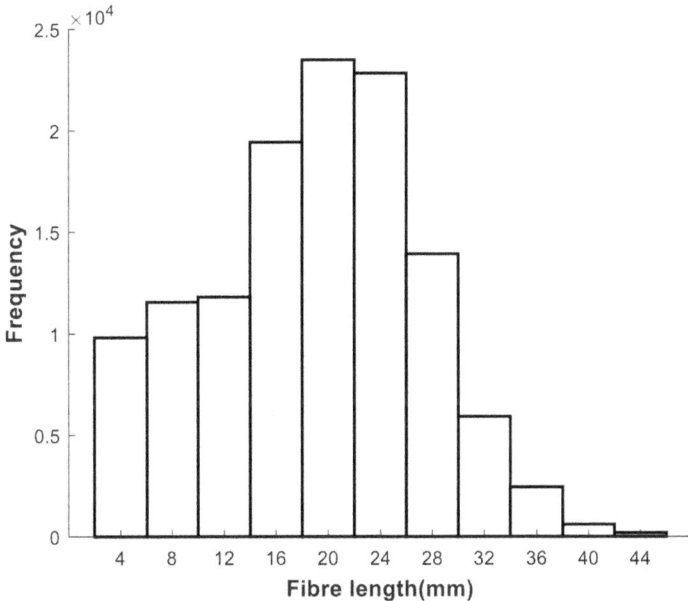

FIGURE 12.4 Distribution of fibre length after breakage using first method.

12.4   MATLAB® CODING

```
clc
clear
close all
format short g
global m
NI=[128 3184 4224 15920 23096 27200 17128 8060 3384 912 288];
% NI is the initial frequency of fibre length
m=length(NI);
dt=0.01;
h=4;
Lmax=m*h;
t1=0.01:dt:1;
R=1;
N1=100;
ns=300;
fn=zeros(length(t1),ns,m);
% formation of matrix B
B=zeros(m,m)
for i=1:m
    for j=1:m
        if (j>i)
        B(i,j)=1;
        elseif (i==j)
            B(i,j)=-fix(i/2);
        else
            B(i,j)=0;
        end
    end
end
B=(1/m)*B;
for j=1:ns % ns is the number of sample paths
randn('state',j);
v1=zeros(m,m);
N=zeros(1,m);
%initial frequency
for i=1:m
N(i)=NI(i);
end
%
    for i=1:m
        dw(i,:)=sqrt(dt)*randn(1,N1);
    end
    for n=2:length(t1)
        for n1=1:m
            Winc(n1)=sum(dw(R*(n-1)+1:R*n));
        end
        for k=1:m
            for l=1:k-1
                s(k,l)=1/k;
```

```
                    for i=1:m
                        if (i==k)
                            dN(i)=-1;
                        elseif (i==l||i==k-1)
                            dN(i)=1;
                        else
                            dN(i)=0;
                        end
                    end
                c=dN'*dN;
                v1=v1+c*N(k)*(1/m);
                end
            end
        N=N+(B*N')'*dt+(sqrtm(v1)*Winc')'
        for i=1:length(N)
            if N(i)>=0
                fn(n,j,:)=N;
            else
            fn(n,j,:)=0;
            break
            end
        end
    end
end
figure
plot(t1,fn(:,:,end))
y1=mean(fn(end,:,end))%y is the average number of fibres in
the last group
sumI=0;
for i=1:m
    sumI=sumI+NI(i)*i*h;
end
avlengthI=sumI/sum(NI);%avlengthI is the average length of
fibres in initial stage
sum1=0;
for i=1:m
    sum1=sum1+N(i)*i*h;
end
avlength=sum1/sum(N);%avlength is the final average length of
fibres
for i=1:m
    avnum(i)=mean(fn(end,:,i));
end
avnum; %avnum is the vector representing average number of
fibers in different classes
figure, bar(h:h:m*h,avnum,1,'w','LineWidth',1.5)
xlim([0 48])
xlabel('Fibre length(mm)','FontSize',12,'FontWeight','bold')
ylabel('Frequency','FontSize',12,'FontWeight','bold')
set(gcf,'Color',[1,1,1])
box off
```

```
%Equivalent SDE-2nd method
for j=1:ns % ns is the number of sample paths
randn('state',j);
N=zeros(1,m);
    for i=1:m
        N(i)=NI(i);
    end
    for n=2:length(t1)
        sum2=zeros(1,m);
        for k=1:m
            for l=1:k-1
                dw=sqrt(dt)*randn(1,N1);
                Winc=sum(dw(R*(n-1)+1:R*n));
                for i=1:m
                    if (i==k)
                        dN(i)=-1;
                    elseif (i==l||i==k-1)
                        dN(i)=1;
                    else
                        dN(i)=0;
                    end
                end
                sum2=sum2+dN*sqrt(N(k)*(1/m))*Winc;
            end
        end
        N=N+(B*N')'*dt+sum2;
        for i=1:length(N)
            if N(i)>=0
                gn(n,j,:)=N;
            else
                gn(n,j,:)=0;
            break
            end
        end
    end
end
figure
plot(t1,gn(:,:,end))
y2=mean(gn(end,:,end))%y2 is the average number of fibres in
the last group
sum3=0;
for i=1:m
    sum3=sum3+N(i)*i*h;
end
avlength_sde=sum3/sum(N);
for i=1:m
    avnum2(i)=mean(gn(end,:,i));
end
avnum2;
figure, bar(h:h:m*h,avnum2,1,'w','LineWidth',1.5)
xlim([0 48])
```

```
xlabel('Fibre length(mm)','FontSize',12,'FontWeight','bold')
ylabel('Frequency','FontSize',12,'FontWeight','bold')
set(gcf,'Color',[1,1,1])
box off
figure, bar(h:h:m*h,NI,1,'w','LineWidth',1.5)
xlim([0 48])
xlabel('Fibre length(mm)','FontSize',12,'FontWeight','bold')
ylabel('Frequency','FontSize',12,'FontWeight','bold')
box off
set(gcf,'Color',[1,1,1])
avlengthI
avlength
avlength_sde
avnum
avnum2
```

Exercises

12.1 The transition probabilities of the fibres from the transient states in a carding machine (Figures 12.1 and 12.2) are $P_{12} = 0.6$, $P_{24} = 0.55$, and $P_{45} = 0.2$. Find

    i.   the average number of revolutions a fibre recycles in the cylinder before it is transferred to the doffer.

    ii. the average number of times the fibre passes through the state E_2 before it is absorbed.

12.2 There are 3 length groups of fibres with each group having 100 fibres initially and $\mu = 1$. Find the expression for the variance covariance matrix under the assumptions used in Section 12.3.3. Also find the variance covariance matrix for the initial number of fibres in each group. Obtain a square matrix B such that $BB^T = B^T B = V\left(N(0)\right)$ where $N(0) = \left(N_1(0), N_2(0), N_3(0)\right)$.

[Hint: Suppose $\lambda_1, \lambda_2, \lambda_3$ are the eigen values of $V\left(N(0)\right)$. Then,

$$V\left(N(0)\right) = Pdiag\left(\lambda_i\right)P^{-1}$$

where the columns of the matrix P are the eigen vectors of the matrix $V\left(N(0)\right)$. Thus, $\sqrt{V\left(N(0)\right)} = Pdiag\left(\sqrt{\lambda_i}\right)P^{-1} = B$].

Appendix A: Statistical Tables

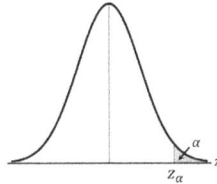

TABLE A1

Areas in the Tail of the Standard Normal Distribution

z_α	0.00	0.01	0.02	0.03	0.04	0.05	0.06	0.07	0.08	0.09
0.0	0.5000	0.4960	0.4920	0.4880	0.4840	0.4801	0.4761	0.4721	0.4681	0.4641
0.1	0.4602	0.4562	0.4522	0.4483	0.4443	0.4404	0.4364	0.4325	0.4286	0.4247
0.2	0.4207	0.4168	0.4129	0.4090	0.4052	0.4013	0.3974	0.3936	0.3897	0.3859
0.3	0.3821	0.3783	0.3745	0.3707	0.3669	0.3632	0.3594	0.3557	0.3520	0.3483
0.4	0.3446	0.3409	0.3372	0.3336	0.3300	0.3264	0.3228	0.3192	0.3156	0.3121
0.5	0.3085	0.3050	0.3015	0.2981	0.2946	0.2912	0.2877	0.2843	0.2810	0.2776
0.6	0.2743	0.2709	0.2676	0.2643	0.2611	0.2578	0.2546	0.2514	0.2483	0.2451
0.7	0.2420	0.2389	0.2358	0.2327	0.2296	0.2266	0.2236	0.2206	0.2177	0.2148
0.8	0.2119	0.2090	0.2061	0.2033	0.2005	0.1977	0.1949	0.1922	0.1894	0.1867
0.9	0.1841	0.1814	0.1788	0.1762	0.1736	0.1711	0.1685	0.1660	0.1635	0.1611
1.0	0.1587	0.1562	0.1539	0.1515	0.1492	0.1469	0.1446	0.1423	0.1401	0.1379
1.1	0.1357	0.1335	0.1314	0.1292	0.1271	0.1251	0.1230	0.1210	0.1190	0.1170
1.2	0.1151	0.1131	0.1112	0.1093	0.1075	0.1056	0.1038	0.1020	0.1003	0.0985
1.3	0.0968	0.0951	0.0934	0.0918	0.0901	0.0885	0.0869	0.0853	0.0838	0.0823
1.4	0.0808	0.0793	0.0778	0.0764	0.0749	0.0735	0.0721	0.0708	0.0694	0.0681
1.5	0.0668	0.0655	0.0643	0.0630	0.0618	0.0606	0.0594	0.0582	0.0571	0.0559
1.6	0.0548	0.0537	0.0526	0.0516	0.0505	0.0495	0.0485	0.0475	0.0465	0.0455
1.7	0.0446	0.0436	0.0427	0.0418	0.0409	0.0401	0.0392	0.0384	0.0375	0.0367
1.8	0.0359	0.0351	0.0344	0.0336	0.0329	0.0322	0.0314	0.0307	0.0301	0.0294
1.9	0.0287	0.0281	0.0274	0.0268	0.0262	0.0256	0.0250	0.0244	0.0239	0.0233
2.0	0.0228	0.0222	0.0217	0.0212	0.0207	0.0202	0.0197	0.0192	0.0188	0.0183
2.1	0.0179	0.0174	0.0170	0.0166	0.0162	0.0158	0.0154	0.0150	0.0146	0.0143
2.2	0.0139	0.0136	0.0132	0.0129	0.0125	0.0122	0.0119	0.0116	0.0113	0.0110
2.3	0.0107	0.0104	0.0102	0.0099	0.0096	0.0094	0.0091	0.0089	0.0087	0.0084
2.4	0.0082	0.0080	0.0078	0.0075	0.0073	0.0071	0.0069	0.0068	0.0066	0.0064
2.5	0.0062	0.0060	0.0059	0.0057	0.0055	0.0054	0.0052	0.0051	0.0049	0.0048
2.6	0.0047	0.0045	0.0044	0.0043	0.0041	0.0040	0.0039	0.0038	0.0037	0.0036
2.7	0.0035	0.0034	0.0033	0.0032	0.0031	0.0030	0.0029	0.0028	0.0027	0.0026
2.8	0.0026	0.0025	0.0024	0.0023	0.0023	0.0022	0.0021	0.0021	0.0020	0.0019
2.9	0.0019	0.0018	0.0018	0.0017	0.0016	0.0016	0.0015	0.0015	0.0014	0.0014
3.0	0.0013	0.0013	0.0013	0.0012	0.0012	0.0011	0.0011	0.0011	0.0010	0.0010

Note: The entries in this Table are the Values of $\alpha = P(Z > z_\alpha)$.

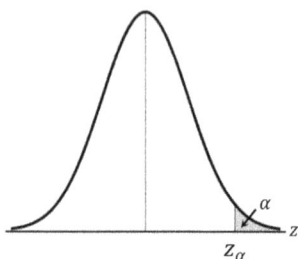

TABLE A2
Standard Normal Variates for Given Probability Values

α	z_α	α	z_α
0.5	0	0.029	1.8957
0.45	0.1257	0.028	1.9110
0.4	0.2533	0.027	1.9268
0.35	0.3853	0.026	1.9431
0.3	0.5244	0.025	1.9600
0.25	0.6745	0.024	1.9774
0.2	0.8416	0.023	1.9954
0.15	1.0364	0.022	2.0141
0.1	1.2816	0.021	2.0335
0.05	1.6449	0.02	2.0537
0.049	1.6546	0.019	2.0749
0.048	1.6646	0.018	2.0969
0.047	1.6747	0.017	2.1201
0.046	1.6849	0.016	2.1444
0.045	1.6954	0.015	2.1701
0.044	1.7060	0.014	2.1973
0.043	1.7169	0.013	2.2262
0.042	1.7279	0.012	2.2571
0.041	1.7392	0.011	2.2904
0.04	1.7507	0.01	2.3263
0.039	1.7624	0.009	2.3656
0.038	1.7744	0.008	2.4089
0.037	1.7866	0.007	2.4573
0.036	1.7991	0.006	2.5121
0.035	1.8119	0.005	2.5758
0.034	1.8250	0.004	2.6521
0.033	1.8384	0.003	2.7478
0.032	1.8522	0.002	2.8782
0.031	1.8663	0.001	3.0902
0.03	1.8808	0.0005	3.2905

Note: The entries in this Table are the values of z_α for the given values of α.

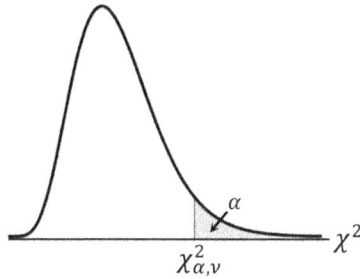

TABLE A3
Values of $\chi^2_{\alpha,v}$ for Chi-square Distribution

v	$\alpha = 0.995$	$\alpha = 0.99$	$\alpha = 0.975$	$\alpha = 0.96$	$\alpha = 0.95$	$\alpha = 0.9$
1	0.000	0.000	0.001	0.003	0.004	0.016
2	0.010	0.020	0.051	0.082	0.103	0.211
3	0.072	0.115	0.216	0.300	0.352	0.584
4	0.207	0.297	0.484	0.627	0.711	1.064
5	0.412	0.554	0.831	1.031	1.145	1.610
6	0.676	0.872	1.237	1.492	1.635	2.204
7	0.989	1.239	1.690	1.997	2.167	2.833
8	1.344	1.646	2.180	2.537	2.733	3.490
9	1.735	2.088	2.700	3.105	3.325	4.168
10	2.156	2.558	3.247	3.697	3.940	4.865
11	2.603	3.053	3.816	4.309	4.575	5.578
12	3.074	3.571	4.404	4.939	5.226	6.304
13	3.565	4.107	5.009	5.584	5.892	7.042
14	4.075	4.660	5.629	6.243	6.571	7.790
15	4.601	5.229	6.262	6.914	7.261	8.547
16	5.142	5.812	6.908	7.596	7.962	9.312
17	5.697	6.408	7.564	8.288	8.672	10.085
18	6.265	7.015	8.231	8.989	9.390	10.865
19	6.844	7.633	8.907	9.698	10.117	11.651
20	7.434	8.260	9.591	10.415	10.851	12.443
21	8.034	8.897	10.283	11.140	11.591	13.240
22	8.643	9.542	10.982	11.870	12.338	14.041
23	9.260	10.196	11.689	12.607	13.091	14.848
24	9.886	10.856	12.401	13.350	13.848	15.659
25	10.520	11.524	13.120	14.098	14.611	16.473
26	11.160	12.198	13.844	14.851	15.379	17.292
27	11.808	12.879	14.573	15.609	16.151	18.114
28	12.461	13.565	15.308	16.371	16.928	18.939
29	13.121	14.256	16.047	17.138	17.708	19.768
30	13.787	14.953	16.791	17.908	18.493	20.599

(*Continued*)

TABLE A3 (Continued)
Values of $\chi^2_{\alpha,v}$ for Chi-square Distribution

v	$\alpha = 0.1$	$\alpha = 0.05$	$\alpha = 0.04$	$\alpha = 0.025$	$\alpha = 0.01$	$\alpha = 0.005$
1	2.706	3.841	4.218	5.024	6.635	7.879
2	4.605	5.991	6.438	7.378	9.210	10.597
3	6.251	7.815	8.311	9.348	11.345	12.838
4	7.779	9.488	10.026	11.143	13.277	14.860
5	9.236	11.070	11.644	12.833	15.086	16.750
6	10.645	12.592	13.198	14.449	16.812	18.548
7	12.017	14.067	14.703	16.013	18.475	20.278
8	13.362	15.507	16.171	17.535	20.090	21.955
9	14.684	16.919	17.608	19.023	21.666	23.589
10	15.987	18.307	19.021	20.483	23.209	25.188
11	17.275	19.675	20.412	21.920	24.725	26.757
12	18.549	21.026	21.785	23.337	26.217	28.300
13	19.812	22.362	23.142	24.736	27.688	29.819
14	21.064	23.685	24.485	26.119	29.141	31.319
15	22.307	24.996	25.816	27.488	30.578	32.801
16	23.542	26.296	27.136	28.845	32.000	34.267
17	24.769	27.587	28.445	30.191	33.409	35.718
18	25.989	28.869	29.745	31.526	34.805	37.156
19	27.204	30.144	31.037	32.852	36.191	38.582
20	28.412	31.410	32.321	34.170	37.566	39.997
21	29.615	32.671	33.597	35.479	38.932	41.401
22	30.813	33.924	34.867	36.781	40.289	42.796
23	32.007	35.172	36.131	38.076	41.638	44.181
24	33.196	36.415	37.389	39.364	42.980	45.559
25	34.382	37.652	38.642	40.646	44.314	46.928
26	35.563	38.885	39.889	41.923	45.642	48.290
27	36.741	40.113	41.132	43.195	46.963	49.645
28	37.916	41.337	42.370	44.461	48.278	50.993
29	39.087	42.557	43.604	45.722	49.588	52.336
30	40.256	43.773	44.834	46.979	50.892	53.672

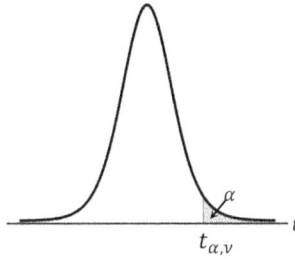

TABLE A4
Values of $t_{\alpha,\nu}$ for t-Distribution

ν	$\alpha = 0.1$	$\alpha = 0.05$	$\alpha = 0.025$	$\alpha = 0.01$	$\alpha = 0.005$
1	3.078	6.314	12.706	31.821	63.657
2	1.886	2.920	4.303	6.965	9.925
3	1.638	2.353	3.182	4.541	5.841
4	1.533	2.132	2.776	3.747	4.604
5	1.476	2.015	2.571	3.365	4.032
6	1.440	1.943	2.447	3.143	3.707
7	1.415	1.895	2.365	2.998	3.499
8	1.397	1.860	2.306	2.896	3.355
9	1.383	1.833	2.262	2.821	3.250
10	1.372	1.812	2.228	2.764	3.169
11	1.363	1.796	2.201	2.718	3.106
12	1.356	1.782	2.179	2.681	3.055
13	1.350	1.771	2.160	2.650	3.012
14	1.345	1.761	2.145	2.624	2.977
15	1.341	1.753	2.131	2.602	2.947
16	1.337	1.746	2.120	2.583	2.921
17	1.333	1.740	2.110	2.567	2.898
18	1.330	1.734	2.101	2.552	2.878
19	1.328	1.729	2.093	2.539	2.861
20	1.325	1.725	2.086	2.528	2.845
21	1.323	1.721	2.080	2.518	2.831
22	1.321	1.717	2.074	2.508	2.819
23	1.319	1.714	2.069	2.500	2.807
24	1.318	1.711	2.064	2.492	2.797
25	1.316	1.708	2.060	2.485	2.787
26	1.315	1.706	2.056	2.479	2.779
27	1.314	1.703	2.052	2.473	2.771
28	1.313	1.701	2.048	2.467	2.763
29	1.311	1.699	2.045	2.462	2.756
30	1.310	1.697	2.042	2.457	2.750

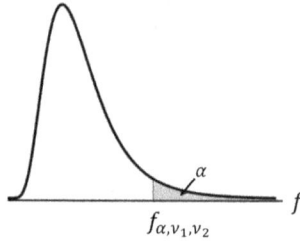

f_{α,v_1,v_2}

TABLE A5

(a) Values of f_{α,v_1,v_2} for F-Distribution at $\alpha = 0.05$

v_2	1	2	3	4	5	6	7	8	9	10	11	12	13	14	15	20	24	30
1	161	200	216	225	230	234	237	239	241	242	243	244	245	245	246	248	249	250
2	18.5	19.0	19.2	19.2	19.3	19.3	19.4	19.4	19.4	19.4	19.4	19.4	19.4	19.4	19.4	19.4	19.5	19.5
3	10.1	9.6	9.3	9.1	9.0	8.9	8.9	8.8	8.8	8.8	8.8	8.7	8.7	8.7	8.7	8.7	8.6	8.6
4	7.7	6.9	6.6	6.4	6.3	6.2	6.1	6.0	6.0	6.0	5.9	5.9	5.9	5.9	5.9	5.8	5.8	5.7
5	6.6	5.8	5.4	5.2	5.1	5.0	4.9	4.8	4.8	4.7	4.7	4.7	4.7	4.6	4.6	4.6	4.5	4.5
6	6.0	5.1	4.8	4.5	4.4	4.3	4.2	4.1	4.1	4.1	4.0	4.0	4.0	4.0	3.9	3.9	3.8	3.8
7	5.6	4.7	4.3	4.1	4.0	3.9	3.8	3.7	3.7	3.6	3.6	3.6	3.6	3.5	3.5	3.4	3.4	3.4
8	5.3	4.5	4.1	3.8	3.7	3.6	3.5	3.4	3.4	3.3	3.3	3.3	3.3	3.2	3.2	3.2	3.1	3.1
9	5.1	4.3	3.9	3.6	3.5	3.4	3.3	3.2	3.2	3.1	3.1	3.1	3.0	3.0	3.0	2.9	2.9	2.9
10	5.0	4.1	3.7	3.5	3.3	3.2	3.1	3.1	3.0	3.0	2.9	2.9	2.9	2.9	2.8	2.8	2.7	2.7
11	4.8	4.0	3.6	3.4	3.2	3.1	3.0	2.9	2.9	2.9	2.8	2.8	2.8	2.7	2.7	2.6	2.6	2.6
12	4.7	3.9	3.5	3.3	3.1	3.0	2.9	2.8	2.8	2.8	2.7	2.7	2.7	2.6	2.6	2.5	2.5	2.5
13	4.7	3.8	3.4	3.2	3.0	2.9	2.8	2.8	2.7	2.7	2.6	2.6	2.6	2.6	2.5	2.5	2.4	2.4
14	4.6	3.7	3.3	3.1	3.0	2.8	2.8	2.7	2.6	2.6	2.6	2.5	2.5	2.5	2.5	2.4	2.3	2.3
15	4.5	3.7	3.3	3.1	2.9	2.8	2.7	2.6	2.6	2.5	2.5	2.5	2.4	2.4	2.4	2.3	2.3	2.2
20	4.4	3.5	3.1	2.9	2.7	2.6	2.5	2.4	2.4	2.3	2.3	2.3	2.2	2.2	2.2	2.1	2.1	2.0
24	4.3	3.4	3.0	2.8	2.6	2.5	2.4	2.4	2.3	2.3	2.2	2.2	2.2	2.1	2.1	2.0	2.0	1.9
30	4.2	3.3	2.9	2.7	2.5	2.4	2.3	2.3	2.2	2.2	2.1	2.1	2.1	2.0	2.0	1.9	1.9	1.8

v_1

TABLE A5

(b) Values of f_{α,v_1,v_2} for F-Distribution at $\alpha = 0.025$

										v_1								
v_2	1	2	3	4	5	6	7	8	9	10	11	12	13	14	15	20	24	30
1	648	799	864	900	922	937	948	957	963	969	973	977	980	983	985	993	997	1001
2	38.5	39.0	39.2	39.2	39.3	39.3	39.4	39.4	39.4	39.4	39.4	39.4	39.4	39.4	39.4	39.4	39.5	39.5
3	17.4	16.0	15.4	15.1	14.9	14.7	14.6	14.5	14.5	14.4	14.4	14.3	14.3	14.3	14.3	14.2	14.1	14.1
4	12.2	10.6	10.0	9.6	9.4	9.2	9.1	9.0	8.9	8.8	8.8	8.8	8.7	8.7	8.7	8.6	8.5	8.5
5	10.0	8.4	7.8	7.4	7.1	7.0	6.9	6.8	6.7	6.6	6.6	6.5	6.5	6.5	6.4	6.3	6.3	6.2
6	8.8	7.3	6.6	6.2	6.0	5.8	5.7	5.6	5.5	5.5	5.4	5.4	5.3	5.3	5.3	5.2	5.1	5.1
7	8.1	6.5	5.9	5.5	5.3	5.1	5.0	4.9	4.8	4.8	4.7	4.7	4.6	4.6	4.6	4.5	4.4	4.4
8	7.6	6.1	5.4	5.1	4.8	4.7	4.5	4.4	4.4	4.3	4.2	4.2	4.2	4.1	4.1	4.0	3.9	3.9
9	7.2	5.7	5.1	4.7	4.5	4.3	4.2	4.1	4.0	4.0	3.9	3.9	3.8	3.8	3.8	3.7	3.6	3.6
10	6.9	5.5	4.8	4.5	4.2	4.1	3.9	3.9	3.8	3.7	3.7	3.6	3.6	3.6	3.5	3.4	3.4	3.3
11	6.7	5.3	4.6	4.3	4.0	3.9	3.8	3.7	3.6	3.5	3.5	3.4	3.4	3.4	3.3	3.2	3.2	3.1
12	6.6	5.1	4.5	4.1	3.9	3.7	3.6	3.5	3.4	3.4	3.3	3.3	3.2	3.2	3.2	3.1	3.0	3.0
13	6.4	5.0	4.3	4.0	3.8	3.6	3.5	3.4	3.3	3.2	3.2	3.2	3.1	3.1	3.1	2.9	2.9	2.8
14	6.3	4.9	4.2	3.9	3.7	3.5	3.4	3.3	3.2	3.1	3.1	3.1	3.0	3.0	2.9	2.8	2.8	2.7
15	6.2	4.8	4.2	3.8	3.6	3.4	3.3	3.2	3.1	3.1	3.0	3.0	2.9	2.9	2.9	2.8	2.7	2.6
20	5.9	4.5	3.9	3.5	3.3	3.1	3.0	2.9	2.8	2.8	2.7	2.7	2.6	2.6	2.6	2.5	2.4	2.3
24	5.7	4.3	3.7	3.4	3.2	3.0	2.9	2.8	2.7	2.6	2.6	2.5	2.5	2.5	2.4	2.3	2.3	2.2
30	5.6	4.2	3.6	3.2	3.0	2.9	2.7	2.7	2.6	2.5	2.5	2.4	2.4	2.3	2.3	2.2	2.1	2.1

TABLE A5

(c) Values of f_{α,v_1,v_2} for F-Distribution at $\alpha = 0.01$

									v_1									
v_2	1	2	3	4	5	6	7	8	9	10	11	12	13	14	15	20	24	30
1	4052	4999	5403	5625	5764	5859	5928	5981	6022	6056	6083	6106	6126	6143	6157	6209	6235	6261
2	98.5	99.0	99.2	99.2	99.3	99.3	99.4	99.4	99.4	99.4	99.4	99.4	99.4	99.4	99.4	99.4	99.5	99.5
3	34.1	30.8	29.5	28.7	28.2	27.9	27.7	27.5	27.3	27.2	27.1	27.1	27.0	26.9	26.9	26.7	26.6	26.5
4	21.2	18.0	16.7	16.0	15.5	15.2	15.0	14.8	14.7	14.5	14.5	14.4	14.3	14.2	14.2	14.0	13.9	13.8
5	16.3	13.3	12.1	11.4	11.0	10.7	10.5	10.3	10.2	10.1	10.0	9.9	9.8	9.8	9.7	9.6	9.5	9.4
6	13.7	10.9	9.8	9.1	8.7	8.5	8.3	8.1	8.0	7.9	7.8	7.7	7.7	7.6	7.6	7.4	7.3	7.2
7	12.2	9.5	8.5	7.8	7.5	7.2	7.0	6.8	6.7	6.6	6.5	6.5	6.4	6.4	6.3	6.2	6.1	6.0
8	11.3	8.6	7.6	7.0	6.6	6.4	6.2	6.0	5.9	5.8	5.7	5.7	5.6	5.6	5.5	5.4	5.3	5.2
9	10.6	8.0	7.0	6.4	6.1	5.8	5.6	5.5	5.4	5.3	5.2	5.1	5.1	5.0	5.0	4.8	4.7	4.6
10	10.0	7.6	6.6	6.0	5.6	5.4	5.2	5.1	4.9	4.8	4.8	4.7	4.6	4.6	4.6	4.4	4.3	4.2
11	9.6	7.2	6.2	5.7	5.3	5.1	4.9	4.7	4.6	4.5	4.5	4.4	4.3	4.3	4.3	4.1	4.0	3.9
12	9.3	6.9	6.0	5.4	5.1	4.8	4.6	4.5	4.4	4.3	4.2	4.2	4.1	4.1	4.0	3.9	3.8	3.7
13	9.1	6.7	5.7	5.2	4.9	4.6	4.4	4.3	4.2	4.1	4.0	4.0	3.9	3.9	3.8	3.7	3.6	3.5
14	8.9	6.5	5.6	5.0	4.7	4.5	4.3	4.1	4.0	3.9	3.9	3.8	3.7	3.7	3.7	3.5	3.4	3.3
15	8.7	6.4	5.4	4.9	4.6	4.3	4.1	4.0	3.9	3.8	3.7	3.7	3.6	3.6	3.5	3.4	3.3	3.2
20	8.1	5.8	4.9	4.4	4.1	3.9	3.7	3.6	3.5	3.4	3.3	3.2	3.2	3.1	3.1	2.9	2.9	2.8
24	7.8	5.6	4.7	4.2	3.9	3.7	3.5	3.4	3.3	3.2	3.1	3.0	3.0	2.9	2.9	2.7	2.7	2.6
30	7.6	5.4	4.5	4.0	3.7	3.5	3.3	3.2	3.1	3.0	2.9	2.8	2.8	2.7	2.7	2.5	2.5	2.4

TABLE A6
The Values of Studentized Range at $\alpha = 0.05$

Degrees of Freedom, v	Number of Treatments (k)							
	3	4	5	6	7	8	9	10
1	26.98	32.82	37.08	40.41	43.12	45.40	47.36	49.07
2	8.33	9.80	10.88	11.73	12.43	13.03	13.54	13.99
3	5.91	6.83	7.50	8.04	8.48	8.85	9.18	9.46
4	5.04	5.76	6.29	6.71	7.05	7.35	7.60	7.83
5	4.60	5.22	5.67	6.03	6.33	6.58	6.80	7.00
6	4.34	4.90	5.31	5.63	5.90	6.12	6.32	6.49
7	4.17	4.68	5.06	5.36	5.61	5.82	6.00	6.16
8	4.04	4.53	4.89	5.17	5.40	5.60	5.77	5.92
9	3.95	4.42	4.76	5.02	5.24	5.43	5.60	5.74
10	3.88	4.33	4.65	4.91	5.12	5.30	5.46	5.60
11	3.82	4.26	4.57	4.82	5.03	5.20	5.35	5.49
12	3.77	4.20	4.51	4.75	4.95	5.12	5.26	5.40
13	3.73	4.15	4.45	4.69	4.88	5.05	5.19	5.32
14	3.70	4.11	4.41	4.64	4.83	4.99	5.13	5.25
15	3.67	4.08	4.37	4.60	4.78	4.94	5.08	5.20
16	3.65	4.05	4.33	4.56	4.74	4.90	5.03	5.15
17	3.63	4.02	4.30	4.52	4.71	4.86	4.99	5.11
18	3.61	4.00	4.28	4.49	4.67	4.82	4.96	5.07
19	3.59	3.98	4.25	4.47	4.65	4.79	4.92	5.04
20	3.58	3.96	4.23	4.45	4.62	4.77	4.90	5.01
21	3.57	3.94	4.21	4.42	4.60	4.74	4.87	4.98
22	3.55	3.93	4.20	4.41	4.58	4.72	4.85	4.96
23	3.54	3.91	4.18	4.39	4.56	4.70	4.83	4.94
24	3.53	3.90	4.17	4.37	4.54	4.68	4.81	4.92
25	3.52	3.89	4.15	4.36	4.53	4.67	4.79	4.90
26	3.51	3.88	4.14	4.35	4.51	4.65	4.77	4.88
27	3.51	3.87	4.13	4.33	4.50	4.64	4.76	4.86
28	3.50	3.86	4.12	4.32	4.49	4.63	4.75	4.85
29	3.49	3.85	4.11	4.31	4.48	4.61	4.73	4.84
30	3.49	3.85	4.10	4.30	4.46	4.60	4.72	4.82

Note: The entries in this Table are the values of $q_{0.05,k,v}$.

Appendix B: MATLAB®
Coding for Statistical Tables

MATLAB® CODING FOR TABLE A1

```
clc
clear
close all
format short
z=0:0.01:3.09;
alpha=1-normcdf(z,0,1);
[z' alpha'];
j=1;
for i=1:10:length(z)
P(j,:)=alpha(i:i+9);
j=j+1;
end
P
```

MATLAB® CODING FOR TABLE A2

```
clc
clear
close all
format short
alpha=[0.5:-0.05:0.1 0.05:0.-0.001:0.001 0.0005];
length(alpha)
z=norminv(1-alpha,0,1);
Table=[alpha' z']
```

MATLAB® CODING FOR TABLE A3

```
clc
clear
close all
format short
for i=1:30
alpha1=[0.995 0.99 0.975 0.96 0.95 0.9];
alpha2=[0.1 0.05 0.04 0.025 0.01 0.005];
chi1(i,:)=chi2inv(1-alpha1,i);
chi2(i,:)=chi2inv(1-alpha2,i);
end
Table1=chi1
Table2=chi2
```

MATLAB® CODING FOR TABLE A4

```
clc
clear
close all
format short
for i=1:30
alpha=[0.1 0.05 0.025 0.01 0.005];
t(i,:)=tinv(1-alpha,i);
end
Table=t
```

MATLAB® CODING FOR TABLE A5

```
clc
clear
close all
format short
alpha1=0.05;
alpha2=0.025;
alpha3=0.01;
for i=1:18
    nu2=[1:15 20 24 30];
    nu1(i)=nu2(i);
    F1(:,i)=finv(1-alpha1,nu1(i),nu2);
    F2(:,i)=finv(1-alpha2,nu1(i),nu2);
    F3(:,i)=finv(1-alpha3,nu1(i),nu2);
end
Table1=F1
Table2=F2
Table3=F3
```

Appendix C: Answers to Exercises

CHAPTER 2:

2.1 mean = 111.2, median = 60.5, mode = 110

2.2 mean = 35.85, median = 36.6667, mode = 38.1481

2.3 50.30, 20.93

2.4 12.199

CHAPTER 3:

3.3 $\frac{4}{7}$

3.4 $\frac{4}{7}$

3.5 i. $\frac{1}{102}$, ii. $\frac{33}{221}$

3.6 0.304

3.7 0.5

3.8 $k = 0.1$ i. 0.35, ii. 0.65 iii. $F(x) = 0$ if $x < -2$; $F(x) = 0.1$ if $-2 \leq x < -1$; $F(x) = 0.35$ if $-1 \leq x < 0$; $F(x) = 0.55$ if $0 \leq x < 1$; $F(x) = 0.70$ if $1 \leq x < 2$; $F(x) = 1$ if $2 \leq x < \infty$.

3.10 $\frac{e^{3t}-1}{3t}$, mean = $\frac{3}{2}$, variance = $\frac{3}{4}$

3.11 i. 0.1

ii.

x	0	1	2
$P(X = x)$	0.395	0.29	0.315

y	0	1	2	3
$P(Y = y)$	0.275	0.12	0.205	0.4

iii. 0.25

iv.

y	0	1	2	3
$P(Y = y \mid X = 1)$	0.517	0.172	0.137	0.172

3.12 i. $k = 2$

ii. $g(u,v) = 2$, $v < u < 1$; $0 < v < 1$
   $= 0$ elsewhere

iii. $h(u) = 2u$, $0 < u < 1$
   $= 0$ elsewhere

CHAPTER 4:

4.1 $\left(\frac{1}{2}\right)^5 = 0.0313$

4.2 0.1152

4.3 0.0729

4.4 0.4305

4.5 0.2205

4.6 i. 0.6065;   ii. 0.0902;   iii. 0.9098;   iv. 0.0821

4.7 0.3132

4.8 $\dfrac{\binom{10}{3} \times \binom{40}{9}}{\binom{50}{12}} = 0.2703$

4.9 $\dfrac{\binom{5}{2} \times \binom{10}{3}}{\binom{15}{5}} = 0.3996$

CHAPTER 5:

5.1 i. 0.9938;   ii. 0.9878;   iii. 0.3944

5.2 i. 0.1359;   ii. 0.0139;   iii. 0.9192;   iv. 0.9332;   v. 0.0968

5.3 i. 144;   ii. 131;   iii. 155

5.4 i. 0.0912;   ii. 0.6267;   iii. 0.0478

5.5 0.0478

5.6 mean = 66.017, standard deviation = 3.88

5.7 mean = 2447.17, standard deviation = 208.56

5.8 841

5.9 i. 0.819;   ii. 0.1326;   iii. 0.0484;   iv. 0.7347

5.10 i. 0.9895;   ii. 0.9

5.11 i. 0.1577;   ii. 0.2057

5.12 i. 0.084;   ii. 0.7568

5.13 0.965

5.14 i. 0.2835;   ii. 0.2498;   iii. 0.2636

5.15 i. 0.1084;   ii. 0.5889

5.16 i. 0.0146;   ii. 0.6189

CHAPTER 6:

6.1 0.0423

6.2 0.0217

6.3 0.2217

6.7 3
6.8 144
6.9 278
6.10 (19.1427, 20.0573), 188
6.11 (25.0577, 26.1423), (24.8873, 26.3127), 81
6.12 267
6.13 (2276.5, 2523.5), (2225.7, 2574.3), 18
6.14 (0.4205, 0.4635), (0.4111, 0.4729), 15
6.15 (0.4844, 0.5346), (0.4765, 0.5425)
6.16 (0.0846, 0.1288)
6.17 (4.0452, 4.5548)
6.18 (4.1319, 5.4281)
6.19 (0.0204, 0.1496)
6.20 2.5164
6.21 (226.32, 2144.50)
6.22 8.052
6.23 (0.3931, 6.3720)

CHAPTER 7:

7.1 $z = 5.77$; the null hypothesis must be rejected.
7.2 $z = 2.11$; the null hypothesis cannot be rejected.
7.3 Less than 189.02 gf
7.4 Greater than 227.21 gf
7.5 $t = 1.45$; the null hypothesis cannot be rejected.
7.6 $t = 1.66$; the null hypothesis cannot be rejected.
7.7 $t = -0.73$; the null hypothesis cannot be rejected.
7.8 $z = 1.78$; the null hypothesis cannot be rejected.
7.9 $z = 2.41$; the null hypothesis must be rejected.
7.10 $t = -4.33$; the null hypothesis must be rejected.
7.11 $t = -0.614$; the null hypothesis cannot be rejected.
7.12 $z = 2.6$; the null hypothesis must be rejected.
7.13 $z = -0.57$; the null hypothesis cannot be rejected.
7.14 $\chi^2 = 18.65$; the null hypothesis cannot be rejected.
7.15 $\chi^2 = 7.95$; the null hypothesis cannot be rejected.
7.16 $f = 0.18$; the null hypothesis cannot be rejected.
7.17 $f = 1.63$; the null hypothesis cannot be rejected.
7.18 $\chi^2 = 6.59$; the null hypothesis cannot be rejected.
7.19 $\chi^2 = 4.15$; the null hypothesis must be rejected.
7.20 $\chi^2 = 85.25$; the null hypothesis must be rejected.
7.21 $\chi^2 = 1.58$; the null hypothesis cannot be rejected.
7.22 $\chi^2 = 9.08$; the null hypothesis must be rejected.

CHAPTER 8:

8.1

Source of Variation	Sum of Square	Degrees of Freedom	Mean Square	f
Between gauge lengths	3.7658	2	1.8829	64.5
Error	0.3503	12	0.0292	
Total	4.1160	14		

Differences are significant.

8.2

Source of Variation	Sum of Square	Degrees of Freedom	Mean Square	f
Between strain rates	32.0853	2	16.0426	111.22
Error	1.7309	12	0.1442	
Total	33.8162	14		

Differences are significant.

8.3 In both the exercises, multiple comparisons between two groups in all combinations are significantly different.

8.4

Source of Variation	Sum of Square	Degrees of Freedom	Mean Square	f
Between machines	4.0032	3	1.3344	2.59
Between shifts	0.3614	2	0.1807	0.35
Error	3.0931	6	0.5155	
Total	7.4577	11		

i. Differences between machines are not significant.
ii. Differences between shifts are not significant.

8.5

Source of Variation	Sum of Square	Degrees of Freedom	Mean Square	f
Between treatments	1.4708	2	0.7354	18.41
Between blocks	0.1440	4	0.0360	0.9
Error	0.3196	8	0.0399	
Total	1.9344	14		

i. Differences between treatments are significant.
ii. Differences between blocks are not significant.

8.6 Means of treatment A and treatment C are significantly different.
Means of treatment B and treatment C are significantly different.
Means of treatment A and treatment B are not significantly different.
Multiple comparisons of block means are not significant.

8.7

Source of Variation	Sum of Square	Degrees of Freedom	Mean Square	f
Detachment settings	11.8144	2	5.9072	145.66
Batt weights	1.4478	2	0.7239	17.85
Interaction	0.1222	4	0.0306	0.75
Error	0.3650	9	0.0406	
Total	13.7494	17		

Differences due to detachment settings are significant.
Differences due to batt weights are significant.
The interaction between detachment setting and batt weight is not significant.

8.8

Source of Variation	Sum of Square	Degrees of Freedom	Mean Square	f
Blend ratios	37.1696	2	18.5848	151.6
Yarn counts	4.7052	2	2.3526	19.19
Interaction	0.7304	4	0.1826	1.49
Error	2.2067	18	0.1226	
Total	44.8119	26		

Differences due to blend ratios are significant.
Differences due to yarn counts are significant.
The interaction between blend ratio and yarn count is not significant.

CHAPTER 9:

9.1 $r = 0.9544$

9.2 $r = 0.9064$

9.3 $r_s = 0.7381$

9.4 $r_s = 0.9292$

9.5 $\hat{y} = -0.0402x + 20.6107$

9.6 $\hat{y} = 0.1153x + 16.9857$

9.7 $\hat{y} = -0.0445x^2 + 1.0373x + 0.532$

9.8 $\hat{y} = 0.0682x^2 - 1.6242x + 10.8667$

9.9 $\hat{y} = 0.0758x_1 - 0.0781x_2 + 0.9325$

9.10 $\hat{y} = 0.1167x_1 + 0.5333x_2 - 8.2222$

9.11 $\hat{y} = 121.78 - 15.5x_1 - 6.67x_2 + 1.83x_1^2 - 5.67x_2^2$ (the coded levels are used for x_i)

9.12 $\hat{y} = 0.04 - 0.02x_1 - 0.03x_2 + 0.02x_3 + 0.01x_1x_2 - 0.01x_1x_3 - 0.01x_2x_3 + 0.01x_2^2$ (the coded levels are used for x_i)

9.13 $\hat{y} = 113.33 + 2.38x_1 - 13.13x_2 - 5.75x_3 + x_1x_2 - 3.25x_1x_3 + 1.75x_2x_3 + 14.83x_1^2 - 16.17x_2^2 - 1.92x_3^2$ (the coded levels are used for x_i)

9.14

Term	Regression Coefficient	Standard Error of Coefficient	t-value	Probability
Constant	113.33	6.43	17.63	0.00*
x_1	2.38	3.94	0.60	0.57
x_2	−13.13	3.94	−3.33	0.02*
x_3	−5.75	3.94	−1.46	0.20
x_1x_2	1.00	5.57	0.18	0.86
x_1x_3	−3.25	5.57	−0.58	0.58
x_2x_3	1.75	5.57	0.31	0.77
x_1^2	14.83	5.79	2.56	0.05*
x_2^2	−16.17	5.79	−2.79	0.04*
x_3^2	−1.92	5.79	−0.33	0.75

The terms x_2, x_1^2 and x_2^2 are significant at 95% confidence level.

CHAPTER 10:

10.1

Source of Variation	Sum of Square	Degrees of Freedom	Mean Square	f
Spindle speed	20.5275	3	6.8425	114.57
Days	0.2325	3	0.0775	1.3
Error	0.5375	9	0.05972	
Total	21.2975	15		

For spindle speed, as $6.8425 > 3.9$, the null hypothesis is rejected. Therefore, there is statistical evidence that there are significant differences in yarn breakage rate due to spindle speed. For days, as $0.0775 < 3.9$, the null hypothesis cannot be rejected. Therefore, there are no significant differences in yarn breakage rate studied on different days.

10.2

Source of Variation	Sum of Square	Degrees of Freedom	Mean Square	f
Suppliers	68.187	3	22.7292	12.54
Days	0.687	3	0.2292	0.13
Equipment	28.687	3	9.5625	5.28
Error	10.875	6	1.8125	
Total	108.438	15		

For suppliers, as $22.7292 > 4.8$, the null hypothesis is rejected. Therefore, there is statistical evidence that there are significant differences in reflectance degree of cotton drill fabrics bleached with bleaching agents supplied by different suppliers. For the block day, $f_{days} (0.13) < 4.8$, the null hypothesis cannot be rejected. Therefore, there is no significant difference in reflectance degree due to bleaching applied on different days. For the block equipment, $f_{equipment} (5.28) > 4.8$, the null hypothesis is rejected. Therefore, there is statistical evidence that there are significant differences in reflectance degree of cotton drill fabrics bleached in different equipment.

10.3

Source	Sum of Squares	Degree of Freedom	Mean Squares	f
A	210.125	1	210.125	1681
B	153.125	1	153.125	1225
C	0.125	1	0.125	1
AB	36.125	1	36.125	289
AC	0.125	1	0.125	1
BC	10.125	1	10.125	81
ABC	0.125	1	0.125	
Total	409.875	7		

The f values of the factors A, B, A × B and B × C are more than their corresponding critical values $f_{0.05,1,1} = 4.48$ and hence, the null hypothesis is rejected; therefore, we do have statistical evidence that there are significant differences thermal conductivity of single jersey knitted fabric due to A, B, A × B and B × C. For C and A × C, as their f values are less than their corresponding critical values, the null hypothesis cannot be rejected; therefore, there is no sufficient statistical evidence to conclude that C and A × C have influence on thermal conductivity of single jersey knitted fabric.

10.4

Source	Sum of Squares	Degree of Freedom	Mean Squares	f
A	49	1	49	8.83
B	36	1	36	6.49
C	2.25	1	2.25	0.41
D	90.25	1	90.25	16.26
AB	2.25	1	2.25	0.41
AC	4	1	4	0.72
AD	4	1	4	0.72
BC	1	1	1	0.18
BD	1	1	1	0.18
CD	0.25	1	0.25	0.05
Error	27.25	5	5.55	
Total	217.75	15		

The f values of the factors A and D are more than their corresponding critical values ($f_{0.05,1,5} = 6.6$) and hence, the null hypothesis is rejected. Therefore, we do have statistical evidence that there are significant differences in the yarn imperfections due to A and D. Now, for B, C, A × B, A × C, A × D, B × C, B × D and C × D, as their f values are less than their corresponding critical values, the null hypothesis cannot be rejected; therefore, there is no sufficient statistical evidence to conclude that B, C, A × B, A × C, A × D, B × C, B × D and C × D have influence on imperfection of 30 Ne cotton yarn.

10.5 i. $\alpha = 2.378$; Levels = (−2.378, −1, 0, +1, +2.378);   ii. $\alpha = 2.828$; Levels = (−2.828, −1, 0, +1, +2.828)

10.6

Run No.	x_1	x_2	x_3	x_4	x_5	Run No.	x_1	x_2	x_3	x_4	x_5
1	−1	−1	0	0	0	24	0	+1	+1	0	0
2	−1	+1	0	0	0	25	−1	0	0	−1	0
3	+1	−1	0	0	0	26	−1	0	0	+1	0
4	+1	+1	0	0	0	27	+1	0	0	−1	0
5	0	0	−1	−1	0	28	+1	0	0	+1	0
6	0	0	−1	+1	0	29	0	0	−1	0	−1
7	0	0	+1	−1	0	30	0	0	−1	0	+1
8	0	0	+1	+1	0	31	0	0	+1	0	−1
9	0	−1	0	0	−1	32	0	0	+1	0	+1
10	0	−1	0	0	+1	33	−1	0	0	0	−1
11	0	+1	0	0	−1	34	−1	0	0	0	+1
12	0	+1	0	0	+1	35	+1	0	0	0	−1
13	−1	0	−1	0	0	36	+1	0	0	0	+1
14	−1	0	+1	0	0	37	0	−1	0	−1	0
15	+1	0	−1	0	0	38	0	−1	0	+1	0
16	+1	0	+1	0	0	39	0	+1	0	−1	0
17	0	0	0	−1	−1	40	0	+1	0	+1	0
18	0	0	0	−1	+1	41	0	0	0	0	0
19	0	0	0	+1	−1	42	0	0	0	0	0
20	0	0	0	+1	+1	43	0	0	0	0	0
21	0	−1	−1	0	0	44	0	0	0	0	0
22	0	−1	+1	0	0	45	0	0	0	0	0
23	0	+1	−1	0	0	46	0	0	0	0	0

CHAPTER 11:

11.1 0.616, 0.2622

11.2 $n = 319$, $c = 10$

11.3 $n = 117$, $c = 7$

11.4 AOQ = 0.0083, ATI = 7012

11.5 $n = 13$, $\bar{x}_L = 65.276$

11.6 $n = 15$, $\bar{x}_U = 7.7373$

11.7 $n = 9$, $t = 0.5476$

11.8 $n = 20$, $t = 1.2733$

11.9 Warning limits for averages = (19.172, 20.451)
    Action limits for averages = (18.819, 20.805)
    Warning limits for ranges = (0.4579, 2.9088)
    Action limits for ranges = (0, 3.6158)
    Process is out of control.

11.10 Warning limits for averages = (49.726, 50.316)
Action limits for averages = (49.5607, 50.4817)
Warning limits for ranges = (0.1, 1.129)
Action limits for ranges = (0, 1.425)
Process is in control.

11.11 Warning limits = (0, 4.58)
Action limits = (0, 6.12)
Process is in control.

11.12 Warning limits = (0.0236, 0.1009)
Action limits = (0.0013, 0.1232)
Process is in control.

11.13 Warning limits = (0.04, 7.8)
Action limits = (0, 10.03)
Process is in control.

11.14 Warning limits = (0, 5.76)
Action limits = (0, 7.58)
Process is in control.

CHAPTER 12:

12.1 i. 5;   ii. 5.45

Bibliography

Alam, M. S., Majumdar. A. and Ghosh., A. 2019. Role of Fibre, Yarn and Fabric Parameters on Bending and Shear Behaviour of Plain Woven Fabrics. *Indian Journal of Fibre and Textile Research*, 44, 9–15.

Allen, E. J., Allen, L. J. S., Arciniega, A. and Greenwood, P. E. 2008. Construction of Equivalent Stochastic Differential Equation Models. *Stochastic Analysis and Applications*, 26, 274–297.

Box, G. E. P. and Behnken, D. W. 1960. Some New Three Level Designs for the Study of Quantitative Variables. *Technometrics*, 2, 455–475.

Box, G. E. P. and Wilson, K. B. 1951. On the Experimental Attainment of Optimum Conditions. *Journal of the Royal Statistical Society, Series B*, 13, 1–45.

Dean, A., Voss, D., and Draguljic, D. 2014. Design and Analysis of Experiments, 2nd Edition, Springer, New York.

Ghosh, A., Mal, P. and Majumdar A. 2019. Advanced Optimization and Decision-Making Techniques in Textile Manufacturing. CRC Press, Boca Raton, FL.

Ghosh, A., Mal, P., Majumdar, A. and Banerjee, D. 2016. Analysis of Knitting Process Variables and Yarn Count Influencing the Thermo-Physiological Comfort Properties of Single Jersey and Rib Fabrics. *Journal of The Institute of Engineers (India): Series E*, 97(2), 89–98.

Ghosh, A., Mal, P., Majumdar, A. and Banerjee, D. 2017. An Investigation on Air and Thermal Transmission Through Knitted Fabric Structures Using Taguchi Method. *Autex Research Journal*, 17(2), 152–163.

Monfort, F. 1962. Carding as Markovian Process, *Journal of the Textile Institute Transactions*, 53(8), T379–T393.

Montgomery, D. C. 2019. Design and Analysis of Experiments, 10th Edition, John Wiley & Sons, Inc., Hoboken, NJ.

Prabhu, N. U. 1966. Stochastic Process, Basic Theory and Its Applications. The Macmillan Company, New York.

Ross, P. J. 1996. Taguchi Techniques for Quality Engineering. 2nd Edition, McGraw Hill, New York, 329–335.

Roy, R. K. 2001. Design of Experiments Using the Taguchi Approach. John Wiley & Sons, Inc, New York, USA.

Singh, A. and Swani, N. M. 1973. A Quantitative Analysis of the Carding Action by the Flats and Doffer in a Revolving-Flat Card, *The Journal of the Textile Institute*, 64(3), 115–123.

Taguchi, G., Chowdhury, S. and Wu, Y. 2005. Taguchi's Quality Engineering Handbook. John Wiley & Sons, Inc., Hoboken, NJ, 1654–1662.

Suggested Further Reading

Anderson, T. W. 2010. An Introduction to Multivariate Statistical Analysis, 3rd Edition, John Wiley & Sons, Inc., Hoboken, NJ.

Devore, J., Farnum, M., and Doi, J. 2003. Applied Statistics for Engineers and Scientists, 3rd Edition, Cengage Learning, Stamford, CT.

Johnson, R., Miller, I., and Freund, J. E. 2017. Miller & Freund's Probability and Statistics for Engineers, 9th Edition, Pearson, London.

Metcalfe, A., Green, D., Greenfield, T., Mansor, M., Smith, A., and Tuke, J. 2019. Statistics in Engineering with Examples in MATLAB® and R, 2nd Edition, CRC Press, Boca Raton, FL.

Miller I., and Miller, M. 2004. John E. Freund's Mathematical Statistics with Applications, 7th Edition, Pearson, New York.

Montgomery, D. C., and Runger, G. C. 2003. Applied Statistics and Probability for Engineers, 3rd Edition, John Wiley & Sons, Inc., Hoboken, NJ.

Montgomery, D. C., and Runger, G. C. 2009. Introduction to Statistical Quality Control, 6th Edition, John Wiley & Sons, Inc., Hoboken, NJ.

Rohatgi, V. K., and Saleh, A. K. E. 2006. An Introduction to Probability and Statistics, 2nd Edition, John Wiley & Sons, Inc. Hoboken, NJ.

Index

For Product Safety Concerns and Information please contact our EU representative GPSR@taylorandfrancis.com
Taylor & Francis Verlag GmbH, Kaufingerstraße 24, 80331 München, Germany